Monografie Matematyczne
Instytut Matematyczny Polskiej Akademii Nauk (IMPAN)

Volume 72
(New Series)

Founded in 1932 by
S. Banach, B. Knaster, K. Kuratowski,
S. Mazurkiewicz, W. Sierpinski, H. Steinhaus

Volumes 31–62 of the series
Monografie Matematyczne were published by
PWN – Polish Scientific Publishers, Warsaw

Adam Osękowski

Sharp Martingale and Semimartingale Inequalities

 Birkhäuser

Adam Osękowski
Institute of Mathematics
University of Warsaw
Warsaw
Poland

ISBN 978-3-0348-0369-4   ISBN 978-3-0348-0370-0 (eBook)
DOI 10.1007/978-3-0348-0370-0
Springer Basel Heidelberg New York Dordrecht London

Library of Congress Control Number: 2012945527

Mathematics Subject Classification (2010): Primary: 60G42, 60G44, 60G46; secondary: 31B05, 31B20, 42B25, 46B20, 46C15

Printed on acid-free paper

Springer Basel AG is part of Springer Science+Business Media (www.birkhauser-science.com)

To the memory of my Father

Contents

Preface

The purpose of this monograph is to present a unified approach to a certain class of semimartingale inequalities, which have their roots at some classical problems in harmonic analysis. Preliminary results in this direction were obtained by Burkholder in 60s and 70s during his work on martingale transforms and geometry of UMD Banach spaces. The rapid development in the field occurred after the appearance of a large paper of Burkholder in 1984, in which he described a powerful method to handle martingale inequalities and used it to obtain a number of interesting results. Since then, the method has been extended considerably in many directions and successfully implemented in the study of related problems in various areas of mathematics.

The literature on the subject is very large. One of the objectives of this exposition is to put most of the existing results together, explain in detail the underlying concepts and point out some connections and similarities. This book contains also a number of new results as well as some open problems, which, we hope, will stimulate the reader's further interest in this field. The recent applications of the above results in the theory of quasiregular mappings (with deep implications in geometric function theory), Fourier multipliers as well as their connections to rank-one convexity and quasiconvexity indicate the need of further developing this area.

Acknowledgment

I would like to express my gratitude to Professor Stanisław Kwapień for many discussions and encouragement, which motivated me greatly during the work on the monograph. I am also indebted to Professor Quanhua Xu at Université de Franche-Comté in Besançon, France; this is where the research on the subject was initiated. Last but not least, I wish also to thank my whole family, especially my wife Joanna, for love and constant support.

This work was financially supported by MNiSW Grant N N201 422139.

Chapter 1

Introduction

Inequalities for semimartingales appear in both measure-based and noncommutative probability theory, where they play a distinguished role, and have numerous applications in many areas of mathematics. Before we introduce the necessary probabilistic background, let us start with a related classical problem which interested many mathematicians during the first part of the 20th century. The question is: how does the size of a periodic function control the size of its conjugate? To be more specific, let f be a trigonometric polynomial of the form

$$f(\theta) = \frac{a_0}{2} + \sum_{k=1}^{N} \left(a_k \cos(k\theta) + b_k \sin(k\theta) \right), \qquad \theta \in [0, 2\pi),$$

with real coefficients $a_0, a_1, a_2, \ldots, a_N, b_1, b_2, \ldots, b_N$. The polynomial conjugate to f is defined by

$$g(\theta) = \sum_{k=1}^{N} \left(a_k \sin(k\theta) - b_k \cos(k\theta) \right), \qquad \theta \in [0, 2\pi).$$

The problem can be stated as follows. For a given $1 \leq p \leq \infty$, is there a universal constant C_p (that is, not depending on the coefficients or the number N) such that

$$||g||_p \leq C_p ||f||_p ? \tag{1.1}$$

Here $||f||_p$ denotes the L_p-norm of f, which is given by $\left[\int_0^{2\pi} |f(\theta)|^p \frac{d\theta}{2\pi} \right]^{1/p}$ when p is finite, and equals the essential supremum of f over $[0, 2\pi)$ when $p = \infty$. This question is very easy when $p = 2$: the orthogonality of the trigonometric system implies that the inequality holds with the constant $C_2 = 1$. Furthermore, this value is easily seen to be optimal. What about the other values of p? As shown by M. Riesz in [179] and [180], when $1 < p < \infty$, the estimate does hold with some absolute $C_p < \infty$; for $p = 1$ or $p = \infty$, the inequality does not hold with any finite constant. The best value of C_p was determined by Pichorides [171] and Cole (unpublished; see Gamelin [79]): $C_p = \cot(\pi/(2p^*))$ is the optimal choice, where $p^* = \max\{p, p/(p-1)\}$.

One may consider a non-periodic version of the problem above. To formulate the statement, we need to introduce the Hilbert transform on the real line. For $1 \leq p < \infty$, let $f \in L^p(\mathbb{R})$ and put

$$Hf(x) = \lim_{\varepsilon \to 0} \frac{1}{\pi} \int_{|y| \geq \varepsilon} \frac{f(x-y)}{y} dy.$$

This limit can be shown to exist almost everywhere. The transform H is the non-periodic analogue of the harmonic conjugate operator and satisfies the following L^p bound: if $1 < p < \infty$, then

$$\|Hf\|_p \leq C_p \|f\|_p,$$

where C_p is the same constant as in (1.1): see [180] and [205].

Riesz's inequality has been extended in many directions. A significant contribution is due to Calderón and Zygmund [41], [42], who obtained the following result concerning singular integral operators. They showed that for a large class of kernels $K : \mathbb{R}^n \setminus \{0\} \to \mathbb{C}$, the limit

$$Tf(x) = \lim_{\varepsilon \to 0} \int_{|y| \geq \varepsilon} f(x-y)K(y)dy$$

exists almost everywhere if $f \in L^p(\mathbb{R}^n)$ for some $1 \leq p < \infty$ and

$$\|Tf\|_p \leq C_p \|f\|_p$$

when $1 < p < \infty$. Here the constant C_p may be different from the one in (1.1), but it depends only on K and p.

We may ask analogous questions in the martingale setting. First let us introduce the necessary notation: suppose that $(\Omega, \mathcal{F}, \mathbb{P})$ is a probability space, filtered by $(\mathcal{F}_n)_{n \geq 0}$, a non-decreasing family of sub-σ-fields of $\mathcal{F}$. Assume that $f = (f_n)_{n \geq 0}$, $g = (g_n)_{n \geq 0}$ are adapted discrete-time martingales taking values in $\mathbb{R}$. Then $df = (df_n)_{n \geq 0}$, $dg = (dg_n)_{n \geq 0}$, the difference sequences of f and g, respectively, are defined by $df_0 = f_0$ and $df_n = f_n - f_{n-1}$ for $n \geq 1$, and similarly for dg. In other words, we have the equalities

$$f_n = \sum_{k=0}^{n} df_k \quad \text{and} \quad g_n = \sum_{k=0}^{n} dg_k, \qquad n = 0, 1, 2, \ldots.$$

Let us now formulate the corresponding version of Riesz's inequality. The role of a conjugate function in the probabilistic setting is played by a ± 1-transform, given as follows. Let $\varepsilon = (\varepsilon_k)_{k \geq 0}$ be a deterministic sequence of signs: $\varepsilon_k \in \{-1, 1\}$ for each k. If f is a given martingale, we define the transform of f by ε as

$$g_n = \sum_{k=0}^{n} \varepsilon_k df_k, \qquad n = 0, 1, 2, \ldots.$$

Of course, such a sequence is again a martingale. A fundamental result of Burk-holder [17] asserts that for any f, ε and g as above we have

$$\|g\|_p \leq C_p \|f\|_p, \qquad 1 < p < \infty, \tag{1.2}$$

for some finite C_p depending only on p. Here and below, $\|f\|_p$ denotes the pth mo-ment of f, given by $\sup_n (\mathbb{E}|f_n|^p)^{1/p}$ when $0 < p < \infty$ and $\|f\|_\infty = \sup_n \operatorname{essup}|f_n|$. This result is a starting point for many extensions and refinements and the purpose of this monograph is to present a systematic and unified approach to this type of problems. Let us first say a few words about the proof of (1.2). The initial approach of Burkholder used a related weak-type estimate and some standard interpolation and duality arguments. Then Burkholder refined his proof and invented a method which can be used to study general estimates for a much wider class of processes. Roughly speaking, the technique reduces the problem of proving a given inequal-ity to the existence of a certain special function or, in other words, to finding the solution to a corresponding boundary value problem. This is described in detail in Chapter 2 below, and will be illustrated on many examples in Chapter 3. The method can also be used in the case when the dominating process f is a sub- or supermartingale and, after some modifications, allows also to establish general inequalities for maximal and square functions. Another very important feature of the approach is that it enables us to derive the optimal constants: this, except for the elegance, provides some additional insight into the structure of the extremal processes and often leads to some further implications.

Coming back to the martingale setting described above, we shall be particu-larly interested in the following estimates.

(i) moment inequalities (or strong type (p, p) inequalities): see (1.2),
(ii) weak-type (p, p) inequalities:

$$\|g\|_{p,\infty} \leq c_{p,p} \|f\|_p, \qquad 1 \leq p < \infty,$$

where $\|g\|_{p,\infty} = \sup_{\lambda>0} \lambda \big(\mathbb{P}(\sup_n |g_n| \geq \lambda)\big)^{1/p}$ denotes the weak pth norm of g,

(iii) logarithmic estimates:

$$\|g\|_1 \leq K \sup_n \mathbb{E}|f_n| \log |f_n| + L(K),$$

(iv) tail and Φ-inequalities:

$$\mathbb{P}(\sup_n |g_n| \geq \lambda) \leq P(\lambda), \qquad \sup_n \mathbb{E}\Phi(|g_n|) \leq C_\Phi,$$

under the assumption that $\|f\|_\infty \leq 1$.

We shall study these and other related problems in the more general setting in which the transforming sequence ε is predictable and takes values in $[-1, 1]$. Here by predictability we mean that each term ε_n is measurable with respect to $\mathcal{F}_{(n-1)\vee 0}$ (in particular, it may be random): this does not affect the martingale property of the sequence g. This assumption will be further relaxed to the case when g is a martingale differentially subordinate to f, which amounts to saying that for any $n \geq 0$ we have $|dg_n| \leq |df_n|$ almost surely. We will succeed in determining the optimal constants in most of the aforementioned estimates.

The next challenging problem is to study the above statements in the vector case, when the processes f, g take values from a certain separable Banach space $\mathcal{B}$. It turns out that when we deal with a Hilbert space, the passage from the real case is typically quite easy and does not require much additional effort. However, the extension of a given inequality to a non-Hilbert space setting is a much more difficult problem and the question about the optimal constants becomes hopeless in general. These and related martingale results are dealt with in Chapter 3. Sections 3.1–3.10 concern Hilbert-space-valued processes, while the next two treat the more general case in which the martingales take values in a separable Banach space $\mathcal{B}$. The final section of that chapter is devoted to some applications of these results to the Haar system on $[0, 1)$.

Chapter 4 contains analogous results in the case when the dominating process f is a sub- or a supermartingale. The notion of differential subordination, which is a very convenient condition in the martingale setting, becomes too weak and needs to be strengthened. We do this by imposing an extra conditional subordination and assume that the terms $|\mathbb{E}(dg_n|\mathcal{F}_{n-1})|$ are controlled by $|\mathbb{E}(df_n|\mathcal{F}_{n-1})|$, $n = 1, 2, \ldots$. This domination, called strong differential subordination, has a very natural counterpart in the theory of Itô processes and stochastic integrals, and is sufficient for our purposes.

In Chapter 5 we show how to extend the above inequalities to continuous-time processes and stochastic integrals. In fact, some of the estimates studied there can be regarded as a motivation to the results of Chapter 3 and Chapter 4. Consider the following example. Let $X = (X_t)_{t\geq 0}$ be a continuous-time real-valued martingale and let $H = (H_t)_{t\geq 0}$ be a predictable process taking values in $[-1, 1]$. As usual, we assume that the trajectories of both these processes are sufficiently regular, that is, are right-continuous and have limits from the left. Suppose that Y is the Itô integral of H with respect to X: that is, we have

$$Y_t = H_0 X_0 + \int_{0+}^{t} H_s \mathrm{d}X_s$$

for $t \geq 0$. Then the pair (X, Y) is precisely the continuous analogue of a martingale and its transform by a predictable process bounded in absolute value by 1. We may ask questions, similar to (i)–(iv) above, concerning the comparison of the sizes of X and Y. As we shall see, some approximation theorems allow one to carry over the inequalities from the discrete-time setting to that above, without changing the constants. In fact, we shall prove much more. The notions of differential subordination and strong differential subordination can be successfully generalized

to the continuous-time case, and we shall present an appropriate modification of
Burkholder's method, which can be applied to study the corresponding estimates
in this wider setting. In one of the final sections, we apply the results to obtain
some interesting estimates for harmonic functions on Euclidean domains, which
can be regarded as extensions of Riesz's inequality.

In Chapter 6 we continue the line of research started in Chapter 5. Namely,
we investigate there continuous-time processes under (strong) differential subordination, but this time we impose the additional orthogonality assumption. The
martingales of this type turn out to be particularly closely related to periodic
functions and their conjugates, mentioned at the beginning. The final section of
that chapter is devoted to the description of the connections between these two
settings.

Chapter 7 deals with another important class of estimates: the maximal ones.
To be more specific, one can ask questions similar to (i)–(iv) above in the case
when f (or g, or both) is replaced by the corresponding maximal function $|f|^* =
\sup_{n \geq 0} |f_n|$ or one-sided maximal function $f^* = \sup_{n \geq 0} f_n$. For example, we shall
present the proof the classical Doob's maximal inequality

$$\|f^*\|_p \leq \frac{p}{p-1}\|f\|_p, \qquad 1 < p \leq \infty,$$

where f is a nonnegative submartingale, as well as a number of related weak-type
and logarithmic estimates. See Section 7.2.

Maximal inequalities involving two processes arise naturally in many situations and are of particular interest when the corresponding non-maximal versions
are not valid. For instance, there is no universal $C_1 < \infty$ such that $\|g\|_1 \leq C_1\|f\|_1$
for all real-valued martingales f and their ± 1-transforms g; on the other hand,
Burkholder [35] showed that for such f and g we have

$$\|g\|_1 \leq 2.536\ldots\| \, |f|^*\|_1,$$

and the bound is sharp. For other related results, see Sections 7.3–7.8.

Problems of this type can be successfully investigated using appropriate modification of Burkholder's method, which was developed in [35]. However, it should
be stressed here that the corresponding boundary value problems become tricky
due to the increase of the dimension. That is, the special functions we search
for depend on three or four variables (each of the terms $|f|^*$, $|g|^*$ ... involves an
additional parameter). Furthermore, in contrast with the non-maximal setting,
the passage to the Hilbert-space-valued setting does require extra effort and the
inequalities for Banach-space-valued semimartingales become even more difficult.
The next interesting feature is that, unlike in the non-maximal case, the additional assumption on the continuity of paths does affect the optimal constants
and Burkholder's method needs further refinement. See Sections 7.9–7.12.

Chapter 8 is the final part of the exposition and is devoted to the study of
square function inequalities. Recall that if $f = (f_n)_{n \geq 0}$ is an adapted sequence,

then its square function is given by

$$S(f) = \left(\sum_{k=0}^{\infty} |df_k|^2 \right)^{1/2}.$$

Again, we may ask about sharp moment, weak type and other related estimates between f and $S(f)$. In addition, we can consider maximal inequalities which involve a martingale, its square and maximal function. There are also continuous-time analogues of such estimates (the role of the square function is played by the square bracket or the quadratic covariance process), but they can be easily reduced to the discrete-time bounds by means of standard approximation. The problems of such type are classical and have numerous extensions in various areas of mathematics: see bibliographical notes at the end of Chapter 8. It turns out that many interesting estimates can be immediately deduced from related results concerning Hilbert-space-valued differentially subordinated martingales (see Chapter 2 below), but this approach does not always allow to keep track of the optimal constants. To derive the sharp estimates, one may apply an appropriate modification of Burkholder's method, invented in [38]. However, the corresponding boundary value problems are difficult in general and this technique has been successfully implemented only in a few of cases: see Sections 8.2, 8.3 and 8.4.

The situation becomes a bit easier when we restrict ourselves to conditionally symmetric martingales. Recall that f is conditionally symmetric when for each $n \geq 1$, the conditional distributions of df_n and $-df_n$ given $\mathcal{F}_{n-1}$ coincide. For example, consider the so-called *dyadic martingales*: take the probability space to be $([0,1], \mathcal{B}(0,1), |\cdot|)$ and put $f_n = \sum_{k=0}^{n} a_k h_k$, where $h = (h_k)_{k \geq 0}$ is the Haar system and $a_1, a_2, \ldots$ are coefficients, which may be real or vector valued. Then the corresponding boundary value problem can often (but not always) be dealt with by solving the heat equation on a part of the domain of the special function. See Sections 8.2, 8.3, 8.5 and 8.6.

The final part of Chapter 8 concerns the conditional square function. It is another classical object in the martingale theory, given by

$$s(f) = \left(\sum_{k=0}^{\infty} \mathbb{E}(|df_k|^2 | \mathcal{F}_{(k-1) \vee 0}) \right)^{1/2}.$$

We shall show how to modify Burkholder's method so that it yields estimates involving f and $s(f)$. The approach can be further extended to imply related estimates for sums of nonnegative random variables and their predictable projections: see Wang [199] and the author [134], but we shall not include this here.

A few words about the organization of the monograph. Typically, each section starts with the statement of a number of theorems, which contain results on semimartingale inequalities. In most cases, the proof of a given theorem is divided into three separate parts. In the first step we make use of Burkholder's method and establish the inequality contained in the statement. The next part deals with the optimality of the constants appearing in this estimate. In the final part we present some intuitive arguments which lead to the discovery of the special function, used in the first step. Wherever possible, we have tried to present new proofs and new reasoning which, as we hope, throws some additional light on the structure of the problems. We also did our best to keep the exposition as self-contained as possible, and omit argumentation or its part only when a similar reasoning has been presented earlier. Each chapter concludes with a section containing historical and bibliographical notes concerning the results studied in the preceding sections as well as some material for further reading.

Chapter 2

Burkholder's Method

We start by introducing the main tool which will be used in the study of semi-martingale inequalities. For the sake of clarity, in this chapter we focus on the description of the method only for discrete-time martingales. The necessary modifications, leading to inequalities for wider classes of processes, will be presented in the further parts of the monograph.

2.1 Description of the technique

2.1.1 Inequalities for ± 1-transforms

Burkholder's method relates the validity of a certain given inequality for semi-martingales to a corresponding boundary value problem, or, in other words, to the existence of a special function, which has appropriate concave-type properties. To start, let us assume that $(\Omega, \mathcal{F}, \mathbb{P})$ is a probability space, which is filtered by $(\mathcal{F}_n)_{n\geq0}$, a non-decreasing family of sub-σ-fields of $\mathcal{F}$. Consider adapted simple martingales $f = (f_n)_{n\geq0}$, $g = (g_n)_{n\geq0}$ taking values in $\mathbb{R}$, with the corresponding difference sequences $(df_n)_{n\geq0}$, $(dg_n)_{n\geq0}$, respectively. Here by simplicity of f we mean that for any nonnegative integer n the random variable f_n takes a finite number of values and there is a deterministic integer N such that $f_N = f_{N+1} = f_{N+2} = \cdots$.

Let $D = \mathbb{R} \times \mathbb{R}$ and let $V : D \to \mathbb{R}$ be a function, not necessarily Borel or even measurable. Let $x,\ y \in \mathbb{R}$ be fixed and denote by $M(x,y)$ the class of all pairs (f,g) of simple martingales f and g starting from x and y, respectively, such that $dg_n \equiv df_n$ or $dg_n \equiv -df_n$ for any $n \geq 1$. Here the filtration may vary as well as the probability space, unless it is assumed to be non-atomic. Suppose that we are interested in the numerical value of

$$U^0(x,y) = \sup\{\mathbb{E}V(f_n, g_n)\}, \tag{2.1}$$

where the supremum is taken over $M(x,y)$ and all nonnegative integers n. Of course, there is no problem with measurability or integrability of $V(f_n, g_n)$, since the sequences f and g are simple. Note that the definition of U^0 can be rewritten in the form

$$U^0(x,y) = \sup_{(f,g)\in M(x,y)} \{\mathbb{E}V(f_\infty, g_\infty)\},$$

where f_∞ and g_∞ stand for the pointwise limits of f and g (which exist due to the simplicity of the sequences). This is straightforward: for any $(f,g) \in M(x,y)$ and any nonnegative integer n, we have $(f_n, g_n) = (\overline{f}_\infty, \overline{g}_\infty)$, where the pair $(\overline{f}, \overline{g}) \in M(x,y)$ is just (f,g) stopped at time n.

In most cases, we will try to provide some upper bounds for U^0, either on the whole domain D, or on its part. The key idea in the study of such a problem is to introduce a class of special functions. The class consists of all $U : D \to \mathbb{R}$ satisfying the following conditions 1° and 2°:

1° (Majorization property) For all $(x,y) \in D$,

$$U(x,y) \geq V(x,y). \tag{2.2}$$

2° (Concavity-type property) For all $(x,y) \in D$, $\varepsilon \in \{-1,1\}$ and any $\alpha \in (0,1)$, $t_1, t_2 \in \mathbb{R}$ such that $\alpha t_1 + (1-\alpha)t_2 = 0$, we have

$$\alpha U(x+t_1, y+\varepsilon t_1) + (1-\alpha)U(x+t_2, y+\varepsilon t_2) \leq U(x,y). \tag{2.3}$$

Using a straightforward induction argument, we can easily show that the condition 2° is equivalent to the following: for all $(x,y) \in D$, $\varepsilon \in \{-1,1\}$ and any simple mean-zero variable d we have

$$\mathbb{E}U(x+d, y+\varepsilon d) \leq U(x,y). \tag{2.4}$$

To put it in yet another words, (2.3) amounts to saying that the function U is *diagonally concave*, that is, concave along the lines of slope ± 1.

The interplay between the problem of bounding U^0 from above and the existence of a special function U satisfying 1° and 2° is described in the two statements below, Theorem 2.1 and Theorem 2.2.

Theorem 2.1. *Suppose that U satisfies 1° and 2°. Then for any simple f and g such that $dg_n \equiv df_n$ or $dg_n \equiv -df_n$ for $n \geq 1$ we have*

$$\mathbb{E}V(f_n, g_n) \leq \mathbb{E}U(f_0, g_0), \qquad n = 0, 1, 2, \ldots. \tag{2.5}$$

In particular, this implies

$$U^0(x,y) \leq U(x,y) \qquad \text{for all } x, y \in \mathbb{R}. \tag{2.6}$$

Proof. The key argument is that the process $(U(f_n, g_n))_{n\geq 0}$ is an $(\mathcal{F}_n)$-supermartingale. To see this, note first that all the variables are integrable, by the simplicity of f and g. Fix $n \geq 1$ and observe that

$$\mathbb{E}[U(f_n, g_n)|\mathcal{F}_{n-1}] = \mathbb{E}[U(f_{n-1} + df_n, g_{n-1} + dg_n)|\mathcal{F}_{n-1}].$$

An application of (2.4) conditionally on $\mathcal{F}_{n-1}$, with $x = f_{n-1}$, $y = g_{n-1}$ and $d = df_n$ yields the supermartingale property. Thus, by 1°,

$$\mathbb{E}V(f_n, g_n) \leq \mathbb{E}U(f_n, g_n) \leq \mathbb{E}U(f_0, g_0) \tag{2.7}$$

and the proof is complete. □

Therefore, we have obtained that $U^0(x, y) \leq \inf U(x, y)$, where the infimum is taken over all U satisfying 1° and 2°. The remarkable feature of the approach is that the reverse inequality is also valid. To be more precise, we have the following statement.

Theorem 2.2. *If U^0 is finite on D, then it is the least function satisfying 1° and 2°.*

Proof. The fact that U^0 satisfies 1° is immediate: the deterministic constant pair (x, y) belongs to $M(x, y)$. To prove 2°, we will use the so-called "splicing argument". Take $(x, y) \in D$, $\varepsilon \in \{-1, 1\}$ and α, t_1, t_2 as in the statement of the condition. Pick pairs (f^j, g^j) from the class $M(x + t_j, y + \varepsilon t_j)$, $j = 1, 2$. We may assume that these pairs are given on the Lebesgue probability space $([0, 1], \mathcal{B}([0, 1]), |\cdot|)$, equipped with some filtration. By the simplicity, there is a deterministic integer T such that these pairs terminate before time T. Now we will "glue" these pairs into one using the number α. To be precise, let (f, g) be a pair on $([0, 1], \mathcal{B}([0, 1]), |\cdot|)$, given by $(f_0, g_0) \equiv (x, y)$ and

$$(f_n, g_n)(\omega) = (f_{n-1}^1, g_{n-1}^1)(\omega/\alpha), \qquad \text{if } \omega \in [0, \alpha),$$

and

$$(f_n, g_n)(\omega) = (f_{n-1}^2, g_{n-1}^2)\left(\frac{\omega - \alpha}{1 - \alpha}\right), \qquad \text{if } \omega \in [\alpha, 1),$$

when $n = 1, 2, \ldots, T$. Finally, we let $df_n = dg_n \equiv 0$ for $n > T$. Then it is straightforward to check that f, g are martingales with respect to the natural filtration and $(f, g) \in M(x, y)$. Therefore, by the very definition of U^0,

$$U^0(x, y) \geq \mathbb{E}V(f_T, g_T)$$

$$= \int_0^\alpha V(f_{T-1}^1, g_{T-1}^1)\left(\frac{\omega}{\alpha}\right) d\omega + \int_\alpha^1 V(f_{T-1}^2, g_{T-1}^2)\left(\frac{\omega - \alpha}{1 - \alpha}\right) d\omega$$

$$= \alpha \mathbb{E}V(f_\infty^1, g_\infty^1) + (1 - \alpha)\mathbb{E}V(f_\infty^2, g_\infty^2).$$

Taking supremum over the pairs (f^1, g^1) and (f^2, g^2) gives

$$U^0(x, y) \geq \alpha U^0(x + t_1, y + \varepsilon t_1) + (1 - \alpha)U^0(x + t_2, y + \varepsilon t_2),$$

which is 2°. To see that U^0 is the least special function, simply look at (2.6). □

The above two facts give the following general method of proving inequalities for ± 1-transforms. Let $V : D \to \mathbb{R}$ be a given function and suppose we are interested in showing that

$$\mathbb{E}V(f_n, g_n) \leq 0, \qquad n = 0, 1, 2, \ldots, \tag{2.8}$$

for all simple f, g, such that $dg_n \equiv df_n$ or $dg_n \equiv -df_n$ for all n (in particular, also for $n = 0$).

Theorem 2.3. *The inequality* (2.8) *is valid if and only if there exists* $U : D \to \mathbb{R}$ *satisfying* $1°$, $2°$ *and the* initial condition

$3°$ $U(x, y) \leq 0$ *for all* x, y *such that* $y = \pm x$.

Proof. If there is a function U satisfying $1°$, $2°$ and $3°$, then (2.8) follows immediately from (2.5), since $3°$ guarantees that the term $\mathbb{E}U(f_0, g_0)$ is nonpositive. To get the reverse implication, we use Theorem 2.2: as we know from its proof, the function U^0 satisfies $1°$ and $2°$. It also enjoys $3°$, directly from the definition of U^0 combined with the inequality (2.8). The only thing which needs to be checked is the finiteness of U^0, which is assumed in Theorem 2.2. Since $U^0 \geq V$, we only need to show that $U^0(x, y) < \infty$ for every (x, y). The condition $3°$, which we have already established, guarantees the inequality on the diagonals $y = \pm x$. Suppose that $|x| \neq |y|$ and let (f, g) be any pair from $M(x, y)$. Consider another martingale pair (f', g'), which starts from $((x + y)/2, (x + y)/2)$ and, in the first step, moves to (x, y) or to (y, x). If it jumped to (y, x), it stops; otherwise, we determine (f', g') by the assumption that the conditional distribution of $(f'_n, g'_n)_{n \geq 1}$ coincides with the (unconditional) distribution of $(f_n, g_n)_{n \geq 0}$. We easily check that g' is a ± 1-transform of f', and hence, for any $n \geq 1$,

$$0 \geq \mathbb{E}V(f'_n, g'_n) = \frac{1}{2}V(y, x) + \frac{1}{2}\mathbb{E}V(f_{n-1}, g_{n-1}).$$

Consequently, taking supremum over f, g and n gives $U^0(x, y) \leq -V(y, x)$ and we are done. $\square$

Remark 2.1. Suppose that V has the symmetry property

$$V(x, y) = V(-x, y) = V(x, -y) \qquad \text{for all } x, y \in \mathbb{R}. \tag{2.9}$$

Then we may replace $3°$ by the simpler condition

$3°'$ $U(0, 0) \leq 0$.

In other words, if there is U which satisfies the conditions $1°$, $2°$ and $3°'$, then there is $\bar{U}$ which satisfies $1°$, $2°$ and $3°$. To prove this, fix U as in the previous sentence. By (2.9), the functions $(x, y) \mapsto U(-x, y)$, $(x, y) \mapsto U(x, -y)$ and $(x, y) \mapsto U(-x, -y)$ also enjoy the properties $1°$, $2°$ and $3°'$, and hence so does $\bar{U}$ given by

$$\bar{U}(x, y) = \min\{U(x, y), U(-x, y), U(x, -y), U(-x, -y)\}, \quad x, y \in \mathbb{R}.$$

But this function satisfies $3°$: indeed, by $2°$,

$$\bar{U}(x, \pm x) = \frac{\bar{U}(x, x) + \bar{U}(-x, -x)}{2} \leq \bar{U}(0, 0) \leq 0,$$

for any $x \in \mathbb{R}$.

The approach described above concerns only real-valued processes. Furthermore, the condition of being a ± 1-transform is quite restrictive. There arises the natural question whether the methodology can be extended to a wider class of martingales and we will shed some light on it.

2.1.2 Inequalities for general transforms of Banach-space-valued martingales

Let us start with the following vector-valued version of Theorem 2.3. The proof is the same as in the real case and is omitted. Let $\mathcal{B}$ be a Banach space.

Theorem 2.4. *Let $V : \mathcal{B} \times \mathcal{B} \to \mathbb{R}$ be a given function. The inequality*

$$\mathbb{E}V(f_n, g_n) \leq 0$$

holds for all n and all pairs (f, g) of simple $\mathcal{B}$-valued martingales such that g is a ± 1-transform of f if and only if there exists $U : \mathcal{B} \times \mathcal{B} \to \mathbb{R}$ satisfying the following three conditions.

$1°$ $U \geq V$ *on $\mathcal{B} \times \mathcal{B}$.*

$2°$ *For all $x, y \in \mathcal{B}$, $\varepsilon \in \{-1, 1\}$ and any $\alpha \in (0, 1)$, $t_1, t_2 \in \mathcal{B}$ such that $\alpha t_1 + (1 - \alpha)t_2 = 0$, we have*

$$\alpha U(x + t_1, y + \varepsilon t_1) + (1 - \alpha)U(x + t_2, y + \varepsilon t_2) \leq U(x, y).$$

$3°$ $U(x, \pm x) \leq 0$ *for all $x \in \mathcal{B}$.*

In the previous situation, we had $dg_n = v_n df_n$, $n = 0, 1, 2, \ldots$, where each v_n was deterministic and took values in the set $\{-1, 1\}$. Now let us consider the more general situation in which the sequence v is simple, predictable and takes values in $[-1, 1]$. Recall that predictability means that each v_n is measurable with respect to $\mathcal{F}_{(n-1)\vee 0}$ and, in particular, this allows random terms. The corresponding version of Theorem 2.4 can be stated as follows. We omit the proof, it requires no new ideas.

Theorem 2.5. *Let $V : \mathcal{B} \times \mathcal{B} \to \mathbb{R}$ be a given function. The inequality*

$$\mathbb{E}V(f_n, g_n) \leq 0$$

holds for all n and all f, g as above if and only if there exists $U : \mathcal{B} \times \mathcal{B} \to \mathbb{R}$ satisfying the following three conditions.

$1°$ $U \geq V$ on $\mathcal{B} \times \mathcal{B}$.

$2°$ *For all* $(x, y) \in \mathcal{B} \times \mathcal{B}$, *any deterministic* $a \in [-1, 1]$ *and any* $\alpha \in (0, 1)$, $t_1, t_2 \in \mathcal{B}$ *such that* $\alpha t_1 + (1 - \alpha)t_2 = 0$ *we have*

$$\alpha U(x + t_1, y + at_1) + (1 - \alpha)U(x + t_2, y + at_2) \leq U(x, y).$$

$3°$ $U(x, y) \leq 0$ *for all* x, $y \in \mathcal{B}$ *such that* $y = ax$ *for some* $a \in [-1, 1]$.

Let us make here some important observations.

Remark 2.2. (i) Condition $2°$ of Theorem 2.5 extends to the following inequality: for all x, $y \in \mathcal{B}$, any deterministic $a \in [-1, 1]$ and any simple mean zero $\mathcal{B}$-valued random variable d we have

$$\mathbb{E}U(x + d, y + ad) \leq U(x, y).$$

(ii) Condition $2°$ can be rephrased as follows: for any $x, y, h \in \mathcal{B}$ and $a \in [-1, 1]$, the function $G = G_{x,y,h,a} : \mathbb{R} \to \mathbb{R}$ given by $G(t) = U(x + th, y + tah)$ is concave.

(iii) Arguing as in Remark 2.1, we can prove the following statement. If V satisfies $V(x, y) = V(-x, y) = V(x, -y)$ for all x, $y \in \mathcal{B}$, then we may replace the above initial condition $3°$ by

$3°'$ $U(0, 0) \leq 0$.

It is worth mentioning here that ± 1 transforms usually are the extremal sequences in the above class of transforms. To be more precise, we have the following decomposition.

Theorem 2.6. *Let* g *be the transform of a* $\mathcal{B}$-*valued martingale* f *by a real-valued predictable sequence* v *uniformly bounded in absolute value by* 1. *Then there exist* $\mathcal{B}$-*valued martingales* $F^j = (F_n^j)_{n \geq 0}$ *and Borel measurable functions* $\phi_j : [-1, 1] \to \{-1, 1\}$ *such that for* $j \geq 1$ *and* $n \geq 0$,

$$f_n = F_{2n+1}^j, \qquad \text{and} \qquad g_n = \sum_{j=1}^{\infty} 2^{-j}\phi_j(v_0)G_{2n+1}^j,$$

where G^j *is the transform of* F^j *by* $\varepsilon = (\varepsilon_k)_{k \geq 0}$ *with* $\varepsilon_k = (-1)^k$.

Proof. First we consider the special case when each v_n takes values in the set $\{-1, 1\}$. Let

$$D_{2n} = \frac{1 + v_0 v_n}{2} d_n,$$

$$D_{2n+1} = \frac{1 - v_0 v_n}{2} d_n.$$

Then $D = (D_n)_{n \geq 0}$ is a martingale difference sequence with respect to its natural filtration. Indeed, for even indices,

$$\mathbb{E}(D_{2n} | \sigma(D_0, D_1, \ldots, D_{2n-1})) = \mathbb{E}\left[\frac{1 + v_0 v_n}{2} \mathbb{E}(d_n | \mathcal{F}_{n-1}) \Big| \sigma(D_0, \ldots, D_{2n-1})\right] = 0.$$

Here $(\mathcal{F}_n)_{n \geq 0}$ stands for the original filtration. Furthermore, we have $D_{2n} = 0$ or $D_{2n+1} = 0$ for all n, so

$$\mathbb{E}(D_{2n+1} | \sigma(D_0, \ldots, D_{2n})) = \mathbb{E}(D_{2n+1} 1_{\{D_{2n}=0\}} | \sigma(D_0, \ldots, D_{2n}))$$
$$= \mathbb{E}(D_{2n+1} 1_{\{D_{2n}=0\}} | \sigma(D_0, \ldots, D_{2n-1})) = 0.$$

Now, let F be the martingale determined by D and let G be its transform by ε. By the definition of D we have $d_n = D_{2n} + D_{2n+1}$ and $v_0 v_n d_n = D_{2n} - D_{2n+1}$, so $f_n = F_{2n+1}$ and $g_n = v_0 G_{2n+1}$.

In the general case when the terms v_n take values in $[-1, 1]$, note that there are Borel measurable functions $\phi_j : [-1, 1] \to \{-1, 1\}$ satisfying

$$t = \sum_{j=1}^{\infty} 2^{-j} \phi_j(t), \qquad t \in [-1, 1].$$

Now, for any j, consider the sequence $v^j = (\phi_j(v_n))_{n \geq 0}$, which is predictable with respect to the filtration $(\mathcal{F}_n)_{n \geq 0}$. By the previous special case there is a martingale F^j and its transform G^j satisfying

$$f_n = F^j_{2n+1},$$
$$\sum_{k=0}^{n} \phi_j(v_k) = \phi_j(v_0) G^j_{2n+1}.$$

It suffices to multiply both sides by 2^{-j} and sum the obtained equalities to get the claimed decomposition. $\qquad \square$

2.1.3 Differential subordination

Now we shall introduce another very important class of martingale pairs. It is much wider than that considered in the previous two subsections and allows many interesting applications. Let $\mathcal{B}$ be a given separable Banach space with the norm $|\cdot|$.

Definition 2.1. Suppose that f, g are martingales taking values in $\mathcal{B}$. Then g is *differentially subordinate* to f, if for any $n = 0, 1, 2, \ldots$,

$$|dg_n| \leq |df_n|$$

with probability 1.

If g is a transform of f by a predictable sequence bounded in absolute value by 1, then, obviously, g is differentially subordinate to f. Another very important example is related to martingale square function. Suppose that f takes values in a given separable Banach space and let g be $\ell^2(\mathcal{B})$-valued process, defined by $dg_n = (0, 0, \ldots, 0, df_n, 0, \ldots)$, $n = 0, 1, 2, \ldots$ (where the difference df_n appears on the nth place). Let us treat f as an $\ell^2(\mathcal{B})$-valued process, via the embedding $f_n \sim (f_n, 0, 0, \ldots)$. Then, obviously, g is differentially subordinate to f and f is differentially subordinate to g. However,

$$||g_n||_{\ell^2(\mathcal{B})} = \left(\sum_{k=0}^{n} |df_k|^2 \right)^{1/2}$$

is the square function of f. Thus, any inequality valid for differentially subordinate martingales with values in $\ell^2(\mathcal{B})$ leads to a corresponding estimate for the square function of a $\mathcal{B}$-valued martingale. This observation will be particularly efficient when $\mathcal{B}$ is a separable Hilbert space.

Let us formulate the version of Burkholder's method when the underlying domination is the differential subordination of martingales. Let $V : \mathcal{B} \times \mathcal{B} \to \mathbb{R}$ be a given Borel function. Consider $U : \mathcal{B} \times \mathcal{B} \to \mathbb{R}$ such that

1° $U(x, y) \geq V(x, y)$ for all $x, y \in \mathcal{B}$,

2° there are Borel $A, B : \mathcal{B} \times \mathcal{B} \to \mathcal{B}^*$ such that for any $x, y \in \mathcal{B}$ and any $h, k \in \mathcal{B}$ with $|k| \leq |h|$, we have

$$U(x + h, y + k) \leq U(x, y) + \langle A(x, y), h \rangle + \langle B(x, y), k \rangle.$$

3° $U(x, y) \leq 0$ for all $x, y \in \mathcal{B}$ with $|y| \leq |x|$.

Theorem 2.7. *Suppose that U satisfies 1°, 2° and 3°. Let f, g be $\mathcal{B}$-valued martingales such that g is differentially subordinate to f and*

$$\mathbb{E}|V(f_n, g_n)| < \infty, \quad \mathbb{E}|U(f_n, g_n)| < \infty$$
$$\mathbb{E}\big(|A(f_n, g_n)||df_{n+1}| + |B(f_n, g_n)||dg_{n+1}|\big) < \infty, \tag{2.10}$$

for all $n = 0, 1, 2, \ldots$. Then

$$\mathbb{E}V(f_n, g_n) \leq 0 \tag{2.11}$$

for all $n = 0, 1, 2, \ldots$.

Proof. Note that this result goes beyond the scope of Burkholder's method described so far, since the processes f, g are no longer assumed to be simple. This is why we have assumed the Borel measurability of U, V, A and B; this is also why we have imposed condition (2.10): it guarantees the integrability of the random variables appearing below. However, the underlying idea is the same: we show that for f, g as above the process $(U(f_n, g_n))_{n \geq 0}$ is a supermartingale. To prove this, we use $2°$ to obtain, for any $n \geq 1$,

$$U(f_n, g_n) \leq U(f_{n-1}, g_{n-1}) + \langle A(f_{n-1}, g_{n-1}), df_n \rangle + \langle B(f_{n-1}, g_{n-1}), dg_n \rangle$$

with probability 1. By (2.10), both sides above are integrable. Taking the conditional expectation with respect to $\mathcal{F}_{n-1}$ yields

$$\mathbb{E}(U(f_n, g_n)|\mathcal{F}_{n-1}) \leq U(f_{n-1}, g_{n-1})$$

and, consequently,

$$\mathbb{E}V(f_n, g_n) \leq \mathbb{E}U(f_n, g_n) \leq \mathbb{E}U(f_0, g_0) \leq 0.$$

This completes the proof. $\qquad\square$

Remark 2.3. (i) Condition $2°$ seems quite complicated. However, if U is of class C^1, it is easy to see that the only choice for A and B is to take the partial derivatives U_x and U_y, respectively. Then $2°$ is equivalent to saying that

$2°'$ for any x, y, h, $k \in \mathcal{B}$ with $|k| \leq |h|$, the function $G = G_{x,y,h,k} : \mathbb{R} \to \mathbb{R}$, given by

$$G(t) = U(x + th, y + tk),$$

is concave.

In a typical situation, U is piecewise C^1, and then $2°'$ still implies $2°$: one takes $A(x, y) = U_x(x, y)$, $B(x, y) = U_y(x, y)$ for (x, y) at which U is differentiable and, for remaining points, one defines A and B as appropriate limits of U_x and U_y.

(ii) Condition $2°'$ can be simplified further. Obviously, it is equivalent to

$$G''(t) \leq 0 \tag{2.12}$$

at the points where G is twice differentiable, and

$$G'(t-) \leq G'(t+) \tag{2.13}$$

for the remaining t. However, the family $(G_{x,y,h,k})_{x,y,h,k}$ enjoys the following translation property:

$$G_{x,y,h,k}(t + s) = G_{x+th,y+tk,h,k}(s) \quad \text{for all } s, t.$$

Hence, it suffices to check (2.12) and (2.13) for $t = 0$ only (but, of course, for all appropriate x, y, h and k).

2.2 Further remarks

Now let us make some general observations, some of which will be frequently used in the later parts of the monograph.

(i) The technique can be applied in the situation when the pair (f, g) takes values in a set D different from $\mathcal{B} \times \mathcal{B}$. For example, one can work in $D = \mathbb{R}_+ \times \mathbb{R}$ or $D = \mathcal{B} \times [0, 1]$, and so on. This does not require any substantial changes in the methodology; one only needs to ensure that all the points (x, y), $(x + t_i, y + \varepsilon t_i)$, and so on, appearing in the statements of $1°$, $2°$ and $3°$, belong to the considered domain D.

(ii) A remarkable feature of Burkholder's method is its efficiency. Namely, if we know a priori that a given estimate

$$\mathbb{E}V(f_n, g_n) \leq 0, \qquad n = 0, 1, 2, \ldots,$$

or, more generally,

$$\mathbb{E}V(f_n, g_n) \leq c, \qquad n = 0, 1, 2, \ldots,$$

is valid, then it can be established using the above approach. In particular, the technique can be used to derive the optimal constants in the inequalities under investigation.

(iii) Formula (2.1) can be used to narrow the class of functions in which we search for the suitable majorant. Here is a typical example. Suppose we are interested in showing the strong-type inequality

$$\mathbb{E}|g_n|^p \leq C^p \mathbb{E}|f_n|^p, \qquad n = 0, 1, 2, \ldots,$$

for all real martingales f and their ± 1-transforms g. This corresponds to the choice $V(x, y) = |y|^p - C^p|x|^p$, $x, y \in \mathbb{R}$. We have that V is homogeneous of order p and this property carries over to the function U^0. It follows from the fact that $(f, g) \in M(x, y)$ if and only if $(\lambda f, \lambda g) \in M(\lambda x, \lambda y)$ for any $\lambda > 0$. Thus, we may search for U in the class of functions which are homogeneous of order p. This reduces the dimension of the problem. Indeed, we need to find an appropriate function u of only one variable and then let $U(x, y) = |y|^p u(|x|/|y|)$ for $|y| \neq 0$ and $U(x, 0) = c|x|^p$ for some c. As another example, suppose that V satisfies the symmetry condition $V(x, y) = V(x, -y)$ for all x, y. Then we may search for U in the class of functions which are symmetric with respect to the second variable.

(iv) A natural way of showing that the constant in a given inequality is the best possible is to construct appropriate examples. However, this can be shown by the use of the reverse implication of Burkholder's method. That is, one assumes the validity of an estimate with a given constant C and then exploits the properties $1°$, $2°$ and $3°$ of the function U^0 to obtain the lower bound for C. This approach is often much simpler and less technical, and will be frequently used in the considerations below.

(v) Suppose that we want to establish the inequality $\mathbb{E}V(f_n, g_n) \leq 0$, $n = 0, 1, 2, \ldots$ for some class of pairs (f, g). As we have seen above, we have to find a corresponding special function. Assume that we have been successful and found an appropriate U. Does it have to coincide with U^0? In other words, is the special function uniquely determined? In general the answer is no: typically there are many functions satisfying $1°$, $2°$ and $3°$. In fact, as we shall see, it may happen that the careful choice of one of them is a key to avoid many complicated calculations. However, the formula defining U^0 is usually a good point to start the search from: see below.

(vi) One might expect that the constants in the martingale inequalities under differential subordination are larger than those in the estimates for ±1-transforms: indeed, the differential subordination is a much weaker condition. However, that is not exactly the case, at least in the non-maximal setting. More precisely, we shall see that in general the constants are the same even when we work with Hilbert-space-valued processes; on the other hand, the constants do differ when we leave the Hilbert-space setting.

This remark is related to the approach we use throughout. Namely, if we want to establish a given inequality for Hilbert-space-valued processes, we first try to solve the corresponding boundary value problem for ±1-transforms of real-valued martingales. If we are successful, we interpret the absolute values appearing in the formula for the special function as the corresponding norm in a Hilbert space and try to verify the conditions $1°$, $2°$ and $3°$. A similar reasoning will be conducted in the more general semimartingale setting.

(vii) Burkholder's method is closely related to the theory of boundary value problems. We shall illustrate this in the simplest setting in $\mathbb{R}^2$, but it will be clear how to get more complicated modifications. Suppose that $D \subseteq \mathbb{R}^2$ is a given set which is diagonally convex: that is, any section of D of slope ±1 is convex. Assume in addition that D is of the form $B \cup C$, where C is a nonempty set (for example: $D = B \cup \partial B$ or $D = \emptyset \cup D$). Let $\beta : C \to \mathbb{R}$ be a given function. The problem is: assume that there is a finite diagonally concave function U on D such that $U \geq \beta$ on C; what is the least such function? There is also a dual problem for diagonally convex majorants.

To deal with such a problem, let $M(x, y)$ denote the class of all simple martingales (f, g) starting from (x, y), taking values in D, satisfying $df_n = dg_n$ or $df_n \equiv -dg_n$ for all $n \geq 1$, and such that their pointwise limit (f_∞, g_∞) has all its values in C. Assume that $M(x, y)$ is nonempty for any $(x, y) \in D$. Let

$$U_\beta(x, y) = \sup\{\mathbb{E}\beta(f_\infty, g_\infty) : (f, g) \in M(x, y)\},$$
$$L_\beta(x, y) = \inf\{\mathbb{E}\beta(f_\infty, g_\infty) : (f, g) \in M(x, y)\}.$$

An argument similar to that used in the proof of Theorems 2.1 and 2.2 leads to the following result, which can be regarded as a probabilistic answer to the boundary value problem above.

Theorem 2.8. *The function U_β is the least diagonally concave function on D which majorizes β on C, provided at least one such function exists. The function L_β is the greatest diagonally biconvex function on D which minorizes β on C, provided at least one such function exists.*

Though the boundary value problems described above are different from those appearing in the classical boundary value theory, there are many similarities and connections. For example, diagonal concavity corresponds to superharmonicity. If (f,g) belongs to the class $M(x,y)$ just defined above and U is diagonally concave, then $(U(f_n, g_n))_{n \geq 1}$ is a supermartingale. This is analogous to the classical result of Doob concerning the composition of superharmonic functions with Brownian motion. For further connections to the classical boundary value theory, see Chapter 6.

(viii) Burkholder's method can be generalized to a much wider setting, in which the differential subordination is replaced by an abstract domination. We shall describe the extension for real-valued martingales. Suppose that $\ll$ is a relation, given on the pairs (d, e) of mean-zero simple random variables, depending only on their common distribution, such that $0 \ll 0$. The relation admits its conditional version $\ll_\mathcal{G}$ for any sub-σ-field $\mathcal{G}$ of $\mathcal{F}$. We will say that f $\ll$-dominates g or that g is $\ll$-dominated by f, if $dg_n \ll_{\mathcal{F}_{n-1}} df_n$ for all $n = 1, 2, \ldots$ (note that $n \neq 0$).

Let $D = \mathbb{R} \times \mathbb{R}$ and let $V : D \to \mathbb{R}$ be a function, not necessarily Borel or even measurable. Let $x, y \in \mathbb{R}$ be fixed and denote by $M(x, y)$ the class of all pairs (f, g) of simple f, g starting from x, y, respectively, such that g is $\ll$-dominated by f. We have that $M(x, y)$ is nonempty, as it contains the deterministic constant pair (x, y). Suppose that our goal is to establish the estimate

$$\mathbb{E}V(f_n, g_n) \leq 0, \qquad n = 0, 1, 2, \ldots, \qquad (2.14)$$

for all simple martingales f, g such that g is $\ll$-dominated by f and $|g_0| \leq |f_0|$ almost surely. The appropriate versions of $1°$, $2°$ and $3°$ can be stated as follows.

$1°$ $U \geq V$ on D,

$2°$ for all $(x, y) \in D$ and any simple random variables d, e such that $e \ll d$, we have $\mathbb{E}U(x + d, y + e) \leq U(x, y)$,

$3°$ $U(x, y) \leq 0$ for all (x, y) such that $|y| \leq |x|$.

Then the argumentation from the previous section yields the following result.

Theorem 2.9. *If there is U satisfying $1°$, $2°$ and $3°$, then the inequality (2.14) is valid. On the other hand, if (2.14) holds and the function*

$$U^0(x, y) = \sup\{\mathbb{E}V(f_n, g_n) : (f, g) \in M(x, y), \, n = 0, 1, 2, \ldots\}$$

is finite, then it is the least function satisfying $1°$, $2°$ and $3°$.

Clearly, the differential subordination corresponds to the relation $e \ll d$ iff $|e| \leq |d|$ almost surely. There are also other interesting and natural examples of relations. In Chapter 3 we shall encounter the so-called *weak domination* of martingales; see also the bibliographical notes at the end of that chapter.

2.3 Integration method

The final part of this short chapter is devoted to a simple, but a very powerful enhancement of Burkholder's method. This argument, if applicable, makes the computations much easier to handle. Fix a function $V : \mathbb{R}^2 \to \mathbb{R}$ and suppose that our goal is to establish the estimate

$$\mathbb{E}V(f_n, g_n) \leq 0, \qquad n = 0, 1, 2, \ldots, \tag{2.15}$$

for all simple real-valued martingales f, g such that g is a ± 1-transform of f. The idea is to find first a "simple" function $u : \mathbb{R}^2 \to \mathbb{R}$, which enjoys the corresponding conditions 2° and 3°. The next step is to take a kernel $k : [0, \infty) \to [0, \infty)$ such that

$$\int_0^\infty k(t) \big| u(x/t, y/t) \big| \mathrm{d}t < \infty \qquad \text{for all } x, y \in \mathbb{R},$$

and to define $U : \mathbb{R}^2 \to \mathbb{R}$ by

$$U(x, y) = \int_0^\infty k(t) u(x/t, y/t) \mathrm{d}t. \tag{2.16}$$

Since f, g are simple and for any $t > 0$, g/t is a ± 1-transform of f/t, we may use 2°, 3° and Fubini's theorem to obtain

$$\mathbb{E}U(f_n, g_n) \leq \mathbb{E}U(f_0, g_0) \leq 0.$$

If the kernel k and the function u were chosen so that the majorization $U \geq V$ holds, then (2.15) follows.

This can be used also to study inequalities for martingales and their differential subordinates. To see this, assume that V, u and k are as above (of course, here 2° and 3° are the versions corresponding to differential subordination). To repeat the above reasoning, we need an argument which will justify the use of Fubini's theorem. So, assume that f, g satisfy the integrability property

$$\mathbb{E} \int_0^\infty k(t) \big| u(f_n/t, g_n/t) \big| \mathrm{d}t < \infty$$

for all n. Then, as before, if we chose u and k appropriately, we obtain the chain of inequalities

$$\mathbb{E}V(f_n, g_n) \leq \mathbb{E}U(f_n, g_n) \leq \mathbb{E}U(f_0, g_0) \leq 0, \qquad n = 0, 1, 2, \ldots.$$

The above argument can also be used in a different manner, as a convenient tool to avoid complicated calculations. Consider a typical situation: we want to establish a given inequality of the form $\mathbb{E}V(f_n, g_n) \leq 0$, $n = 0, 1, 2, \ldots$, say, for ± 1-transforms. Some arguments and observations lead to a candidate U for the special function and the next step is to verify the corresponding conditions $1°$, $2°$ and $3°$. Usually the proof of the concavity property is quite elaborate, especially if we work in the Hilbert-space-valued setting. To avoid this problem, one may try to find a representation (2.16) for some appropriate kernel k and a function u (for which the verification of $2°$ is relatively simple). This will be illustrated by many examples below.

2.4 Notes and comments

Section 2.1. The technique described above has its roots at Burkholder's works from early 80s, though some preliminary results in this direction can be found in the papers [13] by Bollobás, [19] by Burkholder and [53] by Cox. The boundary value problems (in the non-classical sense described above) appear for the first time in [20] and [21] in the study of geometric properties of UMD Banach spaces; see also later papers [23], [26] and [37] by Burkholder. The seminal paper [24] contains the deep results concerning the method for real-valued martingales and is in fact the first exposition in which the approach was used to derive optimal constants in various estimates (see the end of Chapter 3 for details). For the refinement and simplification of the technique, the reader is referred to the survey [32] by Burkholder. The generalization of the method to a general domination $\ll$ defined on the difference sequences $(df_n)_{n\geq 0}$, $(dg_n)_{n\geq 0}$, as well as many examples and applications, can be found in the monograph [112] by Kwapień and Woyczyński. Burkholder's method and the notion of differential subordination have been partially extended to the non-commutative setting: see [128].

We would like to mention here another technique, which is very closely related to Burkholder's method. This is the so-called Bellman's method, which also rests on the construction of an appropriate special function. The technique has been used very intensively mostly in analysis, in the study of Carleson embedding theorems, BMO estimates, square function inequalities, bounds for maximal operators, estimates for A_p weights and many other related results. See, e.g., Dindoš and Wall [68], Nazarov and Treil [119], Nazarov, Treil and Volberg [120], Petermichl and Wittwer [176], Slavin and Vasyunin [186], Vasyunin [193], Vasyunin and Volberg [194], [195], Wittwer [201], [202], [203] and references therein.

Section 2.2. The material presented there is a combination of various remarks and observations from the literature on the subject. In particular, see [24] and [32].

Section 2.3. The integration method was introduced by the author in his Ph.D. thesis during the study of the estimates for weakly dominated martingales: see [124]. Then it was successively investigated in subsequent papers (see [125], [126], [137], [145] and [146]).

Chapter 3

Martingale Inequalities in Discrete Time

This part of the monograph contains a study of inequalities for discrete-time martingales, both in the Hilbert-space and Banach-space-valued setting. It is worth stressing here that the setting of Hilbert-space-valued differentially subordinate martingales has been studied intensively and, essentially, the sharp versions of all the crucial estimates arising in this context have been successfully proved. On the other hand, much is to be done in the non-Hilbert case, in which *no* optimal constants are known.

3.1 Weak type estimates, general case

3.1.1 Formulation of the results

We start from weak type inequalities for martingales and their differential subordinates, taking values in a given separable Hilbert space $\mathcal{H}$. For the sake of the reader's convenience, we have decided to split the reasoning into two parts: first we deal with the case in which the weak and the strong norms are of the same order; then we present the results in the general setting.

Theorem 3.1. *Assume that f, g are $\mathcal{H}$-valued martingales such that g is differentially subordinate to f.*

(i) *If $0 < p < 1$, then the inequality*

$$||g||_{p,\infty} \leq c_{p,p}||f||_p \tag{3.1}$$

does not hold in general with any finite $c_{p,p}$, even if we assume that f is real valued and g is its ± 1-transform.

(ii) *If $1 \leq p \leq 2$, then (3.1) holds with*

$$c_{p,p} = \left(\frac{2}{\Gamma(p+1)}\right)^{1/p}.$$

The constant is the best possible even if f is assumed to be real valued and g is its ± 1-transform.

(iii) *If $p > 2$, then (3.1) holds with*

$$c_{p,p} = \left(\frac{p^{p-1}}{2}\right)^{1/p}.$$

The constant is the best possible even if f is assumed to be real valued and g is its ± 1-transform.

Next, we turn to the case of different orders. Let

$$c_{p,q} = \begin{cases} \infty & \text{if } q > p \text{ or } 0 < q \leq p < 1, \\ 1 & \text{if } 0 \leq q \leq 2 \leq p < \infty, \\ \left(\frac{2}{\Gamma(p+1)}\right)^{1/p} & \text{if } 0 < q \leq p, 1 \leq p < 2, \\ 2^{1/p-2/q} q^{(p-1)/p} \left(\frac{p-q}{p-2}\right)^{(p-1)(p-q)/(pq)} & \text{if } 2 < q \leq p < \infty. \end{cases}$$

Theorem 3.2. *Assume that f, g are $\mathcal{H}$-valued martingales such that g is differentially subordinate to f. Then for any $0 < p, q < \infty$ we have*

$$||g||_{q,\infty} \leq c_{p,q}||f||_p \tag{3.2}$$

and the constant is the best possible. It is already the best possible if f is assumed to be real valued and g is its ± 1-transform.

3.1.2   Proof of Theorem 3.1, the underlying concept

The estimate (3.1) can be investigated using Burkholder's method. First, by homogeneity, it reduces to the inequality

$$\mathbb{P}(|g|^* \geq 1) \leq c_{p,p}^p||f||_p^p. \tag{3.3}$$

If $p \geq 1$, the bound above can be further simplified to the form

$$\mathbb{P}(|g_n| \geq 1) \leq c_{p,p}^p \mathbb{E}|f_n|^p, \qquad n = 0, 1, 2, \ldots, \tag{3.4}$$

or $\mathbb{E}V(f_n, g_n) \leq 0$, where $V(x, y) = 1_{\{|y| \geq 1\}} - c_{p,p}^p|x|^p$. To justify this, we make use of the following stopping time argument. For a fixed $\varepsilon \in (0, 1)$, introduce

$\tau = \inf\{n \geq 0 : |g_n| \geq 1 - \varepsilon\}$. Clearly, we have that

$$\mathbb{P}(|g|^* \geq 1) \leq \mathbb{P}(|g_n| \geq 1 - \varepsilon \text{ for some } n) = \mathbb{P}(|g_{\tau \wedge n}| \geq 1 - \varepsilon \text{ for some } n)$$

$$= \mathbb{P}\left(\bigcup_{n \geq 0} \{|g_{\tau \wedge n}| \geq 1 - \varepsilon\}\right) = \lim_{n \to \infty} \mathbb{P}(|g_{\tau \wedge n}| \geq 1 - \varepsilon), \tag{3.5}$$

since the events $\{|g_{\tau \wedge n}| \geq 1 - \varepsilon\}$ are nondecreasing. It is evident that the martingale $(g_{\tau \wedge n}/(1 - \varepsilon))_{n \geq 0}$ is differentially subordinate to $f/(1 - \varepsilon)$, so applying (3.4) to these sequences gives

$$\mathbb{P}(|g_{\tau \wedge n}| \geq 1 - \varepsilon) \leq \frac{c_{p,p}^p}{(1 - \varepsilon)^p} \mathbb{E}|f_n|^p \leq \frac{c_{p,p}^p}{(1 - \varepsilon)^p} ||f||_p^p, \qquad n = 0, 1, 2, \ldots.$$

This yields (3.3), because of (3.5) and the fact that ε was arbitrary.

We consider the cases $0 < p < 1$, $p = 1$, $1 < p < 2$ and $p > 2$ separately. When $p = 2$, the claim is trivial: it follows immediately from Chebyshev's inequality and

$$||g_n||_2^2 = \mathbb{E} \sum_{k=0}^{n} |dg_k|^2 \leq \mathbb{E} \sum_{k=0}^{n} |df_k|^2 = ||f_n||_2^2, \qquad n = 0, 1, 2, \ldots,$$

with equality attained for constant martingales $f = g = (1, 1, 1, \ldots)$.

3.1.3 Proof of Theorem 3.1, the case $0 < p < 1$

The inequality does not hold with any finite constant, even for ± 1-transforms of real martingales. This will be clear from the following example. Let M, $N \geq 2$ be integers and set $\delta = 1/(2N)$. Consider independent centered random variables $X_1, X_2, \ldots$ such that $X_{2n} \in \{-\delta, M - \delta\}$ and $X_{2n-1} \in \{\delta, -M\}$ with probability 1, $n = 1, 2, \ldots$. Introduce the stopping time $\tau = \inf\{n : |X_n| \neq \delta\}$. Set $f_0 = g_0 \equiv 0$,

$$df_n = (-1)^{n+1} dg_n = X_n 1_{\{\tau \geq n\}}, \qquad n = 1, 2, \ldots, N,$$

and $df_n = dg_n \equiv 0$ for $n > 2N$. Then df, dg are martingale difference sequences and g is a ± 1 transform of f. It follows from the construction that $g^* \geq 1$ almost surely. Indeed, we have $g_k = k\delta$ for $k = 1, 2, \ldots, 2N$ on $[\tau > 2N]$; so in particular, $g_{2N} = 1$ on this set. If $\tau \leq 2N$, then $g_\tau = g_{\tau-1} + dg_\tau \leq (\tau - 1)\delta - (M - \delta) \leq 1 - M \leq -1$. Therefore, g satisfies $||g||_{p,\infty} \geq 1$. On the other hand, if $1 \leq k \leq 2N$, then $|f_k| \in \{0, \delta\}$ on $\{\tau > k\}$ and $|f_k| = M$ on $\{\tau \leq k\}$. Therefore,

$$\mathbb{E}|f_k|^p \leq \delta^p + M^p \mathbb{P}(\tau \leq k) \leq \delta^p + M^p \mathbb{P}(\tau \leq 2N) = \delta^p + M^p \left[1 - \left(\frac{M - \delta}{M + \delta}\right)^N\right].$$

The latter expression can be made arbitrarily small if M and N are taken sufficiently large. To see this, fix $\varepsilon > 0$ and take M such that $M(1 - \exp(-1/M)) \leq 2$

and $M^{p-1} \leq \varepsilon$. Then, take N such that $\delta^p = (2N)^{-p} \leq \varepsilon$ and

$$\left(\frac{M-\delta}{M+\delta}\right)^N = \left(1 - \frac{2\delta}{M+\delta}\right)^N \geq \exp\left(-\frac{1}{M+\delta}\right) - \frac{\varepsilon}{M^p}.$$

For those M and N, we have

$$\delta^p + M^p \left[1 - \left(\frac{M-\delta}{M+\delta}\right)^N\right] \leq \varepsilon + M^{p-1} \cdot M \left[1 - \exp\left(-\frac{1}{M+\delta}\right)\right] + \varepsilon \leq 4\varepsilon.$$

Thus, for this choice of M and N, we have $\|f\|_p \leq (4\varepsilon)^{1/p}$ and therefore the weak type estimate does not hold with any finite $c_{p,p}$.

3.1.4 Proof of Theorem 3.1, the case $p = 1$

The case $p = 1$, as we will see below, plays a distinguished role. Therefore we have decided to present it separately, though the situation for $p \in (1, 2]$ is quite similar.

Proof of (3.4) with $c_{1,1} = 2$. Let $V : \mathcal{H} \times \mathcal{H} \to \mathbb{R}$ be given by $V(x, y) = 1_{\{|y| \geq 1\}} - 2|x|$ and introduce $U_1 : \mathcal{H} \times \mathcal{H} \to \mathbb{R}$ by

$$U_1(x, y) = \begin{cases} |y|^2 - |x|^2 & \text{if } |x| + |y| < 1, \\ 1 - 2|x| & \text{if } |x| + |y| \geq 1. \end{cases} \tag{3.6}$$

We will show that U_1 satisfies the conditions $1°$, $2°$ and $3°$. To show the majorization $V \leq U_1$, observe that this is trivial if $|x| + |y| \geq 1$; if $|x| + |y| < 1$, then

$$U_1(x, y) = |y|^2 - |x|^2 \geq -|x|^2 \geq -2|x| = V(x, y).$$

Now we turn to $2°$. We shall show that for any x, y, h, $k \in \mathcal{H}$ satisfying $|h| \geq |k|$ we have

$$U_1(x + h, y + k) \leq U_1(x, y) + A(x, y) \cdot h + B(x, y) \cdot k, \tag{3.7}$$

where

$$A(x, y) = \begin{cases} -2x & \text{if } |x| + |y| < 1, \\ -2x' & \text{if } |x| + |y| \geq 1, \end{cases}$$

$$B(x, y) = \begin{cases} 2y & \text{if } |x| + |y| < 1, \\ 0 & \text{if } |x| + |y| \geq 1. \end{cases}$$

Here we have used the notation $x' = x/|x|$ for $x \neq 0$ and $x' = 0$ for $x = 0$. The choice for A and B is almost unique, since U_1 is piecewise C^1: thus we are forced to take $A = U_{1x}$ and $B = U_{1y}$ on $\{(x, y) : |x| + |y| \neq 1, |x| \neq 0\}$. To prove (3.7), we start with the observation that $U_1(x, y) \leq 1 - 2|x|$ on $\mathcal{H} \times \mathcal{H}$. This is clear if

$|x| + |y| \geq 1$, and in the remaining case we have $|y|^2 - |x|^2 < (1 - |x|)^2 - |x|^2 = 1 - 2|x|$. Therefore, if $|x| + |y| \geq 1$, we can write

$$U_1(x + h, y + k) \leq 1 - 2|x + h| \leq 1 - 2|x| - 2x' \cdot h,$$

which is (3.7). If $|x| + |y| < 1$ and $|x + h| + |y + k| \leq 1$, then

$$U_1(x + h, y + k) = U(x, y) + 2y \cdot k - 2x \cdot h + |k|^2 - |h|^2,$$

and (3.7) follows from the condition $|h| \geq |k|$. It remains to prove the estimate in the case $|x| + |y| < 1$, $|x + h| + |y + k| > 1$; then it reads

$$1 - 2|x + h| \geq |y|^2 - 2(y \cdot k) - |x|^2 + 2(x \cdot h). \tag{3.8}$$

If $|x + h| \leq 1$, then

$$(1 - |x + h|)^2 < |y + k|^2 \leq |y|^2 + 2(y \cdot k) + |h|^2,$$

which is (3.8). If $|x + h| > 1$, then $|h| \geq |x + h| - |x| > 1 - |x| \geq |y|$ and

$$(|x+h|-1)^2 \leq (|x|+|h|-1)^2 \leq (|h|-|y|)^2 = |y|^2-2|y||h|+|h|^2 \leq |y|^2-2(y\cdot k)+|h|^2$$

and (3.8) follows. Finally, the initial condition 3° is obvious; by the symmetry of V it suffices to verify the inequality $U_1(0,0) \leq 0$ (see Remark 2.1). Therefore, U_1 has the desired properties and hence the weak type estimate holds. □

Sharpness. This is simple: put $f_0 = g_0 \equiv 1/2$ and let $df_1 = -dg_1$ be a centered random variable taking the values $-1/2$ and $3/2$. Finally, set $df_2 = df_3 = \cdots \equiv 0$, $dg_2 = dg_3 = \cdots \equiv 0$. Then we have $|g_1| \equiv 1$ and $||f||_1 = ||f_2||_1 = 1/2$, so both sides of (3.1) are equal. □

Now we will explain how one obtains the formula (3.6) above.

On the search of the suitable majorant. There are two objects to be determined: the a priori unknown optimal value of the constant $c_{1,1}$ and the corresponding special function. A typical first step is to study the given problem in the simplest setting: let us assume that f and g are real valued and g is a ± 1 transform of f. To gain some intuition about the sought-for special function, we start with U^0 given by (2.1). That is, let

$$U^0(x, y) = \sup\{\mathbb{P}(|g_n| \geq 1) - c\mathbb{E}|f_n|\},$$

where the supremum is taken over all n and all $(f, g) \in M(x, y)$ (recall that $M(x, y)$ consists of $\mathbb{R}^2$-valued simple martingales (f, g) starting from (x, y) such that $dg_k \equiv df_k$ or $dq_k \equiv -df_k$ for all $k = 1, 2, \ldots$). By Theorem 2.3, the function U^0 satisfies 1°, 2° and 3°. Thus, we arrived at the following boundary value problem: find the explicit formula for the least function enjoying these three conditions.

In the three steps below, we shall derive the explicit form of U^0.

Step 1. The case $|y| \geq 1$. Our first observation is that

$$U^0(x,y) = 1 - c|x| \qquad \text{if } |y| \geq 1. \tag{3.9}$$

Indeed, if $(f, g) \in M(x, y)$, then $\mathbb{P}(|g_n| \geq 1) \leq 1$ and $\mathbb{E}|f_n| \geq |x|$ for any n. This gives the inequality in one direction; letting $f \equiv x$ and $g \equiv y$ (or using 1°) yields the reverse estimate.

Step 2. The case $|x| + |y| \geq 1$. Next we show that

$$U^0(x,y) = 1 - c|x| \qquad \text{if } |x| + |y| \geq 1. \tag{3.10}$$

To prove this, it suffices to deal with the case $|x| + |y| \geq 1 > |y|$. Furthermore, we may assume $x > 0$, $y \geq 0$, since $U^0(x, y) = U^0(-x, y) = U^0(x, -y)$ for all x, y: see Section 2.2. Reasoning as in the previous step, we obtain $U^0(x, y) \leq 1 - c|x|$. On the other hand, consider the following element of $M(x, y)$: let $df_1 = -dg_1$ be a centered random variable taking values $-x$ and $y + 1$ only and set $df_n = dg_n \equiv 0$ for $n \geq 2$. We have that $f \geq 0$, so $\mathbb{E}|f_n| = x$ for all n, and, in addition, g_1 takes values in the set $\{-1, x+y\}$. This implies the reverse inequality $U^0(x, y) \geq 1 - c|x|$ and yields (3.10).

Step 3. The case $|x| + |y| < 1$. It remains to find U^0 on the set $K = \{(x, y) : |x| + |y| < 1\}$. For (x, y) lying in K, take a centered random variable d for which $|x + d| + |y + d| = 1$ with probability 1. It is easy to verify that almost surely, d takes the values $d_- = (1 - x - y)/2$ and $d_+ = (-1 - x - y)/2$, with probabilities $p_- = (1 + x + y)/2$ and $p_+ = (1 - x - y)/2$, respectively. Hence, by 2° and (3.10), we obtain

$$U^0(x, y) \geq p_- U^0(x + d_-, y + d_-) + p_+ U^0(x + d_+, y + d_+)$$

$$= 1 - c\left[\left|x + \frac{1 - x - y}{2}\right| \cdot \frac{1 + x + y}{2} + \left|x + \frac{-1 - x - y}{2}\right| \cdot \frac{1 - x - y}{2}\right]$$

$$= 1 - \frac{c}{2}(1 + |x|^2 - |y|^2).$$

Now apply the initial condition to get

$$c \geq 2 \tag{3.11}$$

and the inequality above takes form

$$U^0(x, y) \geq |y|^2 - |x|^2.$$

Now we *conjecture* that we have equality above. This leads to the function given in (3.6), at least in the case $\mathcal{H} = \mathbb{R}$. Interpreting $|\cdot|$ as the norm in the Hilbert space, we can treat the obtained function U_1 as a function on $\mathcal{H} \times \mathcal{H}$. □

Remark 3.1. Note that in the search above, we have presented an alternative proof of the fact that the constant 2 is the best possible even for ± 1-transforms of real martingales: see (3.11). This type of approach has been already mentioned in Remark (iv) in Section 2.2.

Remark 3.2. When $\mathcal{H} = \mathbb{R}$, the function U_1 is optimal in the sense that it coincides with the corresponding U^0; this can be quite easily extracted from the above considerations. But when $\mathcal{H}$ is at least two-dimensional, is this still the case? No. Burkholder [26] showed that the optimal choice is given by

$$\overline{U}(x, y) = \begin{cases} 1 - (1 + 2|x|^2 - 2|y|^2 + |x + y|^2|x - y|^2)^{1/2} & \text{if } |x + y| \vee |x - y| < 1, \\ 1 - 2|x| & \text{if } |x + y| \vee |x - y| \geq 1. \end{cases}$$

A DUAL FUNCTION TO U_1. Before we proceed, let us introduce the function $U_\infty : \mathcal{H} \times \mathcal{H} \to \mathbb{R}$ by

$$U_\infty(x, y) = \begin{cases} 0 & \text{if } |x| + |y| \leq 1, \\ (|y| - 1)^2 - |x|^2 & \text{if } |x| + |y| > 1. \end{cases} \tag{3.12}$$

This function, as we shall see below, can be considered as a dual function to U_1. We shall prove the following fact.

Lemma 3.1. *The function U_∞ has the properties* $2°$ *and* $3°$.

Proof. This can be done in the similar manner as before. We show that

$$U_\infty(x + h, y + k) \leq U_\infty(x, y) + A(x, y) \cdot h + B(x, y) \cdot k, \tag{3.13}$$

where

$$A(x, y) = \begin{cases} 0 & \text{if } |x| + |y| < 1, \\ -2x & \text{if } |x| + |y| \geq 1, \end{cases}$$

$$B(x, y) = \begin{cases} 0 & \text{if } |x| + |y| < 1, \\ 2y - 2y' & \text{if } |x| + |y| \geq 1. \end{cases}$$

It can be easily verified that $U_\infty(x, y) \geq (|y| - 1)^2 - |x|^2$ on $\mathcal{H} \times \mathcal{H}$. Hence, if $|x| + |y| \geq 1$, then

$$U_\infty(x + h, y + k) \leq (|y + k| - 1)^2 - |x + h|^2$$
$$= (|y| - 1)^2 - |x|^2 - 2[|y \mid k| - |y| + (y \cdot k) - (x \cdot h)] + |k|^2 - |h|^2$$
$$\leq U_\infty(x, y) + A(x, y) \cdot h + B(x, y) \cdot k,$$

due to the condition $|h| \geq |k|$. If $|x| + |y| < 1$ and $|x + h| + |y + k| \leq 1$, then (3.13) is obvious; both sides are equal to 0. Finally, let $|x| + |y| < 1$ and $|x + h| + |y + k| > 1$. We must show that $(|y + k| - 1)^2 - |x + h|^2 < 0$. If $|y + k| \leq 1$, then this is equivalent to $|x + h| + |y + k| > 1$, the inequality we have assumed. If $|y + k| > 1$, then $|k| \geq |y + k| - |y| > 1 - |y| > |x|$ and

$$(|y + k| - 1)^2 \leq (|y| + |k| - 1)^2 \leq (|k| - |x|)^2 \leq (|h| - |x|)^2 \leq |x + h|^2.$$

Finally, to prove 3°, we use (3.13) and obtain

$$U_\infty(x, y) \leq U_\infty(0, 0) + A(0, 0) \cdot x + B(0, 0) \cdot y = 0,$$

whenever $|y| \leq |x|$. This yields the claim.   □

Let us prove here an auxiliary fact, to be needed later.

Lemma 3.2. *For any* $x, y \in \mathcal{H}$ *we have*

$$|U_1(x, y)| \leq U_1(0, y) - U_1(x, 0) \tag{3.14}$$

and

$$|U_\infty(x, y)| \leq 1 + |x|^2 + |y|^2. \tag{3.15}$$

Proof. We may assume that $\mathcal{H} = \mathbb{R}$. First we show (3.14). If $|x| + |y| \leq 1$, then the estimate is trivial: $||y|^2 - |x|^2| \leq |y|^2 + |x|^2$. If $|x| \geq 1$, then the inequality holds because $|U_1(x, y)| = 2|x| - 1 = -U_1(x, 0)$. Finally, if $1 < |x| + |y| < 1 + |y|$, then

$$U_1(0, y) - U_1(x, 0) \geq U_1(0, 1 - |x|) - U_1(x, 0) = 2|x|^2 - 2|x| + 1 \geq |1 - 2|x|| = U_1(x, y).$$

Now we turn to (3.15): it is clear if $|x| + |y| \leq 1$, while for remaining x, y,

$$|U_\infty(x, y)| \leq (|y| - 1)^2 + |x|^2 \leq 1 + |x|^2 + |y|^2$$

and we are done.   □

3.1.5   Proof of Theorem 3.1, the case $1 < p < 2$

We shall prove a slightly more general result. Let Φ be an increasing convex function on $[0, \infty)$ such that Φ is twice differentiable on $(0, \infty)$, Φ' is strictly concave, $\lim_{t \to \infty} \Phi''(t) = 0$ and $\Phi(0) = \Phi'(0+) = 0$. This is satisfied by the function $\Phi(t) = t^p$, $1 < p < 2$, but also, for example, by $\Phi(t) = t \log(t + 1)$ and $\Phi(t) = t - \log(t + 1)$.

Theorem 3.3. *If* f, g *are* $\mathcal{H}$*-valued martingales and* g *is differentially subordinate to* f, *then, for all* $\lambda > 0$,

$$\mathbb{P}(g^* \geq \lambda) \leq C_\Phi \sup_n \mathbb{E}\Phi(|f_n|/\lambda), \tag{3.16}$$

where

$$C_\Phi = 2 \left(\int_0^\infty \Phi(s) e^{-s} ds \right)^{-1}. \tag{3.17}$$

The constant is the best possible. It is the best possible even if f *is real valued and* g *is assumed to be its* ± 1 *transform.*

Note that the choice $\Phi(t) = t^p$, $1 < p < 2$, yields the estimate (3.1).

Proof of (3.16). It suffices to show the estimate for $\lambda = 1$. With no loss of generality, we may and do assume that

$$\sup_n \mathbb{E}\Phi(|f_n|) < \infty, \tag{3.18}$$

since otherwise there is nothing to prove. By the stopping time argument, we reduce the inequality (3.16) to the bound $\mathbb{E}V_\Phi(f_n, g_n) \leq 0$, $n = 0, 1, 2, \ldots$, where

$$V_\Phi(x, y) = 1_{\{|y| \geq 1\}} - C_\Phi \Phi(|x|).$$

Let us exploit the integration method with u given by (3.6) and the kernel

$$k(t) = \frac{C_\Phi}{2} t^2 \left[\Phi''(t-1) - e^{t-1} \int_{t-1}^{\infty} e^{-s} \Phi''(s) ds \right] 1_{\{t \geq 1\}}.$$

Note that $k \geq 0$, since Φ'' is nonincreasing. Lengthy but straightforward computations show that $U_\Phi : \mathcal{H} \times \mathcal{H} \to \mathbb{R}$ given by

$$U_\Phi(x, y) = \int_0^{\infty} k(t) u(x/t, y/t) dt$$

admits the following explicit formula: if $|x| + |y| \leq 1$, then $U_\Phi(x, y) = |y|^2 - |x|^2$; if $|x| + |y| > 1$, then

$$U_\Phi(x, y) = 1 - |y| C_\Phi \Phi(|x| + |y| - 1) - C_\Phi(1 - |y|) e^{|x|+|y|-1} \int_{|x|+|y|-1}^{\infty} e^{-s} \Phi(s) ds.$$

Now, let us show the majorization $U_\Phi \geq V_\Phi$. Note that it suffices to establish this for $\mathcal{H} = \mathbb{R}$ and x, $y \geq 0$, since U_Φ and V_Φ depend on x and y via their norms $|x|$ and $|y|$. For a fixed x, the function $U_\Phi(x, \cdot)$ is nondecreasing as a function of $y \in [0, \infty)$. This follows directly from the definition of U_Φ and the fact that u also has this property. Consequently, it is enough to prove the majorization for $y \in \{0, 1\}$. If $y = 1$ this is trivial: we have $U_\Phi(x, 1) = V_\Phi(x, 1)$ for all x. If $y = 0$, then for $x > 1$,

$$\frac{d}{dx}(U_\Phi(x, 0) - V_\Phi(x, 0)) = C_\Phi e^{x-1} \left[e^{1-x} \Phi'(x) - \int_{x-1}^{\infty} e^{-s} \Phi'(s) ds \right]$$

$$= C_\Phi e^{x-1} \int_{x-1}^{\infty} e^{-s} (\Phi'(s+1) - \Phi'(s) - \Phi''(s+1)) ds,$$

which is nonnegative, by the mean value property. Therefore it remains to show that $U_\Phi(x, 0) \geq V_\Phi(x, 0)$ for $x \in [0, 1]$. This is equivalent to

$$F(x) := C_\Phi \Phi(x) - x^2 \geq 0$$

and follows from the equality $F(0) = F'(0+) = 0$, the strict concavity of F' and the fact that $F'(1) = U_{\Phi x}(1, 0) - V_{\Phi x}(1, 0) \geq 0$. Consequently, the majorization holds. Therefore, all that is left is to check the integrability which allows the use of Fubini's theorem in the inequalities

$$\mathbb{E}U_\Phi(f_n, g_n) \leq \mathbb{E}U_\Phi(f_0, g_0) \leq 0.$$

Precisely, we shall show that for any $n \geq 0$,

$$\mathbb{E} \int_0^\infty k(t)|u(f_n/t, g_n/t)|dt < \infty.$$

By (3.14), we may write

$$\int_0^\infty k(t)|u(x/t, y/t)|dt \leq U_\Phi(0, y) - U_\Phi(x, 0) \leq U_\Phi(0, y) - V_\Phi(x, 0).$$

Note that $-\mathbb{E}V_\Phi(f_n, 0) < \infty$, due to the assumption (3.18). It remains to prove that $\mathbb{E}U_\Phi(0, g_n) < \infty$ for all n. If $|y| > 1$, then

$$U_\Phi(0, y) \leq 1 + C_\Phi|y| \left[e^{|y|-1} \int_{|y|-1}^\infty e^{-s}\Phi(s)ds - \Phi(|y| - 1) \right]$$

$$\leq 1 + |y| + C_\Phi|y| \left[\Phi(|y|) - \Phi(|y| - 1) \right],$$

where the latter passage is equivalent to the inequality $U_\Phi(y, 0) \geq V_\Phi(y, 0)$, which we have already established. By the convexity of Φ, we have

$$|y| \left[\Phi(|y|) - \Phi(|y| - 1) \right] \leq |y| \left[\Phi(|y|) - \frac{|y| - 1}{|y|}\Phi(|y| - 1) \right] \leq \Phi(2|y| - 1) \leq \Phi(2|y|).$$

Therefore, all we need is the inequality $\mathbb{E}\Phi(2|g_n|) < \infty$. Since Φ' is concave with $\Phi'(0+) = 0$, we have $\Phi'(t) \geq \Phi'(2t)/2$ and integrating this from 0 to r gives $\Phi(2r) \leq 4\Phi(r)$ for any $r > 0$. Consequently,

$$\Phi(r + s) \leq \Phi(2r) + \Phi(2s) \leq 4\Phi(r) + 4\Phi(s) \qquad \text{for } r, s \geq 0. \tag{3.19}$$

This gives the desired integrability, because, by the differential subordination and the triangle inequality,

$$\mathbb{E}\Phi(2|g_n|) \leq \mathbb{E}\Phi\left(2\sum_{k=0}^n |dg_k| \right)$$

$$\leq 4\mathbb{E}\Phi\left(\sum_{k=0}^n |df_k| \right) \leq 4\mathbb{E}\Phi\left(2\sum_{k=0}^n |f_k| \right) \leq 16\mathbb{E}\Phi\left(\sum_{k=0}^n |f_k| \right).$$

It suffices to use (3.19) several times and apply (3.18) at the end. $\square$

Sharpness of (3.16). Fix $\delta \in (0, 1)$ and let (f, g) be a Markov martingale that satisfies the following conditions:

   (i) $(f_0, g_0) \equiv (1/2, 1/2)$.
   (ii) The state $(1/2, 1/2)$ leads to $(0, 1)$ or $(1, 0)$.

(iii) The state of the form $(1 + 2k\delta, 0)$ $(k = 0, 1, 2, \ldots)$ leads to $(2k\delta, 1)$ or $(1 + 2k\delta + \delta, -\delta)$.

(iv) The state of the form $(1 + 2k\delta + \delta, -\delta)$ $(k = 0, 1, 2, \ldots)$ leads to $(2k\delta + 2\delta, -1)$ or $(1 + (2k + 2)\delta, 0)$.

(v) The states lying on the lines $y = \pm 1$ are absorbing.

We do not specify the probabilities in the steps (ii)–(iv), since they are uniquely determined by the fact that (f, g) is a martingale. It is easy to check that g is a ± 1-transform of f. Moreover, we see that f is nonnegative; we will need this later. It can be readily verified that the pointwise limits f_∞, g_∞ exist, with $|g_\infty| = 1$ almost surely and $f_\infty \in \{0, 2\delta, 4\delta, \ldots\}$. Moreover,

$$\mathbb{P}(f_\infty = 0) = \mathbb{P}\left(df_1 = -\frac{1}{2}\right) + \mathbb{P}(df_2 = -1) = \frac{1 + 2\delta}{2 + 2\delta}$$

and, for $k \geq 1$,

$$\mathbb{P}(f_\infty = 2k\delta) = \mathbb{P}(df_1 = 1/2, df_2 = \delta, \ldots, df_{2k} = \delta, df_{2k+1} = \delta - 1)$$
$$+ \mathbb{P}(df_1 = 1/2, df_2 = \delta, \ldots, df_{2k+1} = \delta, df_{2k+2} = -1)$$
$$= \frac{\delta(1-\delta)^{k-1}}{2(1+\delta)^k} + \frac{\delta(1-\delta)^k}{2(1+\delta)^{k+1}} = \frac{\delta}{(1+\delta)^2} \cdot \left(1 - \frac{2\delta}{1+\delta}\right)^{k-1}.$$

Therefore, since Φ vanishes at 0,

$$\mathbb{E}\Phi(|f_\infty|) = \mathbb{E}\Phi(f_\infty) = \frac{\delta}{(1+\delta)^2} \sum_{k=1}^{\infty} \Phi(2k\delta) \left(1 - \frac{2\delta}{1+\delta}\right)^{k-1}.$$

Note that we have the elementary inequality

$$e^{-2\delta(1+\delta)} \geq 1 - \frac{2\delta}{1+\delta} \geq e^{-2\delta},$$

valid for δ sufficiently close to 0. Thus, if we take δ small enough, we may make $\mathbb{E}\Phi(|f_\infty|)$ arbitrarily close to $\frac{1}{2} \int_0^\infty \Phi(t) e^{-t} dt = C_\Phi^{-1}$. This completes the proof. $\qquad \square$

Remark 3.3. In [24], Burkholder uses a slightly different function $\overline{U}_\Phi$ to prove (3.16). Namely, he takes $\overline{U}_\Phi = V_\Phi$ on $\{(x, y) : |y| \geq 1\}$ and $\overline{U}_\Phi = U_\Phi$ on the complement of this set. In fact, it is the least special function, provided $\mathcal{H} = \mathbb{R}$.

On the search of the suitable majorant. Let us show how to obtain Burkholder's function defined in the above remark. As before, our first step is to find the function in the case $\mathcal{H} = \mathbb{R}$ and when g is assumed to be a ± 1-transform of f. Suppose that the best constant in the inequality equals c. We write down the formula (2.1):

$$U^0(x, y) = \sup\{\mathbb{P}(|g_n| \geq 1) - c\mathbb{E}\Phi(|f_n|)\},$$

where the supremum is taken over all n and all $(f, g) \in M(x, y)$. To solve the corresponding boundary value problem, first observe that U^0 satisfies the symmetry condition

$$U^0(x, y) = U^0(-x, y) = U^0(x, -y) \qquad \text{for all } x, y \in \mathbb{R} \qquad (3.20)$$

and hence it suffices to determine it in the first quadrant $[0, \infty) \times [0, \infty)$.

Step 1. *The case* $|y| \geq 1$. Note that for these y we have

$$U^0(x, y) = 1 - c\Phi(|x|). \qquad (3.21)$$

Indeed, for all f, g as above we may write $\mathbb{P}(|g_n| \geq 1) \leq 1$ and $\mathbb{E}\Phi(|f_n|) \geq \Phi(|x|)$, where the latter follows from Jensen's inequality. This gives the estimate in one direction, and the choice of constant f and g yields the reverse.

Step 2. *Key assumptions.* Here we make some conjectures on the function on the set $\{(x, y) : |y| < 1\}$. We leave the formal definition of U^0 and base our search on the two assumptions below; thus we may no longer use the notation U^0 and write U instead.

The first condition concerns regularity of the special function.

(A1) U is continuous on $\{(x, y) : |y| \leq 1\}$ and of class C^1 in the interior of this set.

Then the symmetry condition (3.20) implies

$$U_x(0, y) = U_y(x, 0) = 0 \qquad \text{for } x \in \mathbb{R} \text{ and } y \in (-1, 1). \qquad (3.22)$$

Introduce the notation $A(t) = U(t, 1) = 1 - c\Phi(t)$, $B(t) = U(t, 0)$ and $C(t) = U(0, t)$. The next assumption is the key one. Sometimes we shall refer to this type of conditions as to "structural assumptions".

(A2) The function U satisfies

$$\begin{aligned} U(x, y) &= yA(x + y - 1) + (1 - y)B(x + y) & \text{if } x + y \geq 1 \geq y \geq 0, \\ U(x, y) &= \frac{y}{x + y}C(x + y) + \frac{x}{x + y}B(x + y) & \text{if } x, y \geq 0, \, x + y < 1. \end{aligned}$$
$$(3.23)$$

In other words, when restricted to $[0, \infty) \times [0, 1]$, U is linear along the line segments of slope -1.

Step 3. *The case* $|y| < 1$. Now we will show that the conditions (A1) and (A2) lead to Burkholder's function described in Remark 3.3. Note that we may assume that $U(0, 0) = 0$: if this is not the case, replace U by the larger function $U - U(0, 0)$. Using (3.22) and (3.23), we obtain the differential equations

$$C'(y) = \frac{C(y) - B(y)}{y} \qquad \text{for } y \in [0, 1), \qquad (3.24)$$

$$B'(x) = \frac{B(x) - C(x)}{x} \qquad \text{for } x \in [0, 1), \qquad (3.25)$$

and

$$B'(x) = B(x) - A(x - 1) \qquad \text{for } x \geq 1. \qquad (3.26)$$

By (3.24) and (3.25) and the condition $B(0) = C(0) = U(0,0) = 0$ we have that $B(x) = -C(x)$ on $[0, 1]$. Plugging this into (3.24) yields $C(y) = ay^2$ for all $y \in [0, 1]$ and some fixed $a \in \mathbb{R}$. Since $C(1) = A(0) = 1$, we see that $a = 1$ and we obtain the formulas for B and C on $[0, 1]$. Now, by the second equality in (3.23), we get

$$U(x, y) = |y|^2 - |x|^2 \qquad \text{if } |x| + |y| \leq 1.$$

Similarly, solving (3.26) (recall that $A(t) = 1 - c\Phi(t)$) gives

$$B(x) = 1 - ce^x \int_x^\infty e^{-t}\Phi(t - 1)dt.$$

By the continuity of B at 1, we see that c must be equal to C_Φ, given by (3.17). Finally, by the first equation in (3.23), we get that if $1 - |x| < |y| \leq 1$,

$$U(x, y) = 1 - |y|C_\Phi\Phi(|x| + |y| - 1) - C_\Phi(1 - |y|)e^{|x|+|y|-1} \int_{|x|+|y|-1}^\infty e^{-s}\Phi(s)ds.$$

It remains to treat the absolute values in the formulas above as the norms in $\mathcal{H}$ to obtain the candidate for the function U defined on $\mathcal{H} \times \mathcal{H}$.

3.1.6 Proof of Theorem 3.1, the case $p \geq 2$

Here the situation is entirely different.

Proof of (3.4). Obviously, we may restrict ourselves to f which are bounded in L^p. Then $||df_n||_p < \infty$ for any n and hence, by differential subordination, $||dg_n||_p < \infty$ and $||g_n||_p < \infty$ for all n. Let $V_p(x, y) = 1_{\{|y| \geq 1\}} - p^{p-1}|x|^p/2$. As in the previous case, we shall use integration argument, but this time u is the dual function, given by (3.12). Set

$$k(t) = \frac{p^p(p-1)^{2-p}(p-2)}{4}t^{p-1}1_{\{0 \leq t \leq 1-p^{-1}\}}.$$

A little calculation shows that

$$U_p(x, y) = \int_0^\infty k(t)u(x/t, y/t)dt$$

is given by the following formula: if $|x| + |y| \leq 1 - p^{-1}$, then

$$U_p(x, y) = \frac{1}{2}\left(\frac{p}{p-1}\right)^{p-1}(|y| - (p-1)|x|)(|x| + |y|)^{p-1},$$

while for remaining (x, y),

$$U_p(x,y) = \frac{p^2}{4}\left[|y|^2 - |x|^2 - \frac{2(p-2)|y|}{p} + \frac{(p-1)^2(p-2)}{p^3}\right].$$

By (3.15),

$$\int_0^\infty k(t)|u(x/t, y/t)|\mathrm{d}t \leq \tilde{c}(1 + |x|^2 + |y|^2),$$

for all x, y and some positive constant $\tilde{c}$ depending only on p. Hence the use of the integration argument is permitted: for any n, the random variables f_n and g_n belong to L^p, so Fubini's theorem implies that $\mathbb{E}U_p(f_n, g_n) \leq 0$. Therefore, all we need is the majorization property $U_p \geq V_p$ and, as before, it suffices to establish it for $\mathcal{H} = \mathbb{R}$. Consider the function F given by

$$F(s) = \frac{1}{2}\left(\frac{p}{p-1}\right)^{p-1}(1 - ps) + \frac{p^{p-1}}{2}s^p, \qquad s \in [0, 1].$$

This function is nonnegative: indeed, it is convex and satisfies

$$F((p-1)^{-1}) = F'((p-1)^{-1}) = 0.$$

This gives the majorization for $|x| + |y| \leq 1 - p^{-1}$, since for these x, y it is equivalent to $F(|x|/(|x| + |y|)) \geq 0$. The next step is to show that $U_p(x, y) \geq V_p(x, y)$ for $|y| \geq 1$. For fixed x, the function $U_p(x, \cdot)$ increases on $[1, \infty)$, so it suffices to establish the bound for $y = 1$. That is, we must prove that

$$(p|x|)^p - 1 \geq \frac{p}{2}((p|x|)^2 - 1), \tag{3.27}$$

which is an immediate consequence of the mean value property of the convex function $t \mapsto t^{p/2}$. It remains to show the majorization for (x, y) satisfying $|x| + |y| \geq 1 - p^{-1}$ and $|y| < 1$. The inequality can be rewritten in the form

$$\frac{p^2}{4}\left[|y|^2 - |x|^2 - \frac{2(p-2)|y|}{p} + \frac{(p-1)^2(p-2)}{p^3}\right] \geq -\frac{p^{p-1}|x|^p}{2}$$

which is valid for all real x, y, not only for those satisfying the above restrictions. Indeed, observe that as a function of $|y|$, the left-hand side attains its minimum for $|y| = 1 - 2/p$ and for such y, we again arrive at (3.27). This completes the proof. $\qquad\square$

Remark 3.4. The original special function invented by Suh [189] was much more complicated; on the other hand, her function is the least majorant leading to the weak type estimate. We will present a detailed description of the steps which lead to the discovery of this function, just after the proof of the optimality of $c_{p,p}$.

Sharpness. Consider the following example. Let N be a fixed positive integer and let $\delta > 0$ be uniquely determined by the equation

$$\delta(1 + 2\delta)^N = 1 - \frac{1}{p}. \tag{3.28}$$

Let $(X_n)_{n=0}^{2N+1}$ be a sequence of independent random variables, with the distribution given as follows. Let $X_0 \equiv \delta$,

$$\mathbb{P}\left(X_{2n-1} = \frac{p-2}{p-1}\right) = 1 - \mathbb{P}\left(X_{2n-1} = 1 + \delta\right) = \frac{(p-1)\delta}{1 + (p-1)\delta},$$

$$\mathbb{P}\left(X_{2n} = \frac{(p-2)(1+2\delta)}{(p-1)(1+\delta)}\right) = 1 - \mathbb{P}\left(X_{2n} = \frac{1+2\delta}{1+\delta}\right) = \frac{(p-1)\delta}{1 + 2\delta},$$

for $n = 1, 2, \ldots, N$, and, finally,

$$\mathbb{P}\left(X_{2N+1} = \frac{p(1+\delta)}{p-1}\right) = 1 - \mathbb{P}\left(X_{2N+1} = \frac{p-2}{p-1}\right) = \frac{1}{2 + p\delta}.$$

A straightforward verification gives that $\mathbb{E}X_n = 1$ for $n = 1, 2, \ldots, 2N + 1$. Introduce a stopping time $\tau = \inf\{n \geq 1 : X_n < 1\} \wedge (2N + 1)$. Then by Doob's optional sampling theorem, the sequences f, g defined by

$$g_n = -\delta + X_0 X_1 \cdots X_{\tau \wedge n}, \qquad df_n = (-1)^n dg_n, \quad n = 0, 1, 2, \ldots$$

are martingales and g is a ± 1-transform of f. The paths of the martingale g are of two types:

a) g increases for a number of steps, then it decreases and stops;
b) g increases and reaches 1 at time $2N + 1$.

We will show that when δ is sufficiently small (or, rather, when N is sufficiently large), then the ratio $\mathbb{P}(g_{2N+1} \geq 1)/\mathbb{E}|f_{2N+1}|^p$ can be made arbitrarily close to $p^{p-1}/2$. To this end, denote

$$Q = \frac{1 + \delta(3 - p)}{(1 + (p-1)\delta)(1 + 2\delta)} = 1 - \frac{2(p-1)\delta(\delta + 1)}{1 + (p+1)\delta + 2(p-1)\delta^2} \tag{3.29}$$

and observe that

$$\mathbb{P}(g_{2N+1} \geq 1) = \mathbb{P}(X_1 > 1, X_2 > 1, \ldots, X_{2N+1} > 1) = \frac{Q^N}{2 + p\delta}. \tag{3.30}$$

Now let us derive $\mathbb{E}|f_{2N+1}|^p$. Directly from the definition of τ and the variables X_k we infer that

$$\mathbb{P}(\tau = 2k + 1) = \frac{(p-1)Q^k \delta}{1 + (p-1)\delta}, \quad k = 0, 1, \ldots, N - 1,$$

$$\mathbb{P}(\tau = 2k) = \frac{(p-1)Q^k \delta}{1 + \delta(3 - p)}, \quad k = 1, 2, \ldots, N.$$

Furthermore, the distribution of $|f_{2N+1}|$ is given as follows:

$$|f_{2N+1}|1_{\{\tau=2k+1\}} = \frac{\delta(1+2\delta)^k}{p-1}1_{\{\tau=2k+1\}}, \qquad k = 0, 1, \ldots, N-1,$$

$$|f_{2N+1}|1_{\{\tau=2k\}} = \frac{\delta(1+2\delta)^k}{p-1}1_{\{\tau=2k\}}, \qquad k = 1, 2, \ldots, N,$$

$$\mathbb{P}(|f_{2N+1}| = p^{-1} + \delta, \tau = 2N+1) = \frac{Q^N}{2+p\delta},$$

$$\mathbb{P}(|f_{2N+1}| = p^{-1}, \tau = 2N+1) = \frac{Q^N(1+p\delta)}{2+p\delta}.$$

Thus,

$$\mathbb{E}|f_{2N+1}|^p = Q^N(I + II + III + IV),$$

where

$$I = \frac{(p-1)\delta}{1+(p-1)\delta} \sum_{k=0}^{N-1} Q^{k-N} \left(\frac{\delta(1+2\delta)^k}{p-1}\right)^p,$$

$$II = \frac{(p-1)\delta}{1+\delta(3-p)} \sum_{k=1}^{N} Q^{k-N} \left(\frac{\delta(1+2\delta)^k}{p-1}\right)^p,$$

$$III = \frac{(p^{-1}+\delta)^p}{2+p\delta},$$

$$IV = \frac{p^{-p}(1+p\delta)}{2+p\delta}.$$

Now we let $N \to \infty$ (so $\delta \to 0$). We have that

$$I = \frac{1}{(p-1)^{p-1}} \cdot \frac{\delta}{Q(1+2\delta)^p - 1} \cdot \frac{\delta^p((Q(1+2\delta)^p)^N - 1)}{Q^N}.$$

It follows from the second equality in (3.29) that $Q = 1 - 2(p-1)\delta + O(\delta^2)$ and hence $Q(1+2\delta)^p = 1 + 2\delta + O(\delta^2)$. Thus, letting $\delta \to 0$ in each of the three fractions above, we get, by (3.28),

$$\lim_{\delta \to 0} I = \frac{1}{(p-1)^{p-1}} \cdot \frac{1}{2} \cdot (1-p^{-1})^p = \frac{p-1}{2p^p}.$$

It is easy to see that the limit of II is the same, while

$$\lim_{\delta \to 0} III = \lim_{\delta \to 0} IV = \frac{1}{2p^p}.$$

It suffices to combine this with (3.30) to get

$$\lim_{\delta \to 0} \frac{\mathbb{P}(g_{2N+1} \geq 1)}{\mathbb{E}|f_{2N+1}|^p} = \frac{p^{p-1}}{2},$$

which completes the proof.   □

On the search of the suitable majorant. Fix $p > 2$, assume that $\mathcal{H} = \mathbb{R}$ and suppose that the weak type inequality holds with some constant c. Let

$$U^0(x, y) = \sup\{\mathbb{P}(|g_n| \geq 1) - c^p \mathbb{E}|f_n|^p\},$$

where the supremum is taken over all n and all pairs $(f, g) \in M(x, y)$. As previously, we will use the notation U instead of U^0, since the final formula for the solution of the corresponding boundary value problem will be obtained after a number of assumptions.

Step 1. The case $|y| \geq 1$. For such y, we repeat the argumentation from the previous case and conclude that

$$U(x, y) = 1 - c^p |x|^p.$$

Step 2. A special curve. The following intuitive observation turns out to be very important. Let x be a large real number and let $y \in (-1, 1)$. Suppose we are interested in determining $U^0(x, y)$. To do this we need, loosely speaking, to find such f, g and n, for which $\mathbb{P}(|g_n| \geq 1)$ is large and $\mathbb{E}|f_n|^p$ is relatively small. However, the "gain" we can get from the first term is at most 1. This may not be enough to compensate the "loss" coming from the growth of the pth moment of f (at least if $|x|$ is sufficiently large). This is a consequence of the fact that the second derivative of $x \mapsto x^p$ grows to infinity as $x \to \infty$: this is where the condition $p > 2$ plays a role. In other words, if $y \in (-1, 1)$ and $|x|$ is large, it is natural to conjecture that the best pair $(f, g) \in M(x, y)$ is the constant one: hence $U^0(x, y) = -c^p |x|^p$. This suggests introducing the following assumption:

(A1)   There is an nondecreasing function $\gamma : [b, \infty) \to [0, 1]$ of class C^1 such that if $|y| \leq \gamma(|x|)$, then $U(x, y) = -c^p |x|^p$.

Some experimentation leads to

(A2)   We have $b = 0$ and $\gamma(0) = 0$.

See Figure 3.1 below.

Step 3. Further assumptions. Let $A(y) = U(0, y)$ for $y \in \mathbb{R}$. We impose the following regularity condition on U.

(A3)   The function U is continuous on $\mathbb{R} \times [-1, 1]$ and of class C^1 in the interior of this set.

By Remark (iii) in Section 2.2, we may restrict ourselves to the functions satisfying

$$U(x, y) = U(-x, y) = U(x, -y) \quad \text{for all } x, y \in \mathbb{R}. \tag{3.31}$$

By (A3), this gives

$$U_x(0, y) = 0 \quad \text{for } y \geq 0. \tag{3.32}$$

It remains to determine U on the set $J = \{(x, y) : x > 0, \gamma(x) < y < 1\}$.

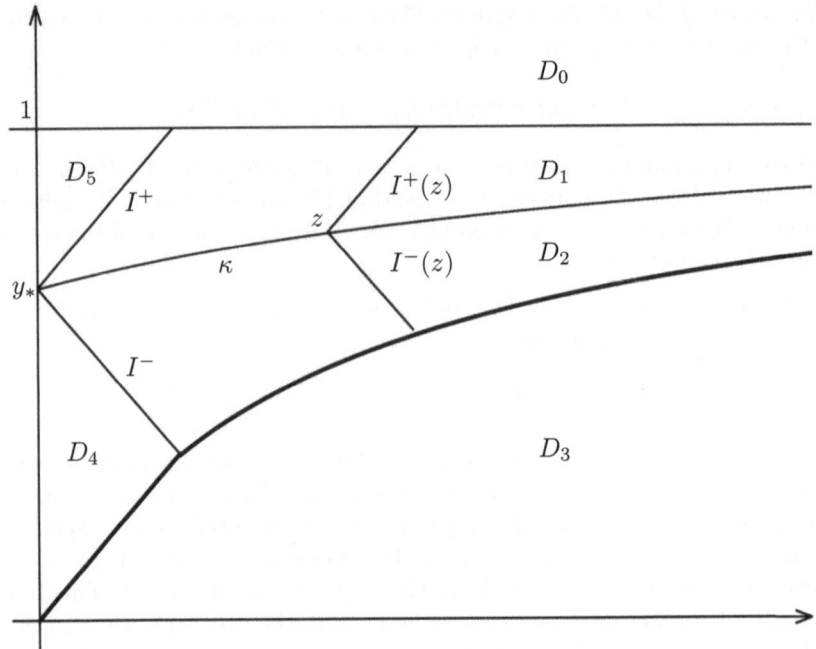

Figure 3.1: Various parameters appearing during the search. The bold curve is the graph of the function γ.

The key structural assumption on U is the following.

(A4)   There is $y_* \in [0,1]$ such that for $(x,y) \in J$,

$$U(x,y) = \frac{x}{x+t}U(x+t, y-t) + \frac{t}{x+t}A(x+y) \qquad \text{if } x+y \le y_*, \quad (3.33)$$

where $t = t(x,y)$ is the unique positive number satisfying $y - t = \gamma(x+t)$, and

$$U(x,y) = \frac{x}{x+1-y}U(x+1-y, 1) + \frac{1-y}{x+1-y}A(y-x) \qquad \text{if } -x+y \ge y_*. \quad (3.34)$$

Condition (A4) forces U to be linear along line segments of slope -1 contained in D_4 and line segments of slope 1 contained in D_5 (the regions D_4 and D_5 are as in Figure 3.1 and will be formally defined below).

  Step 4. A lower bound for c. It turns out that the assumptions (A1)–(A4) imply that $c^p \ge p^{p-1}/2$. To prove this, we start with the observation that by (A3) and (A4) we have, for x satisfying $x + \gamma(x) \le y_*$ and $t \in [0, x]$,

$$U(x - t, \gamma(x) + t) = U(x, \gamma(x)) + (-U_x(x, \gamma(x)) + U_y(x, \gamma(x)))t,$$

so, by (A1) and (A2),

$$U(x - t, \gamma(x) + t) = -c^p x^p + pc^p x^{p-1} t. \tag{3.35}$$

Take $t = x$ and differentiate both sides. We get

$$U_y(0, \gamma(x) + x)(\gamma'(x) + 1) = p(p-1)c^p x^{p-1}.$$

On the other hand, differentiate in (3.35) over t, let $t = x$ and use (3.32) to obtain

$$U_y(0, \gamma(x) + x) = pc^p x^{p-1}.$$

The two equations above give $\gamma'(x) = p - 2$; thus, by (A2), $\gamma(x) = (p-2)x$ for x satisfying $x + \gamma(x) \le y_*$.

By (3.33) and (3.34), the function U is linear on the line segments $I^{\pm}$ of slope ± 1 such that $(0, y^*) \in I^{\pm} \subset J$. Combining this with the symmetry condition $U(x, y) = U(-x, y)$, we get that the function

$$F(t) := U\left(\frac{y_*}{p-1} - t, \frac{(p-2)y^*}{p-1} + t\right)$$

is linear on $[0, 1 - (p-2)y_*/(p-1)]$. Thus

$$U(y^* - 1, 1) = F\left(1 - \frac{(p-2)y_*}{p-1}\right) = F(0) + F'(0)\left(1 - \frac{(p-2)y^*}{p-1}\right),$$

which can be rewritten in the form

$$c^p = \left[(1 - y_*)^p - y_*^p(p-1)^{2-p} + py_*^{p-1}(p-1)^{1-p}\right]^{-1}.$$

It remains to note the right-hand side, as a function of $y_* \in [0, 1]$, attains its minimum $p^{p-1}/2$ at $y_* = 1 - p^{-1}$. This leads to

(A5) $\quad c^p = p^{p-1}/2$ and $y_* = 1 - p^{-1}$.

Step 5. A final assumption. The next key observation is that the segments I^+ and I^-, introduced in the previous step (see also Figure 3.1), have the same length. This suggests the final assumption (A6) below, for whose formulation we need some notation. Introduce the curve

$$\kappa = \left\{\left(x - \frac{1 - \gamma(x)}{2}, \gamma(x) + \frac{1 - \gamma(x)}{2}\right) : x \ge p^{-1}\right\}$$

(for a better understanding of κ, see the geometric properties of $I^+(z)$ and $I^-(z)$ below). Let $D_1 \subset J$ be the closed set bounded by the lines $y = 1$, $-x + y = y_*$ and the curve κ; let $D_2 \subset J$ be the closed set bounded by the line $x + y = y_*$, the curve κ and the graph of γ (see Figure 3.1). Note that D_1 and D_2 have the following property. Take any $z \in \kappa$ and let $I^+(z) \subset D_1$ (respectively, $I^-(z) \subset D_2$)

denote the maximal line segment of slope 1 (respectively, -1), which has z as one of its endpoints. Then $I^+(z)$ and $I^-(z)$ have the same length; so, in a sense, κ divides the set

$$\{(x,y) : y_* - x \le y \le y_* + x, \ \gamma(x) \le y \le 1\}$$

into two halves. Our final assumption can be stated as follows.

(A6)   We assume that
$$U \text{ is linear on each } I^+(z), \ z \in \kappa \tag{3.36}$$

and
$$U \text{ is linear on each } I^-(z), \ z \in \kappa. \tag{3.37}$$

Step 6. A formula for γ. So far, we have derived that

$$\gamma(x) = (p-2)x \qquad \text{for } x \in [0, p^{-1}]$$

and we need to determine this function on the remaining part of the positive half-line. By virtue of (A3) and (3.37), the equation (3.35) is valid for all $x \ge p^{-1}$ and $t \in [0, (1 - \gamma(x))/2]$. This enables us to find the expression of $U_x(z) + U_y(z)$ for any $z \in \kappa$: if $z = (x - (1 - \gamma(x))/2, \gamma(x) + (1 - \gamma(x))/2)$, then

$$U_x(z) + U_y(z) = -pc^p x^{p-1} + \frac{p(p-1)c^p x^{p-2}(1 - \gamma(x))}{1 + \gamma'(x)}.$$

On the other hand, by (A3) and (3.36), this must be equal to

$$\frac{U(x, 1) - U(z)}{\frac{1 - \gamma(x)}{2}}.$$

After some easy manipulations and plugging $c^p = p^{p-1}/2$, this yields

$$\gamma'(x) + 1 = \frac{p^p(p-1)}{4} x^{p-2}(1 - \gamma(x))^2.$$

Standard arguments (cf. [189]) give the existence of a unique $\gamma : (p^{-1}, \infty) \to [0, 1]$ satisfying $\gamma(p^{-1}+) = 1 - 2/p$; then $\gamma(p^{-1}+) = p - 2$ and we obtain the desired function γ.

Step 7. The formula for U. Now we can put all the things together: the equations (3.33), (3.34), (3.36) and (3.37) yield the candidate U, the one invented by Suh. Let $D_0 = \{(x, y) : x > 0, \ y \ge 1\}$ and recall D_1 and D_2 introduced in Step 5. Moreover, let

$$D_3 = \{(x, y) : x \ge 0, \ 0 \le y \le \gamma(x)\},$$
$$D_4 = \{(x, y) : x \ge 0, \ \gamma(x) \le y \le -x + y_*\},$$
$$D_5 = \{(x, y) : \mathbb{R}_+ \times \mathbb{R}_+ \setminus (D_0 \cup D_1 \cup D_2 \cup D_3 \cup D_4)\}$$

(see Figure 3.1). One can check that the above considerations yield the following formula for U. Let G be the inverse to the function $x \mapsto x + \gamma(x)$. Then

$$
U(x, y) = \begin{cases}
1 - \frac{p^{p-1}}{2} x^p & \text{on } D_0, \\
1 - \frac{2(1-y)}{1-\gamma(x-y+1)} & \\
\quad - \frac{1}{2} p^{p-1}(x-y+1)^{p-1}\big(x-(p-1)(1-y)\big) & \text{on } D_1, \\
\frac{p^{p-1}}{2}(G(x+y))^{p-1}\big((p-1)G(x+y)-px\big) & \text{on } D_2, \\
-\frac{p^{p-1}}{2} x^p & \text{on } D_3, \\
\frac{1}{2}\left(\frac{p}{p-1}\right)^{p-1}(x+y)^{p-1}(y-(p-1)x) & \text{on } D_4, \\
\frac{p^{p-1}}{2}(1+x-y)^{p-1}\left(\frac{1-y}{p-1}-x\right) - \frac{p^2(1-y)}{2(p-1)} + 1 & \text{on } D_5.
\end{cases}
\tag{3.38}
$$

The description of U is completed by the condition (3.31). One can now check that the function satisfies the conditions 1°, 2° and 3°. However, this requires a large amount of work and patience; for details, see [189].

3.1.7 Proof of Theorem 3.2

First, let us exclude the trivial cases. For $q > p$ the inequality (3.2) is obvious and its sharpness can be proved, for example, by taking $f = g = (\xi, \xi, \ldots)$, for an appropriate random variable ξ. The case $0 < q \leq p < 1$ is also easy: (3.2) is clear and to see that no finite constant will do, consider the example constructed in the proof of Theorem 3.1 for $0 < p < 1$. Next, if $1 \leq p \leq 2$, then (3.2) and the optimality of $c_{p,q}$ follow directly from (3.1) and the example appearing in the proof of the sharpness of this estimate. Finally, when $q \leq 2 \leq p$, then

$$\|g\|_{q,\infty} \leq \|g\|_{2,\infty} \leq \|g\|_2 \leq \|f\|_2 \leq \|f\|_p,$$

and equality can be attained: take $f = g \equiv 1$.

Thus, we are left with the non-trivial case $2 < q < p$ (the choice $2 < q = p$ is covered by Theorem 3.1). At the first sight, Burkholder's approach is not applicable here, since the inequality (3.2) does not seem to be of the form $\mathbb{E}V(f_n, g_n) \leq c$. To overcome this difficulty, we prove a larger family of related estimates: for any $2 < p < \infty$ and any $\gamma \in [0, 1 - 2/p]$,

$$
\mathbb{P}(|g_n| \geq 1) \leq \frac{p^{p-1}(1-\gamma)^p}{2}\mathbb{E}|f_n|^p + \frac{1}{4}\frac{\gamma^p p^{p-1}}{(p-2)^{p-1}}, \qquad n = 0, 1, 2, \ldots, \tag{3.39}
$$

for all $\mathcal{H}$-valued martingales f, g such that g is differentially subordinate to f. Having proved this, we will deduce (3.2) by picking the optimal value of γ and exploiting the stopping time argument (which will allow us to replace g by $|g|^*$). Observe that (3.39) are of the form in which Burkholder's method can be used. introduce $V_{p,\gamma} : \mathcal{H} \times \mathcal{H} \to \mathbb{R}$ by

$$V_{p,\gamma}(x, y) = 1_{\{|y| \geq 1\}} - \frac{p^{p-1}(1-\gamma)^p}{2}|x|^p.$$

To define the corresponding special function, consider the auxiliary parameters

$$a = a_{p,\gamma} = \frac{\gamma}{(1-\gamma)(p-2)}, \qquad b = b_{p,\gamma} = 1 - \frac{1}{p(1-\gamma)}. \tag{3.40}$$

It is easy to see that $a \le b$. Next, define $k = k_{p,\gamma} : [0,\infty) \to [0,\infty)$ by

$$k(r) = \frac{1}{4} p^p (p-1)^{2-p} (p-2)(1-\gamma)^3 (\gamma + (1-\gamma)r)^{p-3} r^2\, 1_{[a,b]}(r) \tag{3.41}$$

and introduce $R_{p,\gamma} : \mathcal{H} \times \mathcal{H} \to \mathbb{R}$ by the formula

$$R_{p,\gamma}(x,y) = \frac{1}{4} \frac{\gamma^{p-2}(1-\gamma)^2 p^p}{(p-2)^{p-2}}(|y|^2 - |x|^2) + \frac{1}{4} \frac{\gamma^p p^{p-1}}{(p-2)^{p-1}}.$$

The special function $U_{p,\gamma}$ corresponding to (3.39) is given by

$$U_{p,\gamma}(x,y) = \int_0^\infty k(r) U_\infty(x/r, y/r)\mathrm{d}r + R_{p,\gamma}(x,y). \tag{3.42}$$

We now turn to the proof of (3.39). Pick f, g as in the statement of this estimate. Of course, we may assume that $f_n \in L^p$, since otherwise there is nothing to prove. Then $f_n \in L^2$ and, as we have already seen, $||g_n||_2 \le ||f_n||_2$; consequently,

$$\mathbb{E}R_{p,\gamma}(f_n, g_n) \le \frac{1}{4} \frac{\gamma^p p^{p-1}}{(p-2)^{p-1}}.$$

Next, by virtue of (3.15), there is a constant C depending only on γ and p such that

$$\int_0^\infty k(r)\, |U_\infty(x/r, y/r)|\, \mathrm{d}r \le C(1 + |x|^2 + |y|^2),$$

so the integration method is applicable. Combining it with the above upper bound for $\mathbb{E}R_{p,\gamma}(f_n, g_n)$, we obtain

$$\mathbb{E}U_{p,\gamma}(f_n, g_n) \le \frac{1}{4} \frac{\gamma^p p^{p-1}}{(p-2)^{p-1}}$$

and hence to get (3.39), it suffices to show that $U_{p,\gamma} \ge V_{p,\gamma}$. This majorization is just a matter of tedious and lengthy, but rather straightforward computations (see [160] for details).

To deduce (3.2), observe that the stopping time argument applied to (3.39) gives

$$\mathbb{P}(|g|^* \ge 1) \le \frac{p^{p-1}(1-\gamma)^p}{2} ||f||_p^p + \frac{1}{4} \frac{\gamma^p p^{p-1}}{(p-2)^{p-1}}. \tag{3.43}$$

Of course, we may assume that $||f||_p \ne 0$. If $||f||_p \le 1/2$, put

$$\gamma = \left(1 + (p-2)^{-1}(2||f||_p^p)^{-1/(p-1)}\right)^{-1}.$$

Then $\gamma \leq 1 - 2/p$ and

$$1 - \gamma = \left(1 + (p-2)(2\|f\|_p^p)^{1/(p-1)}\right)^{-1}.$$

Plugging this into (3.43) yields

$$\begin{aligned}
\mathbb{P}(|g|^* \geq 1) &\leq \frac{p^{p-1}\|f\|_p^p}{2\left(1 + (p-2)(2\|f\|_p^p)^{1/(p-1)}\right)^{p-1}} \\
&= \frac{p^{p-1}\|f\|_p^{p-q}}{2\left(1 + (p-2)(2\|f\|_p^p)^{1/(p-1)}\right)^{p-1}} \cdot \|f\|_p^q.
\end{aligned} \tag{3.44}$$

To analyze the factor in front of $\|f\|_p^q$, we define the function $G : (0, \infty) \to \mathbb{R}$ by

$$G(t) = \frac{p^{p-1}t^{1-q/p}}{2\left(1 + (p-2)(2t)^{1/(p-1)}\right)^{p-1}}.$$

A straightforward analysis shows that the maximum of G is equal to $c_{p,q}^q$. Thus,

$$\left(\mathbb{P}(|g|^* \geq 1)\right)^{1/q} \leq G(\|f\|_p^p)^{1/q}\|f\|_p \leq c_{p,q}\|f\|_p. \tag{3.45}$$

We have proved this under the assumption $\|f\|_p \leq 1/2$, but (3.45) is valid for all f. Indeed, suppose that $\|f\|_p > 1/2$ and use the weak type $(2, 2)$ bound to get

$$\mathbb{P}(|g|^* \geq 1) \leq \|f\|_2^2 \leq \|f\|_p^2 = \left[c_{p,q}^{-q}\|f\|_p^{2-q}\right] \cdot c_{p,q}^q\|f\|_p^q.$$

However, the expression in the square brackets can be shown to be smaller than 1. Hence (3.45) holds, and it yields (3.2) by standard homogenization.

Sharpness. Let $2 < q < p$ be fixed. To prove the optimality of $c_{p,q}$ one could proceed as previously and construct appropriate examples. However, the calculations turn out to be quite involved; there is a simpler approach which rests on the following modification of Theorem 2.2. For any $(x, y) \in \mathbb{R}^2$ and $t \in [0, 1]$, let $M(x, y, t)$ denote the class of all pairs (f, g) of simple real-valued martingales with $(f_0, g_0) \equiv (x, y)$, satisfying $dg_n \equiv df_n$ or $dg_n \equiv -df_n$ for each $n \geq 1$ and the further assumption $\mathbb{P}(|g_\infty| \geq 1) \geq t$. Note that the class $M(x, y)$, introduced in Chapter 2, coincides with $M(x, y, 0)$ in this new notation. Introduce the function $U^0 : \mathbb{R} \times \mathbb{R} \times [0, 1] \to \mathbb{R}$ by

$$U^0(x, y, t) = \inf\left\{\|f\|_p^p : (f, g) \in M(x, y, t)\right\}.$$

Of course, $U^0(x, y, t) \geq |x|^p$. Note that this estimate can be reversed if $|y| \geq 1$ or $t = 0$, since then the constant pair (x, y) belongs to $M(x, y, t)$. Thus,

$$U^0(x, y, t) = |x|^p \qquad \text{if } |y| \geq 1 \text{ or } t = 0. \tag{3.46}$$

Furthermore, by the use of the "splicing argument", we get that for any $x, y \in \mathbb{R}$, $v \in \mathbb{R}$ and $\varepsilon \in \{-1, 1\}$,

$$\text{the function } G(t) = U^0(x + tv, y + t\varepsilon v, t) \text{ is convex on } [0, 1]. \tag{3.47}$$

Now recall the parameters a and b defined in (3.40). We will need the following further properties of U^0, listed in the two lemmas below.

Lemma 3.3.

(i) *If $t \leq 1/2$, then*

$$U^0(a/2, a/2, t) \leq \frac{1}{2} U^0(0, a, 2t) + \frac{1}{2} \left(\frac{\gamma}{(1-\gamma)(p-2)} \right)^p. \tag{3.48}$$

(ii) *We have*

$$U^0(0, b, 1/2) \leq (p(1-\gamma))^{-p}. \tag{3.49}$$

(iii) *Assume that $\gamma \in (0, 1 - 2/p)$, $\delta \in (0, (p-1)^{-1})$, $y > 0$ and put*

$$\lambda_{p,\delta} = \frac{1 - (p-1)\delta}{1 + (p-1)\delta}.$$

Then for any $t \in [0, \lambda_{p,q}]$,

$$U^0(0, y, t) \leq \lambda_{p,\delta} U^0 \left(0, y + 2 \left(y + \frac{\gamma}{1-\gamma} \right) \delta, t\lambda_{p,\delta}^{-1} \right) + (1 - \lambda_{p,\delta}) \left[\frac{\gamma + (1-\gamma)y}{(p-1)(1-\gamma)} \right]^p.$$

Proof. The claim follows from (3.46) and (3.47). We have

$$U^0(a/2, a/2, t) \leq \frac{1}{2} U^0(0, a, 2t) + \frac{1}{2} U^0(a, 0, 0) = \frac{1}{2} U^0(0, a, 2t) + \frac{1}{2} \left(\frac{\gamma}{(1-\gamma)(p-2)} \right)^p$$

and

$$U^0(0, b, 1/2) \leq \frac{1}{2} U^0(b - 1, 2b - 1, 0) + \frac{1}{2} U^0(1 - b, 1, 1) = (1 - b)^p = (p(1 - \gamma))^{-p}.$$

To check the third estimate, let $\kappa = y + \gamma/(1 - \gamma)$ and note that

$$U^0(0, y, t) \leq \frac{(p-1)\delta}{(p-1)\delta + 1} U^0 \left(\frac{\kappa}{p-1}, y - \frac{\kappa}{p-1}, 0 \right)$$

$$+ \frac{1}{(p-1)\delta + 1} U^0 \left(-\kappa\delta, y + \kappa\delta, ((p-1)\delta + 1)t \right)$$

$$= \frac{(p-1)\delta}{(p-1)\delta + 1} \cdot \left(\frac{\kappa}{p-1} \right)^p$$

$$+ \frac{1}{(p-1)\delta + 1} U^0 \left(-\kappa\delta, y + \kappa\delta, ((p-1)\delta + 1)t \right).$$

Similarly,

$$
U^0\big(-\kappa\delta, y+\kappa\delta, ((p-1)\delta+1)t\big) \le (p-1)\delta \cdot U^0\left(-\frac{\kappa}{p-1}, y-\frac{\kappa}{p-1}+2\kappa\delta, 0\right)
$$
$$
+ (1-(p-1)\delta) \cdot U^0\left(0, y+2\kappa\delta, t\lambda_{p,\delta}^{-1}\right)
$$
$$
= (p-1)\delta\left(\frac{\kappa}{p-1}\right)^p
$$
$$
+ (1-(p-1)\delta)U^0\left(0, y+2\kappa\delta, t\lambda_{p,\delta}^{-1}\right).
$$

Combining these two estimates gives the desired bound. $\qquad\square$

Lemma 3.4. *For any $\gamma \in (0, 1-2/p)$ we have*

$$
U^0\left(a/2, a/2, \frac{1}{4}\left(\frac{p\gamma}{p-2}\right)^{p-1}\right) \le \frac{1}{2}\left(\frac{\gamma}{(1-\gamma)(p-2)}\right)^{p-1}. \tag{3.50}
$$

Proof. If we set $F(s,t) = U^0(0, s-\gamma/(1-\gamma), t)$, then the inequality from (iii) can be rewritten in the more convenient form

$$
F(s,t) \le \lambda_{p,\delta} F(s(1+2\delta), t\lambda_{p,\delta}^{-1}) + (1-\lambda_{p,\delta})\left(\frac{s}{p-1}\right)^p,
$$

where $s = y + \gamma/(1-\gamma)$. Hence, by induction, we have

$$
F(s,t) \le \lambda_{p,\delta}^n F(s(1+2\delta)^n, t\lambda_{p,\delta}^{-n}) + (1-\lambda_{p,\delta})\left(\frac{s}{p-1}\right)^p \frac{[\lambda_{p,\delta}(1+2\delta)^p]^n - 1}{\lambda_{p,\delta}(1+2\delta)^p - 1} \tag{3.51}
$$

for all $s > \gamma/(1-\gamma)$, $\delta \in (0, (p-1)^{-1})$, any positive integer n and any $t \le \lambda_{p,\delta}^n$. Fix a large integer n and let s, δ and t be given by

$$
s = a + \frac{\gamma}{1-\gamma} = \frac{(p-1)\gamma}{(p-2)(1-\gamma)}, \qquad (1+2\delta)^n = \frac{b+\gamma/(1-\gamma)}{a+\gamma/(1-\gamma)} = \frac{p-2}{p\gamma}
$$

and $t = \lambda_{p,\delta}^n/2$. Note that if n is sufficiently large, then $\delta < (p-1)^{-1}$, so we may insert these parameters into (3.51) and let n go to ∞. We have

$$
\lim_{n\to\infty} \lambda_{p,\delta}^n = \lim_{n\to\infty}\left(1 - \frac{2(p-1)\delta}{1+(p-1)\delta}\right)^n = \left(\frac{p\gamma}{p-2}\right)^{p-1},
$$
$$
\lim_{n\to\infty} \frac{1-\lambda_{p,\delta}}{\lambda_{p,\delta}(1+2\delta)^p - 1} = \lim_{\delta\to 0} \frac{2(p-1)\delta}{(1-(p-1)\delta)(1+2\delta)^p - 1 - (p-1)\delta} = p-1
$$

and, by (3.47), the function $U^0(0, a, \cdot) : (0,1) \to \mathbb{R}$ is continuous. Therefore, in the limit, (3.51) becomes

$$U^0\left(0, a, \frac{1}{2}\left(\frac{p\gamma}{p-2}\right)^{p-1}\right)$$

$$\leq \left(\frac{p\gamma}{p-2}\right)^{p-1} U^0\left(0, b, \frac{1}{2}\right) + (p-1)\left(\frac{\gamma}{(1-\gamma)(p-2)}\right)^p\left(\frac{p-2}{p\gamma} - 1\right)$$

$$\leq \left(\frac{\gamma}{p-2}\right)^{p-1}(1-\gamma)^{-p}\left(1 - \frac{p-1}{p-2}\gamma\right),$$

where to get the last inequality we used (3.49). Plugging this estimate into (3.48) gives (3.50). □

Now we can easily show the sharpness of (3.2) and (3.39). Let us first deal with (3.39). Fix $\varepsilon > 0$ and $\gamma \in (0, 1 - 2/p)$. Then, by (3.50) and the definition of U^0, there is a pair (f, g) of simple martingales starting from $(a/2, a/2)$ such that $dg_n \equiv df_n$ or $dg_n \equiv -df_n$ for all $n \geq 1$, and

$$\|f\|_p^p \leq \frac{1}{2}\left(\frac{\gamma}{(1-\gamma)(p-2)}\right)^{p-1} + \varepsilon \quad \text{and} \quad \mathbb{P}(|g_\infty| \geq 1) \geq \frac{1}{4}\left(\frac{p\gamma}{p-2}\right)^{p-1}.$$

Therefore, g is a ± 1-transform of f and

$$\mathbb{P}(|g_\infty| \geq 1) - \frac{p^{p-1}(1-\gamma)^p}{2}\|f\|_p^p \geq \frac{1}{4}\frac{\gamma^p p^{p-1}}{(p-2)^{p-1}} - \frac{p^{p-1}(1-\gamma)^p}{2}\varepsilon$$

and hence (3.43) is sharp, since ε was arbitrary. The case $\gamma \in \{0, 1 - 2/p\}$ follows easily by passing to the limit. Finally, to deal with (3.2), pick $\gamma = 1 - q/p$ and observe that

$$\frac{\mathbb{P}(|g_\infty| \geq 1)}{\|f\|_p^q} \geq \frac{1}{4}\left(\frac{p-q}{p-2}\right)^{p-1} \cdot \left[\frac{1}{2}\left(\frac{p-q}{q(p-2)}\right)^{p-1} + \varepsilon\right]^{-q/p}.$$

However, the right-hand side converges to $c_{p,q}^q$ when $\varepsilon \to 0$. This proves the optimality of the constant $c_{p,q}$ in the weak type (p, q)-estimate. □

On the search of the suitable majorant, $2 < q < p$. Suppose that $\mathcal{H} = \mathbb{R}$ and we want to find the special function which leads to (3.2). The first observation, already mentioned above, is that the estimate is not of appropriate form from our point of view; by homogenization arguments, we are led to study estimates of the form

$$\mathbb{P}(|g_n| \geq 1) \leq \alpha \mathbb{E}|f_n|^p + \beta. \tag{3.52}$$

In view of the weak type inequalities (3.1), it suffices to consider $\alpha \in (0, p^{p-1}/2)$, and the question is: for any such α, what is the least possible value $\beta(\alpha)$ of β and what is the special function which leads to the corresponding sharp estimate?

Of course, we take $V(x, y) = 1_{\{|y| \geq 1\}} - \alpha |x|^p$. It is natural to try to modify Suh's special function U given by (3.38). A little thought and experimentation suggest to consider the following scaling of this object. Let $\gamma \in (0, 1 - 2p^{-1})$ be a fixed parameter and, for any $x \in \mathbb{R}$ and $y \geq 0$, take

$$u(x, y) = U\left(x(1 - \gamma), y(1 - \gamma) + \gamma\right).$$

Extend u to the whole $\mathbb{R}^2$ by setting $u(x, y) = u(x, -y)$ for $x \in \mathbb{R}$, $y < 0$. The function should have the necessary concavity property, but, unfortunately, we have $\lim_{y \downarrow 0} u_y(x, y) > 0$ when $|x| \leq \gamma/((1-\gamma)(p-2))$ (that is, when $(x(1-\gamma), \gamma) \in D_4$, where D_4 is as in Figure 3.1). To overcome this problem, we modify u on the square

$$\{(x, y) \in \mathbb{R}^2 : |x| + |y| \leq \gamma/((1 - \gamma)(p - 2))\},$$

putting $u(x, y) = \kappa_1(|y|^2 - |x|^2) + \kappa_2$ there. The parameters κ_1, κ_2 depend only on p and γ, and we determine them by requiring that u is continuous: we obtain

$$\kappa_1 = \frac{p^p \gamma^{p-2}(1 - \gamma)^2}{4(p - 2)^{p-2}} \qquad \text{and} \qquad \kappa_2 = \frac{p^{p-1} \gamma^p}{4(p - 2)^{p-1}}.$$

Quite miraculously, the modified u has the required concavity property (one easily verifies that the partial derivatives match appropriately at the boundary of the square). What inequality does it yield? Looking at $u(\cdot, 1)$ and $u(0, 0)$, we see that u should correspond to (3.52) with

$$\alpha = \frac{p^{p-1}(1 - \gamma)^p}{2} \qquad \text{and} \qquad \beta = \frac{1}{4} \frac{\gamma^p p^{p-1}}{(p - 2)^{p-1}},$$

i.e., we have obtained the candidate $u_{p,\gamma}$ for special function leading to (3.39). Observe that in the proof above, we have used slightly different objects $U_{p,\gamma}$. The reason why we apply the integration method is clear: it enables us to avoid many technicalities involved in the analysis of Suh's function. How did we get the kernel k? First, observe that there is no nonnegative k for which

$$u_{p,\gamma}(x, y) = \int_0^\infty k(r) U_\infty(x/r, y/r) \mathrm{d}r$$

on the whole $\mathbb{R}^2$. A little experimentation leads to the slightly different equation

$$u_{p,\gamma}(x, y) = \int_0^\infty k(r) U_\infty(x/r, y/r) \mathrm{d}r + \kappa_1(|y|^2 - |x|^2) + \kappa_2, \tag{3.53}$$

but still there is no k for which the identity holds true on $\mathbb{R}^2$. Fortunately, it suffices to require the validity of this identity on an appropriate part of $\mathbb{R}^2$: assuming (3.53) on the set

$$\{(x, y) : |x| + |y| \leq b, \ |y| + \gamma/(1 - \gamma) \geq (p - 2)|x|\} \cup \{(1 - b, 1)\},$$

implies that k is given by (3.41).

Finally, we would like to stress here that it can be shown that

$$u_{p,\gamma}(x, y) = \sup\left\{\mathbb{P}(|g_n| \geq 1) - \frac{p^{p-1}(1-\gamma)^p}{2}\mathbb{E}|f_n|^p : (f, g) \in M(x, y)\right\};$$

in other words, the above $u_{p,\gamma}$ is the least special function leading to (3.39). $\square$

3.2   Weak type estimates for nonnegative f

Now we will study the above weak type estimates under the assumption that the dominating martingale is nonnegative. It can be easily seen from the considerations in the general case (see the sharpness of (3.16)), that the optimal constants do not change for $1 \leq p \leq 2$. It turns out that they do change for remaining p.

3.2.1   Formulation of the result

We will prove the following fact.

Theorem 3.4. *Suppose f is a nonnegative martingale and g is an $\mathcal{H}$-valued martingale which is differentially subordinate to f. Then, for $p \in (0, \infty)$,*

$$||g||_{p,\infty} \leq c^+_{p,p}||f||_p, \tag{3.54}$$

where

$$(c^+_{p,p})^p = \begin{cases} 2 & \text{if } p \in (0, 1), \\ 2/\Gamma(p+1) & \text{if } p \in [1, 2], \\ \frac{p^p}{2^p(p-1)} & \text{if } p > 2. \end{cases}$$

The constant is the best possible, even if $\mathcal{H} = \mathbb{R}$ and g is assumed to be a ± 1-transform of f.

If $p \in (0, 1)$, then the inequality holds for any nonnegative supermartingale f and any g which is strongly differentially subordinate to f. This will be shown in the next chapter and we refer the reader there. Therefore it remains to show the weak type inequality for $p > 2$, which, as in the general case, turns out to be quite a challenging problem.

3.2.2   Proof of Theorem 3.4 for $p > 2$

Proof of (3.54). It suffices to show the estimate for $f \in L^p$; then g_n is also p-integrable, $n = 0, 1, 2, \ldots$. By the stopping time argument and homogeneity, the weak type inequality is equivalent to

$$\mathbb{E}V^+_p(f_n, g_n) \leq 0, \qquad n = 0, 1, 2, \ldots, \tag{3.55}$$

where $V_p^+ : [0, \infty) \times \mathcal{H} \to \mathbb{R}$ is given by $V_p^+(x, y) = 1_{\{|y| \geq 1\}} - (c_{p,p}^+)^p x^p$. Recall the function U_∞ from Subsection 3.1.4 and let, for a fixed parameter $s \in (1, \infty)$ and $x \geq 0$, $y \in \mathcal{H}$,

$$U_{\infty,s}(x, y) = U_\infty \left(\frac{s-1}{s} \left(x - \frac{1}{s-1}, 0, 0, \ldots \right), \frac{s-1}{s} y \right) \tag{3.56}$$

$$= \begin{cases} 0 & \text{if } (x, y) \in D_s, \\ (\frac{s-1}{s})^2 (|y|^2 - x^2) - \frac{2(s-1)}{s} |y| + \frac{2(s-1)}{s^2} x + \frac{s^2-1}{s^2} & \text{if } (x, y) \notin D_s, \end{cases}$$

where $D_s = \{(x, y) \in [0, \infty) \times \mathcal{H} : |x - \frac{1}{s-1}| + |y| \leq \frac{s}{s-1}\}$. Now let $U_p^+ : [0, \infty) \times \mathcal{H} \to \mathbb{R}$ be given by

$$U_p^+(x, y) = \int_0^\infty k(t) U_{\infty,p-1}(x/t, y/t) dt,$$

where

$$k(t) = \frac{p^p(p-1)^2}{4(p-2)^{p-1}} t^{p-1} 1_{[0,1-2/p]}(t).$$

To write the explicit formula for U_p^+, consider the following subsets of $[0, \infty) \times \mathcal{H}$:

$$D_0 = (\mathbb{R}_+ \times \mathcal{H}) \setminus (D_0 \cup D_1),$$

$$D_1 = \left\{ (x, y) : (p-1)x \leq |y| < x + 1 - \frac{2}{p} \right\},$$

$$D_2 = \left\{ (x, y) : |y| < \min\{1 - x, (p-1)x\} \right\}.$$

Then

$$U_p^+(x, y) = \begin{cases} \frac{p^2(|y|^2 - x^2)}{4} - \frac{p(p-2)|y|}{2} + \frac{p(p-2)x}{2(p-1)} + \left(\frac{p-2}{2}\right)^2 & \text{on } D_0, \\ \frac{p^p}{2(p-1)(p-2)^{p-2}} x(|y| - x)^{p-1} & \text{on } D_1, \\ \frac{1}{p-1}(x + |y|)^{p-1} \left[(p-1)|y| - \frac{p^2-2p+2}{2} x \right] & \text{on } D_2. \end{cases}$$

We turn to the proof of (3.55). Clearly, $U_{\infty,p-1}$ inherits property $2°$: hence if a martingale g is $\mathcal{H}$-valued and differentially subordinate to a nonnegative martingale f, then for all $n \geq 0$,

$$\mathbb{E} U_{\infty,p-1}(f_n, g_n) \leq \mathbb{E} U_{\infty,p-1}(f_0, g_0).$$

Furthermore, it is easy to see that $U_{\infty,p-1}(x, y) \leq 0$ for $|y| \leq x$, so the latter expectation is nonpositive. Since $|U_{\infty,p-1}(x, y)| \leq c(1 + x^2 + |y|^2)$ for some absolute c and all $x \geq 0$, $y \in \mathcal{H}$ (see (3.15)), we have that

$$\int_0^\infty k(t) |U_{\infty,p-1}(x/t, y/t)| dt \leq \tilde{c}(1 + x^2 + |y|^2)$$

and hence the use of Fubini's theorem (and the integration argument) is permitted. Consequently, to complete the proof it suffices to show the majorization $U_p^+ \geq V_p^+$.

With no loss of generality we may assume that $\mathcal{H} = \mathbb{R}$ and, due to the symmetry of U_p^+ and V_p^+ with respect to the x-axis, we are allowed to consider $y \geq 0$ only. We distinguish two cases.

The case $x \in [0, 2/p]$. Then it is easy to derive that the function $U_p(x, \cdot)$ is decreasing on $[0, (p-2)x/2]$ and increasing on $[(p-2)x/2, \infty)$. Therefore it is enough to establish the majorization for $y = (p-2)x/2$ and $y = 1$; for the first value of y we have $U_p^+(x, y) = V_p^+(x, y)$, while for the second,

$$V_p^+(x, 1) = 1 - c_{p,p}^p x^p \leq 1 - \frac{p^2 x^2}{4} + \frac{p(p-2)x}{2(p-1)} = U_p^+(x, 1).$$

Here in the second passage we have used the inequality

$$u^{p-1} - 1 \geq (p-1)(u-1), \qquad \text{with } u = px/2. \tag{3.57}$$

The case $x \in (2/p, \infty)$. Then $U_p(x, \cdot)$ is decreasing on $[0, 1 - 2/p]$ and increasing on $[1 - 2/p, \infty)$, so it suffices to prove the majorization for $y = 1 - 2/p$ and $y = 1$. The second possibility has been already checked above; furthermore, $U_p^+(x, 1 - 2/p) \geq V_p^+(x, 1 - 2/p)$ is again equivalent to (3.57).

This completes the proof.   □

Sharpness of (3.54), $p > 2$. This can be obtained by considering a slight modification of the following example, similar to that studied in the general setting. Let $\delta > 0$, $x \in (0, 1/p)$ be numbers satisfying

$$x \left(1 + \frac{2\delta}{p} \right)^N = \frac{1}{p} \tag{3.58}$$

for certain integer $N = N(\delta, x)$. It is clear that we may choose δ and x to be arbitrarily small. Consider a two-dimensional Markov martingale (f, g), which is uniquely determined by the following properties.

(i) $f_0 = x$, $g_0 = (p-1)x$.

(ii) We have $df_n = (-1)^{n+1} dg_n$ for $n = 1, 2, \ldots$.

(iii) If (f_n, g_n) lies on the line $y = (p-1)x$ and $f_n < 1/p$, then in the next step it moves either to the line $x = 0$, or to the line $y = (p-1+\delta)x/(1+\delta)$, $n = 0, 1, 2, \ldots$.

(iv) If (f_n, g_n) lies on the line $y = (p-1+\delta)x/(1+\delta)$, then in the next step it moves either to the line $y = (p-1)x$, or to the line $y = (p-2)x/2$, $n = 0, 1, 2, \ldots$.

(v) If $(f_n, g_n) = (1/p, 1 - 1/p)$ (which may happen only if $n = 2N$), then (f_{n+1}, g_{n+1}) equals either $(0, 1)$, or $(2/p, 1 - 2/p)$.

(vi) The states on the line $x = 0$ and $y = (p-2)x/2$ are absorbing.

For a detailed calculation, we refer the interested reader to [133].   □

On the search of the suitable majorant. A reasoning similar to that presented above in the general setting leads to the special function from [133]. We omit the details. □

3.3 L^p estimates

We turn to the comparison of moments of differentially subordinate martingales.

3.3.1 Formulation of the result

Theorem 3.5. *Assume that f, g are $\mathcal{H}$-valued martingales such that g is differentially subordinate to f.*

(i) *If $0 < p \leq 1$, then the inequality*

$$\|g\|_p \leq C_{p,p}\|f\|_p \qquad (3.59)$$

does not hold in general with any finite $C_{p,p}$, even if $\mathcal{H} = \mathbb{R}$ and g is assumed to be a ± 1-transform of f.

(ii) *If $1 < p < \infty$, then the inequality (3.59) holds with $C_{p,p} = p^* - 1$, where $p = \max\{p, p/(p-1)\}$. The constant is the best possible.*

3.3.2 Proof of Theorem 3.5 for $0 < p \leq 1$

Assume first that $p < 1$. If the moment inequality were valid, then the weak type estimate $\|g\|_{p,\infty} \leq C_{p,p}\|f\|_p$ would hold and we have already proved in Subsection 3.1.3 above that this is impossible.

It remains to show that the moment inequality fails to hold also for $p = 1$. Consider the sequence $(X_n)_{n \geq 0}$ of independent mean 1 random variables, with the distribution described as follows: $X_0 \equiv 1$ and for $n \geq 1$, the variable X_n takes values in the set $\{0, 1 + 1/n\}$. Let f, g be two martingales defined by $f_n = X_0 X_1 X_2 \cdots X_n$ and $dg_n = (-1)^n df_n$, $n = 0, 1, 2, \ldots$. Since f is nonnegative, we have $\|f\|_1 = \mathbb{E}f_0 = 1$. Furthermore, as one easily checks, for any $n \geq 1$,

$$\begin{aligned}
\mathbb{P}(|g_\infty| = 2n) &= \mathbb{P}(g_{2n-1} = 2n) + \mathbb{P}(g_{2n} = -2n) \\
&= \mathbb{P}(f_{2n-2} = 2n - 1, X_{2n-1} = 0) + \mathbb{P}(f_{2n-1} = 2n, X_{2n} = 0) \\
&= \frac{1}{2n-1} \cdot \frac{1}{2n} + \frac{1}{2n} \cdot \frac{1}{2n+1} = \frac{2}{(2n-1)(2n+1)}.
\end{aligned}$$

This implies $\|g\|_1 = \infty$ and hence (3.59) does not hold with any finite constant.

3.3.3 Proof of Theorem 3.5 for $1 < p \leq 2$

The inequality (3.59) is trivial for $p = 2$, so in the proof below we assume that $p < 2$.

Proof of (3.59). With no loss of generality we may assume that $||f||_p < \infty$. It suffices to show that for any nonnegative integer n we have

$$\mathbb{E}V_p(f_n, g_n) \leq 0, \qquad n = 0, 1, 2, \ldots,$$

where $V_p(x, y) = |y|^p - C_{p,p}^p |x|^p$. The well-known Burkholder function corresponding to this problem is

$$U_p(x, y) = p^{2-p}(|y| - (p-1)^{-1}|x|)(|x| + |y|)^{p-1}. \tag{3.60}$$

We shall prove the majorization $U_p \geq V_p$. Observe that, by homogeneity, it suffices to show it for $|x| + |y| = 1$. Then the substitution $s = |x| \in [0, 1]$ transforms the desired inequality into

$$p^{2-p}\left(1 - \frac{p}{p-1}s\right) - (1-s)^p + \frac{s^p}{(p-1)^p} \geq 0.$$

Denoting the left-hand side by $F(s)$, we derive that $F((p-1)/p) = F'((p-1)/p) = 0$. Furthermore,

$$F''(s) = -p(p-1)(1-s)^{p-2} + \frac{ps^{p-2}}{(p-1)^{p-1}},$$

hence there is $s_0 \in (0, 1)$ such that F is concave on $(0, s_0)$ and convex on $(s_0, 1]$. Finally, we have

$$F''((p-1)/p) = p^{4-p}(p-1)^{-1}(2-p) > 0.$$

Combining these facts yields that $F \geq 0$ on $[0, 1]$ and $1°$ is established. We next turn to conditions $2°$ and $3°$. In his original approach, Burkholder verified the concavity property directly. We will provide a different, simpler proof, which exploits the integration method. Namely, one can verify that the following remarkable formula is valid: for any $x, y \in \mathcal{H}$,

$$U_p(x, y) = \frac{p^{3-p}(p-1)(2-p)}{2} \int_0^\infty t^{p-1}u(x/t, y/t)dt,$$

where u is given by (3.6). Thus both $2°$ and $3°$ follow, which in turn yields (3.59), if only we can justify the use of Fubini's theorem. To do this, we apply (3.14) to obtain

$$\int_0^\infty t^{p-1}|u(x/t, y/t)|dt \leq U_p(0, y) - U_p(x, 0) = \frac{p^{2-p}}{p-1}(|x|^p + (p-1)|y|^p)$$

and it suffices to use the fact that $f_n, g_n \in L^p$ for any $n \geq 0$, by virtue of the differential subordination. This completes the proof. $\qquad \square$

Sharpness. For almost all the estimates studied so far, we have proved the optimality of the constants by providing appropriate examples. As already announced (see Remark (iv) in Section 2.2), there is a different method, based on Theorem 2.2. For the sake of convenience, we present both approaches below and discuss the similarities between them.

The first approach. This is essentially taken from [24] (see also [25]): we will construct appropriate examples of f and g and derive directly that the ratio $\|g\|_p/\|f\|_p$ can be arbitrarily close to $(p-1)^{-1}$.

To this end, let $p' \in (p, 2)$, $\delta \in (0, 1)$ and consider the Markov martingale (f, g) such that

(i) We have $(f_0, g_0) = (\frac{1+\delta}{2}, \frac{1+\delta}{2})$ almost surely.

(ii) The state $(\frac{1+\delta}{2}, \frac{1+\delta}{2})$ leads to the point $(1, \delta)$ or to $(\frac{(p'-1)(1+\delta)}{p'}, \frac{1+\delta}{p'})$.

(iii) The state (x, y) satisfying $y = \delta x$ leads to $(\frac{(p'-1)(1+\delta)}{p'} x, \frac{1+\delta}{p'} x)$ or to $(\frac{1+\delta}{1-\delta} x, -\delta \frac{1+\delta}{1-\delta} x)$.

(iv) The state (x, y) satisfying $y = -\delta x$ leads to $(\frac{(p'-1)(1+\delta)}{p'} x, -\frac{1+\delta}{p'} x)$ or to $(\frac{1+\delta}{1-\delta} x, \delta \frac{1+\delta}{1-\delta} x)$.

(v) The states on the lines $y = \pm(p'-1)^{-1} x$ are absorbing.

It can be readily verified that g is a ± 1 transform of f; furthermore, it is clear from the construction that f is nonnegative and $g_\infty = (p'-1)^{-1} f_\infty$ almost surely. For any $N \geq 1$ we have

$$
\begin{aligned}
\mathbb{E}|g_N|^p &= \mathbb{E}|g_N|^p 1_{\{g_N \neq (p'-1)^{-1} f_N\}} + \mathbb{E}|g_N|^p 1_{\{g_N = (p'-1)^{-1} f_N\}} \\
&\geq (p'-1)^{-p} \mathbb{E}|f_N|^p 1_{\{g_N = (p'-1)^{-1} f_N\}} \\
&= (p'-1)^{-p} \left[\|f_N\|_p^p - \mathbb{E}|f_N|^p 1_{\{g_N \neq (p'-1)^{-1} f_N\}} \right].
\end{aligned}
\tag{3.61}
$$

Note that the event $\{g_N \neq (p'-1)^{-1} f_N\}$ means that $|g_n| = \delta f_n$ for $n = 1, 2, \ldots, N$ and hence

$$
\mathbb{P}(g_N \neq (p'-1)^{-1} f_N) = \frac{(1+\delta)(2-p')}{2(1+\delta-p'\delta)} \cdot \left[\frac{(1-\delta)(1-(p'-1)\delta)}{(1-\delta)(1-(p'-1)\delta)+2p'\delta} \right]^{N-1},
$$

where the first factor is just the probability of passing from $(\frac{1+\delta}{2}, \frac{1+\delta}{2})$ to $(1, \delta)$ at the first step and the expression in the square brackets is equal to the conditional probability $\mathbb{P}(|g_n| = \delta f_n \mid |g_{n-1}| = \delta f_{n-1})$, $n = 2, 3, \ldots, N$. Furthermore, on the set $\{g_N \neq (p'-1)^{-1} f_N\}$ we have that $f_1 = 1$ and $f_{n+1}/f_n = (1+\delta)/(1-\delta)$ almost surely, $n = 1, 2, \ldots, N-1$, and consequently, $f_N = (\frac{1+\delta}{1-\delta})^{N-1}$. Therefore

$$
\begin{aligned}
\mathbb{E}|f_N|^p 1_{\{g_N \neq (p'-1)^{-1} f_N\}} &= \left(\frac{1+\delta}{1-\delta} \right)^{(N-1)p} \mathbb{P}(g_N \neq (p'-1)^{-1} f_N) \\
&= \frac{(1+\delta)(2-p')}{2(1+\delta-p'\delta)} \cdot F(\delta)^{N-1},
\end{aligned}
\tag{3.62}
$$

where

$$F(\delta) = \left(\frac{1+\delta}{1-\delta}\right)^p \cdot \frac{(1-\delta)(1-(p'-1)\delta)}{(1-\delta)(1-(p'-1)\delta)+2p'\delta},$$

and a direct computation shows that $F(0+) = 1$, $F'(0+) = 2(p - p') < 0$, so $F(\delta) < 1$ if δ is sufficiently small. Now we may proceed as follows. Fix $\varepsilon > 0$ and choose N so that

$$\mathbb{E}|f_N|^p 1_{\{g_N \neq (p'-1)^{-1}f_N\}} \leq \varepsilon \left(\frac{1+\delta}{2}\right)^p = \varepsilon ||f_0||_p^p \leq \varepsilon ||f_N||_p^p.$$

Such N exists in view of (3.62). Plugging this into (3.61) yields

$$||g_N||_p^p \geq \frac{1-\varepsilon}{(p'-1)^p} ||f_N||_p^p.$$

Since $\varepsilon > 0$ and $p' > p$ were arbitrary, the constant $(p-1)^{-1}$ is indeed the best possible in (3.59).

The second approach. This method is entirely different and exploits Theorem 2.2. Suppose that the best constant in the moment inequality (3.59), to be valid for any real martingale f and its ± 1 transform g, equals β_p. Let us write down the formula (2.1) for $U^0 : \mathbb{R} \times \mathbb{R} \to \mathbb{R}$ in our setting:

$$U^0(x, y) = \sup\{\mathbb{E}|g_n|^p - \beta_p^p \mathbb{E}|f_n|^p\}, \tag{3.63}$$

where the supremum is taken over all n and all pairs $(f, g) \in M(x, y)$. The function U^0 satisfies $1°$, $2°$ and $3°$; moreover, directly from (3.63), we have that it is homogeneous of order p:

$$U^0(\lambda x, \pm \lambda y) = |\lambda|^p U^0(x, y), \qquad x, y \in \mathbb{R}, \lambda \neq 0.$$

Thus the function $w : \mathbb{R} \to \mathbb{R}$ given by

$$w(x) = U^0(x, 1-x) \tag{3.64}$$

is concave and satisfies

$$w(x) = U^0(x, 1-x) = U^0(x, x-1) = (2x-1)^p w\left(\frac{x}{2x-1}\right) \qquad \text{for } x > 1 \tag{3.65}$$

and

$$w((\beta_p + 1)^{-1}) \geq \left(1 - (\beta_p + 1)^{-1}\right)^p - \beta_p^p(\beta_p + 1)^{-p} = 0. \tag{3.66}$$

Using (3.65), we have, for $x > 1$,

$$\frac{w(x) - w(1)}{x-1} + (2x-1)^{p-1} \frac{w(1) - w\left(\frac{x}{2x+1}\right)}{1 - x/(2x+1)} = 2w(1)\frac{(2x-1)^p - 1}{(2x-1) - 1}.$$

Now let $x \downarrow 1$ to get $w'(1+) + w'(1-) = 2pw(1)$; both one-sided derivatives exist, due to the concavity of w. Thus, by (3.66) and the concavity again,

$$\frac{\beta_p + 1}{\beta_p} w(1) \geq \frac{w(1) - w((\beta_p + 1)^{-1})}{1 - (\beta_p + 1)^{-1}}$$

$$\geq w'(1-) \geq \frac{w'(1-) + w'(1+)}{2} = pw(1), \tag{3.67}$$

or $\beta_p \geq (p-1)^{-1}$, since $w(1)$ is strictly negative. The latter follows from concavity of w, (3.66) and the estimate $w(0) = U^0(0, 1) \geq 1$. $\qquad\square$

Remark 3.5. There is an interesting question about the analogies between the two proofs above and we shall provide an intuitive answer. Suppose that β_p is the best constant and let U^0 be as in the second proof. Let D^+ and D^- denote the angles $\{(x, y) : 0 < y < (p-1)^{-1}x\}$ and $\{(x, y) : -(p-1)^{-1}x < y < 0\}$, respectively. A careful inspection shows that in both approaches the lower bound for the constant β_p is obtained by exploiting the concavity of U^0 along the line segments of slope -1 lying in D^+ and the line segments of slope 1 lying in D^-. This is evident in the second proof: look at (3.67). To see that this happens also in the first approach, note that our exemplary pair (f, g) moves along these line segments (or rather their endpoints, up to a small constant $\delta > 0$). The lower bound for β_p is in fact obtained by calculating $EV(f_N, g_N)$ and noting that this expectation is nonpositive. Put the sequence $(\mathbb{E}U^0(f_n, g_n))_{n=0}^N$ in between:

$$EV(f_N, g_N) \leq \mathbb{E}U^0(f_N, g_N) \leq \mathbb{E}U^0(f_{N-1}, g_{N-1}) \leq \cdots \leq \mathbb{E}U^0(f_0, g_0) \leq 0,$$

which is allowed by Burkholder's method. Consequently, we see that the inequality $EV(f_N, g_N) \leq 0$ relies on

$$\mathbb{E}(U^0(f_n, g_n)|\mathcal{F}_{n-1}) \leq U^0(f_{n-1}, g_{n-1}), \qquad n = 1, 2, \ldots, N,$$

which, in turn, makes use of the concavity mentioned above.

On the search of the suitable majorant. We will use (or rather continue) the argumentation from the second approach. So, suppose that the optimal constant in the moment estimate for real martingales and their ± 1-transforms equals β_p. Let U^0 be defined by (3.63) and introduce w by (3.64); clearly, by homogeneity, we will be done if we find an appropriate candidate for w on $[0, 1]$. Following the reasoning presented above, we come to the assumption

(A1) $\beta_p = (p-1)^{-1}$.

Then equality must hold throughout (3.67) and thus w is linear on the interval $((\beta_p + 1)^{-1}, 1) = (1 - 1/p, 1)$: we have $w(x) = ax - b$ there. In addition (see the last passage in (3.67)) we obtain $w'(1-) = w'(1+)$, which, by (3.65), gives $b = (p-1)a/p$. Moreover, we have

$$w(x) \geq (1 - x)^p - (p-1)^{-p} x^p,$$

and the two sides are equal when $x = (p-1)/p$. This enforces the equality of the derivatives of both sides at this point:

$$a = w'\left(\frac{p-1}{p}+\right) = -\frac{p^{3-p}}{p-1}.$$

Thus, in view of homogeneity of U^0, we obtain the candidate U given by (3.60) on the set $\{(x,y) : |y| \leq (p-1)|x|\}$. We complete the search by

(A2)   The formula (3.60) is valid for all x, $y \in \mathbb{R}$.

An important remark is in order. There is no uniqueness of the special function: the considerations above do not say anything about U outside the set $\{(x,y) : |y| \leq (p-1)|x|\}$ and there are many alternative versions of the assumption (A2). For example,

(A2') $U(x,y) = |y|^p - (p-1)^{-p}|x|^p$

also leads to a right candidate; in fact this choice yields the optimal (that is, the least) special function.   □

3.3.4   Proof of Theorem 3.5 for $p > 2$

Here the calculations are similar, so we will be brief.

Proof of (3.59). It suffices to establish the inequality for $f \in L^p$; then $g_n \in L^p$ for each n. Let U_p, $V_p : \mathcal{H} \times \mathcal{H} \to \mathbb{R}$ be given by

$$U_p(x,y) = p^{2-p}(p-1)^{p-1}(|y| - (p-1)|x|)(|x| + |y|)^{p-1},$$
$$V_p(x,y) = |y|^p - (p-1)^p|x|^p.$$

It can be easily verified that U_p satisfies 1°: simply repeat the reasoning from the case $1 < p < 2$. Furthermore, we have the identity

$$U_p(x,y) = \int_0^\infty k(t)u(x/t, y/t)\mathrm{d}t,$$

where

$$k(t) = \frac{p^{3-p}(p-1)^p(p-2)}{2}t^{p-1}$$

and u is given by (3.12). Using (3.15) we show that

$$\int_0^\infty k(t)|u(x/t, y/t)|\mathrm{d}t \leq c(|x|^p + |y|^p),$$

for a certain absolute constant c and all x, $y \in \mathcal{H}$. Hence, Fubini's theorem can be used and the result follows via the integration method.   □

Sharpness. This can be established by a reasoning similar to that in the case $1 < p < 2$. However, we take the opportunity here to present a different approach, based on duality. Fix $p > 2$ and suppose that the optimal constant in (3.59) equals β_p. Take a real martingale $f \in L^q$ and its ± 1-transform g: $dg_n = \varepsilon_n df_n$, $n = 0, 1, 2, \ldots$. Here $q = p/(p-1)$ is the harmonic conjugate to p. Fix n and let $\xi = |g_n|^{q-1} \mathrm{sgn}\, g_n$, $\xi_k = \mathbb{E}(\xi | \mathcal{F}_k)$, $k = 0, 1, 2, \ldots, n$. We have that $(\xi_k)_{k=0}^n \in L^p$ and $\|\xi\|_p = \|g_n\|_q^{q/p}$. Furthermore,

$$\|g_n\|_q^q = \mathbb{E} g_n \xi = \mathbb{E}\left(\sum_{k=0}^n \varepsilon_k df_k\right)\left(\sum_{k=0}^n d\xi_k\right) = \mathbb{E}\sum_{k=0}^n \varepsilon_k df_k d\xi_k$$
$$= \mathbb{E}\left(\sum_{k=0}^n df_k\right)\left(\sum_{k=0}^n \varepsilon_k d\xi_k\right),$$

since $\mathbb{E} df_k d\xi_\ell = 0$ for $k \neq \ell$. Let us use Hölder's inequality and (3.59) applied to the finite martingale $(\xi_k)_{k=0}^n$ and its ± 1-transform by the sequence $(\varepsilon_k)_{k=0}^n$. We obtain

$$\|g_n\|_q^q \leq \|f_n\|_q \left\|\sum_{k=0}^n \varepsilon_k d\xi_k\right\|_p \leq \|f_n\|_q \cdot \beta_p \|\xi\|_p = \beta_p \|f_n\|_q \|g_n\|_q^{q/p},$$

whence $\|g_n\|_q \leq \beta_p \|f_n\|_q$. But we have already shown that the best constant in the L^q inequality equals $(q-1)^{-1} = p - 1$. This yields $\beta_p \geq p - 1$ and we are done.   □

On the search of the suitable majorant. This can be done essentially in the same manner as in the case $1 < p < 2$. The details are left to the reader.   □

3.4   Estimates for moments of different order

The next problem we shall study concerns the best constants in the inequalities comparing pth and qth moments of f and g:

$$\|g\|_p \leq C_{p,q} \|f\|_q, \quad 1 \leq p, q < \infty.$$

It is easy to see that this cannot hold with a finite $C_{p,q}$ if $p > q$: simply take a constant pair $(f, g) = (\xi, \xi)$ with $\xi \in L^p \setminus L^q$. The case $p = q$ has been already studied in the previous section. Thus, in what follows, we assume that $p < q$.

3.4.1   Formulation of the result

Before we state the result, we need to introduce some auxiliary objects. Consider the differential equation

$$p(2-p)h'(x) + p = q(q-1)x^{q-2}h(x)^{2-p}, \quad x > 0. \tag{3.68}$$

To stress the dependence on p and q, we will write $(3.68)_{p,q}$. We have the following fact, proved in [146].

Lemma 3.5.

(i) *Let* $1 \leq p < q < 2$. *Then there exists a unique increasing solution* $h : [0, \infty) \to [0, \infty)$ *of* $(3.68)_{p,q}$ *satisfying* $h(0) > 0$ *and* $h'(t) \to 0$, $h(t) \to \infty$, *as* $t \to \infty$.

(ii) *Let* $2 < p < q < \infty$. *Then there exists a unique increasing solution* $h : [0, \infty) \to [0, \infty)$ *of* $(3.68)_{q,p}$ *satisfying* $h(0) > 0$ *and* $h'(t) \to 0$, $h(t) \to \infty$, *as* $t \to \infty$.

The solution h from the above lemma is the key to define the best constants and the corresponding special functions. Put

$$
L_{p,q} = \begin{cases} \frac{1}{2}(2-p)h(0)^p & \text{if } 1 \leq p < q < 2, \\ \frac{1}{2}(q-2)h(0)^q & \text{if } 2 < p < q < \infty \end{cases} \tag{3.69}
$$

and, for $1 \leq p < q < \infty$,

$$
C_{p,q} = \begin{cases} L_{p,q}^{(q-p)/pq} \left(\frac{q-p}{p}\right)^{1/q} \left(\frac{q}{q-p}\right)^{1/p} & \begin{array}{l} \text{if } 1 \leq p < q < 2 \\ \text{or } 2 < p < q < \infty, \end{array} \\ 1 & \text{otherwise.} \end{cases}
$$

Now we are ready to formulate the main results of this section.

Theorem 3.6. *Let* $1 \leq p < q < \infty$ *and let* f, g *be* $\mathcal{H}$*-valued martingales such that* g *is differentially subordinate to* f. *Then*

$$
||g||_p \leq C_{p,q}||f||_q. \tag{3.70}
$$

Furthermore, if $1 \leq p < q < 2$ *or* $2 < p < q < \infty$, *then*

$$
||g||_p^p \leq ||f||_q^q + L_{p,q}. \tag{3.71}
$$

Both inequalities are sharp even if f *is real valued and* g *is assumed to be its* ± 1*-transform.*

Two remarks are in order. We see that the inequality (3.70) cannot be directly rewritten in the form $\mathbb{E}V(f_n, g_n) \leq 0$, $n = 0, 1, 2, \ldots$. This is why we study the estimate (3.71), which is of the right form and yields (3.70) after a standard homogenization procedure (see below). We will prove the above result only for $1 \leq p \leq 2$, for the remaining set of parameters p the reasoning is similar. Consult [146] for details.

3.4.2 Proof of Theorem 3.6 for $1 \leq p \leq 2$

First, note that if $q \geq 2$, then (3.70) is trivial: $||g||_p \leq ||g||_2 \leq ||f||_2 \leq ||f||_q$ and, obviously, equality may hold. Thus, we restrict ourselves to $q \in (p, 2)$.

Proof of (3.70) *and* (3.71). The inequality (3.71) can be rewritten in the form $\mathbb{E}V_{p,q}(f_n, g_n) \leq L_{p,q}$, $n = 0, 1, 2, \ldots$, where $V_{p,q}(x, y) = |y|^p - |x|^q$. The special function $U_{p,q}$ will be defined using the integration method. Let h be the solution to $(3.68)_{p,q}$ coming from Lemma 3.5 and let $H : [h(0), \infty) \to [0, \infty)$ be the inverse to $t \mapsto t + h(t)$. Now, let u be given by (3.6) and define the kernel $k_{p,q}$ by

$$k_{p,q}(t) = \frac{p(2-p)}{2} h(H(t))^{p-3} h'(H(t)) H'(t) t^2 1_{\{t > h(0)\}}.$$

Then the function

$$U_{p,q}(x, y) = \int_0^\infty k_{p,q}(t) u(x/t, y/t) \mathrm{d}t + L_{p,q} \tag{3.72}$$

has the explicit formula

$$U_{p,q}(x, y) = p \frac{|y|^2 - |x|^2}{2h(0)^{2-p}} + L_{p,q} \tag{3.73}$$

if $|x| + |y| \leq h(0)$, and

$$
\begin{aligned}
U_{p,q}(x, y) = p|y| h(H(|x| + |y|))^{p-1} - (p-1) h(H(|x| + |y|))^p \\
- H(|x| + |y|)^q - q H(|x| + |y|)^{q-1} (|x| - H(|x| + |y|)),
\end{aligned} \tag{3.74}
$$

if $|x| + |y| > h(0)$.

Let us establish the majorization property. Since the proof is quite long and elaborate, we have decided to put it in a separate lemma.

Lemma 3.6. *For any* $(x, y) \in \mathcal{H} \times \mathcal{H}$ *we have* $U_{p,q}(x, y) \geq V_{p,q}(x, y)$.

Proof. Clearly, we may restrict ourselves to $\mathcal{H} = \mathbb{R}$. For the reader's convenience, the proof is split into several steps.

Step 1. Reduction to the set $|x| + |y| \geq h(0)$. Fix x such that $|x| < h(0)$. Then the function $y \mapsto U_{p,q}(x, y) - V_{p,q}(x, y)$ is nonincreasing on $[0, h(0) - |x|]$ (by the standard analysis of the derivative). Consequently,

$$U_{p,q}(x, y) - V_{p,q}(x, y) \geq U_{p,q}(x, 1 - |x|) - V_{p,q}(x, 1 - |x|)$$

and it suffices to establish the majorization for $|x| + |y| \geq h(0)$.

Step 2. The case $p = 1$. The majorization follows from the convexity of the function $t \mapsto t^q$: indeed, the inequality $U_{p,q}(x, y) \geq V_{p,q}(x, y)$ is equivalent to

$$|x|^q - H(|x| + |y|)^q - q H(|x| + |y|)^{q-1} \cdot (|x| - H(|x| + |y|)) \geq 0.$$

Therefore, till the end of the proof, we assume that $p > 1$.

Step 3. *Reduction to the case* $y = 0$. Fix $r \geq h(0)$ and suppose that $|x| + |y| = r$. If we denote $s = |y|$, then the inequality $U_{p,q}(x, y) \geq V_{p,q}(x, y)$ can be rewritten in the form

$$F(s) = psh(H(r))^{p-1} - (p-1)h(H(r))^p$$
$$- H(r)^q - qH(r)^{q-1}h(H(r)) - s^p + (r-s)^q \geq 0. \quad (3.75)$$

We have $F(h(H(r))) = F'(h(H(r))) = 0$. Furthermore, the second derivative of F, equal to $F''(s) = -p(p-1)s^{p-2} + q(q-1)(r-s)^{q-2}$, is negative on $(0, s_0)$ and positive on (s_0, r) for some $s_0 \in (0, r)$. Therefore, to show (3.75), it suffices to prove that $F(0) \geq 0$, or $U_{p,q}(x, 0) \geq V_{p,q}(x, 0)$.

Step 4. *Proof of* $U_{p,q}(x, 0) \geq V_{p,q}(x, 0)$ *for large* $|x|$. Since $q < 2$, we have for s large enough,

$$\frac{s^q - (H(s))^q - q(H(s))^{q-1}h(H(s))}{h(H(s))^p} \geq \frac{q(q-1)}{2}s^{q-2}h(H(s))^{2-p}$$

$$= \frac{q(q-1)}{2}H(s)^{q-2}h(H(s))^{2-p} \cdot \left(\frac{s}{H(s)}\right)^{q-2}.$$

But, by (3.68),

$$\frac{q(q-1)}{2}H(s)^{q-2}h(H(s))^{2-p} \geq \frac{p}{2}$$

and, by de l'Hospital rule,

$$\frac{s}{H(s)} = 1 + \frac{h(H(s))}{H(s)} \to 1, \quad \text{as } s \to \infty.$$

Since $p/2 > p - 1$, we see that

$$\frac{s^q - (H(s))^q - q(H(s))^{q-1}h(H(s))}{h(H(s))^p} \geq p - 1$$

for large s. This is equivalent to $U_{p,q}(x, 0) \leq V_{p,q}(x, 0)$ with $|x| = s$.

Step 5. $U_{p,q}(x, 0) \geq V_{p,q}(x, 0)$: *the general case.* Suppose the inequality does not hold for all x. Let T denote the largest t satisfying $U_{p,q}(t, 0) = V_{p,q}(t, 0)$ (its existence is guaranteed by the continuity of $U_{p,q}$ and $V_{p,q}$ and the previous step). By the considerations above, $U_{p,q} \geq V_{p,q}$ on the set $|x| + |y| \geq T$. Consider the processes $f = (f_0, f_1, f_2)$, $g = (g_0, g_1, g_2)$ on a probability space $([0, 1], \mathcal{B}([0, 1]), |\cdot|)$ such that $(f_0, g_0) \equiv (T, 0)$ and

$$df_1 = dg_1 = \delta\chi_{[0,t_0]} - h(H(T))\chi_{(t_0,1]},$$
$$df_2 = -dg_2 = \delta\chi_{[0,t_1]} - (h(H(T)) - \delta)\chi_{(t_1,t_0]},$$

where

$$t_0 = \frac{h(H(T))}{h(H(T)) + \delta}, \quad t_1 = t_0 \cdot \frac{h(H(T)) - \delta}{h(H(T))} = \frac{h(H(T)) - \delta}{h(H(T)) + \delta}.$$

It is straightforward to check that f and g are martingales and g is differentially subordinate to f. Note that we have $U_{p,q}(f_2, g_2) \geq V_{p,q}(f_2, g_2)$ almost surely, as $|f_2| + |g_2| \geq T$ with probability 1. By the integration method, we may write

$$-T^q = U_{p,q}(T, 0) = \mathbb{E}U_{p,q}(f_0, g_0) \geq \mathbb{E}U_{p,q}(f_2, g_2) \geq \mathbb{E}V_{p,q}(f_2, g_2)$$
$$= \frac{2\delta}{h(H(T)) + \delta}(h(H(T))^p - H(T)^q) - \frac{h(H(T)) - \delta}{h(H(T)) + \delta}(T + 2\delta)^q,$$

which is equivalent to

$$\frac{(T + 2\delta)^q - T^q}{2\delta} - \frac{(T + 2\delta)^q}{h(H(T)) + \delta} + \frac{H(T)^q}{h(H(T)) + \delta} \geq \frac{h(H(T))^p}{h(H(T)) + \delta}.$$

Letting $\delta \to 0$ and multiplying throughout by $h(H(T))$ yields

$$h(H(T))^p \leq H(T)^q - T^q - qT^{q-1}h(H(T)). \tag{3.76}$$

But we have

$$H(T)^q - T^q - qT^{q-1}h(H(T)) \leq T^q - H(T)^q - qH(T)^{q-1}h(H(T)). \tag{3.77}$$

To see this, substitute $H(T) = a > 0$, $T - H(T) = b > 0$ and observe that (3.77) becomes

$$2[(a + b)^q - a^q] - [qa^{q-1} + q(a + b)^{q-1}]b \geq 0.$$

Now calculate the derivative of the left-hand side with respect to b: it is equal to

$$q(a + b)^{q-1} - qa^{q-1} - q(q - 1)(a + b)^{q-2}b \geq 0,$$

due to mean value theorem. Thus (3.77) follows. To conclude the proof, observe that the inequalities (3.76) and (3.77) give

$$h(H(T))^p \leq T^q - H(T)^q - qH(T)^{q-1}h(H(T))$$

and since $p < 2$, this implies

$$(p - 1)h(H(T))^p < T^q - H(T)^q - qH(T)^{q-1}h(H(T)),$$

a contradiction: the inequality above is equivalent to $U_{p,q}(T, 0) < V_{p,q}(T, 0)$. $\qquad \square$

Now we turn to the proof of (3.71). We may assume that $||f||_q < \infty$, which implies $||g||_p \leq ||g||_q < \infty$, by Burkholder's inequality (3.59). The use of the integration method is permitted: this is due to (3.14) and the bound $H(t) + h(t) \leq c(1 + t)$ coming from (3.5) and the definition of H. We obtain $\mathbb{E}V_{p,q}(f_n, g_n) \leq U_{p,q}(0, 0) = L_{p,q}$ for any n, which implies (3.71). To deduce (3.70) from this estimate, observe that for any $\lambda > 0$, the martingale $g \cdot \lambda^{1/(q-p)}$ is differentially subordinate to $f \cdot \lambda^{1/(q-p)}$. Consequently,

$$||g||_p^p \leq \lambda||f||_q^q + \frac{L_{p,q}}{\lambda^{p/(q-p)}}.$$

It can be easily checked that the right-hand side, as a function of λ, attains its minimum for

$$\lambda = \left(\frac{p}{q-p} \cdot \frac{L_{p,q}}{||f||_q^q} \right)^{(q-p)/q}$$

and the minimum is equal to

$$\left[L_{p,q}^{(q-p)/pq} \cdot \left(\frac{q-p}{p} \right)^{1/q} \left(\frac{q}{q-p} \right)^{1/p} ||f||_q \right]^p = [C_{p,q}||f||_q]^p.$$

This completes the proof of (3.70).   □

Sharpness. Let $T > h(0)$ be a fixed number to be specified later, pick a positive integer N and put $\delta = (T - h(0))/(2N)$. Consider a Markov martingale (f, g), determined uniquely by the following six conditions.

  (i) It starts from $(h(0)/2, h(0)/2)$ almost surely.
  (ii) From $(h(0)/2, h(0)/2)$ it moves either to $(h(0), 0)$, or to $(0, h(0))$.
  (iii) For $h(0) \le x < T$, the state $(x, 0)$ leads to $(H(x), h(H(x)))$ or to $(x + \delta, -\delta)$.
  (iv) For $h(0) \le x < T$, the state $(x + \delta, -\delta)$ leads to $(H(x + 2\delta), -h(H(x + 2\delta)))$ or to $(x + 2\delta, 0)$.
  (v) For $x \ge T$, the states $(x, 0)$ are absorbing.
  (vi) All the states lying on the curves $y = \pm h(x)$ are absorbing.

Let us split the reasoning into a few steps.

   Step 1. It is easy to see from the conditions (i)–(vi) that (f, g) converges almost surely. Even more, we have $df_{2N+2} = df_{2N+3} = \cdots \equiv 0$. Let us compute $\mathbb{E}U_{p,q}(f_{2N+1}, g_{2N+1})$, or rather relate it to $\mathbb{E}V_{p,q}(f_{2N+1}, g_{2N+1})$. To begin, note that the process (f_n, g_n) moves from $(x, 0)$ to $(x + 2\delta, 0)$ in two steps with probability

$$\left(1 - \frac{\delta}{x - H(x) + \delta} \right) \left(1 - \frac{\delta}{x + 2\delta - H(x + 2\delta)} \right).$$

Hence the probability that (f_n, g_n) ever reaches $(T, 0)$ equals

$$p_\delta = \frac{1}{2} \prod_{n=0}^{N-1} \left(1 - \frac{\delta}{x_{2n+1} - H(x_{2n+1}) + \delta} \right) \left(1 - \frac{\delta}{x_{2n+1} + 2\delta - H(x_{2n+1} + 2\delta)} \right).$$

In the next lemma we study the limit behavior of p_δ, as $\delta \to 0$.

Lemma 3.7. *We have*

$$\lim_{\delta \to 0} p_\delta = \frac{1}{2} \exp(-\Phi'(H(T))).$$

Proof. Indeed,

$$\frac{1}{2} \exp \left[\sum_{k=0}^{N-1} \left(\frac{\delta}{x_{2n+1} - H(x_{2n+1}) + \delta} + \frac{\delta}{x_{2n+1} + 2\delta - H(x_{2n+1} + 2\delta)} \right) \right] \cdot p_\delta^{-1} \to 1$$

as $\delta \to 0$, and the Riemann sum

$$\sum_{k=0}^{N-1} \left(\frac{\delta}{x_{2n+1} - H(x_{2n+1}) + \delta} + \frac{\delta}{x_{2n+1} + 2\delta - H(x_{2n+1} + 2\delta)} \right)$$

converges, as $\delta \to 0$, to

$$-\int_{h(0)}^{T} \frac{1}{x - H(x)} dx = -\int_{h(0)}^{T} \frac{1}{h(H(x))} dx.$$

The substitution $y = H(x)$ (so $x = y + h(y)$) transforms the integral to

$$-\int_{0}^{H(T)} \frac{1 + h'(y)}{h(y)} dy = -\int_{0}^{H(T)} \Phi''(y) dy = -\Phi'(H(T)).$$

This yields the claim. $\qquad\square$

Step 2. The next observation is that g is a ± 1 transform of f. Hence, the sequence $(\mathbb{E}U_{p,q}(f_n, g_n))$ is nonincreasing: see Theorem 2.1. The key fact is that this sequence is almost constant if δ is small. Roughly speaking, this follows from the fact that the pair (f, g) moves along line segments of slope $1/-1$ when it is below/above the x-axis and $U_{p,q}$ is linear on such segments. The precise statement is the following (see [146] for the proof).

Lemma 3.8. *There is a function* $R : [0,1] \to \mathbb{R}$ *satisfying* $R(\delta)/\delta \to 0$ *as* $\delta \to 0$ *and for which*

$$\mathbb{E}U_{p,q}(f_{2N+1}, g_{2N+1}) \geq U_{p,q}(h(0)/2, h(0)/2) - NR(\delta).$$

Step 3. *Sharpness of* (3.71). On the set $\{(x, y) : y = \pm h(x)\}$ the functions $U_{p,q}$ and $V_{p,q}$ coincide. The variable (f_{2N+1}, g_{2N+1}) belongs to this set unless $(f_{2N+1}, g_{2N+1}) = (T, 0)$, which occurs with probability p_δ. Hence we may write, by Lemma 3.8,

$$\mathbb{E}V_{p,q}(f_{2N+1}, g_{2N+1})$$
$$= \mathbb{E}U_{p,q}(f_{2N+1}, g_{2N+1}) + (V_{p,q}(T, 0) - U_{p,q}(T, 0))p_\delta$$
$$\geq U_{p,q}(h(0)/2, h(0)/2) - NR(\delta) + (V_{p,q}(T, 0) - U_{p,q}(T, 0))p_\delta.$$

Let N go to ∞. Then $NR(\delta) \to 0$; furthermore, by the majorization $U_{p,q} \geq V_{p,q}$ and Lemma 3.7,

$$0 \leq (V_{p,q}(T, 0) - U_{p,q}(T, 0))p_\delta \to -\frac{T^q - (H(T))^q - q(H(T))^{q-1}(T - H(T))}{2 \exp(q(H(T))^{q-1})}.$$

$$(3.78)$$

Now we proceed as follows. We have $h'(t) \to 0$ as $t \to \infty$, so $H(T)/T \to 1$ as $T \to \infty$. Thus, for a fixed $\varepsilon > 0$ we choose such a T, that the expression appearing as the limit on the right is smaller than ε. Keeping this T fixed, we may choose an N such that $NR(\delta) < \varepsilon$ and the expression in the middle of (3.78) is smaller than 2ε. In consequence, we obtain

$$||g||_p^p - ||f||_q^q = \mathbb{E}V_{p,q}(f_{2N+1}, g_{2N+1}) \geq U_{p,q}(h(0)/2, h(0)/2) - 3\varepsilon.$$

However, $U_{p,q}(h(0)/2, h(0)/2)$ equals $L_{p,q}$ and ε was arbitrary; therefore, (3.71) is sharp.

   Step 4. The optimality of $C_{p,q}$, $1 \leq p < q < 2$. For fixed $\varepsilon > 0$, let f^ε, g^ε be real martingales such that g^ε is a ± 1 transform of f^ε and

$$||g^\varepsilon||_p^p > ||f^\varepsilon||_q^q + L_{p,q} - \varepsilon. \tag{3.79}$$

Let

$$\lambda = \left(\frac{p}{q-p} \cdot \frac{L_{p,q} - \varepsilon}{||f||_q^q} \right)^{(q-p)/q}.$$

The martingale $\tilde{g} = g^\varepsilon \cdot \lambda^{1/(q-p)}$ is a ± 1 transform of $\tilde{f} = f^\varepsilon \cdot \lambda^{1/(q-p)}$. Multiply both sides of (3.79) by $\lambda^{p/(q-p)}$ to obtain

$$||\tilde{g}||_p^p > \left[(L_{p,q} - \varepsilon)^{(q-p)/pq} \cdot \left(\frac{q-p}{p} \right)^{1/q} \left(\frac{q}{q-p} \right)^{1/p} \right]^p ||\tilde{f}||_q^p.$$

It is clear that the expression in the square brackets can be made arbitrarily close to $C_{p,q}$ by a proper choice of ε. This implies $C_{p,q}$ can not be replaced by a smaller constant in (3.70). $\qquad\qquad\qquad\qquad\qquad\qquad\qquad\qquad\qquad\qquad\qquad\qquad\qquad\square$

On the search of the suitable majorant. As usual, we assume that $\mathcal{H} = \mathbb{R}$ and search for the majorant corresponding to the estimate (3.71) for ± 1-transforms. As previously, we consider $1 \leq p < q < 2$. Clearly, we may and do assume that U is symmetric in the sense that $U(x, -y) = U(-x, y) = U(x, y)$, so it suffices to determine it in the first quadrant $[0, \infty)^2$. It is quite natural to consider the limit case $p = q \in (1, 2)$ and look at the properties of the function $U_{p,p}$ (which is given by (3.60)). We see that

   (i) $U_{p,p}$, restricted to $[0, \infty)^2$, is linear along the line segments of slope -1,
   (ii) $(U_{p,p})_y(x, 0) = 0$ for any $x \geq 0$,
   (iii) the set $\{(x,y) \in [0, \infty) \times [0, \infty) : U_{p,p}(x, y) = V_{p,p}(x, y)\}$ is the half-line $y = (p-1)^{-1}x$.

   It is reasonable to assume that for general $1 \leq p < q < 2$, the function U should also enjoy (i) and (ii). Next, if we restrict to any line segment of slope -1 contained in $[0, \infty)^2$, we expect the graph of U to be a line tangent to the graph of $V_{p,q}$. What about the set $\{U = V_{p,q}\}$? One should not expect such a regular answer as for $p = q$, since $V_{p,q}$ is not homogeneous. However, it is natural to guess that the set coincides with the graph of a certain C^1 function h. This leads to the following candidate for U: for any $x \geq 0$ and $t \in [-x, h(x)]$,

$$U(x + t, h(x) - t) = V_{p,q}(x, h(x)) + \left[(V_{p,q})_x(x, h(x)) - (V_{p,q})_y(x, h(x))\right]t.$$

Using the formula for $V_{p,q}$, the above becomes

$$U(x + t, h(x) - t) = h(x)^p - x^q - (qx^{q-1} - ph^{p-1}(x))t. \tag{3.80}$$

Applying the condition (ii) yields the equation (3.68). The substitution of x and y in the place of $x + t$ and $h(x) - t$, respectively, transforms (3.80) into (3.74). To obtain (3.73), take (x, y) from the square $D = \{(x, y) : |x| + |y| < h(0)\}$ and consider a line segment of slope -1, containing (x, y), with the endpoints P_1, P_2 on the boundary of the square. Take $U(x, y)$ as the corresponding convex combination of $U(P_1)$ and $U(P_2)$. This gives the candidate studied above. $\square$

Remark 3.6. Here the use of the integration method significantly reduces the complexity of calculations. A direct verification of the concavity property 2°, especially in the Hilbert-space-valued setting, would require a large amount of work.

3.5 Moment estimates for nonnegative martingales

What happens to the best constant in the inequality (3.59) if f or g is assumed to be nonnegative? This question will be answered in this section.

3.5.1 Formulation of the results

We start with the case when f is a nonnegative martingale and g is an $\mathcal{H}$-valued martingale, differentially subordinate to f. This setting was studied by Burkholder in [36] and here is the main result of that paper.

Theorem 3.7. *If f is a nonnegative martingale and g is an $\mathcal{H}$-valued martingale differentially subordinate to f, then*

$$\|g\|_p \leq C_{p+,p}\|f\|_p, \tag{3.81}$$

where

$$C_{p+,p} = \begin{cases} (p-1)^{-1} & \text{if } p \in (1, 2], \\ p^{1/p}[(p-1)/2]^{(p-1)/p} & \text{if } p \in (2, \infty). \end{cases}$$

The constant $C_{p+,p}$ is the best possible. It is already the best possible if g is assumed to be a nonnegative ± 1 transform of f.

An extension to $p < 1$ will be investigated in Theorem 4.11 of Chapter 4. Now we turn to the case when the dominated martingale g is nonnegative.

Theorem 3.8. *If f is an $\mathcal{H}$-valued martingale and g is a nonnegative martingale differentially subordinate to f, then, for $1 \leq p < \infty$,*

$$\|g\|_p \leq C_{p,p+}\|f\|_p, \tag{3.82}$$

where

$$C_{p,p+} = \begin{cases} 1 & \text{if } p = 1, \\ p^{-1/p}[2/(p-1)]^{(p-1)/p} & \text{if } p \in (1,2), \\ p-1 & \text{if } p \in [2,\infty). \end{cases}$$

The constant $C_{p,p+}$ is the best possible. It is already the best possible if f is assumed to be real valued and g is a nonnegative ± 1 transform of f.

3.5.2   Proof of Theorem 3.7

Proof of (3.81). We only need to consider the case $p > 2$; for $p \in (1,2]$ the constant is the same as in the general setting (see the examples studied in Section 3.3). We shall present an approach based on the integration argument. Recall the function $U_{\infty,s}$ defined in (3.56) and let

$$U_{p,s}(x,y) = \frac{p(p-1)(p-2)s^2}{2(s-1)} \int_0^\infty t^{p-1} U_{\infty,s}(x/t, y/t) dt.$$

We have that if $sx \leq |y|$,

$$U_{p,s}(x,y) = (|y| - x)^{p-1}\big[x(sp - s - 1) + |y|(s - p + 1)\big]$$

and if $sx \geq |y|$,

$$U_{p,s}(x,y) = \left(\frac{s-1}{s+1}\right)^{p-1} (x + |y|)^{p-1}\big[x(s - ps - 1) + |y|(s + p - 1)\big].$$

The function $U_{p,p}$ is the special function corresponding to the inequality (3.81). We shall now show the majorization $U_{p,p}(x,y) \geq |y|^p - C_{p+,p}^p x^p$. By homogeneity, it suffices to prove this for $x + |y| = 1$. Introduce $F : [0,1] \to \mathbb{R}$ by

$$F(|y|) = U_{p,p}(1 - |y|, y) - |y|^p - C_{p+,p}^p(1 - |y|)^p, \qquad y \in \mathcal{H}, \ |y| \leq 1.$$

We easily check that

$$F(1) = F\left(\frac{p-1}{p+1}\right) = F'\left(\frac{p-1}{p+1}\right) = 0$$

and that there is a t_0, with $(p-1)/(p+1) < t_0 < 1$, such that F is convex on $(0, t_0)$ and concave on $(t_0, 1)$. Thus $F \geq 0$ and the majorization follows. It suffices to apply Fubini's theorem to get the claim (the necessary integrability is guaranteed in the usual way).   □

Sharpness of (3.81). We need to deal with the case $p > 2$ only. We will make use of the appropriate version of Theorem 2.2 (which takes into account the fact that both martingales are nonnegative).

Suppose that the best constant in the inequality (3.81), restricted to nonnegative g, equals β. Let $U^0 : [0, \infty) \times [0, \infty) \to \mathbb{R}$ be defined by

$$U^0(x, y) = \sup\{\mathbb{E}g_n^p - \beta^p \mathbb{E}f_n^p\}, \tag{3.83}$$

where the supremum is taken over all n and pairs (f, g) of simple nonnegative martingales which belong to $M(x, y)$. Then U^0 satisfies $1°$, $2°$, $3°$ and, in addition, it is homogeneous of order p. Let δ be a fixed positive number. We start by applying $2°$ to $x = 1$, $y = p$, $\varepsilon = 1$ and centered random variables taking values -1 and δ; we get

$$U^0(1, p) \geq \frac{\delta}{1 + \delta} U^0(0, p - 1) + \frac{1}{1 + \delta} U^0(1 + \delta, p + \delta). \tag{3.84}$$

Next, use $2°$ to $x = 1 + \delta$, $y = p + \delta$, $\varepsilon = -1$ and centered random variables taking values 1 and $-a = -(p - 1)\delta/(p + 1)$ (the number a is chosen to ensure that $p + \delta + a = p(1 + \delta - a)$). We get

$$U^0(1 + \delta, p + \delta) \geq \frac{a}{1 + a} U^0(2 + \delta, p + \delta - 1) + \frac{1}{1 + a} U^0(1 + \delta - a, p + \delta + a)$$

$$= \frac{a}{1 + a} U^0(2 + \delta, p + \delta - 1) + \frac{(1 + \delta - a)^p}{1 + a} U^0(1, p),$$

where in the last line we have used the homogeneity. Combining this with (3.84) and then applying the majorization property we obtain

$$U^0(1, p)\left[1 - \frac{(1 + \delta - a)^p}{(1 + \delta)(1 + a)}\right] \geq \frac{\delta U^0(0, p - 1)}{1 + \delta} + \frac{a U^0(2 + \delta, p + \delta - 1)}{(1 + \delta)(1 + a)}$$

$$\geq \frac{\delta(p - 1)^p}{1 + \delta} + \frac{a[(p + \delta - 1)^p - \beta^p(2 + \delta)^p]}{(1 + \delta)(1 + a)}. \tag{3.85}$$

Dividing throughout by δ and letting $\delta \to 0$ yields

$$0 \geq (p - 1)^p + \frac{p - 1}{p + 1}[(p - 1)^p - \beta^p \cdot 2^p],$$

which is equivalent to $\beta^p \geq p[(p - 1)/2]^{p-1}$. This completes the proof. $\quad\square$

3.5.3 Proof of Theorem 3.8

Proof of (3.82). We can restrict ourselves to the case $1 \leq p < 2$. If $p = 1$, then

$$\|g\|_1 = \mathbb{E}g_0 \leq \mathbb{E}|f_0| \leq \|f\|_1$$

and, clearly, the constant 1 is the best possible. Hence, from now on we assume that $p \in (1, 2)$. The proof is very similar to the one of the preceding theorem, so we will only present the main steps. We use the integration method again, this time with the function $\overline{U}_{1,s} : \mathcal{H} \times [0,\infty) \to \mathbb{R}$ given by

$$\overline{U}_{1,s}(x,y) = U_1\left(\frac{s-1}{s}x, \frac{s-1}{s}\left(y - \frac{1}{s-1}\right)\right) + \frac{2y(s-1)}{s^2} - \frac{1}{s^2}$$

$$= \begin{cases} (\frac{s-1}{s})^2(y^2 - |x|^2) & \text{if } |x| + |y - \frac{1}{s-1}| \leq 1, \\ 1 - \frac{2(s-1)}{s}|x| + \frac{2(s-1)}{s^2}y - \frac{1}{s^2} & \text{if } |x| + |y - \frac{1}{s-1}| > 1. \end{cases}$$

Here $s > 1$ is a fixed parameter. Since U_1 enjoys the concavity property, so does $\overline{U}_{1,s}$ and hence

$$\mathbb{E}\overline{U}_{1,s}(f_n, g_n) \leq \mathbb{E}\overline{U}_{1,s}(f_0, g_0) \leq 0, \tag{3.86}$$

since $\overline{U}_{1,s}(x,y) \leq 0$ if $y \leq |x|$. Introduce $U_{p,s} : \mathcal{H} \times [0,\infty) \to \mathbb{R}$ by

$$U_{p,s}(x,y) = \frac{p(p-1)(2-p)s^2}{2(s-1)} \int_0^\infty t^{p-1} u_{2,s}(x/t, y/t) dt. \tag{3.87}$$

The explicit expression for $U_{p,s}$ is the following: if $|x| \leq sy$, then

$$U_{p,s}(x,y) = \left(\frac{s-1}{s+1}\right)^{p-1}(|x|+y)^{p-1}\big[|x|(-s-p+1) + y(sp-s+1)\big],$$

while for $|x| \geq sy$,

$$U_{p,s}(x,y) = (|x|-y)^{p-1}\big[|x|(p-s-1) + y(s-sp+1)\big].$$

Now one easily checks that

$$U_{p,p}(x,y) \geq C_{p,p+}^{-p}y^p - |x|^p,$$

which, combined with (3.86) and the formula for $U_{p,p}$ yields

$$C_{p,p+}^{-p}||g_n|| \leq ||f_n||_p \leq ||f||_p, \qquad n = 0, 1, 2, \ldots,$$

for all $f \in L^p$. It remains to take the supremum over n and the claim is proved. $\square$

Sharpness of (3.82). We have to deal with the case $p \in (1,2)$ only. Suppose the best constant in the estimate equals β and introduce the function U^0 by (3.83), the supremum being taken over the same class of martingales. We proceed similarly and exploit the properties of this function to get

$$U^0(p,1)\left[1 - \frac{(1+\delta-a)^p}{(1+\delta)(1+a)}\right] \geq \frac{\delta U^0(p-1,0)}{1+\delta} + \frac{aU^0(p+\delta-1, 2+\delta)}{(1+\delta)(1+a)}$$

$$\geq -\frac{\delta\beta^p(p-1)^p}{1+\delta} + \frac{a[(2+\delta)^p - \beta^p(p+\delta-1)^p]}{(1+\delta)(1+a)}.$$

Now dividing by δ and letting $\delta \to 0$ yields the desired estimate $\beta \geq C_{p,p+}$. $\square$

3.6 Logarithmic estimates

3.6.1 Formulation of the result

As we have already seen above, the moment inequality (3.59) does not hold with any finite constant when $p = 1$. A natural question is how to bound the first moment of the dominated martingale g in terms of f. In this section we will provide one of the possible answers to this problem. Typically in such a situation one studies $L \log L$ inequalities of the form

$$||g||_1 \leq K \sup_n \mathbb{E}|f_n| \log |f_n| + L, \tag{3.88}$$

for some universal constants K and L (here and below, we use the convention $0 \log 0 = 0$). There are two questions to be asked:

a) For which K there is a universal $L < \infty$ such that (3.88) holds?

b) Suppose that K is as in a) and let $L(K)$ denote the smallest L in (3.88). What is the numerical value of $L(K)$?

The main result of this section provides the answers to both these questions.

Theorem 3.9. *Let f, g be two martingales taking values in $\mathcal{H}$ such that g is differentially subordinate to f.*

(i) *If $K \leq 1$, then $L(K) = \infty$. This is true even if $\mathcal{H} = \mathbb{R}$ and g is assumed to be a ± 1 transform of f.*

(ii) *If $1 < K < 2$, then*

$$L(K) = \frac{K^2}{2(K-1)} \exp(-K^{-1}). \tag{3.89}$$

The constant $L(K)$ is best possible even if $\mathcal{H} = \mathbb{R}$, f is a nonnegative martingale and g is its ± 1 transform.

(iii) *If $K \geq 2$, then*

$$L(K) = K \exp(K^{-1} - 1). \tag{3.90}$$

The constant is already the best possible if $\mathcal{H} = \mathbb{R}$, f is a nonnegative martingale and g its ± 1 transform.

3.6.2 Proof of Theorem 3.9

Proof of (3.88) *for $K > 1$.* Clearly, we may assume that

$$\sup_n \mathbb{E}|f_n| \log |f_n| < \infty,$$

otherwise there is nothing to prove. All we need is the estimate

$$\mathbb{E}V(f_n, g_n) \leq 0, \qquad n = 0, 1, 2, \ldots,$$

where $V(x, y) = |y| - K|x| \log |x| - L(K)$ for $x, y \in \mathcal{H}$. We use the integration method, with u given by (3.6) and $k(t) = 1_{\{t \geq 1\}}$. We derive that

$$
\begin{aligned}
U(x, y) &= \int_0^\infty k(t) u(x/t, y/t) dt \\
&= \begin{cases} |y|^2 - |x|^2 & \text{if } |x| + |y| \leq 1, \\ -2|x| \log(|x| + |y|) + 2|y| - 1 & \text{if } |x| + |y| > 1. \end{cases}
\end{aligned}
\tag{3.91}
$$

Let, for $1 < K \leq 2$,

$$
\alpha = \alpha_K = \frac{2(K-1)}{K^2} \exp(K^{-1}) \quad \text{and} \quad d = d_K = \frac{K-1}{K} \exp(K^{-1}),
$$

while for $K > 2$, put

$$
\alpha = \alpha_K = \frac{1}{K} \exp(1 - K^{-1}) \quad \text{and} \quad d = d_K = \frac{1}{2} \exp(1 - K^{-1}).
$$

Lemma 3.9. *For any* $(x, y) \in \mathcal{H}$ *we have*

$$
\alpha^{-1} U(xd, yd) \geq V(x, y).
\tag{3.92}
$$

Proof. We will only study the case $1 < K \leq 2$, since the majorization for the remaining values of K can be treated in a similar manner. We start by observing that for $|x| + |y| \leq 1$ we have

$$
|y|^2 - |x|^2 \geq -2|x| \log(|x| + |y|) + 2|y| - 1.
$$

This is straightforward: for any fixed x, the function

$$
s \mapsto s^2 - |x|^2 + 2|x| \log(|x| + s) - 2s + 1
$$

is nonincreasing and equal to 0 for $s = 1 - |x|$. Thus, to establish (3.92), it suffices to show that for all $x, y \in \mathcal{H}$ we have

$$
2|y|d - 2|x|d \log(|x|d + |y|d) - 1 - \alpha(|y| - K|x| \log |x| - L(K)) \geq 0.
$$

Now fix x, substitute $s = |y|$ and denote the left-hand side by $F(s)$. Then F is convex on $[0, \infty)$ and satisfies $F(|x|/(K-1)) = F'(|x|/(K-1)) = 0$. This completes the proof. $\qquad\square$

Thus it remains to show that the use of Fubini's theorem in the integration argument is allowed. We have, by (3.14),

$$
\int_0^\infty k(t) |u(x/t, y/t)| dt \leq U(y, 0) - U(x, 0)
$$

$$
\leq 2|x| + 2|y| + \beta|x| \log |x| + \gamma,
$$

for some absolute constants α, β, $\gamma > 0$. This yields

$$\mathbb{E} \int_1^\infty |u_1(f_n d/t, g_n d/t)| dt < \infty$$

for any $d > 0$. This follows from the assumption $\sup_n \mathbb{E}|f_n| \log |f_n| < \infty$; the proof is complete. □

Sharpness. We consider the cases $K \geq 2$ and $K < 2$ separately.

The case $K \geq 2$. Let f, g be given by $f_n = g_n \equiv \exp(K^{-1} - 1)$, $n = 0, 1, 2, \ldots$. Obviously, g is a ± 1 transform of f and it can be easily verified that the two sides of (3.88) are equal.

The case $K < 2$. This is more involved, and we divide the proof into a few steps.

Step 1. Assume the inequality (3.88) holds with some L; our aim is to show that $L \geq L(K)$. For $x \geq 0$ and $y \in \mathbb{R}$, define

$$U^0(x, y) = \sup \{\mathbb{E}|g_n| - K\mathbb{E}f_n \log f_n\},$$

where the supremum is taken over all n and all pairs $(f, g) \in M(x, y)$ such that f is nonnegative. Then U^0 satisfies $1°$, $2°$ and the following version of the initial condition:

$3°$ $U^0(x, \pm x) \leq L$ for all $x \geq 0$.

Furthermore, the assumption on the sign of f in the definition of U^0 yields the following homogeneity-type property of this function: if $\lambda > 0$, then

$$U^0(\lambda x, \lambda y) = \lambda U^0(x, y) - K\lambda x \log \lambda. \tag{3.93}$$

To see this, let (f, g) be a pair as in the definition of $U^0(x, y)$. We have

$$\begin{aligned}
\lambda \mathbb{E}(|g_n| - Kf_n \log f_n) &= \mathbb{E}[|\lambda g_n| - K\lambda f_n \log(\lambda f_n)] + K\lambda \mathbb{E}f_n \log \lambda \\
&= \mathbb{E}[|\lambda g_n| - K\lambda f_n \log(\lambda f_n)] + K\lambda x \log \lambda \\
&\leq U^0(\lambda x, \lambda y) + K\lambda x \log \lambda,
\end{aligned}$$

as the pair $(\lambda f, \lambda g)$ belongs to $M(\lambda x, \lambda y)$ and $\lambda f \geq 0$. Taking the supremum over all n and all f, g as above, we get

$$U^0(x, y) \leq U^0(\lambda x, \lambda y) + K\lambda x \log \lambda.$$

Switching λ to λ^{-1} and replacing x, y by λx, λy, respectively, yields the reverse estimate.

Step 2. Let A, $B : [0, \infty) \to \mathbb{R}$ be given by $A(x) = U^0(x, 0)$ and

$$B(x) = \frac{x}{K} - (K - 1)x \log \left(\frac{K - 1}{K} x\right). \tag{3.94}$$

Fix $x > 0$, $\delta > 0$ and use $2°$ with $x > 0$, $y = 0$, $\varepsilon = -1$, $t_1 = -x/K$, $t_2 = \delta x$ and $\alpha = K\delta/(1 + K\delta)$ to obtain

$$A(x) \geq \frac{K\delta}{1 + K\delta} U^0 \left(\frac{K-1}{K} x, \frac{x}{K} \right) + \frac{1}{1 + K\delta} U^0(x + \delta x, -\delta)$$

$$\geq \frac{K\delta}{1 + K\delta} B(x) + \frac{1}{1 + K\delta} U^0(x + \delta x, -\delta),$$

where in the second passage we have exploited $1°$. Let us use $2°$ again, this time with $x := x + \delta x$, $y = -\delta$, $\varepsilon = 1$, $t_1 = -(x + (2 - K)\delta x)/K$, $t_2 = \delta x$ and $\alpha = K\delta/(1 + 2\delta)$. Then apply $1°$ to obtain

$$U^0(x + \delta x, -\delta) \geq \frac{K\delta}{1 + 2\delta} B(x + 2\delta) + \frac{1 + 2\delta - K\delta}{1 + 2\delta} A\big(x(1 + 2\delta)\big).$$

Combining the two estimates above and exploiting (3.93) leads to

$$A(x) \geq \frac{K\delta}{1 + K\delta} B(x) + \frac{1}{1 + K\delta} \cdot \frac{K\delta}{1 + 2\delta} B\big(x(1 + 2\delta)\big)$$

$$+ \frac{1}{1 + K\delta} \cdot \frac{1 + 2\delta - K\delta}{1 + 2\delta} \left[(1 + 2\delta) A(x) - K(1 + 2\delta) x \log(1 + 2\delta) \right],$$

or, equivalently,

$$\frac{2(K - 1)\delta}{1 + K\delta} A(x) \geq \frac{K\delta}{1 + K\delta} B(x) + \frac{K\delta}{(1 + K\delta)(1 + 2\delta)} B\big(x(1 + 2\delta)\big)$$

$$- \frac{1 + 2\delta - K\delta}{1 + K\delta} \cdot Kx \log(1 + 2\delta).$$

Divide throughout by 2δ, let $\delta \to 0$ and use continuity of B to obtain

$$(K - 1)A(x) \geq KB(x) - Kx. \tag{3.95}$$

The next move is to use $2°$ with $x := x/2$, $y = x/2$, $\varepsilon = -1$, $t_1 = (K - 2)/(2K)$, $t_2 = x/2$ and $\alpha = K/2$. With an aid of $1°$, this yields

$$U^0(x/2, x/2) \geq \frac{K}{2} B(x) + \frac{2 - K}{K} A(x).$$

However, the left-hand side is not larger than L, in view of $3°$. Thus, combining this with (3.95), we obtain

$$L \geq \frac{K}{2(K - 1)} [B(x) + (K - 2)x]. \tag{3.96}$$

This is valid for all $x \geq 0$. It is straightforward to check that the expression in the square brackets attains its maximum at the point $x_0 = K/((K - 1) \exp(K^{-1}))$. Plugging this into (3.96) gives

$$L \geq \frac{K^2}{2(K - 1)} \exp(-K^{-1}),$$

that is, $L \geq L(K)$. The proof is complete. $\qquad \square$

On the search of the suitable majorant. We shall only deal with the case $1 < K < 2$; for $K \geq 2$, the reasoning is similar. As usual, we start by defining the function U^0, corresponding to the logarithmic estimate for ± 1-transforms. That is, for a given $K > 1$ and any x, $y \in \mathbb{R}$, set $V(x, y) = |y| - K|x| \log|x|$ and

$$U^0(x, y) = \sup\{\mathbb{E}V(f_n, g_n)\},$$

where the supremum is taken over all n and all $(f, g) \in M(x, y)$.

For the sake of convenience, we split the further reasoning into several steps.

Step 1. *A special curve.* Let us gather some overall information on the function U^0. Clearly, it enjoys the symmetry property

$$U^0(x, y) = U^0(-x, y) = U^0(x, -y) \qquad \text{for all } x, y \in \mathbb{R}. \tag{3.97}$$

How does the set $\{U^0 = V\}$ look like? First, note that when $|x| < e^{-1}$ and $y \in \mathbb{R}$, then $U^0(x, y) > V(x, y)$. This can be seen by considering a pair $(f, g) \in M(x, y)$ such that $df_1 = dg_1$ and f_1 takes the values $\pm e^{-1}$: then

$$U^0(x, y) \geq \mathbb{E}V(f_1, g_1) = \mathbb{E}|g_1| - K\mathbb{E}|f_1| \log|f_1| = \mathbb{E}|g_1| + Ke^{-1}$$
$$\geq |y| + Ke^{-1} > |y| - K|x| \log|x| = V(x, y).$$

On the other hand, suppose that $x > e^{-1}$ and y is a fixed positive number. To derive $U^0(x, y)$, we need to find a pair $(f, g) \in M(x, y)$ for which $||g||_1$ is large in comparison with $\mathbb{E}|f_n| \log|f_n|$. However, $||g||_1$ increases only when g gets close to 0; if y is large, the increase of $||g||_1$ will not be compensated by the cost arising from the increase of $\mathbb{E}|f_n| \log|f_n|$. It is reasonable to impose the following assumption (as usual, we pass from the notation U^0 to U).

(A1) There is a function $\gamma : (e^{-1}, \infty) \to \mathbb{R}$ such that $U(x, y) = V(x, y)$ if and only if $|y| \geq \gamma(|x|)$.

How does the function γ look like? It is natural to conjecture that

(A2) There is $x_* > e^{-1}$ such that γ is decreasing on (e^{-1}, x_*) and increasing on (x_*, ∞). Furthermore, γ is of class C^1 on both intervals, $\lim_{x \to e^{-1}+} \gamma(x) = \lim_{x \to \infty} \gamma(x) = \infty$.

Step 2. *Further assumptions on* U. As usual, we impose the following regularity condition:

(A3) U is of class C^1 on $\mathbb{R}^2$.

The final, "structural" assumption is based on quite elaborate experimentation. Let

$D_1 = \{(x, y) \in [0, \infty) \times [0, \infty) : x_* + \gamma(x_*) \leq x + y \leq x + \gamma(x)\}$,
$D_2 = \{(x, y) \in [0, \infty) \times [0, \infty) : x + y < x_* + \gamma(x_*) \text{ and } -x + y \leq -x_* + \gamma(x_*)\}$,
$D_3 = \{(x, y) \in [0, \infty) \times [0, \infty) : -x_* + \gamma(x_*) \leq -x + y \leq -x + \gamma(x)\}$

(see Figure 3.2).

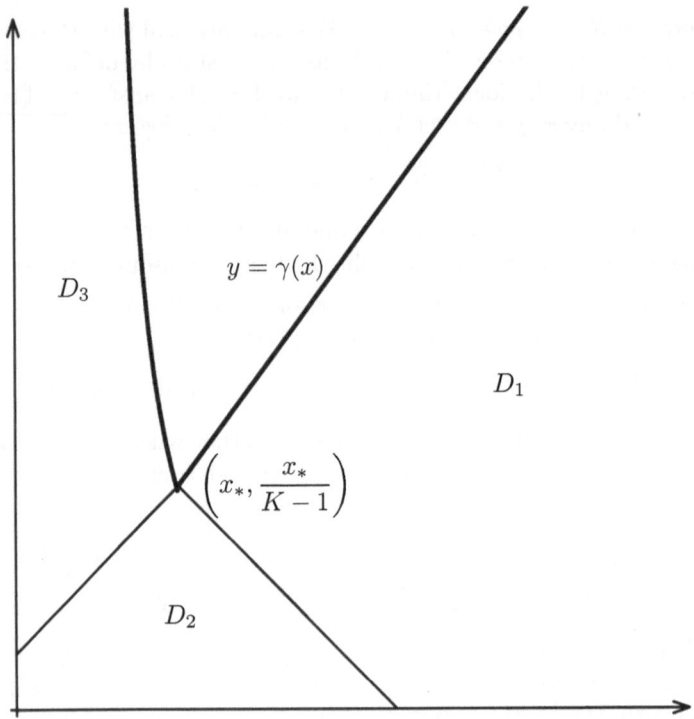

Figure 3.2: The regions D_1–D_3 arising during the search. The bold
curve is the graph of the function γ.

The assumption reads

(A4)   On the set D_1, U is linear along the line segments of slope -1. On $D_2 \cup D_3$,
        U is linear along the line segments of slope 1.

Step 3. *Derivation of* γ *and* U *on* D_1. By (A1), (A3) and (A4) we get that
for all $x \geq x_*$ and $t \in [0, \gamma(x)]$,

$$
\begin{aligned}
U(x+t, \gamma(x)-t) &= U(x, \gamma(x)) + U_x(x, \gamma(x))t - U_y(x, \gamma(x))t \\
&= V(x, \gamma(x)) + V_x(x, \gamma(x))t - V_y(x, \gamma(x))t \qquad (3.98) \\
&= \gamma(x) - Kx\log x - K(1 + \log x)t - t.
\end{aligned}
$$

By the symmetry property (3.97) and (A3), we have $U_y(x, 0) = 0$ for all x, so the
formula (3.98) yields

$$
\gamma'(x) = \frac{K\gamma(x)}{x} - 1.
$$

Solving this differential equation gives $\gamma(x) = \frac{x}{K-1} + cx^K$ for some constant c. It
is easy to see that $c \geq 0$, since otherwise γ would not satisfy (A2). Furthermore,

we must have $c = 0$: for positive c it can be shown that the function U grows too fast, by virtue of (3.98). Thus, $\gamma(x) = \frac{x}{K-1}$ for $x \geq x_*$ and (3.98) implies that

$$U(x,y) = Ky - x - Kx \log\left(\frac{K-1}{K}(x+y)\right) \tag{3.99}$$

provided $(x,y) \in D_1$.

Step 4. Derivation of γ on D_2. For any $x \in [x_*, \frac{K}{K-1}x_*]$, the point $P = (x, \frac{K}{K-1}x_* - x)$ belongs to $\partial D_1 \cap \partial D_2$, so by (A3), (A4) and (3.99) we may write

$$U\left(x - t, \frac{K}{K-1}x_* - x - t\right) = U(P) - [U_x(P) + U_y(P)]t,$$

provided $t \leq \min\{x, \frac{K}{K-1}x_* - x\}$. This, after some tedious but simple calculations, gives

$$U(x,y) = \frac{K-1}{2x_*}(y^2 - x^2) - (K\log x_* + 1)x + \frac{K^2 x_*}{2(K-1)}$$

for $(x,y) \in D_1$. By the symmetry condition (3.97) we have $x_* = \exp(-K^{-1})$: here we make use of the assumption $K \leq 2$, since precisely for these values of K the set D_2 has a nonempty intersection with the y-axis. In conclusion, we have obtained the following expression for U:

$$U(x,y) = \frac{(K-1)\exp(K^{-1})}{2}(y^2 - x^2) + \frac{K^2}{2(K-1)}\exp(-K^{-1}) \qquad \text{on } D_2. \tag{3.100}$$

Step 5. Derivation of γ and U on D_3. We proceed as in Step 3 and obtain, for $x \in (e^{-1}, x^*)$ and $t \in [0, x]$,

$$U(x - t, \gamma(x) - t) = \gamma(x) - Kx\log x - t + K(1 + \log x)t. \tag{3.101}$$

The equation $U_x(0,y) = 0$, valid for $y \geq 0$ by virtue of (3.97) and (A3), implies the following condition on γ:

$$\gamma'(x) = 1 - \frac{1}{1 + \log x}, \qquad x \in (e^{-1}, x_*).$$

However, $\gamma(x_*) = \frac{x_*}{K-1}$, so, by (A2),

$$\gamma(x) = x + \frac{x_*(2-K)}{K-1} + \int_x^{x_*} \frac{1}{1 + \log t}\,dt.$$

This, together with (3.101), allows to derive the formula for U on D_3. Indeed, if $G : (\frac{2-K}{K-1}x_*, \infty) \to (e^{-1}, x_*]$ denotes the inverse of the function $x \mapsto \gamma(x) - x$, then

$$U(x,y) = y - x + G(y - x) - Kx\log G(y - x).$$

To complete the description, apply the symmetry condition (3.97) to obtain the formula for U on the whole plane. In fact, this function is the least majorant (in the real-valued setting). To pass to the Hilbert-space-valued setting, one can proceed as usual and interpret $|\cdot|$ as the corresponding norm. However, we do not know if the obtained function satisfies the conditions $1°$–$3°$. Instead of the quite complicated verification of these, we make use of the integration argument, but this requires an additional effort.

$Step$ $6.$ The $passage$ to the $formula$ (3.91). It is more or less clear that as the "simple" integrand function, we are forced to take u given by (3.6). This follows, for example, from the fact that it is the right choice in the proof of L^p estimates for small p (that is, $1 < p < 2$). However, it is easy to check that no kernel leads to the function U invented above. A natural idea is to try to choose a kernel for which the obtained function agrees with U on a large part of $\mathbb{R}^2$. But which part is important? To answer this, let us inspect the proof of the optimality of the constant $L(K)$. It is clear that we have exploited the properties of U^0 on the set D_1; by (3.99), this suggests to search for k for which

$$\int_0^\infty k(t)u(x/t, y/t)\mathrm{d}t = Ky - x - Kx \log\left(\frac{K-1}{K}(x+y)\right)$$

provided $(x, y) \in D_1$. After some manipulations, this leads precisely to (3.91). □

3.7   Inequalities for bounded martingales

In this section we turn to another important class of inequalities. Namely, we shall assume that the dominating process f satisfies the boundedness condition $||f||_\infty \leq 1$.

3.7.1   Formulation of the results

We will establish two types of results: tail estimates and Φ-inequalities.

Theorem 3.10. *Let f and g be $\mathcal{H}$-valued martingales such that $||f||_\infty \leq 1$ and g is differentially subordinate to f. Then*

$$\mathbb{P}(|g|^* \geq \lambda) \leq P(\lambda) = \begin{cases} 1 & if\ 0 < \lambda \leq 1, \\ \lambda^{-2} & if\ 1 < \lambda \leq 2, \\ e^{2-\lambda}/4 & if\ \lambda > 2. \end{cases} \tag{3.102}$$

The inequality is sharp for each λ, even if f is real and g is assumed to be its ± 1-transform.

Note that in contrast with the previous tail estimates, we cannot assume that $\lambda = 1$, since the condition $||f||_\infty \leq 1$ is not invariant with respect to scaling. In other words, each value of λ requires an independent analysis.

We have also the following one-sided version for real martingales.

Theorem 3.11. *Let f and g be $\mathbb{R}$-valued martingales such that $||f||_\infty \leq 1$ and g is differentially subordinate to f. Then*

$$\mathbb{P}(g^* \geq \lambda) \leq P(\lambda) = \begin{cases} 1 & \text{if } 0 < \lambda \leq 1, \\ (1 - \sqrt{\lambda - 1}/2)^2 & \text{if } 1 < \lambda \leq 2, \\ e^{2-\lambda}/4 & \text{if } \lambda > 2. \end{cases} \quad (3.103)$$

For each $\lambda > 0$, the bound on the right is sharp, even if we assume that g is a ± 1-transform of f.

Now let us formulate the second result. Let Φ be an increasing convex function on $[0, \infty)$, twice differentiable on $(0, \infty)$, such that $\Phi(0) = 0$ and $\Phi'(0+) = 0$.

Theorem 3.12.

(i) *If Φ' is convex, then*

$$\sup_n \mathbb{E}\Phi(|g_n|) \leq \frac{1}{2} \int_0^\infty \Phi(t) e^{-t}\, dt. \quad (3.104)$$

(ii) *If Φ' is concave, then*

$$\sup_n \mathbb{E}\Phi(|g_n|) \leq \Phi(1). \quad (3.105)$$

Both inequalities are sharp, even if f is real and g is assumed to be its ± 1- transform.

3.7.2   Proof of Theorem 3.10

Proof of (3.102). If $0 < \lambda \leq 2$, then the inequality is straightforward. Indeed, for $\lambda \in (0, 1]$ there is nothing to prove, while for $\lambda \in (1, 2]$, by Doob's weak type inequality (see (7.18) below),

$$\mathbb{P}(|g|^* \geq \lambda) \leq \frac{||g||_2^2}{\lambda^2} \leq \frac{||f||_2^2}{\lambda^2} \leq \frac{||f||_\infty^2}{\lambda^2} \leq \frac{1}{\lambda^2}.$$

The main difficulty lies in showing the estimate for $\lambda > 2$. Using the stopping time argument, it suffices to establish the corresponding bound

$$\mathbb{P}(|g_n| \geq \lambda) \leq P(\lambda) \qquad n = 0, 1, 2, \ldots.$$

We may use Burkholder's method: take $V_\lambda(x, y) = 1_{\{|y| \geq \lambda\}}$ for $(x, y) \in S = \{(x, y) \in \mathcal{H} \times \mathcal{H} : |x| \leq 1\}$. In order to introduce the majorant, consider the following subsets of the strip S:

$$D_0 = \{(x,y) \in S : |x| + |y| \leq 1\},$$
$$D_1 = \{(x,y) \in S : 1 < |x| + |y| < \lambda - 1 \text{ and } 0 < |x| < 1\},$$
$$D_2 = \{(x,y) \in S : \lambda - 1 - |x| < |y| < \lambda - 1 + |x| \text{ and } |x| < 1\},$$
$$D_3 = \{(x,y) \in S : \lambda - 1 + |x| < |y| < \sqrt{\lambda^2 - 1 + |x|^2},\}$$
$$D_4 = \{(x,y) \in S : |y| > \sqrt{\lambda^2 - 1 + |x|^2} \text{ and } |x| < 1\}.$$

Define U_λ by

$$U_\lambda(x,y) = \begin{cases} (1 + |y|^2 - |x|^2)e^{2-\lambda}/4 & \text{on } D_0, \\ (1 - |x|)e^{|x|+|y|-\lambda+1}/2 & \text{on } D_1, \\ (1 - |x|^2)((\lambda - |y|)^2 + 1 - |x|^2)^{-1} & \text{on } D_2, \qquad (3.106) \\ 1 - (\lambda^2 - 1 - |y|^2 + |x|^2)(4(\lambda - 1))^{-1} & \text{on } D_3, \\ 1 & \text{on } D_4 \end{cases}$$

and extend it to the whole S by continuity and the condition $U_\lambda(x,y) = 1$ for $|x| = 1$ and $|y| = \lambda$. It is evident that U_λ majorizes V_λ. Now we will show that

$$U_\lambda(x + h, y + k) \leq U_\lambda(x,y) + A(x,y) \cdot h + B(x,y) \cdot k \qquad (3.107)$$

for all $(x,y) \in S$ such that $|x| < 1$ and $|k| \leq |h|$. Here

$$A(x,y) = \begin{cases} -e^{2-\lambda}x/2 & \text{on } D_0, \\ -e^{|x|+|y|-\lambda+1}x/2 & \text{on } D_1, \\ -2(\lambda - |y|)^2[(\lambda - |y|)^2 + 1 - |x|^2]^{-2}x & \text{on } D_2, \\ -x/(2\lambda - 2) & \text{on } D_3, \\ 0 & \text{on } D_4, \end{cases}$$

$$B(x,y) = \begin{cases} e^{2-\lambda}y/2 & \text{on } D_0, \\ e^{|x|+|y|-\lambda+1}(1 - |x|)y'/2 & \text{on } D_1, \\ 2(\lambda - |y|)(1 - |x|^2)[(\lambda - |y|)^2 + 1 - |x|^2]^{-2}y' & \text{on } D_2, \\ y/(2\lambda - 2) & \text{on } D_3, \\ 0 & \text{on } D_4. \end{cases}$$

and A and B are extended to the whole S in such a way that their restrictions to $\overline{D_3}$ and $S \setminus \overline{D_3}$ are continuous. To establish (3.107), fix (x,y), $(h,k) \in \mathcal{H} \times \mathcal{H}$ such that $|x| < 1$ and $|k| \leq |h|$ and consider the function $G = G_{x,y,h,k}$ given on $\{t \in \mathbb{R} : |x + th| \leq 1\}$ by the formula $G(t) = u(x + th, y + tk)$. It suffices to show that G satisfies $G''(0) \leq 0$ if $(x,y) \in D_0 \cup D_1 \cup \cdots \cup D_4$ and $G'(0-) \geq G'(0+)$ otherwise. If $(x,y) \in D_0$, then

$$G''(0) = e^{2-\lambda}(|k|^2 - |h|^2)/2 \leq 0.$$

For $(x, y) \in D_1$, $G''(0)$ equals

$$\frac{e^{|x|+|y|-\lambda+1}}{2} \left[-|x|(x' \cdot h + y' \cdot k)^2 + |k|^2 - |h|^2 - \frac{(|k|^2 - (y' \cdot k)^2)(|x| + |y| - 1)}{|y|} \right]$$

which is also nonpositive. If $(x, y) \in D_2$, then a little calculation yields

$$G''(0) = -2[(\lambda - |y|)^2 1 - |x|^2]^{-3}(I + II + III),$$

where

$$I = (|h|^2 - |k|^2)(\lambda - |y|)^2[(\lambda - |y|)^2 + 1 - |x|^2] \geq 0,$$
$$II = [\{(\lambda - |y|)^2 - (1 - |x|^2)\}y' \cdot k + 2(\lambda - |y|)x \cdot h]^2 \geq 0,$$
$$III = (\lambda - |y|)[(\lambda - |y|)^2 + 1 - |x|^2][|k|^2 - (y' \cdot k)^2][\lambda - |y| - (1 - |x|^2)/|y|] \geq 0.$$

Therefore $G''(0) \leq 0$ in this case as well. If $(x, y) \in D_3$, we have

$$G''(0) = \frac{|k|^2 - |h|^2}{2(\lambda - 1)} \leq 0$$

and, finally, $G''(0) = 0$ on D_4. It suffices to verify the inequality for one-sided derivatives. For example, suppose that $(x, y) \in \partial D_3 \cap \partial D_4$, that is, $|y|^2 = \lambda^2 - 1 + |x|^2$. If $(x + th, y + tk) \in D_4$ for small positive t, then $G'(0+) = 0$ and $G'(0-) < 0$. Similarly, if $(x + th, y + tk) \in D_3$ for small $t > 0$, then $G'(0-) = 0 > G'(0+)$. The remaining inequalities for one-sided derivatives are verified in the same manner. This finishes the proof of (3.107).

The final observation is that if $|y| \leq |x| \leq 1$, then, by the property (3.107) we have just proved,

$$U_\lambda(x, y) \leq U_\lambda(0, 0) + A(0, 0) \cdot x + B(0, 0) \cdot y = P(\lambda).$$

Now we are ready to establish (3.102). Let f, g be as in the statement. We can not use Burkholder's method yet, since we do not have the necessary integrability: the functions A and B blow up when $|x| \to 1$ and $|y| \to \lambda$. To overcome this, fix $\alpha \in (0, 1)$ and apply Theorem 2.7 to the pair $(\alpha f, \alpha g)$. This is allowed, since this pair takes values in the set $\{(x, y) \in \mathcal{H} \times \mathcal{H} : |x| \leq \alpha\}$, on which U_λ, A and B are bounded. We get

$$\mathbb{E}V_\lambda(\alpha f_n, \alpha g_n) \leq \mathbb{E}U_\lambda(\alpha f_0, \alpha g_0) \leq P(\lambda),$$

that is, $\mathbb{P}(|g_n| \geq \lambda/\alpha) \leq P(\lambda)$. Letting $\alpha \to 1$ gives $\mathbb{P}(|g_n| > \lambda) \leq P(\lambda)$; to get the nonstrict inequality on the left, it suffices to use the fact that P is a continuous function of the parameter λ. $\qquad\square$

Sharpness. If $\lambda \leq 1$, then take the constant pair $(f, g) \equiv (1, 1)$. Then f, g have the required properties (that is, they are martingales and g is a ± 1-transform of f) and the two sides of (3.102) are equal. Suppose now that $\lambda \in (1, 2]$ and assume that we find a pair (f, g) of bounded martingales such that $|f_\infty| \equiv 1$, $|g_\infty| \in \{0, \lambda\}$ and g is a ± 1-transform of f. Then

$$\mathbb{P}(|g|^* \geq \lambda) \geq \mathbb{P}(|g_\infty| \geq \lambda) = \frac{||g||_2^2}{\lambda^2} = \frac{||f||_2^2}{\lambda^2} = \frac{1}{\lambda^2}.$$

The desired pair can be defined as follows. First, introduce the sets

$$E_1 = \{(1,0), (1 - \lambda/2, \lambda/2)\} \quad \text{and} \quad E_2 = \{(1, \lambda), (-\lambda/2, -1 + \lambda/2)\}.$$

The process (f, g) is uniquely determined by the conditions $f_0 = g_0 \equiv 1/2$,

$$(f_{2k+1}, g_{2k+1}) \in E_1, \quad (f_{2k+2}, g_{2k+2}) \in E_2, \quad (-f_{2k+3}, -g_{2k+3}) \in E_1$$

and $(-f_{2k+4}, -g_{2k+4}) \in E_2$ for $k = 0, 1, 2, \ldots$.

Finally, we turn to the case $\lambda > 2$. Fix a positive integer $N > \lambda - 2$ and set $\delta = (\lambda - 2)/(2N)$. Consider a sequence $(\xi_n)_{n=1}^{2N}$ of independent centered random variables such that $\xi_1 \in \{-1/2 - \delta, \delta\}$, $\xi_{2k} \in \{-1 + \delta, \delta\}$ and $\xi_{2k+1} \in \{-\delta, 1\}$ for $k = 1, 2, \ldots, N - 1$, and, finally, $\xi_{2N} \in \{-1 + \delta, 1 + \delta\}$. Put

$$F_n = \frac{1}{2} + \sum_{k=0}^{n \wedge (2N)} \xi_k, \quad \text{and} \quad G_n = \frac{1}{2} + \sum_{k=0}^{n \wedge (2N)} (-1)^n \xi_k.$$

Then F, G are martingales such that G is a ± 1-transform of F. Let $\tau = \inf\{n : |F_n| = 1\}$ and put $f_n = F_{\tau \wedge n}$, $g_n = G_{\tau \wedge n}$. Then f is a martingale satisfying $||f||_\infty \leq 1$ and g is its ± 1-transform. It is easy to check that g is nonnegative and reaches λ if $\xi_1 = -1/2 - \delta$, $|\xi_k| = \delta$ for $2 \leq k \leq 2N - 1$ and $\xi_{2N} = 1 + \delta$. Therefore, using the independence of the ξ_j's, we get

$$\mathbb{P}(g^* \geq \lambda) \geq \frac{1}{2 + 2\delta} \cdot (1 - \delta)^{N-1} \cdot (1 + \delta)^{-N+1} \cdot \frac{1 - \delta}{2}.$$

However, since $\delta = (\lambda - 2)/(2N)$, we see that for sufficiently large N, the right-hand side can be made arbitrarily close to

$$\frac{1}{2} \cdot e^{-(\lambda-2)/2} \cdot e^{-(\lambda-2)/2} \cdot \frac{1}{2} = P(\lambda).$$

This completes the proof.   □

On the search of the suitable majorant. Let us first focus on the case $\mathcal{H} = \mathbb{R}$. It turns out that the majorant we obtain in this case will be a bit different from the one above.

Step 1. Initial observations. Fix $\lambda > 2$ and let us use an appropriate version of (2.1). Recall that for any $(x, y) \in [-1, 1] \times \mathbb{R}$, the class $M(x, y)$ consists of martingale pairs (f, g) such that $(f_0, g_0) \equiv (x, y)$, $\|f\|_\infty \leq 1$ and g is a ± 1-transform of f. Define

$$U^0(x, y) = \sup\{\mathbb{P}(|g_n| \geq \lambda) : (f, g) \in M(x, y), \, n = 0, 1, 2, \ldots\}. \qquad (3.108)$$

Obviously, $0 \leq U^0(x, y) \leq 1$ for all $(x, y) \in [-1, 1] \times \mathbb{R}$. If $|y| \geq \lambda$, then considering the constant pair $(f, g) \equiv (x, y)$ gives $U^0(x, y) = 1$. Similarly, if $|y| < \lambda$ but $|x| = 1$, then $M(x, y)$ has only one element, a constant one. This gives $U^0(x, y) = 0$. Finally, we have the symmetry

$$U^0(x, y) = U^0(x, -y) = U^0(-x, y) \qquad (3.109)$$

for all $(x, y) \in [-1, 1] \times \mathbb{R}$. Thus it remains to determine the majorant on the set $C = [0, 1] \times [0, \lambda]$.

Step 2. Key assumptions. Now we will impose some conditions on the majorant. As they do not follow from (3.108), we change the notation and write U instead of U^0, as usually. The general idea, already exploited in the previous estimates, is to decompose C into a finite family $\{C_i\}$ such that on each C_i the majorant U is either linear along all lines of slope -1 or linear along all lines of slope 1. Some experimentation leads to the following decomposition:

$$C_0 = \{(x, y) \in C : x + y \leq 1\},$$
$$C_1 = \{(x, y) \in C : 1 < x + y < \lambda - 1\},$$
$$C_2 = \{(x, y) \in C : \lambda - 1 - x \leq y \leq \lambda - 1 + x\},$$
$$C_3 = \{(x, y) \in C : \lambda - 1 + x < y < 1\},$$
$$C_4 = \{(x, y) \in C : y \geq 1\}.$$

After symmetrization, this leads to a decomposition of $[-1, 1] \times \mathbb{R}$ which is slightly different from the one generated by the sets D_0–D_4 above. As already mentioned, we have

$$U(x, y) = 1 \qquad \text{on } C_4. \qquad (3.110)$$

Next, motivated by some experimentation, we assume that:

(A1)   On C_0, C_1 and C_2 the majorant is linear along the lines of slope -1 and on C_3 it is linear along the lines of slope 1.

(A2) U is of class C^1 on $(-1, 1) \times \mathbb{R}$.

In fact, it will turn out that the differentiability on the whole strip cannot hold (so there is no U satisfying both (A1) and (A2) as well as the previous properties), but the arguments will lead to the right function and it will be clear how to modify the second assumption.

Step 3. The formula for U. Denote $A(x) = U(x, 0)$ and $B(y) = U(0, y)$ for $x \in [0, 1]$ and $y \in [0, \lambda]$. By (A1), we have

$$B(\lambda - 1) = \frac{1}{2}U(-1, \lambda) + \frac{1}{2}U(1, \lambda - 2) = \frac{1}{2}. \qquad (3.111)$$

Furthermore, by (A1), if $(x, y) \in C_1$, then

$$U(x, y) = (1 - x)B(x + y) + xU(x + y - 1, 1) = (1 - x)B(x + y).$$

Now, by (A2) and (3.109), we have $U_x(0, y) = 0$ if $(0, y) \in C_1$, which, by the above, leads to the differential equation $B'(y) = B(y)$. Using (3.111), we get

$$B(y) = e^{y-\lambda+1}/2 \qquad \text{for } 1 \le y \le \lambda - 1 \tag{3.112}$$

and hence

$$U(x, y) = (1 - x)e^{x+y-\lambda+1}/2 \qquad \text{on } C_1. \tag{3.113}$$

Using (A1) we have, for $(x, y) \in C_2$,

$$U(x, y) = \frac{2 - 2x}{-x - y + \lambda + 1}U(1, x + y - 1)$$
$$+ \frac{2 - 2x}{-x - y + \lambda + 1}U\left(\frac{x + y - \lambda + 1}{2}, \frac{x + y + \lambda - 1}{2}\right)$$
$$= \frac{2 - 2x}{-x - y + \lambda + 1}U\left(\frac{x + y - \lambda + 1}{2}, \frac{x + y + \lambda - 1}{2}\right).$$

Using (A1) and (A2) again, we obtain

$$U\left(\frac{x + y - \lambda + 1}{2}, \frac{x + y + \lambda - 1}{2}\right) = \frac{-x - y + 1 - \lambda}{4}U\left(-1, \frac{x + y + \lambda - 3}{2}\right)$$
$$+ \frac{x + y + \lambda - 3}{4}U(1, \lambda)$$
$$= \frac{x + y + \lambda - 3}{4}$$

and plugging this into the previous equation yields

$$U(x, y) = \frac{(1 - x)(x + y - \lambda + 3)}{2(-x - y + \lambda + 1)} \qquad \text{on } C_2. \tag{3.114}$$

Now pick $(x, y) \in C_3$ and use (A1) to obtain

$$U(x, y) = \frac{x}{x - y + \lambda}U(x - y + 1, \lambda) + \frac{\lambda - y}{x - y + \lambda}B(y - x)$$
$$= \frac{x}{x + \lambda - y} + \frac{\lambda - y}{x - y + \lambda}B(y - x).$$

Since $U_x(0, y) = 0$ for $y \in [\lambda - 1, \lambda]$, this leads to the differential equation

$$B'(y) = -\frac{B(y)}{\lambda - y} + \frac{1}{\lambda - y},$$

with the solution of the form $B(y) = c(\lambda - y) + 1$. Since $B(\lambda - 1) = 1/2$, as derived above, we obtain $B(y) = \frac{y-\lambda}{2} + 1$ and hence

$$U(x, y) = \frac{y - \lambda}{2} + 1 \qquad \text{on } C_3. \tag{3.115}$$

Finally, let us take $(x, y) \in C_0$. We have

$$U(x, y) = \frac{x}{x + y} A(x + y) + \frac{y}{x + y} B(x + y).$$

The conditions $U_x(0, y) = U_y(x, 0) = 0$, $x \in [0, 1)$, $y \in [0, 1]$, imply

$$B'(y) = \frac{-A(y) + B(y)}{y} \quad \text{and} \quad A'(x) = \frac{A(x) - B(x)}{x}.$$

Consequently, $A'(s) = -B'(s)$ for $s \in [0, 1)$, so $A(s) = -B(s) + c$ for some constant c. Putting $s = 0$ and using the equality $A(0) = B(0)$ we get that $c = 2B(0)$ and, coming back to the differential equation for B, we get

$$B'(y) = \frac{2B(y) - 2B(0)}{y}.$$

Solving this equation, we get $B(y) = ay^2 + B(0)$ for some constant a. However, B is of class C^1, so, by (3.112), $B(1) = B'(1) = e^{2-\lambda}/2$. This implies $a = B(0) = e^{2-\lambda}/4$ and hence

$$U(x, y) = \frac{e^{2-\lambda}}{4}(-x^2 + y^2 + 1) \qquad \text{on } C_0. \tag{3.116}$$

One can show that the function U above (after extension to the whole $[-1, 1] \times \mathbb{R}$ by (3.109)) coincides with the function U^0. In addition, this function works also for differentially subordinate martingales in the real-valued setting. However, we have done all the computation in the case $\mathcal{H} = \mathbb{R}$: if the dimension of $\mathcal{H}$ over $\mathbb{R}$ is at least 2, then the usual extension, given by $U_\lambda^{\mathcal{H}}(x, y) = U(|x|, |y|)$ does no longer work. For example, take a point $(0, y) \in \mathcal{H} \times \mathcal{H}$ such that $|y| = \sqrt{\lambda^2 - 1 + |x|^2} < \lambda$. We have $U(0, |y|) = \frac{y-\lambda}{2} + 1 < 1$. On the other hand, since $\mathcal{H}$ is at least two-dimensional, there is a norm-one vector $z \in \mathcal{H}$ which is orthogonal to y. For a Rademacher variable ε we have

$$\mathbb{E}U(|z\varepsilon|, |y + z\varepsilon|) = \frac{U(|-z|, |y - z|) + U(|z|, |y + z|)}{2} = 1,$$

since $|y \pm z| = \lambda$. We see that the concavity property $2°$ is not satisfied. However, some experiments based on the example we have just considered lead to the function U_λ given in (3.106). $\qquad \square$

3.7.3   Proof of Theorem 3.11

The case $\lambda \leq 1$ is trivial and the case $\lambda \geq 2$ follows immediately from the preceding theorem, so it suffices to prove the assertion for $\lambda \in (1,2]$. However, we shall present here a new argument which works for all values of λ. Let

$$U^0(x,y) = \sup\{\mathbb{P}(g_n^* \geq \lambda)\},$$

where the supremum is taken over all pairs (f,g) of real martingales starting from (x,y), such that $\|f\|_\infty \leq 1$ and g satisfies $|dg_n| \leq |df_n|$ almost surely for all $n \geq 1$. Note that this class is much larger than $M(x,y)$. For such pairs (f,g) we may write, for any $s > 0$,

$$\mathbb{P}(g_n^* \geq \lambda) = \mathbb{P}((g_n + s)^* \geq \lambda + s)$$

$$= \mathbb{P}(|g_n + s|^* \geq \lambda + s) - \mathbb{P}\left(\inf_{0 \leq k \leq n}(g_k + s) \leq -\lambda - s\right)$$

and if $s \to \infty$, then the second probability in the last line converges to 0. Consequently, from the remarks just after (3.116), we infer that

$$U^0(x,y) = \lim_{s \to \infty} U_{\lambda+s}(x, y + s),$$

where U_λ is given by (3.110), (3.113), (3.114), (3.115), (3.116) and the symmetry condition (3.109). Calculating the limit, we obtain

$$U^0(x,y) = \begin{cases} (1 - |x|)e^{|x| + y - \lambda + 1}/2 & \text{if } (x,y) \in D_1, \\ \frac{(1 - |x|)(3 + |x| + y - \lambda)}{2(\lambda - |x| - y + 1)} & \text{if } (x,y) \in D_2, \\ 1 - (\lambda - y)/2 & \text{if } (x,y) \in D_3, \\ 1 & \text{if } (x,y) \in D_4, \end{cases}$$

where D_1–D_4 are the subsets of $[-1,1] \times \mathbb{R}$ given by

$$D_1 = \{(x,y) : y < \lambda - 1 - |x|\},$$
$$D_2 = \{(x,y) : \lambda - 1 - |x| \leq y < \lambda - 1 + |x|\},$$
$$D_3 = \{(x,y) : \lambda - 1 + |x| \leq y \leq \lambda\},$$
$$D_4 = \{(x,y) : y > \lambda\}.$$

It remains to note that $P(\lambda) = \sup_{|y| \leq |x| \leq 1} U^0(x,y)$: the claim follows.

3.7.4   Proof of Theorem 3.12

It turns out that the inequalities (3.104) and (3.105) are completely different in nature.

Proof of (3.104). As the reader has noticed, the constant from (3.104) appears also in (3.16). In fact the proofs of both these estimates have many similarities, though it should be stressed that the conditions imposed on Φ are entirely different: in (3.16) we considered functions with a strictly concave derivative, while here we assume that Φ' is a convex function on $(0, \infty)$. We keep the notation $S = \{(x, y) \in \mathcal{H} \times \mathcal{H} : |x| \leq 1\}$ from one of the previous subsections, and define

$$D_1 = \{(x, y) \in S : |x| + |y| \leq 1\},$$
$$D_2 = \{(x, y) \in S : |x| + |y| > 1\}.$$

Let $V_\Phi, U_\Phi : \{(x, y) \in \mathcal{H} \times \mathcal{H} : |x| \leq 1\} \to \mathbb{R}$ be given by $V_\Phi(x, y) = \Phi(|y|)$ and

$$U_\Phi(x, y) = \begin{cases} \frac{|y|^2 - |x|^2 + 1}{2} \int_0^\infty e^{-t} \Phi(t) dt & \text{on } D_1, \\ |x|\Phi(|x| + |y| - 1) + (1 - |x|)e^{|x|+|y|} \int_{|x|+|y|}^\infty e^{-t}\Phi(t - 1) dt & \text{on } D_2. \end{cases}$$

A little calculation shows that

$$U_\Phi(x, y) = \int_1^\infty k_\Phi(t) U_\infty(x/t, y/t) dt + \frac{|y|^2 - |x|^2 + 1}{2} \int_0^\infty e^{-t}\Phi(t) dt,$$

where

$$k_\Phi(t) = \frac{t^2}{2}\left[e^t \int_t^\infty e^{-s}\Phi''(s - 1) ds - \Phi''(t - 1)\right].$$

Note that k_Φ is nonnegative, due to the assumptions on Φ. It is easy to see that $U_\Phi(x, y)$ decreases as $|x|$ increases. Consequently, if $z \in \mathcal{H}$ has norm 1, then

$$U_\Phi(x, y) \geq U_\Phi(z, y) = \Phi(|y|) = V_\Phi(x, y).$$

Thus it suffices to show that we have the integrability needed to apply Fubini's theorem. But this is evident: since f is bounded, so is g_n for any n. $\qquad \square$

Sharpness. We repeat the argumentation from the part of Subsection 3.1.5 concerning the sharpness of (3.16). Let f and g be the sequences studied there. Then f is a ± 1-transform of g, g is bounded in absolute value by 1 and $\mathbb{E}\Phi(|f_\infty|)$ can be made arbitrarily close to $\frac{1}{2}\int_0^\infty \Phi(t)e^{-t} dt$. This is exactly what we need. $\qquad \square$

On the search of the suitable majorant. This is part very similar to the search presented in Subsection 3.1.5. We omit the details. $\qquad \square$

Proof of (3.105). Let $\Psi : [0, \infty) \to [0, \infty)$ be given by $\Psi(s) = \Phi(\sqrt{s})$. We easily check that for $s > 0$,

$$\Psi''(s) - \frac{1}{4s^{3/2}}(\Phi''(\sqrt{s})\sqrt{s} - \Phi'(\sqrt{s})) \leq 0,$$

since Φ' is concave and $\Phi'(0+) = 0$. Therefore, we have $\Psi'(1)(s^2 - 1) \geq \Psi(s^2) - \Psi(1)$, or, in terms of Φ,

$$\frac{\Phi'(1)}{2}(s^2 - 1) + \Phi(1) \geq \Phi(s). \tag{3.117}$$

Now we are ready to apply Burkholder's method. Let $V_\Phi(x, y) = \Phi(|y|)$ and

$$U_\Phi(x, y) = \frac{\Phi'(1)}{2}(|y|^2 - |x|^2) + \Phi(1).$$

The majorization $U_\Phi \geq V_\Phi$ is a consequence of (3.117):

$$U_\Phi(x, y) \geq \frac{\Phi'(1)}{2}(|y|^2 - 1) + \Phi(1) \geq V_\Phi(x, y).$$

Obviously, the condition $2°$ is satisfied and $U_\Phi(x, y) \leq \Phi(1)$ for $|y| \leq |x|$. Thus

$$\mathbb{E}V_\Phi(f_n, g_n) \leq \mathbb{E}U_\Phi(f_0, g_0) \leq \Phi(1)$$

and we are done.   □

Sharpness. This is trivial: take $(f, g) \equiv (1, 1)$.   □

On the search of the suitable majorant. As usual, assume that $\mathcal{H} = \mathbb{R}$ and let us restrict ourselves to the case of ± 1-transforms. Let $V(x, y) = \Phi(|y|)$. We may assume that the function U we search for satisfies the symmetry condition

$$U(x, y) = U(x, -y) = U(-x, y).$$

Consequently, it suffices to determine it on $S^+ = [0, 1] \times [0, \infty)$. The key assumptions are the following:

(A1) U is of class C^1 on S,
(A2) U is linear along all line segments of slope 1 contained in the set S^+.

These conditions imply that $U(0, 0) = (U(1, 1) + U(-1, -1))/2 = \Phi(1)$. What is more important, if we modify U outside the diagonals $\{(x, y) : |x| = |y|\}$ (but keep $1°$, $2°$ and $3°$ valid), then the new function will yield the sharp Φ-inequality as well. Now if we look at the functions used in the previous problems, the guess $U(x, y) = a(|y|^2 - |x|^2) + \Phi(1)$ comes quickly into one's mind. This leads to the special function considered above.   □

3.8   Inequalities for nonsymmetric martingale transforms

The next problem we shall study is the following. Assume that f is a martingale and g is its transform by a predictable sequence taking values in $[0, 1]$. We may ask about optimal constants in the corresponding weak, strong and logarithmic estimates. We can also study these inequalities in a wider setting, where the martingale transform is replaced by a weaker assumption which may be regarded as "nonsymmetric differential subordination":

$$|dg_n|^2 \leq df_n \cdot dg_n, \qquad n = 0, 1, 2, \ldots. \tag{3.118}$$

Note that this is equivalent to saying that the martingale $-f/2 + g$ is differentially subordinate to $f/2$.

3.8.1 Formulation of the result

We begin with the result concerning weak-type estimates. Let

$$K_{p,\infty} = \begin{cases} 1 & \text{if } 1 \le p \le 2, \\ \frac{1}{2}\left[\frac{(2c+p-1)^{p-1}}{c+1}\right]^{1/p} & \text{if } p > 2, \end{cases}$$

where $c = c(p) > 1$ is the unique number satisfying

$$c^{p-1} = 2c + 1. \tag{3.119}$$

Theorem 3.13. *Assume that f, g are $\mathcal{H}$-valued martingales satisfying the condition (3.118). Then for any $1 \le p < \infty$ we have*

$$||g||_{p,\infty} \le K_{p,\infty}||f||_p \tag{3.120}$$

and the constant is the best possible. In fact, it is already the best possible if $\mathcal{H} = \mathbb{R}$ and g is a transform of f by a deterministic sequence taking values in $\{0,1\}$.

The moment inequalities turn out to be much more difficult. Let $1 < p < \infty$. In [52], Choi introduced constants c_p, depending only on p, with the following behavior as $p \to \infty$:

$$c_p = \frac{p}{2} + \frac{1}{2}\log\left(\frac{1+e^{-2}}{2}\right) + \frac{[\frac{1}{2}\log\frac{1+e^{-2}}{2}]^2 + \frac{1}{2}\log\frac{1+e^2}{2} - 2(e^2+1)^{-2}}{p} + \cdots.$$

Theorem 3.14. *Assume that f, g are real-valued martingales satisfying (3.118). Then for $1 < p < \infty$,*

$$||g||_p \le c_p||f||_p \tag{3.121}$$

and the constant is the best possible. It is already the best possible if $\mathcal{H} = \mathbb{R}$ and g is a transform of f by a deterministic sequence taking values in $\{0,1\}$.

Finally, let us state the logarithmic estimate.

Theorem 3.15. *Suppose that f, g are $\mathcal{H}$-valued martingales satisfying (3.118). Then for $K > 1/2$,*

$$||g||_1 \le K \sup_n \mathbb{E}|f_n|\log|f_n| + L^{ns}(K), \tag{3.122}$$

where

$$L^{ns}(K) = \begin{cases} K^2(2K-1)^{-1} & \text{if } 1/2 < K < 1, \\ K\exp(K^{-1} - 1) & \text{if } K \ge 1. \end{cases} \tag{3.123}$$

The constant $L^{ns}(K)$ is the best possible if $K \ge 1$. Furthermore, it is of the best order $O((K-1/2)^{-1})$ when $K \to 1/2+$. For $K \le 1/2$ there is no $L^{ns}(K) < \infty$ for which (3.122) holds.

In fact, for $1/2 < K < 1$, the optimal constant L in (3.122) is not smaller than $\frac{K^2}{2K-1}\exp(1 - K^{-1})$, so we are quite close with our choice of $L^{ns}(K)$.

We will only focus on Theorem 3.13 and Theorem 3.15. The proof of the moment inequality (3.121) consists of two parts: (i) one shows the estimate in the case when g is a transform of f and (ii) extends it to martingales satisfying (3.118). The proof of the first part is extremely technical and we have decided not to include it here. We refer the interested reader to Choi's original proof in [52]. Our contribution is the proof of the second part, but this extension is postponed to Chapter 5. Finally, observe that (3.121) concerns real-valued martingales only. The value of the optimal constant for Hilbert-space-valued martingales is unknown (though we strongly believe that it is the same as in the real-valued setting).

3.8.2   Proof of Theorem 3.13

Proof of (3.120), $1 \leq p \leq 2$. Let $V_{p,\infty} : \mathcal{H} \times \mathcal{H} \to \mathbb{R}$ be defined by $V_{p,\infty}(x, y) = 1_{\{|y|\geq 1\}} - |x|^p$. Furthermore, let U_1 be given by (3.6) and put

$$U_{p,\infty}(x, y) = pU_1(x/2, -x/2 + y)$$

$$= \begin{cases} py \cdot (y - x) & \text{if } |x| + |2y - x| < 2, \\ p - p|x| & \text{if } |x| + |2y - x| \geq 2. \end{cases}$$

We have the majorization $U_{p,\infty} \geq V_{p,\infty}$. Indeed, if $|x| + |2y - x| < 2$, then $|y| \leq |x/2| + |y - x/2| < 1$ and, consequently,

$$1_{\{|y|\geq 1\}} - |x|^p = -|x|^p \leq -\frac{p|x|^2}{4} \leq p|y|(|y| - |x|) \leq py \cdot (y - x).$$

On the other hand, if $|x| + |2y - x| \geq 2$, then the majorization follows from the estimate $p - ps \geq 1 - s^p$, valid for all $s \geq 0$, by virtue of the mean-value theorem. Since $-f/2 + g$ is differentially subordinate to $f/2$, Burkholder's method gives

$$\mathbb{E}V_{p,\infty}(f_n, g_n) \leq \mathbb{E}U_{p,\infty}(f_n, g_n) = p\mathbb{E}U_1(f_n/2, -f_n/2 + g_n) \leq 0$$

and we are done. $\square$

Sharpness, $1 \leq p \leq 2$. This is trivial: take $(f, g) \equiv (1, 1)$. $\square$

On the search of the suitable majorant. Let us assume that $\mathcal{H} = \mathbb{R}$ and let us restrict to transforms by deterministic sequences with values in $\{0, 1\}$. Set $V(x, y) = 1_{\{|y|\geq 1\}} - K^p|x|^p$. As in the case of the inequality (3.105), the key *guess* is that the equality holds for the trivial pair $(1, 1)$ (so $K = 1$). If one makes this assumption, then all we need is a majorant of V, which is concave along the lines of slope 0 or 1, and which is equal to 0 on the line segments $\{(x, \pm x) : .x \in [-1, 1]\}$. How to construct such functions? A natural idea is to take a diagonally concave function U

(we considered a lot of them while studying "symmetric" martingale transforms) and write it in skew coordinates, that is, consider $(x, y) \mapsto U(x/2, -x/2+y)$. Then the function has the required concavity, and after scaling and/or translation we may hope that we will obtain the right majorant. The first try is to take the function $U(x, y) = |y|^2 - |x|^2$ and consider its "skewed" version $(x, y) \mapsto ay \cdot (y - x) + b$ for some constants $a > 0$ and $b \in \mathbb{R}$. This does not work: the majorization fails to hold. The second try is to take U_1, and this leads to the function $U_{p,\infty}$ above. □

Proof of (3.120), $p > 2$. This is much more elaborate. Let U_∞ be given by (3.12) and put

$$u(x, y) = U_\infty(x/2, -x/2 + y)$$
$$= \begin{cases} 0 & \text{if } |x| + |2y - x| < 2, \\ \left(|y - \frac{x}{2}| - 1\right)^2 - |\frac{x}{2}|^2 & \text{if } |x| + |2y - x| \geq 2. \end{cases}$$

Next, we introduce the special function $U_{p,\infty} : \mathcal{H} \times \mathcal{H} \to \mathbb{R}$ corresponding to the weak type estimate for $p > 2$. Let $c = c(p)$ be given by (3.119) and let

$$b = b(p) = \frac{2(p-1)}{2c + p - 1}. \tag{3.124}$$

Put

$$U_{p,\infty}(x, y) = \int_0^b t^{p-1} u(2x/t, 2y/t) \mathrm{d}t. \tag{3.125}$$

A brief calculation gives

$$U_{p,\infty}(x, y) = \frac{2}{p(p-1)(p-2)} (|x| + |2y - x|)^{p-1} (|2y - x| - (p-1)|x|)$$

if $|x| + |2y - x| \leq b$, and

$$U_{p,\infty}(x, y) = b^{p-2} \left[\frac{|2y - x|^2 - |x|^2}{p - 2} - \frac{2b|2y - x|}{p - 1} + \frac{b^2}{p} \right]$$

if $|x| + |2y - x| > b$. We will also need the function $V_{p,\infty} : \mathcal{H} \times \mathcal{H} \to \mathbb{R}$, given by

$$V_{p,\infty}(x, y) = \alpha_p(K_{p,\infty}^{-p} 1_{\{|y| \geq 1\}} - |x|^p),$$

where

$$\alpha_p = \frac{2(p-1)^{p-2}}{p(p-2)}.$$

Since $U_{p,\infty}$ satisfies $2°$ and $U_{p,\infty}(0,0) = 0$, the proof of (3.120) will be complete if we establish the majorization $U_{p,\infty} \geq V_{p,\infty}$. This will be done in the three lemmas below.

First, note that it suffices to establish the majorization in the real case and for x, y satisfying $0 \leq x \leq 2y$. To see this, let us (for a moment) write $U_{p,\infty}^{\mathcal{H}}$, $V_{p,\infty}^{\mathcal{H}}$ instead of $U_{p,\infty}$, $V_{p,\infty}$, to indicate the Hilbert space we are working with. For x, $y \in \mathcal{H}$, take $x' = |x|$ and $y' = |x/2| + |x/2 - y|$. Then $0 \leq x' \leq 2y'$, $2y' - x' = |2y - x|$ and $y' \geq |y|$, so

$$U_{p,\infty}^{\mathcal{H}}(x, y) - V_{p,\infty}^{\mathcal{H}}(x, y) \geq U_{p,\infty}^{\mathbb{R}}(x', y') - V_{p,\infty}^{\mathbb{R}}(x', y').$$

This justifies the reduction. Let us consider the cases $y \leq b/2$, $y \in (b/2, 1)$ and $y \geq 1$ separately.

Lemma 3.10. *We have*

$$\frac{2}{p(p-1)(p-2)}(2y)^{p-1}(2y - px) \geq -\frac{2(p-1)^{p-2}}{p(p-2)}x^p. \qquad (3.126)$$

This yields the majorization for $y \leq b/2$.

Proof. The estimate is clear for $x = 0$. If $x > 0$ and we divide both sides by x^p, the inequality takes the form $F_0(2y/x) \geq 0$, where

$$F_0(s) := \frac{2}{p(p-1)(p-2)}s^{p-1}(s - p) + \frac{2(p-1)^{p-2}}{p(p-2)}$$

for $s > 0$. It suffices to note that F_0 is convex and $F_0(p-1) = F_0'(p-1) = 0$. $\square$

Lemma 3.11.

(i) *For all $s \geq 0$,*

$$b^{p-2}\left(-\frac{s}{p-2} + \frac{b^2}{p(p-1)^2}\right) + \frac{2(p-1)^{p-2}}{p(p-2)}s^{p/2} \geq 0. \qquad (3.127)$$

(ii) *We have*

$$b^{p-2}\left[\frac{4y^2 - 4xy}{p-2} - \frac{2b(2y - x)}{p-1} + \frac{b^2}{p}\right] + \frac{2(p-1)^{p-2}}{p(p-2)}x^p \geq 0. \qquad (3.128)$$

This yields the majorization for $y \in (b/2, 1)$.

Proof. (i) Denote the left-hand side of (3.127) by $F_1(s)$. It is evident that the function F_2 is convex on $\mathbb{R}$. In addition, it is straightforward to check that

$$F_1(b^2/(p-1)^2) = F_1'(b^2/(p-1)^2) = 0. \qquad (3.129)$$

The claim follows.

(ii) The partial derivative of the left-hand side of (3.128) with respect to y equals

$$\frac{4b^{p-2}}{p-2}\left(2y - x - \frac{b(p-2)}{p-1}\right).$$

Thus it suffices to verify the estimate for $2y = x + b(p-2)/(p-1)$. Plug this into (3.128) to get the inequality $F_1(x^2) \geq 0$, which has been already proved in (i). $\square$

Lemma 3.12. *If $y \geq 1$, then*

$$b^{p-2}\left[\frac{4y^2 - 4xy}{p-2} - \frac{2b(2y-x)}{p-1} + \frac{b^2}{p}\right] - \frac{2(p-1)^{p-2}}{p(p-2)}\left(\frac{1}{C_p^p} - x^p\right) \geq 0. \quad (3.130)$$

This yields the majorization for $y \geq 1$.

Proof. We divide the proof into three parts.

Step 1. *A reduction.* Denoting the left-hand side of (3.130) by $F_2(x,y)$, we derive that its partial derivative with respect to y is

$$F_{2y}(x,y) = \frac{4b^{p-2}}{p-2}\left(2y - x - \frac{b(p-2)}{p-1}\right).$$

Hence, it suffices to establish the estimate on the line segment

$$H_1 = \left\{(x,y) : y = 1, \, x \geq 0, \, 2y - x - \frac{b(p-2)}{p-1} \geq 0\right\}$$

and the half-line

$$H_2 = \left\{(x,y) : y > 1, \, 2y - x - \frac{b(p-2)}{p-1} = 0\right\}.$$

Step 2. *The segment H_1.* It is obvious that $x \mapsto F_2(x,1)$ is convex on the interval $[0, 2 - b(p-2)/(p-1)]$. After some lengthy, but easy calculations we verify that

$$1 - \frac{b}{2} < 2 - \frac{b(p-2)}{p-1}, \quad \text{and} \quad F_2\left(1 - \frac{b}{2}, 1\right) = F_2'\left(1 - \frac{b}{2}, 1\right) = 0,$$

which yields the estimate on H_1.

Step 3. *The half-line H_2.* Plugging $2y = x + b(p-2)/(p-1)$ into the estimate transforms it into

$$b^{p-2}\left(-\frac{x^2}{p-2} + \frac{b^2}{p(p-1)^2}\right) + \frac{2(p-1)^{p-2}}{p(p-2)}x^p \geq \frac{2(p-1)^{p-2}}{p(p-2)C_p^p}. \quad (3.131)$$

The left-hand side is equal to $F_1(x^2)$, where F_1 was defined in the proof of Lemma 3.11. Note that we have

$$x \geq 2 - \frac{b(p-2)}{p-1} \geq \frac{b}{p-1},$$

the latter being equivalent to $b \leq 2$, which is obvious. Thus, by the convexity of F_1 and (3.129), we get that the left-hand side of (3.131) attains its minimum at $x = 2 - b(p-2)/(p-1)$. However, then the estimate reads $F_2(2 - b(p-2)/(p-1)) \geq 0$, and we have already showed this in the preceding step. $\qquad\square$

Sharpness. Here the calculations are similar to those in the symmetric case, see Subsection 3.1.6. We will not present all the details and only exhibit the following Markov process (f, g). Namely, let $\kappa > 0$ and $\delta = (b - \kappa)/N$, where N is a large positive integer (such that $\kappa + \delta > (p - 1)\delta/2$). Consider the Markov martingale (f, g), uniquely determined by the following conditions:

(i) $(f_0, g_0) \equiv (0, 0)$.

(ii) If $(f_n, g_n) = (0, y)$ with $y < b$, then (f, g) moves either to $(\delta, y + \delta)$, or to the point on the line $y + \kappa = (3 - p)x/2$ (which is uniquely determined by the fact that f and g are martingales).

(iii) If $(f_n, g_n) = (0, b)$, then (f, g) moves either to $(1 - b, 1)$, or to the point on the line $y + \kappa = (3 - p)x/2$.

(iv) If $(f_n, g_n) = (\delta, y)$, $y + \kappa > (p - 1)\delta/2$, then (f, g) moves either to $(0, y)$, or to the point on the line $y + \kappa = (p - 1)x/2$.

(v) All the states on the lines $y + \kappa = (p - 1)x/2$, $y + \kappa = (3 - p)x/2$ and $y = 1$ are absorbing.

Clearly, the pair (f, g) depends on κ and N. It can be verified that

$$\lim_{\kappa \to 0} \lim_{N \to \infty} \frac{\mathbb{P}(g^* \geq 1)}{\|f\|_p^p} = K_{p,\infty}^p,$$

so the constant is indeed the best possible. $\square$

On the search of the suitable majorant. Let

$$V(x, y) = 1_{\{|y| \geq 1\}} - K^p |x|^p.$$

We simply follow the steps from the symmetric case and use the integration method. The "nonsymmetric" analogue of U_∞ is precisely the function u above and we integrate it against the kernel t^{p-1} from 0 to a certain positive number b. In this way we get a family of functions (indexed by b), each of which majorizes V with some $K > 0$ (or rather, αV, for some α, $K > 0$). Now all we need is to pick the "largest" element from the class, namely, the one for which the corresponding constant K is the smallest. It is quite fortunate that this approach leads to a sharp estimate. $\square$

3.8.3   Proof of Theorem 3.15

Proof of (3.122). The proof is very similar to that presented in the symmetric setting. Thus we will be brief and leave most of analogous calculations to the reader. Take $V(x, y) = |y| - K|x| \log |x| - L^{ns}(K)$ and let us write the special function (3.91) from the symmetric case in the skew coordinates (that is, put $x/2$ and $y - x/2$ in place of x and y). We obtain

$$U(x, y) = \begin{cases} y \cdot (y - x) & \text{if } |x| + |2y - x| < 2, \\ |2y - x| - |x| \log \frac{|x| + |2y - x|}{2} - 1 & \text{if } |x| + |2y - x| \geq 2. \end{cases}$$

Let, for $K \in (1/2, 1]$,

$$\alpha = \alpha_K = 2K^{-1} - K^{-2} \quad \text{and} \quad d = d_K = 2 - K^{-1},$$

while for $K > 1$, put

$$\alpha = \alpha_K = K^{-1} \exp(1 - K^{-1}) \quad \text{and} \quad d = d_K = \exp(1 - K^{-1}).$$

One can show that for all $K > 1/2$ and all $x, y \in \mathcal{H}$,

$$\alpha_K^{-1} U(d_K x, d_K y) \geq V(x, y).$$

Furthermore, it can be verified that the use of Fubini's theorem applies. Thus, the result follows. $\qquad \square$

On a lower bound for the constant. If $K \geq 1$, then the equality is attained for the constant pair $(f, g) \equiv (\exp(K^{-1} - 1), \exp(K^{-1} - 1))$. Let us assume that $K \in (1/2, 1)$ and let $L(K)$ denote the best constant in (3.122). Again, the reasoning is similar to that in the symmetric case, so we will only sketch the main steps. Introduce the function $U^0 : [0, \infty) \times \mathbb{R} \to \mathbb{R}$, by

$$U^0(x, y) = \sup\{\mathbb{E}|g_n| - K\mathbb{E}f_n \log f_n\},$$

where the supremum is taken over all n and all pairs (f, g) of simple martingales starting from (x, y) such that f is nonnegative and g satisfies $dg_k = \theta_k df_k$, $k \geq 1$, for some deterministic $\theta_k \in \{0, 1\}$. One can show that

1° $U^0(x, y) \geq |y| - Kx \log x$,
2° the function U^0 is concave along the lines of slope 0 or 1.
3° $U^0(x, x) \leq L(K)$ for all $x \geq 0$.

Furthermore, the restriction to nonnegative f gives the additional homogeneity property:

4° For any $x \geq 0$, $y \in \mathbb{R}$ and $\lambda > 0$,

$$U^0(\lambda x, \lambda y) = \lambda U^0(x, y) - K\lambda \log x.$$

Now we exploit these properties to obtain a lower bound for $L(K)$. By 2° and 1°,

$$U^0\left(1, \frac{1}{2}\right) \geq \frac{2K\delta}{1 + 2K\delta} U^0\left(\frac{2K-1}{2K}, \frac{1}{2}\right) + \frac{1}{1 + 2K\delta} U^0\left(1 + \delta, \frac{1}{2}\right) \quad (3.132)$$

$$\geq \frac{2K\delta}{1 + 2K\delta}\left[\frac{1}{2} - \frac{2K-1}{2}\log\frac{2K-1}{2K}\right] + \frac{1}{1 + 2K\delta} U^0\left(1 + \delta, \frac{1}{2}\right)$$

and, similarly, using 4°,

$$U^0\left(1+\delta,\frac{1}{2}\right)$$
$$\geq \frac{2K\delta}{1+2K\delta}U^0\left(1+\delta-\frac{1}{2K},\frac{1}{2}-\frac{1}{2K}\right) + \frac{1}{1+2K\delta}U^0\left(1+2\delta,\frac{1}{2}+\delta\right)$$
$$\geq \frac{2K\delta}{1+2K\delta}\left[\frac{1}{2K}-\frac{1}{2}-K\left(1+\delta-\frac{1}{2K}\right)\log\left(1+\delta-\frac{1}{2K}\right)\right]$$
$$+\frac{1}{1+2K\delta}(1+2\delta)\left[U^0\left(1,\frac{1}{2}\right)-K\log(1+2\delta)\right].$$

Plug this into (3.132), subtract $U^0(1,1/2)$ from both sides, divide throughout by δ and let δ go to 0. One obtains

$$U^0\left(1,\frac{1}{2}\right) \geq -K\log\frac{2K-1}{2K}-\frac{1}{2}.$$

Now, using 2° again, we get

$$U^0\left(\frac{1}{2},\frac{1}{2}\right) \geq KU^0\left(\frac{2K-1}{2K},\frac{1}{2}\right) + (1-K)U^0\left(1,\frac{1}{2}\right).$$

By 1° and the preceding inequality, we obtain, after some straightforward computations,

$$U^0\left(\frac{1}{2},\frac{1}{2}\right) \geq \frac{2K-1}{2}-\frac{K}{2}\log\frac{2K-1}{2K}.$$

The homogeneity property implies that for $x > 0$,

$$U^0(x,x) \geq (2K-1)x - Kx\log\frac{2K-1}{2K} - Kx\log(2x).$$

The right-hand side, as a function of $x \in (0,\infty)$, attains its maximum at $x_0 = \frac{K}{2K-1}\exp(1-K^{-1})$; using 3°,

$$L(K) \geq U^0(x_0,x_0) = \frac{K^2}{2K-1}\exp(1-K^{-1}).$$

This completes the proof.   □

3.9 On optimal control of martingales

Now let us turn to another type of results. Let f be a real-valued martingale and let g be a transform of f by a predictable process v bounded in absolute value by 1. Then g can be viewed as the result of controlling f by v. Suppose that g satisfies the bound

$$||g||_p \geq 1. \tag{3.133}$$

What can be said about the size of f, for example its qth moment, $q \geq p$? The moment estimates studied in the previous sections give a precise answer to this question: we have that

$$||f||_q \geq C_{p,q}^{-1} \tag{3.134}$$

and the constant $C_{p,q}^{-1}$ is the best possible, that is, cannot be replaced by a larger one. In other words, if f can be controlled so that (3.133) is satisfied, then necessarily (3.134) must hold and this bound cannot be improved.

Let us consider a different condition on g, given in terms of the one-sided maximal function of g:

$$\mathbb{P}(\sup_n g_n \geq 1) \geq t, \tag{3.135}$$

where t is a fixed number from the interval $[0, 1]$. What can now be said about the pth moments of f? The weak type inequalities give the bound $||f||_p \geq t/c_{p,p}$, but in general this is not optimal. For example, if $t = 1$ and $p > 1$, then the optimal bound is equal to 1. To see this, assume first that $\mathbb{P}(f_0 < 1) > 0$. This yields $\mathbb{P}(g_0 < 1) > 0$ and (3.136) implies that g, and hence also f, is not bounded in L^p: in particular $||f||_p \geq 1$. On the other hand, if $f_0 \geq 1$ almost surely, then $||f||_p \geq ||f_0||_p \geq 1$. Clearly, this bound is attained for the constant pair $(1, 1)$.

In fact we will study this problem in a more general setting and allow f, g to start from an arbitrary point of $\mathbb{R} \times \mathbb{R}$. Furthermore, instead of (3.135), we shall consider the condition

$$\mathbb{P}\big(\sup_n g_n \geq 0\big) \geq t, \tag{3.136}$$

which is easier to handle. The results easily carry over to the setting described by (3.135): it suffices to replace g with $g + 1$.

3.9.1 Formulation of the result

For any $1 \leq p < \infty$, let $U_p : \mathbb{R}^2 \to \mathbb{R}$ be given by (3.142), (3.157) or (3.166), depending on whether $p = 1$, $1 < p < 2$ or $p \geq 2$.

Theorem 3.16. *Let $1 \leq p < \infty$ be fixed. For given $x, y \in \mathbb{R}$, consider two martingales f, g satisfying the conditions $f_0 \equiv x$, $g_0 \equiv y$ and $|dg_n| \leq |df_n|$ for all $n \geq 1$. Then*

$$\mathbb{P}(g^* \geq 0) - ||f||_p^p \leq U_p(x, y). \tag{3.137}$$

The inequality is sharp for each p, x and y.

This result enables us to solve the problem formulated at the beginning. Introduce the function $L_1 : \mathbb{R} \times \mathbb{R} \times [0,1] \to [0,\infty)$ by

$$L_1(x,y,t) = \begin{cases} |x| & \text{if } y > (1 - 2t^{-1})|x|, \\ -y - [(y^2 - x^2)(1-t)]^{1/2} & \text{if } y \le (1 - 2t^{-1})|x|. \end{cases}$$

Furthermore, for $p > 1$, let L_p be given by

$$L_p(x,y,t) = \sup_{C>0} \frac{t - U_p(Cx, Cy)}{C^p}. \tag{3.138}$$

Theorem 3.17. *Let f, g be real-valued martingales such that $f_0 \equiv x$, $g_0 \equiv y$ and $|dg_n| \le |df_n|$ for $n \ge 1$. Assume in addition, that g satisfies the one-sided condition (3.136) for a given t. Then for $1 \le p < \infty$,*

$$||f||_p^p \ge L_p(x,y,t) \tag{3.139}$$

and the bound is the best possible. It is the best possible even in the special case when $df_n \equiv dg_n$ or $df_n \equiv -dg_n$ for all $n \ge 1$.

As the second application of Theorem 3.16, we obtain the following "one-sided versions" of the weak-type (p,p) inequalities.

Theorem 3.18. *Let $1 \le p < \infty$. Assume that f, g are real-valued martingales such that g is differentially subordinate to f. Then*

$$\mathbb{P}\Big(\sup_n g_n \ge 1\Big) \le K_{p,\infty} ||f||_p^p, \tag{3.140}$$

where $K_{1,\infty} = 2$,

$$K_{p,\infty} = \frac{\left(\frac{2p}{p-1}\right) \Gamma\left(\frac{p+1}{p}\right)^p}{\Gamma\left(\frac{p-1}{p}\right)^p} \qquad \text{for } 1 < p < 2,$$

and $K_{p,\infty} = p^{p-1}/2$ for $p \ge 2$. The constant $K_{p,\infty}$ is the best possible, even when g is a ± 1-transform of f.

3.9.2 Proofs in the case $p = 1$

Proof of (3.137) for $p = 1$. We use Burkholder's method. By the stopping time argument, the inequality can be rewritten in the form

$$\mathbb{E}V_1(f_n, g_n) \le U_1(x,y), \qquad n = 0, 1, 2, \ldots, \tag{3.141}$$

where $V_1(x, y) = 1_{\{y \geq 0\}} - |x|$ and U_1 is defined as follows. First, consider the subsets D_0, D_1 and D_2 of $\mathbb{R}^2$ given by

$$D_0 = \{(x, y) : y \geq -|x|\},$$
$$D_1 = \{(x, y) : |x| - 2 \leq y < -|x|\},$$
$$D_2 = \mathbb{R}^2 \setminus (D_0 \cup D_1\}.$$

We introduce the function $U_1 : \mathbb{R}^2 \to \mathbb{R}$ by the formula

$$U_1(x, y) = \begin{cases} 1 - |x| & \text{if } (x, y) \in D_0, \\ \frac{1}{4}[(y + 2)^2 - x^2] & \text{if } (x, y) \in D_1, \\ \frac{2|x|}{|x| - y} - |x| & \text{if } (x, y) \in D_2. \end{cases} \tag{3.142}$$

To prove (3.141), it suffices to verify the conditions $1°$ and $2°$ (the initial condition is not required, due to the special form of (3.141) and the fact that f, g are assumed to start from x and y, respectively). The majorization $U_1 \geq V_1$ is straightforward: it is clear on $D_0 \cup D_2$, while on D_1 we use the inequality $y + 2 \geq |x|$ to obtain

$$U_1(x, y) \geq 0 \geq -|x| = V_1(x, y).$$

To prove $2°$, let us introduce functions A_1, $B_1 : \mathbb{R}^2 \to \mathbb{R}$ by

$$A_1(x, y) = \begin{cases} -\text{sgn } x & \text{if } (x, y) \in D_0, \\ -\frac{1}{2}x & \text{if } (x, y) \in D_1, \\ \left(-\frac{2y}{(|x| - y)^2} - 1\right) \text{sgn } x & \text{if } (x, y) \in D_2 \end{cases}$$

and

$$B_1(x, y) = \begin{cases} 0 & \text{if } (x, y) \in D_0, \\ \frac{1}{2}(y + 2) & \text{if } (x, y) \in D_1, \\ \frac{2|x|}{(|x| - y)^2} & \text{if } (x, y) \in D_2. \end{cases}$$

We will show that for any $(x, y) \in \mathbb{R}^2$ and k_1, $k_2 \in \mathbb{R}$ satisfying $|k_1| \geq |k_2|$ we have

$$U_1(x + k_1, y + k_2) \leq U_1(x, y) + A_1(x, y)k_1 + B_2(x, y)k_2. \tag{3.143}$$

This will be done by proving that for any x, y and any $a \in [-1, 1]$ the function $G = G_{x,y,a}$ given by $G(t) = U_1(x + t, y + at)$ is concave. To accomplish this, we will show, as usually, that $G''(0) \leq 0$ for those x, y, a, for which the second derivative exists and $G'(0-) \geq G'(0+)$ for the remaining x, y and a. We derive that

$$G''(0) = \begin{cases} 0 & \text{if } (x, y) \in D_0^\circ, x \neq 0, \\ \frac{1}{2}(a^2 - 1) & \text{if } (x, y) \in D_1^\circ, \\ 4(1 - a)(y - ax)(x - y)^{-3} & \text{if } (x, y) \in D_2^\circ, x \neq 0, \end{cases}$$

where D° denotes the interior of the set D. It is evident that all the expressions are nonpositive. Now, if $(x, y) \in \partial D_0 \cap \partial D_1$, then

$$G'(0-) = \frac{1}{2}[(y + 2)a - x] \geq \frac{1}{2}[-(y + 2) - x] = -1 = G'(0+).$$

If $(x, y) \in \partial D_1 \cap \partial D_2$, then $G'(0-) = G'(0+)$. If $(x, y) \in \partial D_2 \cap \partial D_0$, then

$$G'(0-) = \frac{1 + a}{2x} - 1 \geq -1 = G'(0+).$$

If $x = 0$ and $y \geq 0$, then $G'(0-) = 1 \geq -1 = G'(0+)$. Finally, if $x = 0$ and $y < -2$, then

$$G'(0-) = \frac{2}{y} + 1 \geq -\frac{2}{y} - 1 = G'(0+).$$

This establishes the concavity property 2°. Thus, by Burkholder's argument, (3.141) will be proved once we can show that the random variables $U(f_n, g_n)$, $A(f_n, g_n)df_{n+1}$ and $B(f_n, g_n)dg_{n+1}$ are integrable for any $n = 0, 1, 2, \ldots$. However, the latter is evident: we have that $|U(x, y)| \leq 1 + |x|$ and $|A(x, y)| \leq 1$, $|B(x, y)| \leq 1$ for all $x, y \in \mathbb{R}$. $\square$

Sharpness of (3.137) *for* $p = 1$. For each (x, y), we will exhibit an example for which the two sides of (3.137) are equal. Furthermore, for each such example, we shall derive the explicit formula for

$$P(x, y) = \mathbb{P}(\sup_n g_n \geq 0). \tag{3.144}$$

This will be useful in the further applications of (3.137). We split the reasoning into three parts, corresponding to the sets D_0, D_2 and D_1.

    Step 1. $(x, y) \in D_0$. If $y \geq 0$, then we take the trivial pair $(f, g) \equiv (x, y)$. Then we have equality in (3.137) and, obviously, $P(x, y) = 1$. On the other hand, if $y < 0$, consider independent centered random variables $\xi_1, \xi_2, \ldots$ such that $\xi_k = \pm 2^{k-1}|y|$ for $k = 1, 2, \ldots$ and introduce the stopping time $\tau = \inf\{n : \xi_n > 0\}$. Put

$$f_n = x - (\xi_1 + \xi_2 + \cdots + \xi_{\tau \wedge n})\operatorname{sgn} x, \qquad g_n = y + \xi_1 + \xi_2 + \cdots + \xi_{\tau \wedge n}.$$

It is easy to see that g has the following behavior: it starts from y and in the next step it rises to 0 or drops to $2y$. If it jumped to 0, it stays there forever; if it came to $2y$, then, in the next move, it jumps to 0 (and stops), or drops to $4y$, and so on. Thus, we have that $P(x, y) = 1$. On the other hand, f has constant sign, which is due to the assumption $y \geq -|x|$, and hence $||f||_1 = |x|$, so the two sides of (3.137) are equal.

    Step 2. $(x, y) \in D_2$. Suppose that $x \geq 0$; the case $x < 0$ can be studied likewise. Consider a Markov martingale (f, g) such that

(i) $(f_0, g_0) \equiv (x, y)$,
(ii) $(f_1, g_1) \in \{(0, y - x), ((x - y)/2, (y - x)/2)\}$,

(iii) if $(f_1, g_1) = (0, y - x)$, the process stops,

(iv) if $(f_1, g_1) = ((x - y)/2, (y - x)/2)$, the process evolves according to the rule described in Step 1.

It is easy to compute that

$$P(x, y) = \mathbb{P}((f_1, g_1) = ((x - y)/2, (y - x)/2)) = \frac{2x}{x - y}$$

and since the sign of f does not change, $||f||_1 = x$. Consequently, we obtain equality in (3.137).

Step 3. $(x, y) \in D_1$. Consider a Markov martingale (f, g) such that

(i) $(f_0, g_0) \equiv (x, y)$,

(ii) $df_1 = dg_1$ and (f_1, g_1) jumps to the line $y = -x$ or $y = -x - 2$,

(iii) the process $(f_n, g_n)_{n \geq 2}$ evolves according to the rules described in Step 1 and Step 2.

We easily derive that

$$P(x, y) = \mathbb{P}(f_1 = -g_1) + \mathbb{P}(f_1 = -g_1 - 2, f_2 = g_2)$$
$$= \frac{x + y + 2}{2} - \frac{x + y}{2} \cdot \frac{-x + y + 2}{2} = 1 + \frac{x^2 - y^2}{4}$$

and

$$||f||_1 = \mathbb{E}|f_1| = -y + \frac{x^2 - y^2}{2},$$

so the two sides of (3.137) are equal. Thus $U_1(x, y)$ is indeed the best possible in (3.137) for any choice of the starting point (x, y). $\qquad \square$

Proof of Theorem 3.18 for $p = 1$. Apply Theorem 3.16 to the martingales $2f$ and $2g - 2$, conditionally on $\mathcal{F}_0$. Taking expectation of both sides, we get

$$\mathbb{P}(g^* \geq 1) = \mathbb{P}\left((2g - 2)^* \geq 0\right) \leq ||2f||_1 + \mathbb{E}U_1\left(2f_0, 2g_0 - 2\right).$$

We will show that for all points $(x, y) \in \mathbb{R}^2$ satisfying $|y| \leq |x|$ we have

$$U_1(x, y - 2) \leq 0. \tag{3.145}$$

This will give (3.140). Furthermore, we will prove that we have equality in (3.145) for at least one such point: this will imply the optimality of the constant. We easily check that if $|y| \leq |x|$ and $y + |x| \leq 2$, then

$$U_1(x, y - 2) = \frac{2|x|}{|x| - y + 2} - |x| \leq \frac{2|x|}{|x| - |y| + 2} - |x| \leq 0,$$

with equality iff $|x| = |y|$. On the other hand, if $|y| \leq |x|$ and $y + |x| > 2$, then $|x| > 1$ and consequently $U_1(x, y - 2) = 1 - |x| < 0$. The proof is complete. $\qquad \square$

Proof of (3.139) *for p = 1.* The estimate is trivial if $y \geq -|x|$, so we assume that $y < -|x|$. Let $C > 0$ be an arbitrary constant. Application of (3.137) to the martingales Cf, Cg yields $\mathbb{P}(g^* \geq 0) - C\|f\|_1 \leq U(Cx, Cy)$, which, by (3.136), leads to the bound

$$\|f\|_1 \geq \frac{t - U(Cx, Cy)}{C}. \tag{3.146}$$

We will maximize the right-hand side over C. We have that

$$\frac{t - U(Cx, Cy)}{C} = \begin{cases} -y + (t-1)C^{-1} + \frac{1}{4}C(x^2 - y^2) & \text{if } C(|x| - y) < 2, \\ \left(t - \frac{2|x|}{|x| - y}\right)C^{-1} + |x| & \text{if } C(|x| - y) \geq 2. \end{cases}$$

If $y > (1 - 2t^{-1})|x|$ (or, equivalently, $t < \frac{2|x|}{|x|-y}$), then by a straightforward analysis of the derivative we see that the expression on the right, as a function of C, is increasing on $(0, \infty)$. Letting $C \to \infty$ in (3.146) gives $\|f\|_1 \geq |x|$, which is (3.139). Finally, if $y \leq (1 - 2t^{-1})|x|$, then the right-hand side of (3.146) is increasing on $(0, C_0)$ and nonincreasing on (C_0, ∞), where

$$C_0 = 2\left(\frac{1-t}{y^2 - x^2}\right)^{1/2}.$$

Plugging $C = C_0$ into (3.146) gives (3.139). $\qquad\square$

Sharpness of (3.139). Fix $(x, y) \in \mathbb{R}^2$ and $t \in [0, 1]$. Obviously, if $y \geq 0$, then the constant pair $(f, g) \equiv (x, y)$ gives equality in (4.96); therefore we may and do assume that $y < 0$. Consider the function $C \mapsto P(Cx, Cy)$, where P was introduced in (3.144). If there is $C > 0$ such that $P(Cx, Cy) = t$, then let (f, g) be the corresponding pair starting from (Cx, Cy). Then $(f/C, g/C)$ starts from (x, y), satisfies the appropriate condition for difference sequences, $\mathbb{P}((g/C)^* \geq 0) = t$ and

$$\mathbb{P}(g^* \geq 0) - \|f\|_1 = U_1(Cx, Cy). \tag{3.147}$$

This implies

$$L_1(x, y, t) \leq \|f/C\|_1 = \frac{t - U_1(Cx, Cy)}{C} \leq L_1(x, y, t),$$

which yields the claim. On the other hand, if there is no C such that $P(Cx, Cy) = t$, then, by continuity of P, we see that $P(Cx, Cy) > t$ for all $C > 0$. Take a large C and take the appropriate pair (f, g) starting from (Cx, Cy). We have, by (3.147),

$$L_1(x, y, t) \leq \|f/C\|_1 \leq \frac{1 - U_1(Cx, Cy)}{C}.$$

However, the right-hand side converges to $|x|$ as $C \to \infty$; furthermore, $|x| \leq L_1(x, y, t)$. Thus, taking sufficiently large C, we see that the pair $(f/C, g/C)$ satisfies the following properties: it starts from (x, y), has appropriate difference sequences, $\mathbb{P}((g/C)^* \geq 0) \geq t$ and $\|f/C\|_1$ is arbitrarily close to $L_1(x, y, t)$. This completes the proof. $\qquad\square$

3.9.3 An auxiliary differential equation, $1 < p < 2$

For $p > 1$, the situation becomes much more complicated. We start with the following auxiliary result.

Theorem 3.19. *Let $1 < p < 2$. There is a continuous function $H = H_p : [0, \infty) \to [0, \infty)$ satisfying the differential equation*

$$H'(x) = \frac{p(p-1)}{2} x^{p-2} (H(x) - x)^2 \qquad (3.148)$$

for $x > 0$ and such that

$$H(0) = \frac{\left(\frac{2p}{p-1}\right)^{1/p} \Gamma\left(\frac{p+1}{p}\right)}{\Gamma\left(\frac{p-1}{p}\right)}. \qquad (3.149)$$

Proof. Observe that the equation (3.148) has Riccati's form. Therefore, the substitution

$$j(y) = \exp\left[\int_0^y \frac{p(p-1)}{2} z^{p-2}(z - H(z))\mathrm{d}z\right] \qquad (3.150)$$

transforms (3.148) into

$$yj''(y) + (2 - p)j'(y) - \frac{p(p-1)}{2} y^{p-1} j(y) = 0, \qquad y > 0. \qquad (3.151)$$

Two linearly independent solutions to this equation are given by

$$j_1(y) = y^{(p-1)/2} I_{-1+1/p}(z_0),$$
$$j_2(y) = y^{(p-1)/2} I_{1-1/p}(z_0),$$

where $z_0 = \sqrt{\frac{2(p-1)}{p}} y^p$ and I_α stands for the modified Bessel function of the first kind (see Abramowitz and Stegun [1]). That is, I_α satisfies

$$z^2 I_\alpha''(z) + z I_\alpha'(z) - (z^2 + \alpha^2) I_\alpha(z) = 0 \qquad (3.152)$$

and can be written as

$$I_\alpha(z) = \sum_{k=0}^{\infty} \frac{\left(\frac{z}{2}\right)^{2k+\alpha}}{k! \Gamma(\alpha + k + 1)}, \qquad (3.153)$$

or, in integral form, as

$$I_\alpha(z) = \frac{1}{\pi} \int_0^\pi \exp(z \cos\theta) \cos(\alpha\theta)\mathrm{d}\theta - \frac{\sin(\alpha\pi)}{\pi} \int_0^\infty \exp(-z \cosh t - \alpha t)\mathrm{d}t. \quad (3.154)$$

To see that j_1, j_2 satisfy (3.151), we plug them into the equation and obtain

$$\frac{p^2}{4} y^{\frac{p-3}{2}} \left[z_0^2 I_{\pm(1-1/p)}''(z_0) + z_0 I_{\pm(1-1/p)}'(z_0) - \left((1 - p^{-1})^2 + z_0^2\right) I_{\pm(1-1/p)}(z_0)\right] = 0,$$

which is valid thanks to (3.152). Coming back to (3.148), we take $j = j_1 - j_2$: in view of (3.154), we have

$$j(y) = \frac{2\sin(\pi - \pi/p)}{\pi} y^{(p-1)/2} \int_0^\infty \exp(-z_0 \cosh t) \cosh((1 - 1/p)t) dt > 0 \quad (3.155)$$

and recover H by the formula

$$H(x) = x - \frac{2x^{2-p} j'(x)}{p(p-1)j(x)}. \quad (3.156)$$

To see that (3.149) is satisfied, note that by (3.153) we have the equalities

$$\lim_{x \to 0+} j(x) = -\frac{\left(\frac{p-1}{2p}\right)^{-(p-1)/(2p)}}{\Gamma(1/p)},$$

$$\lim_{x \to 0+} x^{2-p} j'(x) = (p-1) \frac{\left(\frac{p-1}{2p}\right)^{(p-1)/(2p)}}{\Gamma(2 - 1/p)},$$

from which the initial condition follows after simple manipulations. □

We will need the following further properties of H.

Lemma 3.13. *For any $x > 0$ we have $H(x) > x$ and $H'(x) \geq 1$.*

Proof. By (3.156), the first estimate is equivalent to $j'(x) < 0$, since j is positive, as we have already observed above. However, it is clear that j is decreasing: it suffices to look at the integrand in (3.155). To show that $H'(x) \geq 1$, suppose that there is $x_0 > 0$ such that $H'(x_0) < 1$. Let $x_1 = \inf\{x > x_0 : H'(x) = 1\}$, with the convention $\inf \emptyset = \infty$. Then for any $x \in (x_0, x_1)$ we have $0 < H(x) - x < H(x_0) - x_0$ and $x^{p-2} < x_0^{p-2}$, so, by (3.148), $H'(x) < H'(x_0)$. Letting $x \to x_1$ gives $x_1 = \infty$, so $H'(x) < H'(x_0) < 1$ for all $x > x_0$. This contradicts the estimate $H(x) > x$ and completes the proof of $H'(x) \geq 1$. □

3.9.4 Proofs in the case $1 < p < 2$

We start with Theorem 3.16. The idea is the same as in the previous section. Let $V_p : \mathbb{R}^2 \to \mathbb{R}$ be given by $V_p(x, y) = 1_{\{y \geq 0\}} - |x|^p$. To define the corresponding special function $U_p : \mathbb{R}^2 \to \mathbb{R}$, introduce the following subsets of $[0, \infty) \times [0, \infty)$, where $H = H_p$ is the function studied in the preceding subsection and $h = h_p$ stands for its inverse:

$$D_0 = \{(x, y) : y \geq 0\},$$
$$D_1 = \{(x, y) : h(x - y) - y < x, -x \leq y < 0\},$$
$$D_2 = \{(x, y) : h(x - y) < x \leq h(x - y) - y, y - x \leq -H(0)\},$$
$$D_3 = \{(x, y) : x - H(0) < y < -x\},$$
$$D_4 = \{(x, y) : x \leq h(x - y)\}.$$

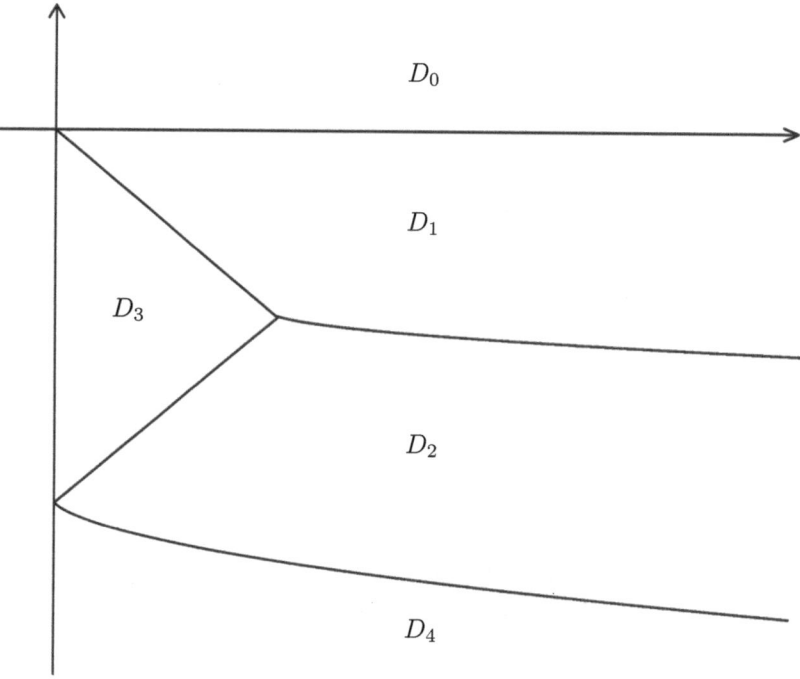

Figure 3.3: The regions D_0–D_4.

The function $U_p : \mathbb{R}^2 \to \mathbb{R}^2$ is given by the equation

$$U_p(x, y) = U_p(-x, y)$$

and

$$U_p(x, y) = \begin{cases} 1 - x^p & \text{if } (x, y) \in D_0, \\ 1 + (x + y)^{p-1}((p - 1)y - x) + \frac{2y}{H(x+y)-x-y} & \text{if } (x, y) \in D_1, \\ [h(x - y)]^{p-1}[(p - 1)h(x - y) - px] & \text{if } (x, y) \in D_2, \\ [(y + H(0))^2 - x^2](H(0))^{-2} & \text{if } (x, y) \in D_3, \\ -x^p & \text{if } (x, y) \in D_4. \end{cases}$$

$$(3.157)$$

Now we will prove that the function U_p satisfies the conditions $1°$ and $2°$. We start with the majorization property.

Lemma 3.14. *We have $V_p(x, y) \le U_p(x, y) \le 1 - |x|^p$ for all $(x, y) \in \mathbb{R}^2$.*

Proof. Since $U_p(x, y) = 1 - |x|^p$ for $y \ge 0$ and $U_p(x, y) = -|x|^p$ for sufficiently small y, it is enough to show that for any fixed x, the function $y \mapsto U_p(x, y)$

is nondecreasing. By symmetry, we will be done if we verify this for $x \geq 0$. If $(x, y) \in D_0^o$, then $U_{py} = 0$. If $(x, y) \in D_1$, then

$$U_{py}(x, y) = \frac{2H(x + y) - 2x}{(H(x + y) - x - y)^2}$$

and we must prove that the numerator is nonnegative. We consider two cases. If $x \leq H(0)$, then $y \geq -x$ (by the definition of D_1) and $H(x + y) \geq H(0) \geq x$. If, conversely, $x > H(0)$, then $x + y > h(x - y)$, or, equivalently, $H(x + y) > x - y$; it remains to note that the right-hand side is larger than x, as needed. Suppose now that $(x, y) \in D_2$. Then it can be derived that

$$U_{py}(x, y) = p(p - 1)(h(x - y))^{p-2} h'(x - y)[x - h(x - y)] \geq 0,$$

since all the factors are nonnegative: see the definition of D_2. If $(x, y) \in D_3$, then $y > -H(0)$ and $U_{py}(x, y) = 2(y + H(0))(H(0))^{-2} > 0$. Finally, $U_{py} = 0$ on D_4, which completes the proof. $\square$

The next step is to establish the concavity property. Introduce the functions $A_p, B_p : \mathbb{R}^2 \to \mathbb{R}$ by $A_p(x, y) = U_{px}(x, y)$ and

$$B_p(x, y) = \begin{cases} U_{py}(x, y) & \text{if } y \neq 0, \\ 0 & \text{if } y = 0. \end{cases}$$

Lemma 3.15. *For any $(x, y) \in \mathbb{R}^2$ and $k_1, k_2 \in \mathbb{R}$ satisfying $|k_2| \leq |k_1|$ we have*

$$U_p(x + k_1, y + k_2) \leq U_p(x, y) + A_p(x, y)k_1 + B_p(x, y)k_2. \qquad (3.158)$$

Proof. We argue as previously and check the concavity of the function $G = G_{x,y,a}$, given by $G(t) = U_p(x + t, y + at)$, where $x \geq 0$, $y \in \mathbb{R}$ and $a \in [-1, 1]$ are fixed parameters. It is straightforward to check that G is of class C^1 except for t such that $y + at = 0$. Thus, by the translation argument, it suffices check the condition $G''(0) \leq 0$ for (x, y) lying in the interiors of the sets D_i and then to verify whether $G'_{x,0,a}(0-) \geq G'_{x,0,a}(0+)$. If (x, y) belongs to the interior of D_0 and $x \neq 0$, then $G''(0) = -p(p - 1)x^{p-2} < 0$. If $(x, y) \in D_1^o$, then a brief computation shows that $G''(0) = I + II$, where

$$I = \frac{2(a^2 - 1)}{(H(x + y) - x - y)^2},$$

$$II = -\frac{2(a + 1)^2(H'(x + y) - 1)(H(x + y) - x + y)}{(H(x + y) - x - y)^3}. \qquad (3.159)$$

Clearly, $I \leq 0$. To show that $II \leq 0$, note that this is equivalent to $x + y > h(x - y)$, and is guaranteed by the definition of D_1. Let $(x, y) \in D_2^o$. Then $G''(0) = I + II$,

where

$$I = \frac{2(a^2 - 1)}{(h(x-y) - x + y)^2},$$

$$II = \frac{2(a-1)^2(1 - h'(x-y))(h(x-y) - x - y)}{(-h(x-y) + x - y)^3}$$

(3.160)

and both terms are nonpositive: this follows from the inequalities

$$h'(x-y) \leq 1, \quad h(x-y) \leq x - y \quad \text{and} \quad h(x-y) \geq x + y,$$

where the first two come from Lemma 3.13 and the latter is imposed in the definition of D_2. Next observe that if $(x,y) \in D_3^o$, then

$$G''(0) = 2(a^2 - 1)(H(0))^{-2} \leq 0.$$

(3.161)

Finally, if $(x,y) \in D_4^o$ and $x \neq 0$, then

$$G''(0) = -p(p-1)x^{p-2} \leq 0.$$

(3.162)

To complete the proof, note that if $a < 0$, then

$$G'_{x,0,a}(0-) = -px^{p-1} \geq -px^{p-1} + \frac{2a}{H(x) - x} = G'_{x,0,a}(0+),$$

while for $a \geq 0$,

$$G'_{x,0,a}(0-) = -px^{p-1} + \frac{2a}{H(x) - x} \geq -px^{p-1} = G'_{x,0,a}(0+).$$

The required properties are satisfied and the claim follows.   □

We will also need the following bound for A_p and B_p.

Lemma 3.16. *There is an absolute constant c such that for any x, $y \in \mathbb{R}$,*

$$|A_p(x,y)| \leq c(1 + |x|^{p-1} + |y|^{p-1}), \qquad |B_p(x,y)| \leq c(1 + |x|^{p-1} + |y|^{p-1}). \quad (3.163)$$

Proof. It suffices to prove the existence of the constant c on each D_i. This is obvious for D_0, D_3 and D_4. Note that if $x \geq 0$, then $A_p(x,y) \leq 0$ and $B_p(x,y) \geq 0$: the first inequality follows from the fact that the function $x \mapsto U_p(x,y)$ is even and concave (see the proof of (3.158)) and the second has been shown in the proof of (3.14). Furthermore, if $(x,y) \in D_1$,

$$A_p(x,y) = p(x+y)^{p-2}(-x + (p-2)y) - \frac{2y}{(H(x+y) - x - y)^2}(H'(x+y) - 1)$$

$$\geq p(x+y)^{p-2}(-x + (p-2)y),$$

$$B_p(x,y) = (p-1)^2(x+y)^{p-2}y + \frac{2}{H(x+y) - x - y} - \frac{2y(H'(x+y) - 1)}{(H(x+y) - x - y)^2}$$

$$\leq \frac{2}{H(x+y) - x - y} - \frac{2yH'(x+y)}{(H(x+y) - x - y)^2}$$

$$\leq \frac{2}{H(0)} - p(p-1)y(x+y)^{p-2},$$

where in the last inequality we have used the fact that $H(x) - x \geq H(0)$ (which follows from $H'(x) \geq 1$) and the differential equation (3.148). Consequently, (3.163) holds on D_1. Finally, if $(x, y) \in D_2$, then

$$
\begin{aligned}
A_p(x, y) &= ph(x - y)^{p-2}[(p-1)(h(x-y) - x)h'(x-y) - h(x-y)] \\
&\geq ph(x-y)^{p-2}[-(p-1)xh'(x-y) - h(x-y)] \\
&\geq px^{p-2}[-(p-1)x - x] = p^2 x^{p-1},
\end{aligned}
$$

where in the second inequality we used $h(x-y) < x$ (see the definition of D_2) and $h'(x-y) < 1$ (because $h = H^{-1}$ and $H'(x) > 1$). Similarly,

$$
\begin{aligned}
B_p(x, y) &= p(p-1)(h(x-y))^{p-2}[x - h(x-y)]h'(x-y) \\
&\leq p(p-1)(h(x-y))^{p-2}xh'(x-y) \leq p(p-1)x^{p-1},
\end{aligned}
$$

and we are done.   □

Proof of (3.137) *for* $1 < p < 2$. Clearly, we may and do assume that $\|f\|_p < \infty$; then also $\|g\|_p < \infty$, in virtue of Burkholder's moment inequality. By the preceding lemmas and Burkholder's method, all we need is the integrability of the random variables $U_p(f_n, g_n)$, $A_p(f_n, g_n)df_{n+1}$ and $B_p(f_n, g_n)dg_{n+1}$ for all $n = 0, 1, 2, \ldots$. By Lemma 3.14, we have $\mathbb{E}|U_p(f_n, g_n)| < \infty$. To deal with the remaining variables, we make use of Lemma 3.16, Hölder's inequality and the condition $df_n, dg_n \in L^p$. This finishes the proof.   □

Sharpness of (3.137) *for* $1 < p < 2$. Compared to the case $p = 1$, this is much more delicate. The reasoning is very similar to that in the proof of the sharpness of (3.70) and (3.71). First, if $(|x|, y) \in D_0 \cup D_4$, the constant pair gives equality in (3.137). For the remaining points (x, y), fix $\delta \in (0, H(0))$ and consider a Markov family with the transitions described as follows.

(i) The states belonging to D_0 and D_4 are absorbing.

(ii) If $(f_0, g_0) = (x, y) \in D_1$, then $df_1 = -dg_1$ and (f, g) goes to the line $y = 0$ or to the set $\partial D_1 \cap \partial D_2$.

(iii) If $(f_0, g_0) = (x, y) \in D_2 \setminus \partial D_1$, then $df_1 = dg_1$ and (f, g) goes to the set $\partial D_2 \cap \partial D_4$ or to $\partial D_1 \cap \partial D_2$.

(iv) If $(f_0, g_0) = (x, y) \in D_3$, then $df_1 = dg_1$ and (f, g) goes to the line $y = -x$ or to $y = -x - H(0)$.

(v) If $(f_0, g_0) = (x, y) \in \partial D_1 \cap \partial D_2$, then $df_1 = dg_1$ and (f, g) goes to $(f_0 + \delta, g_0 + \delta)$ or to $\partial D_2 \cap \partial D_4$.

(vi) If $f_0 < 0$, then $(-f, g)$ evolves according to the rules above.

Note that the moves described in steps (i)–(iv) have the property that $\mathbb{E}U_p(f_1, g_1) = U_p(x, y)$: in each case the process (f, g) moves to endpoints of a line segment along which U_p is linear. On the other hand, since U_p is of class C^1, we have that for $(f_0, g_0) \equiv (x, y) \in \partial D_1 \cap \partial D_2$,

$$\mathbb{E} U_p(f_1, g_1) \geq U_p(x, y) - R\delta^2, \qquad (3.164)$$

where $R = R(x, y)$ is a certain positive constant. Furthermore, since U_p is of class C^2 on D_1, for any $T > H(0)$ there is a finite $R > 0$ depending only on T such that (3.164) holds for all $(x, y) \in \partial D_1 \cap \partial D_2$ with $x \leq T$.

Now let $\varepsilon > 0$, $\delta > 0$ and let (f, g) start from a fixed $(x, y) \in \partial D_1 \cap \partial D_2$. Clearly, (f, g) converges almost surely for any δ. In fact, it is easy to see that there is $T > 0$, independent of δ, such that

$$\mathbb{P}(f^* > T) < \varepsilon. \qquad (3.165)$$

Let $R = R(T)$ be the corresponding constant in (3.164). Now, let $N = \inf\{n : f_n \geq T\}$. If $N < \infty$, then $N = O(1/\delta)$ and, consequently,

$$\mathbb{E} U_p(f_N, g_N) \geq U_p(x, y) - R\delta^2 \cdot N \geq U_p(x, y) - \varepsilon,$$

provided δ is sufficiently small. Therefore,

$$\begin{aligned}
\mathbb{E} V_p(f_N, g_N) &= \mathbb{E} U_p(f_N, g_N) + \mathbb{E}\left[V_p(f_N, g_N) - U_p(f_N, g_N)\right] \\
&\geq U_p(x, y) + \mathbb{E}\left[V_p(f_N, g_N) - U_p(f_N, g_N)\right] - \varepsilon.
\end{aligned}$$

However, $U_p(f_N, g_N) = V_p(f_N, g_N)$ on the set $\{f^* < T\}$ and $U_p(f_N, g_N) < V_p(f_N, g_N) + 1$ on $\{f^* \geq T\}$ (see Lemma 3.14). Thus, by (3.165),

$$\mathbb{E} V_p(f_N, g_N) \geq U_p(x, y) - 2\varepsilon$$

and since $\varepsilon > 0$ was arbitrary, $U_p(x, y)$ is the optimal bound in (3.137) provided $(|x|, y) \in \partial D_1 \cap \partial D_2$. For the remaining points, it suffices to use the fact that (f, g) reaches $D_0 \cup D_4 \cup (\partial D_1 \cap \partial D_2)$ after at most two steps, in which there is no change in $\mathbb{E} U_p(f, g)$. $\qquad \square$

Proof of Theorem 3.18 *for* $1 < p < 2$. Applying Theorem 3.16 to the pair

$$(H(0)f, H(0)g - H(0))$$

we obtain

$$\begin{aligned}
\mathbb{P}(g^* \geq 1) = \mathbb{P}\left((H(0)g - H(0))^* \geq 0\right) \\
\leq \|H(0)f\|^p \ | \ \mathbb{E} U_p\left(II(0)f_0, H(0)g_0 - H(0)\right).
\end{aligned}$$

Thus it suffices to prove that for any point (x, y) such that $|x| \geq |y|$ we have $U_p(x, y - H(0)) \leq 0$, with equality for at least one such point. In the proof of Lemma 3.14 we have shown that for any fixed x the function $y \mapsto U_p(x, y)$ is nondecreasing. Combining this with (3.158) yields

$$\begin{aligned}
U_p(x, y - H(0)) &= U_p(|x|, y - H(0)) \leq U_p(|x|, |x| - H(0)) \\
&= U_p(0, -H(0)) + A_p(0, -H(0))|x| + B_p(0, -H(0))|x| = 0,
\end{aligned}$$

with equality for $x \in [-H(0)/2, H(0)/2]$ and $y = |x|$. $\qquad \square$

Finally, we turn to the result on the control of martingales.

Proof of Theorem 3.17, $1 < p < 2$. We proceed as in the case $p = 1$. Let $C > 0$. By (3.137), applied to the martingales Cf and Cg, we have $C^p||f||_p \geq \mathbb{P}(g^* \geq 0) - U_p(Cx, Cy)$, so, by (3.136),

$$||f||_p^p \geq \frac{t - U_p(Cx, Cy)}{C^p}.$$

This immediately yields (3.139): see (3.138). To see that the bound is optimal, note that if $y \geq 0$, then $U_p(Cx, Cy) = 1 - C^p|x|^p$ and hence $L_p(x, y, t) = |x|$; this is obviously the best possible (take the constant pair). Suppose then that $y < 0$. It is evident from the example studied above that for any $\varepsilon > 0$ there are a $\delta = \delta(\varepsilon) > 0$ and a $C = C(\varepsilon)$ bounded away from 0 such that the appropriate pair (f, g), starting from (Cx, Cy), satisfies $\mathbb{E}V_p(f_\infty, g_\infty) \geq U_p(Cx, Cy) - \varepsilon$ and $\mathbb{P}(g^* \geq 0) = t$. Consequently,

$$||f/C||_p^p \leq \frac{\mathbb{P}(g^* \geq 0) - U_p(Cx, Cy) + \varepsilon}{C^p} = \frac{t - U_p(Cx, Cy)}{C^p} + \frac{\varepsilon}{C^p}$$
$$\leq L_p(x, y, t)^p + \frac{\varepsilon}{C^p}$$

and since ε was arbitrary, the result follows. $\square$

3.9.5   The case $p \geq 2$

As previously, one introduces the function $V_p : \mathbb{R}^2 \to \mathbb{R}$ given by $V_p(x, y) = 1_{\{y \geq 0\}} - |x|^p$. Let U_p be given by

$$U_p(x, y) = \begin{cases} U\left(\left(\frac{2}{p^{p-1}}\right)^{1/p} x, \left(\frac{2}{p^{p-1}}\right)^{1/p} y + 1 \right) & \text{if } y \geq -(p^{p-1}/2)^{1/p}, \\ -|x|^p & \text{if } y < -(p^{p-1}/2)^{1/p}, \end{cases} \quad (3.166)$$

where U is Suh's special function corresponding to the weak type inequality, defined in (3.38). Now one proceeds precisely in the same manner as in the preceding subsection. We omit the details.

3.10 Inequalities for weakly dominated martingales

3.10.1 Weak domination and modification of Burkholder's method

In this section we will study the weak-type, moment and logarithmic inequalities under a different type of domination.

Definition 3.1. Let f, g be two $\mathcal{B}$-valued martingales. We say that g is weakly dominated by f, if for any convex and increasing function ϕ on $[0, \infty)$ we have $\mathbb{E}\phi(|g_0|) \leq \mathbb{E}\phi(|f_0|)$ and

$$\mathbb{E}(\phi(|dg_n|)|\mathcal{F}_{n-1}) \leq \mathbb{E}(\phi(|df_n|)|\mathcal{F}_{n-1}), \qquad n = 1, 2, \ldots,$$

with probability 1.

Clearly, this assumption is less restrictive than the differential subordination. Another important example is the following. We say that martingales f and g are tangent, if for any $n = 0, 1, 2, \ldots$ the conditional distributions of df_n and dg_n given $\mathcal{F}_{n-1}$ coincide (set $\mathcal{F}_{-1}$ to be the trivial σ-field). If f, g satisfy this condition, then g is weakly dominated by f and f is weakly dominated by g.

There is a version of Burkholder's method which works in this new setting. To see this, let $V : \mathcal{B} \times \mathcal{B} \to \mathbb{R}$ be a given Borel function, satisfying the symmetry condition

$$V(x, y) = V(-x, -y) \qquad \text{for all } x, y \in \mathcal{B}. \tag{3.167}$$

Consider a function $U : \mathcal{B} \times \mathcal{B} \to \mathbb{R}$ which enjoys the following three properties:

1° $U(x, y) \geq V(x, y)$ for all $x, y \in \mathcal{B}$,
2° there are Borel functions

$$A : \mathcal{B} \times \mathcal{B} \to \mathcal{B}^*, \quad B : \mathcal{B} \times \mathcal{B} \to \mathcal{B}^* \quad \text{and} \quad \phi : \mathcal{B} \times \mathcal{B} \times [0, \infty) \to \mathbb{R}$$

with $\phi(x, y, \cdot)$ convex and increasing for any fixed x, y, for which the following condition is satisfied. For any $x, y, h, k \in \mathcal{B}$ we have

$$U(x + h, y + k) \leq U(x, y) + \langle A(x, y), h \rangle + \langle B(x, y), k \rangle \\ + \phi(x, y, |k|) - \phi(x, y, |h|). \tag{3.168}$$

3° $U(0, 0) \leq 0$.

Theorem 3.20. *Suppose that U satisfies 1°, 2° and 3°. Let f, g be $\mathcal{B}$-valued martingales such that g is weakly dominated by f. If f and g satisfy the integrability conditions*

$$\mathbb{E}|V(f_n, g_n)| < \infty, \quad \mathbb{E}|U(f_n, g_n)| < \infty,$$
$$\mathbb{E}\big(|A(f_n, g_n)||df_{n+1}| + |B(f_n, g_n)||dg_{n+1}|\big) < \infty, \tag{3.169}$$
$$\mathbb{E}\phi(f_n, g_n, |df_{n+1}|) < \infty$$

for all $n \geq 0$, then

$$\mathbb{E}V(f_n, g_n) \leq 0 \qquad n = 0, 1, 2, \ldots. \tag{3.170}$$

Proof. First, note that we may assume that f and g start from 0. To see this, we consider, with no loss of generality, the underlying probability space $([0,1], \mathcal{B}([0,1]), |\cdot|)$. Let (f', g') be given by

$$(f'_n, g'_n)(\omega) = (f_n, g_n)(2\omega)1_{\{\omega \leq 1/2\}} + (-f_n, -g_n)(2\omega - 1)1_{\{\omega > 1/2\}}.$$

In other words, we have split the probability space into two halves and copied (f, g) into the left half and its reflection $(-f, -g)$ into the right half. Consider a natural filtration of (f', g') and add the trivial σ-field $\mathcal{F}_{-1} = \{\emptyset, [0,1]\}$. Then $(f_{-1}, g_{-1}) \equiv (0, 0)$ and g' is weakly dominated by f'. Finally, f', g' satisfy (3.169) and we have $\mathbb{E}V(f'_n, g'_n) = \mathbb{E}V(f_n, g_n)$, in view of the assumption (3.167).

Now fix $n \geq 0$ and apply 2°, with $x = f_n$, $y = g_n$, $h = df_{n+1}$ and $k = dg_{n+1}$. We get

$$U(f_{n+1}, g_{n+1}) \leq U(f_n, g_n) + \langle A(f_n, g_n), df_{n+1}\rangle + \langle B(f_n, g_n), dg_{n+1}\rangle$$
$$+ \phi(f_n, g_n, |dg_{n+1}|) - \phi(f_n, g_n, |df_{n+1}|).$$

By (3.169), both sides are integrable, so taking the conditional expectation with respect to $\mathcal{F}_n$ yields $\mathbb{E}[U(f_{n+1}, g_{n+1})|\mathcal{F}_n] \leq U(f_n, g_n)$ thanks to the weak domination. This gives the supermartingale property of the sequence $(U(f_n, g_n))_{n \geq 0}$ and hence, by 1° and 3°,

$$\mathbb{E}V(f_n, g_n) \leq \mathbb{E}U(f_n, g_n) \leq \mathbb{E}U(f_0, g_0) = U(0, 0) \leq 0.$$

This completes the proof. $\qquad\qquad\qquad\qquad\qquad\qquad\qquad\qquad\qquad\qquad\quad\square$

3.10.2   Formulation of the results

Let $\mathcal{H}$ be a separable Hilbert space. We start with the weak-type $(1,1)$ estimate.

Theorem 3.21. *Let f, g be two $\mathcal{H}$-valued martingales such that g is weakly dominated by f. Then*

$$||g||_{1,\infty} \leq 2\sqrt{2}||f||_1. \tag{3.171}$$

The next statement concerns moment estimates.

Theorem 3.22. *Let f, g be two $\mathcal{H}$-valued martingales such that g is weakly dominated by f. Then for $1 < p \leq 2$,*

$$\left\|\sqrt{|f|^2 + |g|^2}\right\|_p \leq \sqrt{2}(p-1)^{-1}||f||_p \tag{3.172}$$

and

$$||g||_p \leq \frac{\sqrt{1 + 2p - p^2}}{p - 1}||f||_p. \tag{3.173}$$

Furthermore, if $p > 2$, then

$$||g||_p \leq 1.48(p - 1)||f||_p. \tag{3.174}$$

Finally, we turn to logarithmic estimates. The result can be stated as follows.

Theorem 3.23. *Let f, g be two $\mathcal{H}$-valued martingales such that g is weakly dominated by f. Then for $K > \sqrt{2}$,*

$$\left\|\sqrt{f^2 + g^2}\right\|_1 \leq \sup_{n \geq 0} K\mathbb{E}|f_n| \log|f_n| + L(K), \tag{3.175}$$

where

$$L(K) = \begin{cases} \frac{K^2}{2(K-\sqrt{2})} \exp(-\sqrt{2}/K) & \text{if } K \in (\sqrt{2}, 2\sqrt{2}), \\ 2\sqrt{2}(4 - 2\sqrt{2})^{2\sqrt{2}/K-1} \exp(-\sqrt{2}/K) & \text{if } K \geq 2\sqrt{2}. \end{cases}$$

3.10.3 Proof of Theorem 3.21

We start with some technical results that are needed in the proof of the weak-type estimate. First, let us investigate the properties of the following class of functions. For $0 \leq r < \sqrt{2} - 1$, let $\phi_r : \mathbb{R}_+ \to \mathbb{R}_+$ be defined by

$$\phi_r(s) := \begin{cases} s^2 & \text{if } s \leq r + \sqrt{2} - 1, \\ (2\sqrt{2} - 2r)s + \beta & \text{if } s > r + \sqrt{2} - 1, \end{cases} \tag{3.176}$$

where $\beta = \beta(r) = (r + \sqrt{2} - 1)^2 - (2\sqrt{2} - 2r)(r + \sqrt{2} - 1)$ is a real number such that ϕ_r is continuous. Let

$$\begin{aligned} D &= \{(x, y) \in \mathcal{H} \times \mathcal{H} : |x| \leq \sqrt{2} - \sqrt{|y|^2 + 1}\} \\ &= \{(x, y) \in \mathcal{H} \times \mathcal{H} : \sqrt{2}|x| + \sqrt{|x|^2 + |y|^2} \leq 1\}. \end{aligned} \tag{3.177}$$

Lemma 3.17. *The function ϕ_r enjoys the following properties:*

(i) *It is convex.*

(ii) *If $(x, y) \in D$, then $\phi_{|x|}(s) \geq s^2$ for $s \leq 1 + |y|$.*

Proof. (i) The function ϕ_r is obviously convex on each of the sets $[0, r + \sqrt{2} - 1]$ and $[r + \sqrt{2} - 1, \infty)$. Moreover,

$$(\phi_r)'_-(r + \sqrt{2} - 1) = 2(r + \sqrt{2} - 1) < 2\sqrt{2} - 2r = (\phi_r)'_+(r + \sqrt{2} - 1).$$

(ii) By (i), for every $0 \leq r < \sqrt{2} - 1$ there exists such a real number $\bar{s} = \bar{s}(r) > r + \sqrt{2} - 1$, that $\phi_r(s) < s^2$ if and only if $s > \bar{s}$. Hence it is enough to show that $1 + |y| < \bar{s}$, or check that

$$\phi_{|x|}(1 + |y|) > (1 + |y|)^2.$$

But $1 + |y| \geq 1 > 2\sqrt{2} - 2 > r + \sqrt{2} - 1$, so the above inequality takes the form

$$(|x| + \sqrt{2} - 1)^2 + (2\sqrt{2} - 2|x|)[1 + |y| - (|x| + \sqrt{2} - 1)] \geq (1 + |y|)^2,$$

which is equivalent to

$$(\sqrt{2} - 3|x| - |y|)(2\sqrt{2} - |x| + |y|) \geq 0.$$

Since $2 - \sqrt{2} - |x| + |y| \geq 3 - 2\sqrt{2} + |y| > 0$, it suffices to prove that $3|x| + |y| \leq \sqrt{2}$. But $(x, y) \in D$, hence $3|x| + |y| \leq |y| + 3\sqrt{2} - 3\sqrt{|y|^2 + 1}$. The right-hand side does not exceed $\sqrt{2}$, because, from the Schwarz inequality,

$$|y| + 2\sqrt{2} \leq \sqrt{|y|^2 + 1}\sqrt{1 + 8} = 3\sqrt{|y|^2 + 1}.$$

The proof is complete.   □

Further properties of the functions ϕ_r are contained in the following.

Lemma 3.18. *Assume that* $(x, y) \in D$. *Then*

(i) *For each* $k \in \mathcal{H}$,

$$|y|^2 + 2(y \cdot k) + \phi_{|x|}(|k|) \geq 0. \tag{3.178}$$

(ii) *If* $h \in \mathcal{H}$ *and* $|x + h| \geq \sqrt{2} - 1$, *then*

$$2\sqrt{2}|x + h| - 1 \geq |x + h|^2 + \phi_{|x|}(|h|) - |h|^2. \tag{3.179}$$

(iii) *If* $h, k \in \mathcal{H}$, $|x + h| \leq \sqrt{2} - 1$ *and* $(x + h, y + k) \notin D$, *then*

$$-1 \geq |x + h|^2 - 2\sqrt{2}|x + h| - |y + k|^2 - \phi_{|x|}(|k|) + |k|^2. \tag{3.180}$$

Proof. (i) If $|k| \leq 1 + |y|$, then using Lemma 3.17 (ii) we have $\phi_{|x|}(k) \geq |k|^2$ and the inequality holds. Suppose then that $|k| > 1 + |y|$. We have $|y| + 1 > |x| + \sqrt{2} - 1$ (because $(x, y) \in D$), hence

$$|y|^2 + 2(y \cdot k) + \phi_{|x|}(|k|) \geq |y|^2 - 2|y||k| + (2\sqrt{2} - 2|x|)|k| + \beta(|x|)$$
$$= 2|k|(\sqrt{2} - |x| - |y|) + |y|^2 + \beta(|x|).$$

We have $\sqrt{2} - |x| - |y| \geq 0$, because for $(x, y) \in D$ we have $|y| \leq 1$ and $|x| \leq \sqrt{2} - 1$. The right-hand side is nonnegative for $|k| = 1 + |y|$, so it remains nonnegative for bigger $|k|$.

(ii) Introduce the function $\theta : \mathbb{R} \to \mathbb{R}$ given by $\theta(t) = t^2 - 2\sqrt{2}t + 1$. We must prove that

$$\theta(|x + h|) \leq |h|^2 - \phi_{|x|}(|h|). \tag{3.181}$$

If $|h| \leq |x| + \sqrt{2} - 1$, then $\phi_{|x|}(|h|) = |h|^2$ and

$$\sqrt{2} - 1 \leq |x + h| \leq |x| + |h| \leq |x| + |x| + \sqrt{2} - 1 \leq 3\sqrt{2} - 3 < \sqrt{2} + 1,$$

which yields the desired inequality, because the function θ has two zeros $\sqrt{2} \pm 1$. Suppose then, that $|h| \geq |x| + \sqrt{2} - 1$. Fix x, $|h|$ and let us try to maximize the value $\theta(|x + h|)$. The function θ has minimum at $\sqrt{2}$, hence, because of

$$|h| - |x| \leq |x + h| \leq |x| + |h|,$$

the biggest value of $\theta(|x+h|)$ is equal to $\theta(|h|-|x|)$ or $\theta(|x|+|h|)$. Let us check (3.181) for both $|x+h|=|x|+|h|$ and $|x+h|=|h|-|x|$.

In the first case the inequality (3.179) is equivalent to

$$|x|^2 + 2|x||h| + |h|^2 - 2\sqrt{2}(|x|+|h|) + 1$$
$$\leq |h|^2 - (2\sqrt{2}-2|x|)|h| - (|x|+\sqrt{2}-1)^2 + (2\sqrt{2}-2|x|)(|x|+\sqrt{2}-1),$$

which does not depend on h; all the terms with $|h|$ vanish. On the other hand it is true for $|h|=|x|+\sqrt{2}-1$ (from the preceding case). Hence it holds for any h.

If $|x+h|=|h|-|x|$, then (3.179) takes the form

$$|x|^2 - 2|x||h| + |h|^2 - 2\sqrt{2}(|h|-|x|) + 1$$
$$\leq |h|^2 - (2\sqrt{2}-2|x|)|h| - (|x|+\sqrt{2}-1)^2 + (2\sqrt{2}-2|x|)(|x|+\sqrt{2}-1),$$

or

$$4|x||h| \geq |x|^2 + 2\sqrt{2}|x| + 1 + (|x|+\sqrt{2}-1)^2 - (2\sqrt{2}-2|x|)(|x|+\sqrt{2}-1).$$

It suffices to apply the preceding argument: this inequality holds for $|h|=|x|+\sqrt{2}-1$, so also for h with a bigger norm.

(iii) We have $0 \leq |x+h| \leq \sqrt{2}-1$, hence $\theta(|x+h|) \geq 0$, so there exists a vector $\bar{k} \in \mathcal{H}$ such that $\theta(|x+h|) = |y+\bar{k}|^2$, or, equivalently,

$$-1 = |x+h|^2 - 2\sqrt{2}|x+h| - |y+\bar{k}|^2.$$

This implies $(y+\bar{k}, y+k) \in \partial D$. But $(x+h, y+k) \notin D$, therefore we have $|y+k| > |y+\bar{k}|$. We want to show that

$$|y+k|^2 - |y+\bar{k}|^2 \geq |k|^2 - \phi_{|x|}(|k|). \tag{3.182}$$

If $|k| \leq 1+|y|$, then $\phi_{|x|}(|k|) \geq |k|^2$ (by Lemma 3.17 (ii)) and we are done. If $|k| > 1+|y|$, then $|k| > |x|+\sqrt{2}-1$. Let $k' = k \cdot \frac{1+|y|}{|k|}$. The inequality (3.182) holds for k' (because $|k'|=1+|y|$), so we may write

$$\phi_{|x|}(|k|) - |y+\bar{k}|^2 + |y+k|^2 - |k|^2$$
$$= \phi_{|x|}(|k'|) - |y+\bar{k}|^2 + |y+k'|^2 - |k'|^2 + \phi_{|x|}(|k|) - \phi_{|x|}(|k'|) + 2(y \cdot (k-k'))$$
$$\geq (2\sqrt{2}-2|x|)(|k|-|k'|) - 2|y|(|k|-|k'|)$$
$$= 2(\sqrt{2}-|x|-|y|)(|k|-1-|y|) \geq 0.$$

The proof is complete. $\qquad\square$

Let $\psi : [0,\infty) \to \mathbb{R}$ be given by

$$\psi(s) = \begin{cases} s^2 & \text{if } s \leq 1, \\ 2s-1 & \text{if } s > 1. \end{cases} \tag{3.183}$$

Lemma 3.19. *For any $r \in [0, \sqrt{2} - 1)$ and $s \geq 0$ we have*

$$\psi(s) \leq \phi_r(s) \leq \psi(\sqrt{2}s). \tag{3.184}$$

Proof. We have $\psi = \phi_r$ on $[0, r + \sqrt{2} - 1]$, while for $s > r + \sqrt{2} - 1$,

$$\phi_r'(s) = 2\sqrt{2} - 2r \geq 2 \geq \psi'(s).$$

This proves the left inequality. Let us turn to the right one. Clearly, it holds for $s \leq \min(1/\sqrt{2}, r + \sqrt{2} - 1)$. Furthermore, since the slope of the linear part of ϕ_r does not exceed $2\sqrt{2}$, it suffices to prove the bound for s lying between $1/\sqrt{2}$ and $r + \sqrt{2} - 1$. $\qquad \square$

Proof of (3.171). Let $V_{1,\infty}, U_{1,\infty} : \mathcal{H} \times \mathcal{H} \to \mathbb{R}$ be given by $V_{1,\infty}(x, y) = 1_{\{|y| \geq 1\}} - 2\sqrt{2}|x|$ and

$$U_{1,\infty}(x, y) = \begin{cases} |y|^2 - |x|^2 & \text{if } (x, y) \in D, \\ 1 - 2\sqrt{2}|x| & \text{if } (x, y) \notin D, \end{cases}$$

where D is given by (3.177). We need to verify the conditions 1°, 2°, 3° and the integrability assumption (3.169). To check the majorization $U_{1,\infty}(x, y) \geq V_{1,\infty}(x, y)$, note that it suffices to deal with $(x, y) \in D$. We have $|x| \leq \sqrt{2} - 1$ and

$$|y|^2 - |x|^2 \geq -|x|^2 \geq -(\sqrt{2} - 1)|x| \geq -2\sqrt{2}|x| = 1_{\{|y| > 1\}} - 2\sqrt{2}|x|.$$

We turn to 2°. We shall establish this condition with

$$(A(x, y), B(x, y)) = \begin{cases} (-2x, 2y) & \text{if } (x, y) \in D, \\ (-2\sqrt{2}x', 0) & \text{if } (x, y) \notin D, \end{cases}$$

and

$$\phi(x, y, t) = \begin{cases} \phi_{|x|}(t) & \text{if } (x, y) \in D, \\ 0 & \text{if } (x, y) \notin D. \end{cases}$$

To do this, we start with the observation that $U_{1,\infty}(x, y) \geq 1 - 2\sqrt{x}$ for all x, y. Indeed, we have equality for $(x, y) \notin D$, while for $(x, y) \in D$, by the definition of D, $\sqrt{|y|^2 + 1} \leq \sqrt{2} - |x|$. Squaring both sides of this inequality gives

$$|y|^2 - |x|^2 \geq 1 - 2\sqrt{2}|x|,$$

as claimed. Therefore, if $(x, y) \notin D$, then

$$U_{1,\infty}(x, y) - 2(x' \cdot h) = 1 - 2\sqrt{2}|x| - 2\sqrt{2}(x' \cdot h) \geq 1 - 2\sqrt{2}|x + h|$$
$$\geq U_{1,\infty}(x + h, y + k).$$

Suppose then that (x, y) belongs to D. If $(x+h, y+k) \in D$, then $|h| \leq |x+h| + |x| \leq |x| + \sqrt{2} - 1$ and $|k| \leq |y+k| + |y| \leq 1 + |y|$, so, by Lemma 3.17 (ii), $\phi(|h|) = |h|^2$, $\phi(|k|) \geq |k|^2$ and the inequality (3.168) is true, since

$$|y + k|^2 - |x + h|^2 = |y|^2 - |x|^2 - 2(x \cdot h) + 2(y \cdot k) + |k|^2 - |h|^2.$$

Assume then, that $(x + h, y + k) \notin D$. Then (3.168) takes the form

$$1 - 2\sqrt{2}|x + h| \geq |y|^2 + 2(y \cdot k) + \phi_{|x|}(|k|) - |x|^2 - 2(x \cdot h) - \phi_{|x|}(|h|). \quad (3.185)$$

If $|x + h| \geq \sqrt{2} - 1$, then the above inequality follows immediately from (3.178) and (3.179). If $|x + h| < \sqrt{2} - 1$, then $|h| < |x| + \sqrt{2} - 1$, $\phi_{|x|}(|h|) = |h|^2$ and (3.185) is equivalent to (3.180). This completes the proof of $2°$.

The initial condition $3°$ is obvious. Finally, (3.169) is also very easy, since A and B are bounded, and there is an absolute c such that $\Phi_r(s) \leq c(s + 1)$ for all permissible r and all $s \geq 0$. The inequality (3.171) follows. $\qquad\square$

3.10.4 Proof of Theorem 3.22

First let us show the following auxiliary fact.

Lemma 3.20. *Let* $1 < p < 2$.

(i) *For any nonnegative numbers* a, b *and positive* x_0,

$$(a + b)^{p/2} \geq (1 + x_0)^{p/2-1}a^{p/2} + (x_0^{-1} + 1)^{p/2-1}b^{p/2}. \quad (3.186)$$

(ii) *We have*

$$\min_{x_0 > 0} \left\{ 2^{p/2}(p - 1)^{-p}(1 + x_0)^{1-p/2} - x_0^{1-p/2} \right\} = \left(\frac{\sqrt{1 + 2p - p^2}}{p - 1} \right)^p. \quad (3.187)$$

Proof. (i) After the substitution $x = b/a$, (3.186) takes the form

$$(1 + x)^{p/2} - (1 + x_0)^{p/2-1} - (x_0^{-1} + 1)^{p/2-1}x^{p/2} \geq 0,$$

and the left-hand side, as a function of x, attains its minimum 0 at $x = x_0$ (straightforward analysis of the derivative).

(ii) Differentiating the expression in the parentheses, one gets

$$\left(1 - \frac{p}{2}\right) x_0^{-p/2} \left[\left(\frac{2x_0}{(p - 1)^2(1 + x_0)} \right)^{p/2} - 1 \right],$$

so we see that the minimum is attained at $x_0 = (p-1)^2/(1+2p-p^2)$. Substituting this value into the expression, we get the right-hand side of (3.187). $\qquad\square$

We turn to the proof of the moment inequality for $1 < p < 2$. Clearly, we may and do assume that $\|f\|_p < \infty$, which implies $df_n \in L^p$ for any n. Thus, by the weak domination, $dg_n \in L^p$ and, by the triangle inequality, $g_n \in L^p$. Furthermore, we may assume that f and g start from 0. Let $V_p : \mathcal{H} \times \mathcal{H} \to \mathbb{R}$ be given by

$$V_p(x, y) = (|x|^2 + |y|^2)^{p/2} - (\sqrt{2}(p - 1)^{-1}|x|)^p.$$

The corresponding special function is obtained by the integration argument. Let $u = U_{1,\infty}$ and

$$k(t) = \frac{p^{3-p}(p-1)(2-p)}{2} t^{p-1}.$$

It is easy to derive that

$$U_p(x,y) = \int_0^\infty t^{p-1} u(x/t, y/t)dt$$
$$= p^{2-p}(\sqrt{|x|^2 + |y|^2} - \sqrt{2}(p-1)^{-1}|x|)(\sqrt{2}|x| + \sqrt{|x|^2 + |y|^2})^{p-1}.$$

We easily verify the majorization $U_p \geq V_p$. Moreover, since $|u(x,y)| \leq c(1 + |x|^2 + |y|^2)$ for all $x, y \in \mathcal{H}$ and some universal c, we get that

$$\int_0^\infty k(t)|u(x/t, y/t)|dt < \tilde{c}(1 + |x|^p + |y|^p)$$

and hence

$$\mathbb{E}U_p(f_n, g_n) \leq \mathbb{E}U_p(f_0, g_0) = U_p(0,0) = 0.$$

This yields the inequality (3.172). To establish (3.173), we use Lemma 3.20 (i) with $a = |g_n|^2$, $b = |f_n|^2$ and $x_0 > 0$ to obtain

$$(1 + x_0)^{p/2-1}\mathbb{E}|g_n|^p + (x_0^{-1} + 1)^{p/2-1}\mathbb{E}|f_n|^p \leq \frac{2^{p/2}}{(p-1)^p}\mathbb{E}|f_n|^p,$$

or

$$\mathbb{E}|g_n|^p \leq \left[2^{p/2}(p-1)^{-p}(1 + x_0)^{1-p/2} - x_0^{1-p/2}\right]\mathbb{E}|f_n|^p.$$

An application of (3.187) completes the proof of (3.173).

Finally, a few words about (3.174). This inequality is established in a similar manner. Namely, one constructs a dual function to $U_{1,\infty}$ and uses the integration argument to obtain the corresponding special function U_p. For details, see [124].

3.10.5   Proof of Theorem 3.23

Let $V : \mathcal{H} \times \mathcal{H} \to \mathbb{R}$ be given by $V(x,y) = \sqrt{|x|^2 + |y|^2} - K|x| \log|x| - L(K)$. The special function $U : \mathcal{H} \times \mathcal{H} \to \mathbb{R}$ is given by the integration argument: put $u = U_{1,\infty}$, $k(t) = 1_{[1,\infty)}(t)$ and

$$U(x,y) = \int_0^\infty k(t)u(x/t, y/t)dt$$
$$= \begin{cases} |y|^2 - |x|^2 & \text{if } (x,y) \in D, \\ 2\sqrt{2}|x| \log\left[\sqrt{2}|x| + \sqrt{|x|^2 + |y|^2}\right] - 2\sqrt{|x|^2 + |y|^2} + 1 & \text{if } (x,y) \notin D. \end{cases}$$

Let us introduce the auxiliary constant $t(K) = L(K) \cdot \max\{2\sqrt{2}/K, 1\}$.

We have the following majorization.

Lemma 3.21. *For $K > \sqrt{2}$ and any $x, y \in \mathcal{H}$,*

$$L(K)U(x/t(K), y/t(K)) \geq V(x, y). \tag{3.188}$$

Proof. Substitute $r = |y|$, $s = \sqrt{|x|^2 + |y|^2}$. We will only consider the case $K \in (\sqrt{2}, 2\sqrt{2})$, the argumentation for the remaining values of K is similar. First assume that $(x/t(K), y/t(K)) \notin D$. Then, after some manipulations, the inequality (3.188) takes the form

$$0 \leq -2\sqrt{2}r \log\left(\frac{\sqrt{2}r + s}{t(K)}\right) + 2s + t(K)L(K)^{-1}(Kr\log r - s).$$

Now fix r and consider the right-hand side as a function of $s \in (0, \infty)$. This function is convex and vanishes, along with its derivative, at $s = 2r/(K - \sqrt{2})$. This yields the majorization outside D. Now suppose that $(x/t(K), y/t(K)) \in D$ and substitute $r = |y|$, $s = \sqrt{|x|^2 + |y|^2}$. Then $\sqrt{2}r + s \leq t(K)$ and (3.188) takes the form

$$0 \leq -2r^2 + s^2 + \frac{8L(K)}{K^2}[Kr\log r - s + L(K)]. \tag{3.189}$$

For fixed r, the right-hand side, as a function of s, attains its minimum for $s_0 = \min\{4L(K)/K^2, t(K) - \sqrt{2}r\}$. If $t(K) - \sqrt{2}r \geq 4L(K)/K^2$, or, equivalently, $r \leq \exp(-\sqrt{2}/K)$, then, since $r\log r > 1/e$,

$$F(r) = -2r^2 + s_0^2 + \frac{8L(K)}{K^2}[Kr\log r - s_0 + L(K)]$$

$$= -2r^2 + \frac{8L(K)}{K}r\log r + \frac{4L(K)}{K^2}\exp(-\sqrt{2}/K)(K + \sqrt{2}).$$

We have $F(\exp(-\sqrt{2}/K)) = F'(\exp(-\sqrt{2}/K)-) = 0$ and

$$F''(r) = -4 + \frac{8L(K)}{Kr} > \frac{4K}{K - \sqrt{2}} - 4 > 0.$$

Thus F is nonnegative for $r \leq \exp(-\sqrt{2}/K)$. If $r > \exp(-\sqrt{2}/K)$, then $s_0 = t(K) - \sqrt{2}r$ and $(x/t(K), y/t(K)) \in \partial D$, so (3.189) holds by the continuity of U and the case $(x/t(K), y/t(K)) \notin D$, which we have already considered. $\square$

Proof of inequality (3.175). With no loss of generality, we may restrict ourselves to f satisfying $\mathbb{E}|f_n|\log|f_n| < \infty$ for all n. There are universal constants α, β such that for $x, y \in \mathbb{R}$,

$$|x + y|\log^+|x+y| \leq (|x|+|y|)\log^+(|x|+|y|) \leq \alpha(|x|\log|x|+|y|\log|y|)+\beta. \tag{3.190}$$

In consequence, for any n the random variable $|df_n|\log^+|df_n|$ is integrable and, by the weak domination, so is $|dg_n|\log^+|dg_n|$. The further application of (3.190) gives that $|g_n| \log |g_n| \in L^1$ and, finally, $(|f_n| + |g_n|) \log(|f_n| + |g_n|) \in L^1$. Now, since

$$|u(x,y)| \le (|x|^2 + |y|^2) 1_D(x,y) + (1 + 2\sqrt{2}|y|) 1_{D^c}(x,y),$$

a little calculation shows that

$$\int_1^\infty |u(x/t, y/t)| \mathrm{d}t \le a(|x| + |y|) \log(|x| + |y|) + b,$$

for some universal constants a and b. Therefore we are allowed to use Fubini's theorem and obtain

$$\mathbb{E}U(f_n, g_n) \le \mathbb{E}U(f_0, g_0) = U(0,0) = 0$$

for all n. Apply this to the pair $(f/t(K), g/t(K))$ and combine it with (3.188) to get

$$\mathbb{E}\sqrt{|f_n|^2 + |g_n|^2} \le K\mathbb{E}|f_n| \log |f_n| + L(K).$$

It suffices to take supremum over n to complete the proof.   □

3.11   UMD spaces

3.11.1   The inequalities under differential subordination

There is a natural question whether the estimates studied in the previous sections carry over to the case when the martingales f and g take values in a certain separable Banach space $\mathcal{B}$. As one might expect, this is not true for all Banach spaces. So, there is another problem: provide a geometric characterization of the spaces in which the estimates do hold.

The first result in this direction concerns the inequalities for differentially subordinated martingales. It turns out the only "good spaces" are those isomorphic to Hilbert spaces.

Theorem 3.24. *Let $\mathcal{B}$ be a Banach space.*

(i) *Suppose that there is $1 < p < \infty$ and a universal $\beta = \beta(p, \mathcal{B}) < \infty$ such that*

$$||g||_p \le \beta ||f||_p$$

for any $\mathcal{B}$-valued martingales f, g such that g is differentially subordinate to f. Then $\mathcal{B}$ is isomorphic to a Hilbert space.

(ii) *Suppose that there is a universal $\beta = \beta(\mathcal{B}) < \infty$ such that*

$$||g||_{1,\infty} \le \beta ||f||_1$$

for any $\mathcal{B}$-valued martingales f, g such that g is differentially subordinate to f. Then $\mathcal{B}$ is isomorphic to a Hilbert space.

Proof. (i) Let r_0, r_1, r_2, ... be independent Rademacher variables and take a sequence $(a_k)_{k \geq 0}$ with $a_k \in \mathcal{B}$ for all k. Furthermore, take $b \in \mathcal{B}$ of norm 1 and consider the martingales

$$f_n = \sum_{k=0}^{n} a_k r_k, \quad g_n = \sum_{k=0}^{n} |a_k| r_k b, \qquad n = 0, 1, 2, \ldots.$$

Then g is differentially subordinate to f and f is differentially subordinate to g. Thus, by the Khintchine–Kahane inequality we have, for any n,

$$\left\| \sum_{k=0}^{n} a_k r_k \right\|_p \simeq \left\| \sum_{k=0}^{n} |a_k| r_k b \right\|_p = \left\| \sum_{k=0}^{n} |a_k| r_k \right\|_p \simeq \left\| \sum_{k=0}^{n} |a_k| r_k \right\|_2 = \left(\sum_{k=0}^{n} |a_k|^2 \right)^{1/2}$$

and by the result of Kwapień [110], $\mathcal{B}$ is isomorphic to a Hilbert space.

(ii) Recall that a martingale f is conditionally symmetric if for any $n \geq 1$ the conditional distributions of df_n and $-df_n$ with respect to $\mathcal{F}_{n-1}$ coincide. Using the extrapolation method of Burkholder and Gundy, we see that the finiteness of β implies that for any $1 < p < \infty$ there is β_p such that

$$\|g\|_p \leq \beta_p \|f\|_p$$

whenever f, g are conditionally symmetric martingales such that g is differentially subordinate to f: see the proof of (3.203) below. Thus the claim follows from the examples exhibited in the proof of the first part. $\qquad\square$

3.11.2 UMD spaces and ζ-convexity

By the results of the previous subsection, in order to study moment inequalities for Banach-space-valued martingales, we need to impose a stronger domination on the processes. A natural choice is to consider ± 1-transforms. Let $\mathcal{B}$ be a given separable Banach space and fix $1 < p < \infty$. Let $\beta_p(\mathcal{B})$ be the least extended real number β such that if f is a $\mathcal{B}$-valued martingale and g is its ± 1-transform, then

$$\|g\|_p \leq \beta \|f\|_p. \tag{3.191}$$

Definition 3.2. We say that $\mathcal{B}$ is a UMD space (Unconditional for Martingale Differences) if there is $1 < p < \infty$ such that $\beta_p(\mathcal{B}) < \infty$.

It will be obvious from the geometrical characterization of UMD spaces given below that if $\beta_p(\mathcal{B})$ is finite for some $1 < p < \infty$, then it is finite for all $1 < p < \infty$. It follows from the results of Section 3.3 above that if $\mathcal{H}$ is a Hilbert space, then $\beta_p(\mathcal{H}) = p^* - 1$ for all $1 < p < \infty$. In consequence,

$$\beta_p(\mathcal{B}) \geq \beta_p(\mathbb{R}) = p^* - 1$$

for all Banach spaces $\mathcal{B}$. Moreover, the scalar case implies also that ℓ^p is a UMD space for $p \in (1, \infty)$, because $\beta_p(\ell^p) = p^* - 1$ (simply by integration term-by-term). More generally, we have $\beta_p(\ell^p_{\mathcal{B}}) = \beta_p(\mathcal{B})$, so if $\mathcal{B}$ is a UMD space, then so is $\ell^p_{\mathcal{B}}$ for all $1 < p < \infty$; a similar fact is valid for $L^p_{\mathcal{B}}(0,1)$. On the other hand, the spaces ℓ^1, ℓ^∞, $L^1(0,1)$ and $L^\infty(0,1)$ are not UMD, since UMD spaces are reflexive, even superreflexive (for some references on this subject, see the bibliographical notes at the end of the chapter).

Let us turn to a geometrical characterization of UMD spaces. To formulate it, we need the following auxiliary notion. A function $\zeta : \mathcal{B} \times \mathcal{B} \to \mathbb{R}$ is said to be *biconvex* if for any x, $y \in \mathcal{B}$ both $\zeta(x, \cdot)$ and $\zeta(\cdot, y)$ are convex on $\mathcal{B}$. Furthermore, ζ is biconcave if $-\zeta$ is biconvex.

Definition 3.3. We say that a Banach space $\mathcal{B}$ is ζ-*convex* if there is a biconvex function $\zeta : \mathcal{B} \times \mathcal{B} \to \mathbb{R}$ such that

$$\zeta(x, y) \le |x + y| \qquad \text{if } |x| = |y| = 1, \tag{3.192}$$

and

$$\zeta(0, 0) > 0. \tag{3.193}$$

One of the main results of this section is the following classical result obtained by Burkholder.

Theorem 3.25. *A Banach space $\mathcal{B}$ is UMD if and only if it is ζ-convex.*

Remark 3.7. This result is very much in the spirit of Burkholder's method: a certain martingale inequality is valid if and only if there is a special function with some convex-type properties, satisfying the majorization and the initial condition (the latter two properties are (3.192) and (3.193), respectively). A very interesting feature is that we establish a number of inequalities ((3.191) for $1 < p < \infty$) with the use of a single special function. However, we have already seen such a phenomenon before: the proof of Burkholder's L^p estimates (3.59), $1 < p < 2$, was based on the properties of U_1, given by (3.6), and the integration argument. Here the situation is analogous: as we shall see, the function ζ is closely related to a weak-type $(1,1)$ inequality; the novelty comes from the fact that the integration argument does not seem to work and we use the extrapolation method of Burkholder and Gundy [40].

To gain some understanding about ζ-convexity, let us start with the case $\mathcal{B} = \mathbb{R}$. The function $\zeta = \zeta_{\mathbb{R}}$ must satisfy $\zeta_{\mathbb{R}}(0,0) > 0$ and the condition (3.192) means that

$$\zeta_{\mathbb{R}}(1,1) \le 2, \quad \zeta_{\mathbb{R}}(1,-1) \le 0, \quad \zeta_{\mathbb{R}}(-1,1) \le 0 \quad \text{and} \quad \zeta_{\mathbb{R}}(-1,-1) \le 2.$$

A little experimentation leads to the function $\zeta(x, y) = 1 + xy$. If $\mathcal{B}$ is a Hilbert space, one generalizes the above formula to $\zeta_{\mathcal{H}}(x, y) = 1 + x \cdot y$ (and $\zeta_{\mathcal{H}}(x, y) = 1 + \text{Re}(x \cdot y)$ in the complex case). Note that in both cases, the value at $(0, 0)$ is equal to 1. In general, we have $\zeta(0,0) \leq 1$: to see this, fix x of norm one and use the biconvexity together with (3.192) to obtain

$$\zeta(0,0) \leq \frac{\zeta(x,x) + \zeta(x,-x) + \zeta(-x,x) + \zeta(-x,-x)}{4} \leq 1.$$

In fact, we shall show that if $\mathcal{B}$ is not a Hilbert space, then the strict inequality $\zeta(0,0) < 1$ holds. In other words, a Banach space is a Hilbert space if and only if there is a biconvex ζ satisfying (3.192) such that $\zeta(0,0) = 1$: see Theorem 3.28 below.

Next, we show that the notion of ζ-convexity can be slightly relaxed.

Lemma 3.22. *Suppose ζ is a biconvex function on $\{(x,y) \in \mathcal{B} \times \mathcal{B} : |x| \vee |y| \leq 1\}$, satisfying (3.192). Then there is a biconvex function u on the whole $\mathcal{B} \times \mathcal{B}$ satisfying*

$$u(x,y) \leq |x+y| \qquad \text{if } |x| \vee |y| \geq 1, \tag{3.194}$$

$$u(x,y) = u(y,x) = u(-x,-y) \qquad \text{for all} \quad x, y \in \mathcal{B} \tag{3.195}$$

and

$$\zeta(0,0) \leq u(0,0). \tag{3.196}$$

Proof. We start with the observation that, by (3.192),

$$\zeta(x,y) \leq |x+y| \qquad \text{if } |x| \vee |y| = 1. \tag{3.197}$$

To see this, assume with no loss of generality that $|x| < |y| = 1$ and let $\alpha \in (0,1)$ be such that $|z| = 1$, where $z = \alpha^{-1}(x+y) - y$ (by Darboux property, such α exists). Then we have $x = \alpha z + (1-\alpha)(-y)$ and, by (3.192),

$$\zeta(x,y) \leq \alpha\zeta(z,y) + (1-\alpha)\zeta(-y,y) \leq \alpha|z+y| = |x+y|.$$

Define

$$u(x,y) = \begin{cases} \zeta(x,y) \vee |x+y| & \text{if } |x| \vee |y| < 1, \\ |x+y| & \text{if } |x| \vee |y| \geq 1. \end{cases}$$

This function is biconvex on $\mathcal{B} \times \mathcal{B}$. Indeed, if $|y| \geq 1$, then $u(x,y) = |x+y|$, so $u(\cdot,y)$ is convex on $\mathcal{B}$. If $|y| < 1$, then $u(\cdot,y)$ is locally convex on the complement of the unit sphere of $\mathcal{B}$. If $|y| < 1 = |x|$ and $\alpha_1 x_1 + \alpha_2 x_2 = x$, where $\alpha_1, \alpha_2 \in (0,1)$ satisfy $\alpha_1 + \alpha_2 = 1$, then

$$u(x,y) = |x+y| \leq \alpha_1 |x_1+y| + \alpha_2 |x_2+y| \leq \alpha_1 \zeta(x_1,y) + \alpha_2 \zeta(x_2,y),$$

as needed. Clearly, u satisfies (3.194) and (3.196), directly from its definition. Finally, to guarantee (3.195), replace $u(x,y)$ by $u(x,y) \vee u(y,x) \vee u(-x,-y) \vee u(-y,-x)$, if necessary. $\qquad \square$

We shall also need the following "regularization" lemma.

Lemma 3.23. *Suppose that ζ is a biconvex function, defined on the set $\{(x,y) \in \mathcal{B} \times \mathcal{B} : |x| \vee |y| \leq 1\}$, such that the condition (3.192) holds. Then there is a biconvex function $u : \mathcal{B} \times \mathcal{B} \to \mathbb{R}$ satisfying (3.194), (3.195), the property*

$$u(x, -x) \leq u(0,0) \qquad \text{for all } x \in \mathcal{B} \tag{3.198}$$

and the further condition

$$u(0,0) \geq \frac{\zeta(0,0)}{1+r}, \tag{3.199}$$

where $r \in [0,1]$ is any number for which

$$\sup_{|x| \leq 1} \zeta(x, -x) \leq \sup_{|x| \leq r} \zeta(x, -x).$$

The reason for considering this result is that the biconvex function u coming from Lemma 3.22 need not satisfy the condition (3.198) (which is crucial in our further considerations). We have decided to put this into a separate lemma, since for many classical Banach spaces $\mathcal{B}$, the function $u_{\mathcal{B}}$, the greatest biconvex function satisfying (3.194), does have this property. For the proof of this fact and of Lemma 3.23, consult Burkholder's paper [26].

We come back to Theorem 3.25. In the proof we shall need the following special class of $\mathcal{B} \times \mathcal{B}$-valued martingales. We say that $Z = (Z_n)_{n\geq0} = (X_n, Y_n)_{n\geq0}$ is a *zigzag* martingale (or has the *zigzag* property) if for any $n \geq 1$, either $X_n - X_{n-1} \equiv 0$ or $Y_n - Y_{n-1} \equiv 0$. For example, if f is a $\mathcal{B}$-valued martingale and g is its ±1-transform, then $(f + g, f - g)$ is a zigzag martingale; and conversely, if $Z = (X, Y)$ has the zigzag property, then $g = (X - Y)/2$ is a ±1-transform of $f = (X + Y)/2$. The conditional Jensen inequality implies that if u is a biconcave function on $\mathcal{B} \times \mathcal{B}$ and Z is a simple zigzag martingale, then

$$\mathbb{E}u(Z_n) \leq \mathbb{E}u(Z_{n-1}) \leq \cdots \leq \mathbb{E}u(Z_0). \tag{3.200}$$

In fact, a stronger statement holds true: the process $(u(Z_n))_{n\geq0}$ is a supermartingale. Indeed, for any $n \geq 1$ we have that $Z_n = (X_n, Y_{n-1})$ or $Z_n = (X_{n-1}, Y_n)$. If, say, the first possibility takes place, then

$$\mathbb{E}\big[u(Z_n)|\mathcal{F}_{n-1}\big] = \mathbb{E}\big[u(X_{n-1} + dX_n, Y_{n-1})|\mathcal{F}_{n-1}\big] \leq u(X_{n-1}, Y_{n-1}) = u(Z_{n-1}).$$

In the next step of our study we establish the weak type $(1,1)$ inequality for ±1-transforms.

Theorem 3.26. *Let $u : \mathcal{B} \times \mathcal{B} \to \mathbb{R}$ be a biconvex function satisfying $u(0,0) > 0$, (3.194) and (3.198). Then*

$$\|g\|_{1,\infty} \leq \frac{2}{u(0,0)}\|f\|_1. \tag{3.201}$$

Proof. It suffices to show that for any simple martingale f and its ± 1-transform g we have

$$\mathbb{P}(|g_n| \geq 1) \leq \frac{2}{u(0,0)} ||f||_1.$$

Here Burkholder's method comes into play. Essentially, up to some manipulations, u is the special function corresponding to the above weak type estimate. Let $U, V : \mathcal{B} \times \mathcal{B} \to \mathbb{R}$ be given by

$$U(x,y) = u(0,0) - u(x,y) \quad \text{and} \quad V(x,y) = u(0,0)1_{\{|x|\vee|y|\geq 1\}} - |x+y|.$$

Then U majorizes V. This is evident for $|x| \vee |y| \geq 1$, while for remaining x, y, it follows from (3.198) and the biconvexity of u:

$$\begin{aligned}
u(0,0) - u(x,y) &\geq u(x,-x) - u(x,y) \\
&\geq \lim_{t\to\infty} t^{-1}\big[u(x,-x) - u(x,-x+t(x+y))\big] \\
&\geq |x+y|.
\end{aligned}$$

Now, since U is biconcave and $Z = (X,Y) = (f+g, f-g)$ is zigzag, we may apply (3.200) and obtain

$$\begin{aligned}
u(0,0)\mathbb{P}(|g_n| \geq 1) &= u(0,0)\mathbb{P}(|X_n - Y_n| \geq 2) \\
&\leq u(0,0)\mathbb{P}(|X_n| \vee |Y_n| \geq 1) \\
&\leq \mathbb{E}V(Z_n) + \mathbb{E}|X_n + Y_n| \leq \mathbb{E}U(Z_n) + 2||f_n||_1 \\
&\leq \mathbb{E}U(Z_0) + 2||f_n||_1 \leq 2||f_n||_1.
\end{aligned}$$

Here in the last passage we have used the fact that $X_0 \equiv 0$ or $Y_0 \equiv 0$, and $U(x,0) \leq 0$, $U(0,y) \leq 0$ for all x, $y \in \mathcal{B}$. For example,

$$U(x,0) = \frac{U(x,0) + U(-x,0)}{2} \leq U(0,0) = 0,$$

with a similar argument for $U(0,y)$. $\square$

It seems appropriate to make here the following important observation. Recall that $u_\mathcal{B}$ denotes the greatest biconvex function u on $\mathcal{B} \times \mathcal{B}$ satisfying (3.194). It turns out that $u_\mathcal{B}$ has the following nice probabilistic interpretation.

Lemma 3.24. *For any x, $y \in \mathcal{B}$,*

$$u_\mathcal{B}(x,y) = \inf\{\mathbb{E}|X_\infty + Y_\infty|\}. \tag{3.202}$$

Here the infimum is taken over $\overline{Z}(x,y)$, which consists of all simple zigzag martingales $Z = (X,Y)$ starting from (x,y) and satisfying $|X_\infty - Y_\infty| \geq 2$.

Proof. Denote the right-hand side of (3.202) by $v(x, y)$. Let u be a biconvex function on $\mathcal{B} \times \mathcal{B}$ satisfying (3.194) and suppose that $Z = (X, Y) \in \overline{Z}(x, y)$. Since $|X_\infty| \vee |Y_\infty| \geq 1$, we see that

$$u(x, y) = \mathbb{E}u(Z_0) \leq \mathbb{E}(Z_1) \leq \cdots \leq \mathbb{E}u(Z_\infty) \leq \mathbb{E}|X_\infty + Y_\infty|$$

and hence $u \leq v$. Consequently, this yields the estimate $u_\mathcal{B} \leq v$. To get the reverse estimate, first we use the splicing argument to show that v is biconvex. Thus, it suffices to show that v satisfies the bound (3.194). Suppose that $|x| \vee |y| \geq 1$. If $|x - y| \geq 2$, then $v(x, y) \leq |x + y|$, which can be seen by considering a constant pair $(x, y) \in \overline{Z}(x, y)$. If $|x - y| < 2$ and $|y| \geq 1$ (as we can assume), then $x + y \neq 0$ and hence, for large t,

$$v(x, y) \leq (1 - t^{-1})v(-y, y) + t^{-1}v(-y + t(x + y), y) \leq |x + y|.$$

This completes the proof.   □

We turn to Theorem 3.25. In the result below we establish a part of it.

Theorem 3.27. *Let* $u : \mathcal{B} \times \mathcal{B} \to \mathbb{R}$ *be a biconvex function satisfying* $u(0, 0) > 0$, (3.194) *and* (3.198). *Then*

$$\|g\|_p \leq \frac{324}{u(0, 0)}(p^* - 1)\|f\|_p, \qquad 1 < p < \infty. \tag{3.203}$$

This result will be proved in two steps. First we show the estimate (3.203) in the special case when f and g are dyadic; then we prove the claim in the general setting. See Section 3.13 for the necessary definitions.

Proof of (3.203) *for dyadic martingales.* We make use of the so-called extrapolation method introduced by Burkholder and Gundy. We will establish the following *good λ-inequality*. If f, g are dyadic martingales, then for any $\delta > 0$, $\lambda > 0$ and $\alpha = 8\delta/[u(0, 0)(\beta - 2\delta - 1)]$, $\beta > 2\delta + 1$, we have

$$\mathbb{P}(g^* > \beta\lambda, \ f^* \leq \delta\lambda) \leq \alpha\mathbb{P}(g^* > \lambda). \tag{3.204}$$

Here $f^* = \sup_{n \geq 0} |f_n|$ denotes the maximal function of f. Introduce the random variables

$$\mu = \inf\{n : |g_n| > \lambda\},$$
$$\nu = \inf\{n : |g_n| > \beta\lambda\},$$
$$\sigma = \inf\{n : |f_n| > \delta\lambda \text{ or } |df_{n+1}| > 2\delta\lambda\}.$$

It is easy to see that μ and ν are stopping times. Furthermore, so is σ, since f is dyadic and $|df_{n+1}|$ is measurable with respect to $\mathcal{F}_n$. Thus, $F = (F_n)_{n \geq 0}$, $G = (G_n)_{n \geq 0}$, given by

$$F_n = \sum_{k=0}^n 1_{\{\mu < n \leq \nu \wedge \sigma\}} df_k, \quad G_n = \sum_{k=0}^n \varepsilon_k 1_{\{\mu < n \leq \nu \wedge \sigma\}} df_k, \qquad n = 0, 1, 2, \ldots,$$

are martingales, and so, by (3.201),

$$\begin{aligned}
\mathbb{P}(g^* > \beta\lambda, f^* \le \delta\lambda) &= \mathbb{P}(\mu \le \nu < \infty, \sigma = \infty) \\
&\le \mathbb{P}(G^* > \beta\lambda - 2\delta\lambda - \lambda) \\
&\le \frac{2}{u(0,0)} \frac{\|F\|_1}{\beta\lambda - 2\delta\lambda - \lambda} \\
&\le \alpha\mathbb{P}(g^* > \lambda).
\end{aligned}$$

This implies $\mathbb{P}(g^* > \beta\lambda) \le \mathbb{P}(f^* > \delta\lambda) + \alpha\mathbb{P}(g^* > \lambda)$ and, consequently,

$$\begin{aligned}
\frac{\|g^*\|_p^p}{\beta^p} &= p \int_0^\infty \lambda^{p-1} \mathbb{P}(g^* \ge \beta\lambda) d\lambda \\
&\le p \int_0^\infty \lambda^{p-1} \mathbb{P}(f^* \ge \delta\lambda) d\lambda + \alpha p \int_0^\infty \lambda^{p-1} \mathbb{P}(g^* \ge \lambda) d\lambda \\
&= \frac{\|f^*\|_p^p}{\delta^p} + \alpha\|g^*\|_p^p.
\end{aligned}$$

We may subtract $\alpha\|g^*\|_p^p$ from both sides, since this number is finite, due to simplicity of g. If $\alpha\beta^p < 1$, the estimate becomes

$$\|g\|_p \le \|g^*\|_p \le \frac{\beta}{\delta(1 - \alpha\beta^p)^{1/p}} \|f^*\|_p.$$

Now if we put $\beta = 1 + 1/p$,

$$\delta = \frac{u(0,0)p^p}{8(p+1)^{p+1} + 2p^{p+1}}$$

and use Doob's bound $\|f^*\|_p \le \frac{p}{p-1}\|f\|_p$, $1 < p < \infty$, we arrive at (3.203). $\quad\square$

Proof of (3.203), *general case.* Let $p \in (1, \infty)$ and let

$$\gamma = \frac{324}{u(0,0)}(p^* - 1)$$

be the constant appearing in (3.203). Let $V : \mathcal{B} \times \mathcal{B} \to \mathbb{R}$ be given by

$$V(x,y) = \left|\frac{x-y}{2}\right|^p - \gamma^p \left|\frac{x+y}{2}\right|^p$$

and introduce $U : \mathcal{B} \times \mathcal{B} \to \mathbb{R}$ by

$$U(x,y) = \sup\{\mathbb{E}V(Z_n)\},$$

where the supremum is taken over all n and all simple zigzag martingales $Z = (X, Y)$ starting from (x, y), such that dX_n and dY_n takes at most two non-zero values for each n, each value with some probability. Then U is a majorant of V, $U(x,y) = U(-x,-y)$ for all x, $y \in B$ and $U(0,0) \le 0$, by (3.203) in the dyadic case. Furthermore, using the splicing argument, we get that U is finite (see, e.g., the proof of Theorem 2.3) and biconcave: the latter property follows from the fact that any function which is midconcave and locally bounded from below is automatically concave. Consequently, if f is a simple martingale (not necessarily dyadic) and g is its ± 1-transform, then

$$\begin{aligned}
\mathbb{E}|g_n|^p - \gamma^p \mathbb{E}|f_n|^p &= \mathbb{E} V(f_n + g_n, f_n - g_n) \\
&\le \mathbb{E} U(f_n + g_n, f_n - g_n) \\
&\le \mathbb{E} U(f_0 + g_0, f_0 - g_0) \\
&\le U(0,0) \le 0.
\end{aligned}$$

Here we have used the fact that $U(x,0) \le U(0,0)$ and $U(0,y) \le U(0,0)$, which comes from the biconcavity of U and the symmetry of U. This completes the proof.   □

We move to the remaining half of Theorem 3.25.

If B is UMD, then it is ζ-convex. Assume that there are $1 < p < \infty$ and $\beta < \infty$ such that the L^p inequality (3.191) holds for all B-valued f and their ± 1-transforms g. Arguing as previously, we see that there is a finite biconcave majorant U of the function

$$V(x,y) = \left| \frac{x-y}{2} \right|^p - \beta^p \left| \frac{x+y}{2} \right|^p,$$

satisfying $U(0,0) \le 0$. Put

$$\zeta(x,y) = \frac{2}{1 + \beta^p} \big[1 - U(x,y) \big].$$

Then ζ is biconvex and satisfies $\zeta(0,0) \ge 2/(1+\beta^p) > 0$. It remains to verify the condition (3.192). Suppose that x, y are of norm one and take $t = |x+y|/2$. Then $|x-y|/2 \ge |x| - t = 1 - t$ and

$$\begin{aligned}
\zeta(x,y) &\le \frac{2}{1 + \beta^p} \big[1 - V(x,y) \big] \\
&\le \frac{2}{1 + \beta^p} \big[1 - (1-t)^p + \beta^p t^p \big] \\
&\le 2t = |x+y|,
\end{aligned}$$

where we have used the elementary inequality

$$1 + \beta^p t^p \le (1 + \beta^p) t + (1-t)^p,$$

valid for $t \in [0,1]$ and $\beta \ge p^* - 1$.   □

Finally, we shall establish the aforementioned characterization of Hilbert spaces.

Theorem 3.28. *Suppose that $\mathcal{B}$ is a Banach space for which $\zeta = \zeta_{\mathcal{B}}$ satisfies $\zeta(0,0) = 1$. Then $\mathcal{B}$ is a Hilbert space.*

In the proof, we shall need the following two facts from the theory of convex bodies: see, e.g., [96].

Lemma 3.25. *Suppose that $\mathcal{B}$ is a two-dimensional real Banach space. Then the norm of $\mathcal{B}$ is generated by an inner product if and only if the unit sphere of $\mathcal{B}$ is an ellipse.*

Lemma 3.26. *If γ is a symmetric (about the origin) closed curve in the plane, then there is a unique ellipse of maximal area inscribed in γ, which touches γ in at least four points which are pairwise symmetric.*

Proof of Theorem 3.28. Suppose that $\mathcal{B}$ is not a Hilbert space. We shall find $x \in \mathcal{B}$ and a mean zero simple random variable Y such that $|Y| \geq 1$ almost surely and $\mathbb{E}|x + Y| < 1$. This will yield the claim: by (3.194) we shall obtain $\mathbb{P}(\zeta(x, Y) \leq |x + Y|) = 1$ and hence, by the biconvexity of ζ,

$$\zeta(0,0) \leq \frac{\zeta(x,0) + \zeta(-x,0)}{2} = \zeta(x,0) \leq \mathbb{E}\zeta(x,Y) \leq \mathbb{E}|x + Y| < 1.$$

With no loss of generality, we can assume that $\mathcal{B}$ is two-dimensional. Let $S_{\mathcal{B}}$ denote the unit sphere of $\mathcal{B}$ and let S_0 be the ellipse of maximal area inscribed in $S_{\mathcal{B}}$. Denote by $\|\cdot\|$ the norm induced by S_0; by taking affine transformations, if necessary, we may assume that S_0 is a unit circle. Let $\pm A$ and $\pm C$ be the four contact points guaranteed by Lemma 3.26 and assume that there are no contact points in the interior of the arc AC. After rotation, we may assume that $A = (1, 0)$ and $C = (\cos 2\theta, \sin 2\theta)$, where $\theta \in (0, \pi/4]$. Introduce the third point $D = s(\cos \theta, \sin \theta)$, where $s > 0$ is uniquely determined by the equality $|s(\cos \theta, \sin \theta)| = 1$. Note that $s > 1$, since D is not a contact point.

Now, for a given $t \in (-s, s)$, let

$$x(t) = -t(\cos \theta, \sin \theta)$$

and let $Y : [0, 1) \to \mathcal{B}$ be a simple function given by

$$Y = A1_{[0,p)} + C1_{[p,2p)} - D1_{[2p,1)},$$

where p is chosen to ensure that Y is centered. That is, $p = s/(2(s + \cos \theta))$. Let ϕ, ψ be the functions on $(-s, s)$, given by

$$\begin{aligned}
\phi(s) &= \mathbb{E}|x(t) + Y| \\
&= p|(1 - t\cos\theta, -t\sin\theta)| + p|(\cos 2\theta - t\cos\theta, \sin 2\theta - t\sin\theta)| \\
&\quad + (1 - 2p)\frac{s + t}{s},
\end{aligned}$$

$$\psi(s) = p||(1 - t\cos\theta, -t\sin\theta)|| + p||(\cos 2\theta - t\cos\theta, \sin 2\theta - t\sin\theta)||$$
$$+ (1 - 2p)\frac{s+t}{s}.$$

Then $\phi \le \psi$ with $\phi(0) = \psi(0)$ and, in addition,

$$\psi'(0) = \frac{(1 - s^2)\cos\theta}{s(s + \cos\theta)} < 0,$$

because $s > 1$. Consequently $\psi(t)$, and hence also $\phi(t)$, is negative for some small positive t. It suffices to take $x = x(t)$ to get the claim.   □

The above characterization can be rephrased in the following probabilistic form.

Theorem 3.29. *Suppose that a separable Banach space $\mathcal{B}$ has the following property, for all martingales f, g taking values in $\mathcal{B}$ such that g is a ± 1-transform of f, we have*

$$\mathbb{P}(g^* \ge 1) \le 2||f||_1.$$

Then $\mathcal{B}$ is a Hilbert space.

We shall not establish this result here. However, see the proof of Theorem 4.6 below for a similar reasoning.

3.12   Escape inequalities

Escape inequalities arise naturally in the study of almost sure convergence of martingales, where are used to control counting functions of various types. For example, the upcrossing method of Doob [70] introduces such a function for real martingales. There is also a different one, related to the method of rises, developed by Dubins; this works for nonnegative martingales. Here we will describe a counting function invented by Davis which is dimension-free and allows to study vector-valued martingales f and g.

3.12.1   A counting function

Let $x = (x_n)_{n \ge 0}$ be a sequence in a Banach space $\mathcal{B}$. By Cauchy's criterion, x converges if and only if $C_\varepsilon(x) < \infty$ for all $\varepsilon > 0$, where $C_\varepsilon(x)$ is the number of ε-escapes of x. The counting function C_ε is defined as follows: $C_\varepsilon(x) = 0$ and $\nu_0(x) = 0$ if the set $\{n \ge 0 : |x_n| \ge \varepsilon\}$ is empty. If this set is nonempty, let

$$\nu_0(x) = \inf\{n \ge 0 : |x_n| \ge \varepsilon\}.$$

Now let $C_\varepsilon(x) = 1$ and $\nu_1(x) = \infty$ if $\{n > \nu_0(x) : |x_n - x_{\nu_0(x)}| \ge \varepsilon\}$ is empty. If nonempty, continue as above. If this induction stops at some point, that is,

$\nu_i(x) = \infty$ for some i, set $\nu_j(x) = \infty$ for $j > i$. Then we have that $C_\varepsilon(x) \leq j$ if and only if $\nu_j(x) = \infty$. If f is a $\mathcal{B}$-valued martingale, then $C_\varepsilon(f)$ is the random variable given by $C_\varepsilon(f)(\omega) = C_\varepsilon(f(\omega))$.

3.12.2 Formulation of the results

We start with the escape inequality for Hilbert-space-valued martingales.

Theorem 3.30. *Let f and g be $\mathcal{H}$-valued martingales such that g is differentially subordinate to f. Then*

$$\mathbb{P}(C_\varepsilon(g) \geq j) \leq \frac{2||f||_1}{\varepsilon j^{1/2}}, \qquad j \geq 1, \tag{3.205}$$

and both the constant 2 and the exponent $1/2$ are the best possible. If, in addition, $||f||_1$ is finite, then g converges almost everywhere.

This estimate carries over to martingales which take values in Banach spaces isomorphic to Hilbert spaces (with possible change in the constant). In view of the results of the previous section, it is not so surprising that this bound does not hold for a wider class of Banach spaces. Here is the precise statement.

Theorem 3.31. *Suppose that $\mathcal{B}$ is a Banach space. Then $\mathcal{B}$ is isomorphic to a Hilbert space if and only if there is $\beta = \beta(\mathcal{B})$ such that for any $\mathcal{B}$-valued martingales f and g with g being differentially subordinate to f, we have*

$$\mathbb{P}(C_\varepsilon(g) \geq j) \leq \frac{\beta||f||_1}{\varepsilon j^{1/2}}, \qquad j \geq 1. \tag{3.206}$$

One may impose a stronger domination on the martingale g and study Banach spaces in which the corresponding escape inequality is valid. As one might expect, the restriction to ± 1-transforms leads to UMD spaces.

Theorem 3.32. *Let $\mathcal{B}$ be a Banach space. Then $\mathcal{B}$ is UMD if and only if there are $\alpha > 0$ and $\beta > 0$ such that for any $\mathcal{B}$-valued martingale f and its ± 1-transform g we have*

$$\mathbb{P}(C_\varepsilon(g) \geq j) \leq \frac{\beta||f||_1}{\varepsilon j^\alpha}, \qquad j \geq 1. \tag{3.207}$$

Finally, if we take $g = f$, we obtain the even larger class of superreflexive spaces (see, e.g., [100] and [101] for the necessary background).

Theorem 3.33. *Let $\mathcal{B}$ be a Banach space. Then $\mathcal{B}$ is superreflexive if and only if there are $\alpha > 0$ and $\beta > 0$ such that for any $\mathcal{B}$-valued martingale f we have*

$$\mathbb{P}(C_\varepsilon(f) \geq j) \leq \frac{\beta||f||_1}{\varepsilon j^\alpha}, \qquad j \geq 1. \tag{3.208}$$

Throughout this section we assume that f and g start from 0. This can be done with no loss of generality, by means of a standard argument: see the first lines of the proof of Theorem 3.20.

3.12.3   Proof of Theorem 3.30

We may assume that f is of finite length: there is a deterministic m such that $f_n = f_m$ and $g_n = g_m$ for $n > m$. For $j = 0, 1, 2, \ldots$, let $\tau_j = \nu_j(g) \wedge m$, where, as previously, $\nu_j(g)$ is a random variable defined by $\nu_j(g)(\omega) = \nu_j(g(\omega))$. Then $(\tau_j)_{j\geq 0}$ is a nondecreasing family of stopping times. Let h be the $\ell^2(\mathcal{H})$-valued sequence given as follows: if $n \leq \tau_0$, then $dh_n = (dg_n, 0, 0, \ldots)$; if $\tau_j < n \leq \tau_{j+1}$, then $dh_n = (0, 0, \ldots, 0, dg_n, 0, 0, \ldots)$, where dg_n appears at the $(j+1)$st coordinate. Then h is a martingale with respect to $(\mathcal{F}_n)_{n\geq 0}$ and

$$S_m(g, \tau) = |h_m| = \left(\sum_{j=0}^{\infty} |g_{\tau_{j+1}} - g_{\tau_j}|^2 \right)^{1/2}.$$

By the definition of the family $(\tau_j)_{j\geq 0}$, we see that $|g_{\tau_{j+1}} - g_{\tau_j}| \geq \varepsilon$ if $\tau_{j+1} \leq m$, so $|S_m(g, \tau)|^2 \geq \varepsilon^2 C_\varepsilon(g)$. On the other hand, we see that h is differentially subordinate to f (which can be regarded as an $\ell^2(\mathcal{H})$-valued martingale, by means of the embedding on the first coordinate). Thus

$$\mathbb{P}(C_\varepsilon(g) \geq j) \leq \mathbb{P}(S(g, \tau) \geq \varepsilon j^{1/2}) \leq \frac{2\|f\|_1}{\varepsilon j^{1/2}}. \tag{3.209}$$

To establish the optimality of the constants, let $\mathcal{H} = \ell^2$, fix $j \geq 1$ and consider independent random variables $r_1, r_2, \ldots, r_{2j}$ such that

$$\mathbb{P}(r_{2k+1} = 1/2) = \mathbb{P}(r_{2k+1} = -1/2) = 1/2$$

and

$$\mathbb{P}(r_{2k} = -1) = 1 - \mathbb{P}(r_{2k} = 3) = 3/4.$$

Put $f_0 \equiv 0$ and set

$$df_{2k+1} = (0, 0, \ldots, 0, r_{2k+1}, 0, \ldots), \quad df_{2k+2} = (0, 0, \ldots, 0, r_{2k+1}r_{2k}, 0, \ldots),$$

where in both differences the nonzero term occurs on the kth place, $k = 0, 1, 2, \ldots, j - 1$. Finally, define g by taking $dg_k = (-1)^{k+1} df_k$, $k = 0, 1, 2, \ldots, 2j$. One can easily check that $|g_{2k+2} - g_{2k}| = 1$ for $0 \leq k \leq j$, so $C_1(g) = j$; on the other hand, $\|f\|_1 = j^{1/2}/2$. This completes the proof.

3.12.4   Proof of Theorem 3.31

Proof of (3.206). This is straightforward. Suppose that $\mathcal{B}$ is isomorphic to a Hilbert space $\mathcal{H}$: there exist a linear map T from $\mathcal{B}$ onto $\mathcal{H}$ and strictly positive α_1 and α_2 such that

$$\alpha_1 |Tx| \leq |x| \leq \alpha_2 |Tx|$$

for all $x \in \mathcal{B}$. Let f, g be two $\mathcal{B}$-valued martingales such that g is differentially subordinate to f. We may transfer these to $\mathcal{H}$-valued processes F, G using the formulas

$$F_n(\omega) = \alpha_2 T f_n(\omega), \quad G_n(\omega) = \alpha_1 T g_n(\omega).$$

Then F, G are martingales with respect to $(\mathcal{F}_n)_{n \geq 0}$, which can be verified, for example, by approximating f and g by simple processes. Moreover, G is differentially subordinate to F: indeed,

$$|dG_n| = \alpha_1 |T(dg_n)| \leq |dg_n| \leq |df_n| \leq \alpha_2 |T(df_n)| = |dF_n|.$$

Similarly, we get $C_\varepsilon(g) \leq C_{\alpha_1 \varepsilon/(2\alpha_2)}(G)$ and $|F_n| \leq \alpha_2 \alpha_1^{-1} |f_n|$, so, by Theorem 3.30,

$$\mathbb{P}(C_\varepsilon(g) \geq j) \leq \mathbb{P}(C_{\alpha_1 \varepsilon/(2\alpha_2)}(G) \geq j) \leq \frac{2\|F\|_1}{(\alpha_1 \varepsilon/(2\alpha_2)) j^{1/2}} \leq \frac{4\alpha_2^2 \|f\|_1}{\alpha_1^2 \varepsilon j^{1/2}}.$$

This completes the proof. $\qquad \square$

Escape characterization of spaces isomorphic to Hilbert spaces. Suppose that the inequality (3.206) holds for all $\mathcal{B}$-valued martingales f, g such that g is differentially subordinate to f. Take $\varepsilon = j = 1$ and observe that $\{g^* > 1\} \subseteq \{C_1(g) \geq 1\}$, so

$$\mathbb{P}(g^* > 1) \leq \beta \|f\|_1,$$

which implies $\|g\|_{1,\infty} \leq \beta \|f\|_1$. Thus, by the second part of Theorem 3.24, the claim follows. $\qquad \square$

3.12.5 Proof of Theorem 3.32

Recall that $\beta_p(\mathcal{B})$ is the least extended real number β such that for any $\mathcal{B}$-valued martingale f and its ± 1-transform g,

$$\|g\|_p \leq \beta \|f\|_p.$$

This constant remains unchanged if we allow g to be a transform of f by a certain real predictable sequence v bounded in absolute value by 1. This can be justified by the use of Theorem 2.6 of Chapter 2.

If, in addition, $2 \leq q < \infty$, then we define $\gamma_{p,q}(\mathcal{B})$ to be the least extended real number γ such that

$$\|S(f,q)\|_p \leq \gamma \|f\|_p$$

for all $\mathcal{B}$-valued martingales f, where $S(f,q) = (\sum_{n=0}^\infty |df_n|^q)^{1/q}$. Finally, for any nondecreasing sequence $\tau = (\tau_n)_{n \geq 0}$ of stopping times and any $\mathcal{B}$-valued martingale f, let

$$S(f,q,\tau) = \left(|f_{\tau_0}|^q + \sum_{n=1}^\infty |f_{\tau_n} - f_{\tau_{n-1}}|^q \right)^{1/q}.$$

We start with the following auxiliary fact.

Lemma 3.27. *For any $1 < p < \infty$ and $2 \leq q < \infty$, if f is a $\mathcal{B}$-valued martingale, g is its transform by a certain real-valued predictable sequence bounded in absolute value by 1 and τ is a nondecreasing sequence of stopping times, then*

$$\|S(g, q, \tau)\|_p \leq \beta_p(\mathcal{B})\gamma_{p,q}(\mathcal{B})\|f\|_p. \tag{3.210}$$

Proof. This is straightforward. For any fixed positive integer N, the process $h = (g_{\tau_n \wedge N})_{n \geq 0}$ is a martingale, so

$$\left\| \left(|g_{\tau_0 \wedge N}|^q + \sum_{n=1}^{\infty} |g_{\tau_n \wedge N} - g_{\tau_{n-1} \wedge N}|^q \right)^{1/q} \right\|_p = \|S(h, q)\|_p \leq \gamma_{p,q}(\mathcal{B})\|h\|_p$$

$$\leq \gamma_{p,q}(\mathcal{B})\|g\|_p \leq \beta_p(\mathcal{B})\gamma_{p,q}(\mathcal{B})\|f\|_p.$$

It suffices to let $N \to \infty$ to get the claim. $\square$

Lemma 3.28. *For any $2 \leq q < \infty$, if f is a $\mathcal{B}$-valued martingale, g is its ± 1-transform and τ is a nondecreasing sequence of stopping times, then*

$$\lambda\mathbb{P}(S(g, q, \tau) \geq \lambda) \leq 2\beta_2(\mathcal{B})\gamma_{2,q}(\mathcal{B})\|f\|_1. \tag{3.211}$$

Proof. We divide the reasoning into two steps.

Step 1. *Embedding f into a martingale with small jumps.* By a theorem of Mc-Connell [117], there exist a probability space $(\Omega', \mathcal{F}', \mathbb{P}')$, a $\mathcal{B}$-valued, continuous-path martingale $X = (X_t)_{t \geq 0}$ and a nondecreasing sequence $(\sigma_n)_{n \geq 0}$ of stopping times such that f has the same distribution as $(X_{\sigma_n})_{n \geq 0}$. Let $\mu_0 < \mu_1 < \mu_2 < \cdots$ be stopping times given by

$$\mu_0 = \inf\{t \geq 0 : |X_t| \geq \delta/2\}$$

and, inductively, for $n \geq 1$,

$$\mu_n = \inf\{t \geq 0 : |X_t - X_{\mu_{n-1}}| \geq \delta/2\}.$$

Now, let $0 \leq \rho_0 \leq \rho_1 \leq \rho_2 \leq \cdots$ be the sequence of stopping times obtained by interlacing $(\sigma_n)_{n \geq 0}$ and $(\mu_n \wedge \sigma_\infty)_{n \geq 0}$, where $\sigma_\infty = \lim_{n \to \infty} \sigma_n$. In other words, for each k, the random variable ρ_k is the kth order statistic associated with the random variables $\sigma_0, \sigma_1, \sigma_2, \ldots, \mu_0 \wedge \sigma_\infty, \mu_1 \wedge \sigma_\infty, \mu_2 \wedge \sigma_\infty, \ldots$. Put $F_n = M_{\rho_n}$ for $n \geq 0$. Then F is a martingale satisfying $\|F\|_1 \leq \|f\|_1$, with a difference sequence such that $dF^* \leq \delta$. Since $(\sigma_n)_{n \geq 0}$ are stopping times coming from the embedding f into X, we see that f embeds into F as well: f has the same distribution as $(F_{\eta_n})_{n \geq 0}$, where (η_n) are certain stopping times relative to the filtration with respect to which F is a martingale. The next observation is that since g is a ± 1-transform of f, we have that g has the same distribution as $(\sum_{k=0}^{\eta_n} v_k dF_k)_{n \geq 0}$, where $v = (v_k)_{k \geq 0}$ is a predictable sequence taking values in $\{-1, 1\}$. Let $G_n = \sum_{k=0}^{n} v_k dF_k$. Consequently, there is a sequence $T = (T_n)_{n \geq 0}$ of stopping times such that $(g_{\tau_n})_{n \geq 0}$ has the same distribution as $(G_{T_n})_{n \geq 0}$, and hence $S(g, q, \tau)$, $S(G, q, T)$ also have the same distribution.

Step 2. *Proof of* (3.211). Fix positive constants λ, b, introduce the stopping time

$$R = \inf\{n \geq 0 : |F_n| > b\lambda\}$$

and denote the martingales $(F_{R \wedge n})_{n \geq 0}$, $(G_{R \wedge n})_{n \geq 0}$ by F^R and G^R, respectively. Observe that, by Chebyshev's inequality,

$$\mathbb{P}'(S(G, q, T) \geq \lambda, \, F^* \leq b\lambda) = \mathbb{P}'(S(G^R, q, T) \geq \lambda, \, R = \infty) \leq \lambda^{-p}||S(G^R, q, T)||_p^p,$$

which, by (3.210), can be bounded from above by $\lambda^{-p}\beta_p^p(\mathcal{B})\gamma_{p,q}^p(\mathcal{B})||F^R||_p^p$. By the definition of R and the fact that $dF^* \leq \delta$, we have

$$||F^R||_p^p \leq (b\lambda + \delta)^{p-1}||F^R||_1 \leq (b\lambda + \delta)^{p-1}||F||_1.$$

Putting all the things together, we get

$$\begin{aligned}
\mathbb{P}(S(g, q, \tau) \geq \lambda) &= \mathbb{P}'(S(G, q, T) \geq \lambda) \\
&\leq \mathbb{P}'(S(G, q, T) \geq \lambda, \, F^* \leq b\lambda) + \mathbb{P}'(F^* > b\lambda) \\
&\leq \lambda^{-1}\left[\beta_p^p(\mathcal{B})\gamma_{p,q}^p(\mathcal{B})(b + \delta\lambda^{-1})^{p-1} + b^{-1}\right]||F||_1 \\
&\leq \lambda^{-1}\left[\beta_p^p(\mathcal{B})\gamma_{p,q}^p(\mathcal{B})(b + \delta\lambda^{-1})^{p-1} + b^{-1}\right]||f||_1.
\end{aligned}$$

Letting $\delta \to 0$ yields

$$\mathbb{P}(S(g, q, \tau) \geq \lambda) \leq \lambda^{-1}\left[b^{p-1}\beta_p^p(\mathcal{B})\gamma_{p,q}^p(\mathcal{B}) + b^{-1}\right]||f||_1$$

and putting $p = 2$, $b = \beta_2^{-1}(\mathcal{B})\gamma_{2,q}^{-1}(\mathcal{B})$ gives the desired estimate (3.211). $\quad\square$

Proof of (3.207). Suppose that $\mathcal{B}$ is a UMD space. Then $\beta_p(\mathcal{B})$ is finite for all $1 < p < \infty$. Furthermore, as shown by Maurey [116], $\mathcal{B}$ is superreflexive; by the results of Pisier [173], this implies $\gamma_{p,q}(\mathcal{B}) < \infty$ for some $q \in [2, \infty)$ and all $1 < p < \infty$. It remains to use the inequality (3.211): arguing as in (3.209), we obtain

$$\mathbb{P}(C_{\mathcal{F}}(g) > j) \leq 2\beta_2(\mathcal{B})\gamma_{2,q}(\mathcal{B})||f||_1/(\varepsilon j^{1/q}),$$

which is precisely (3.207), with $\alpha = 1/q$ and $\beta = 2\beta_2(\mathcal{B})\gamma_{2,q}(\mathcal{B})$. $\quad\square$

Escape characterization of UMD spaces. Assume that (3.207) holds for all $\mathcal{B}$-valued martingales f and their ± 1-transforms g. As previously, this implies

$$\lambda\mathbb{P}(g^* \geq \lambda) \leq \beta||f||_1, \qquad \lambda > 0,$$

and, by extrapolation, $||g||_p \leq \beta_p||f||_p$ for some universal β_p, $1 < p < \infty$. Thus, $\mathcal{B}$ is a UMD space. $\quad\square$

3.12.6   Proof of Theorem 3.33

Proof of (3.208). We proceed as in the proof of (3.207) and obtain

$$\mathbb{P}(C_\varepsilon(f) \geq j) \leq 2\gamma_{2,q}(\mathcal{B})\|f\|_1/(\varepsilon j^{1/q}).$$

It suffices to use Pisier's result from [173], which states that the constant $\gamma_{2,q}(\mathcal{B})$ is finite for some q. This completes the proof.   □

Escape characterization of superreflexive spaces. If $\mathcal{B}$ is not superreflexive, then (3.208) fails to hold. To see this, fix positive β, $\varepsilon \in (0,1)$ and pick j so large that $\beta(\varepsilon j^\alpha) < 1$. By results of James [99] and Pisier [173], there is a $\mathcal{B}$-valued martingale f satisfying $\|f\|_1 \leq \|f\|_\infty \leq 1$ and $|df_k| \geq \varepsilon$ almost surely for all $k < j$. Consequently, $\mathbb{P}(C_\varepsilon(f) \geq j) = 1$, which contradicts (3.208).   □

3.13   The dyadic case and inequalities for the Haar system

For many martingale inequalities, the best constants in the dyadic case are different from those of the general case: see, e.g., Chapter 8. However, when we study estimates for ± 1-transforms, the best bounds are the same. First let us introduce the necessary definitions.

Definition 3.4. Suppose that f is a martingale taking values in a certain Banach space $\mathcal{B}$.

(i) We say that f is *dyadic* if $f_0 \equiv b_0 \in \mathcal{B}$ and for any $n \geq 1$ and any nonempty set of the form

$$\{f_0 = b_0, df_1 = b_1, \ldots, df_{n-1} = b_{n-1}\},$$

the restriction of df_n to this set either vanishes identically or has its values in $\{-b_n, b_n\}$ for some $b_n \in \mathcal{B} \setminus \{0\}$.

(ii) We say that f is *conditionally symmetric* if for any $n \geq 1$, the conditional distributions of df_n and of $-df_n$, given $\mathcal{F}_{n-1}$, coincide.

Therefore, if a martingale f is dyadic, it is automatically conditionally symmetric. For example, a sequence of partial sums of Haar series forms a dyadic martingale. To recall the definition of the Haar system $h = (h_0, h_1, h_2, \ldots)$ on $[0,1)$, we shall use the same notation for an interval $[a,b)$ and its indicator function. Let

$$
\begin{aligned}
h_0 &= [0,1), & h_1 &= [0,1/2) - [1/2,1), \\
h_2 &= [0,1/4) - [1/4,1/2), & h_3 &= [1/2,3/4) - [3/4,1), \\
h_4 &= [0,1/8) - [1/8,1/4), & h_5 &= [1/4,3/8) - [3/8,1/2), & \ldots.
\end{aligned}
$$

A classical inequality due to Paley (see [166] and [115]) can be stated as follows: if $1 < p < \infty$, then there is an absolute c_p such that

$$\left\| \sum_{k=0}^{n} \varepsilon_k a_k h_k \right\|_p \leq c_p \left\| \sum_{k=0}^{n} a_k h_k \right\|_p \tag{3.212}$$

for all $a_k \in \mathbb{R}$, $\varepsilon_k \in \{-1, 1\}$ and $n \geq 0$. There are other interesting estimates of this type, including the weak-type, logarithmic and tail bounds. We may also extend these results by allowing the coefficients a_k to be vector valued. What are the best constants in these inequalities? It turns out that they are the same as in the corresponding bounds for ±1-transforms. Here is the precise statement.

Theorem 3.34. *Let $\mathcal{B}$ be a Banach space and let $V : \mathcal{B} \times \mathcal{B} \to \mathbb{R}$ be a given function which is locally bounded from below. The following conditions are equivalent:*

(i) *For any simple $\mathcal{B}$-valued martingale f and its ±1-transform g we have*

$$\mathbb{E}V(f_n, g_n) \leq 0, \qquad n = 0, 1, 2, \ldots. \tag{3.213}$$

(ii) *For any $a_k \in \mathcal{B}$ and $\varepsilon_k \in \{-1, 1\}$, $k = 0, 1, 2, \ldots$, we have*

$$\int_0^1 V\left(\sum_{k=0}^{n} a_k h_k(s), \sum_{k=0}^{n} \varepsilon_k a_k h_k(s) \right) ds \leq 0. \tag{3.214}$$

Proof. The implication (i)$\Rightarrow$(ii) is obvious, since $(\sum_{k=0}^{n} a_k h_k)_{n=0}^{\infty}$ is a martingale and $(\sum_{k=0}^{n} \varepsilon_k a_k h_k)_{n=0}^{\infty}$ is its ±1-transform (we consider the probability space $([0,1), \mathcal{B}(0,1), |\cdot|)$, equipped with the filtration generated by the Haar system). To get the reverse implication, introduce the function $W^0 : \mathcal{B} \times \mathcal{B} \to \mathbb{R}$ given by the formula

$$W^0(x, y) = \sup \left\{ \int_0^1 V\left(x + \sum_{k=1}^{n} a_k h_k(s), y + \sum_{k=1}^{n} \varepsilon_k a_k h_k(s) \right) ds \right\}, \quad x, y \in \mathcal{B}.$$

Here the supremum is taken over all n, all $a_k \in \mathcal{B}$ and all $\varepsilon_k \in \{-1, 1\}$. Letting $n = 0$ in the definition of W^0 gives the majorization $W^0 \geq V$ on $\mathcal{B} \times \mathcal{B}$. Using an appropriate modification of the splicing argument, we can show that W^0 is diagonally mid-concave: that is, for any x, $y\,z \in \mathcal{B}$, the functions $t \mapsto W^0(x + tz, y + tz)$ and $t \mapsto W^0(x + tz, y - tz)$, $t \in \mathbb{R}$, are midpoint concave and hence also concave since they are locally bounded from below. Finally, by (3.214) we get that $W^0(x, \pm x) \leq 0$ for $x \in \mathcal{B}$ and that W^0 is finite (see the proof of Theorem 2.3 in Chapter 2). In other words, we have checked that W^0 satisfies the conditions $1°$, $2°$ and $3°$ of Theorem 2.4 from Chapter 2 and hence (3.213) follows. $\square$

Remark 3.8. Since $(\sum_{k=0}^{n} a_k h_k)_{n=0}^{\infty}$ is a dyadic martingale, the theorem above implies that the best bounds for ±1-transforms are already the best possible in the dyadic case.

Therefore we see that the best constant in Paley's inequality (3.212) is equal to $p^* - 1$, both in the real and the Hilbert-space-valued case. This immediately yields the unconditional constant of the Haar system. Recall that for $1 < p < \infty$ and a given sequence $e = (e_0, e_1, e_2, \dots)$ in real L^p, the *unconditional constant* of e is the least $\beta \in [1, \infty]$ with the property that if n is a nonnegative integer and $a_0, a_1, a_2, \dots, a_n$ are real numbers such that $\| \sum_{k=0}^n a_k e_k \|_p = 1$, then

$$\left\| \sum_{k=0}^n \varepsilon_k a_k e_k \right\|_p \le \beta$$

for all choices of $\varepsilon_k \in \{-1, 1\}$. Theorem 3.34 gives the following.

Theorem 3.35. *The unconditional constant of the Haar system in $L^p(0,1)$ equals $p^* - 1$.*

There is an alternative way to define the unconditional constant, by requiring that the coefficients ε_k take values in $\{0, 1\}$ (see, e.g., James [102]): this corresponds to discarding out some of the summands $a_k e_k$, instead of changing their signs. If we use this definition, then the nonsymmetric version of Theorem 3.34 and the inequality (3.121) imply that the unconditional constant of the Haar system in $L^p(0,1)$ equals c_p, the constant determined by Choi [52] (see Section 3.8 above).

3.14   Notes and comments

Section 3.1. The weak type inequality (3.1), with some absolute constant, appeared for the first time in Burkholder's paper [17]. It was studied in the special case $p = 1$ and when g is a transform of a real-valued martingale f. For other proofs of this result (still with non-optimal constant), consult Davis [56], Garsia [82], Gundy [85], Neveu [121] and Rao [181]. The sharp version of this estimate, still in the real-valued case, can be found in [19]. The general setting, for Hilbert-space-valued processes and under the differential subordination, was established by Burkholder in [24]. This paper contains also the proof of the sharp extension to $p \in (1, 2]$, as well as of the more general Φ-inequality (3.16). The question about the best constant in the case $p > 2$ was open for twenty years and was finally answered by Suh in [189], using the complicated function given by (3.38). The above simplified approach, based on the integration argument, is due to the author (see [151]). Theorem 3.2, which concerns the case of different orders, has been established by the author in [160]. There are also noncommutative versions of the weak-type estimates: see Randrianantonina [177] for the setting of ± 1-transforms and the author's paper [128] for differentially subordinate martingales.

   Section 3.2. The sharp weak type inequality for $p > 2$ in the case when f is nonnegative was established by the author in [133] using a special function that was even more complicated than the one constructed by Suh in (3.38) (though there are many similarities between these two objects). The above approach, using the integration argument, is entirely new.

Section 3.3. The roots of the inequality (3.59) go back to classical results of Marcinkiewicz [115] and Paley [166], who established this bound in the particular case when the martingales are dyadic (the concept of martingale did not appear there, the results being formulated in terms of Haar and Rademacher series). The L^p estimate for real-valued martingales, with some absolute constant, was first proved by Burkholder [17] in the case when g is a transform of f. This follows immediately from the weak type estimate, using interpolation and duality arguments. See also Garsia [82]. The sharp version of the L^p inequality, in the case when the martingales are Hilbert-space-valued and under differential subordination, can be found in Burkholder's paper [24]; see also [22] and [25]. To show the optimality of the constant $p^* - 1$, Burkholder constructed in [24] appropriate examples; the above approach is taken from his later paper [32]. For noncommutative version of the L^p estimates, consult Pisier and Xu [174] and Randrianantonina [177].

Section 3.4. The validity of the strong (q, p) estimate with some constant (not necessarily optimal) follows trivially from the L^p estimates of Section 3.3. The best values of $C_{p,q}$ were determined by the author in [146] and the presentation above comes directly from that paper.

Section 3.5. The L^p estimates for differentially subordinate martingales under the additional assumption that the dominating process is nonnegative were investigated by Burkholder in [36]. The above proof, based on the integration argument, is new; Burkholder verified the condition 2° directly in his original approach. Furthermore, to show the optimality of the constant, he constructed a certain family of Markovian martingales. The case when the dominated process is nonnegative was studied by the author in [125].

Section 3.6. The logarithmic estimates for martingale transforms (with some absolute constants) can be easily obtained from the weak type inequality using interpolation or extrapolation (see, e.g., [17] or [40]). The proof of the sharp version for differentially subordinate martingales is contained in [126]. The author showed the optimality of $L(K)$ by providing appropriate examples: the above approach, based on exploiting the properties of the special function, is new. For some extensions and variations of the logarithmic estimates, see also [146] and [149].

Section 3.7. The inequalities in the case when the dominating process is bounded were studied by Burkholder in [28] and [32].

Section 3.8. The sharp weak type $(1,1)$ inequality for nonsymmetric martingale transforms appeared for the first time in Burkholder's paper [27]. The best constants in the corresponding L^p estimate were determined by Choi [52]. The remaining results of Section 3.8 are due to the author and can be found in [145] and [159].

Section 3.9. Control inequalities connected to the weak type $(1,1)$ estimate for martingale transforms were first studied by Burkholder in [24]; then the results were generalized by Choi [51]. The approach presented above is slightly different from those used there. For example, Burkholder determined the function

$$U(x,y) = \inf\{||f||_1 : \mathbb{P}(|g|^* \geq 1) = 1\},$$

where the infimum is taken over all $(f,g) \in M(x,y)$. Note that under the probability above we have a two-sided maximal bound for g. Analogously, Choi derived the explicit form of

$$U(x,y,t) = \inf\{||f||_1 : (f,g) \in M(x,y), \mathbb{P}(|g|^* \geq 1) = t\}.$$

Both functions easily lead to the estimates

$$||f||_1 \geq L_1(x,y,1) \quad \text{and} \quad ||f||_1 \geq L_1(x,y,t),$$

studied in Theorem 3.17. Our approach, based on the inequality (3.137), is entirely new and also provides the answer in the case when p is strictly larger than 1. There is an interesting related problem of determining the function

$$U_p(x,y,t) = \inf\{||f||_p : (f,g) \in M(x,y), \mathbb{P}(|g|^* \geq 1) = t\},$$

which would give a more precise information on the control. This seems to be open for all $1 < p < \infty$, $p \neq 2$.

Section 3.10. The notion of weak domination is due to Kwapień and Woyczyński [112]. The proofs of the weak type inequality, the L^p estimate $(1 < p < 2)$ and the logarithmic bound presented above are new, though the first two estimates appear (with worse constants) in [124]. For related results concerning tangent and weakly dominated martingales, see Cox and Veraar [55], Hitczenko [89], [91], [92] and Kwapień and Woyczyński [111], [112].

Section 3.11. UMD spaces appear naturally in harmonic analysis. The work of M. Riesz [180] in 1920s on the Hilbert transform and that of Calderón and Zygmund [41], [42] in 1950's on more general singular integral operators lead to a natural question whether the results can be carried over to Banach-space-valued functions. In late 1930's, Bochner and Taylor [12] proved that not all Banach spaces behave well even for the Hilbert transform. Later it was proved that those Banach spaces which are well behaved for the Hilbert transform (HT spaces), are also well behaved for a wide class of singular integrals. A striking and impressive result is that the class of HT spaces coincides with the class of UMD spaces. That a UMD space is HT is due to Burkholder and McConnell: see [23]. The reverse implication was established by Bourgain [16]. This gives rise to another interesting question about providing a geometric characterization of HT spaces. This was answered by Burkholder in [20]. See also the papers [21], [23] and [26] of Burkholder for refinement of this result and more information on the subject. For related martingale inequalities in UMD spaces, see Cox and Veraar [55], Lee [113], [114], McConnell [118].

Section 3.12. Essentially all the material in this section comes from Burkholder's paper [31]. For more information on superreflexive Banach spaces, see Enflo [71], James [99], [100], [101], Maurey [116] and Pisier [172], [173].

Section 3.13. As already mentioned, the inequality (3.212) was established by Paley [166]. In fact, Paley formulated this result in terms of the Walsh system on the Lebesgue unit interval and the above statement is due to Marcinkiewicz [115]. That the unconditional constant of the Haar system in real L^p equals $p^* - 1$, is due to Burkholder: see [22], [24] and [25]. It was then conjectured by Pełczyński [168] that the complex unconditional constant of the Haar basis is also equal to $p^* - 1$ (the definition of the constant is the same as in the real case, one only replaces the signs $\varepsilon_k \in \{-1, 1\}$ by numbers $e^{i\theta_k}$ from the unit circle). This conjecture was proved by Burkholder [30]. The alternative unconditional constant, with transforming coeeficients $\varepsilon_k \in \{0, 1\}$, was derived by Choi in [52].

Chapter 4

Sub- and Supermartingale Inequalities in Discrete Time

4.1 α-strong differential subordination and α-subordination

As we have seen, differential subordination implies many interesting estimates in the martingale setting. However, if we want to extend these inequalities to a wider class of processes, this domination turns out to be too weak. Thus we need more restrictive assumptions, and we propose a convenient one below.

Definition 4.1. Let $\alpha \geq 0$ be a fixed number and suppose that f, g are adapted sequences of integrable $\mathcal{B}$-valued random variables. We say that g is α-strongly differentially subordinate to f if g is differentially subordinate to f and for any $n \geq 1$ we have

$$|\mathbb{E}(dg_n|\mathcal{F}_{n-1})| \leq \alpha|\mathbb{E}(df_n|\mathcal{F}_{n-1})| \qquad (4.1)$$

almost surely.

Note that if f and g are martingales, then the condition (4.1) is automatically satisfied and hence differential subordination and strong differential subordination coincide. As another example, observe that if f is any sequence of integrable random variables and g is its transform by a predictable sequence bounded in absolute value by 1, then g is 1-strongly differentially subordinate to f.

The strong differential subordination is usually imposed when the dominating process f is assumed to be a sub- or supermartingale (then of course f is $\mathbb{R}$-valued, but g may still be vector valued). From now on, we restrict ourselves to these f and turn to the description of the corresponding version of Burkholder's method. Assume that $V : \mathbb{R} \times \mathcal{B} \to \mathbb{R}$ is a given Borel function and consider $U : \mathbb{R} \times \mathcal{B} \to \mathbb{R}$ which satisfies the properties

143

1° $U(x,y) \geq V(x,y)$ for all $(x,y) \in \mathbb{R} \times \mathcal{B}$,

2° there are Borel $A : \mathbb{R} \times \mathcal{B} \to \mathbb{R}$ and $B : \mathbb{R} \times \mathcal{B} \to \mathcal{B}^*$ satisfying

$$\alpha|B| \leq -A \qquad \text{on } \mathbb{R} \times \mathcal{B} \tag{4.2}$$

and such that for any $x,\, h \in \mathbb{R}$ and $y,\, k \in \mathcal{B}$ with $|k| \leq |h|$, we have

$$U(x+h, y+k) \leq U(x,y) + A(x,y)h + \langle B(x,y), k\rangle, \tag{4.3}$$

3° $U(x,y) \leq 0$ for all $x \in \mathbb{R}$, $y \in \mathcal{B}$ such that $|y| \leq |x|$.

Theorem 4.1. *Let V be as above and assume that U satisfies* 1°, 2° *and* 3°. *Suppose that f is a submartingale and let g be a $\mathcal{B}$-valued sequence which is α-strongly subordinate to f. Assume further that $f,\, g$ satisfy the integrability conditions*

$$\mathbb{E}|V(f_n, g_n)| < \infty, \quad \mathbb{E}|U(f_n, g_n)| < \infty,$$
$$\mathbb{E}\big(|A(f_n, g_n)||df_{n+1}| + |B(f_n, g_n)||dg_{n+1}|\big) < \infty,$$

for all $n \geq 0$. Then

$$\mathbb{E}V(f_n, g_n) \leq 0, \qquad n = 0,\, 1,\, 2,\, \dots.$$

Proof. It suffices to show that the process $(U(f_n, g_n))_{n \geq 0}$ is a supermartingale. Fix $n \geq 1$ and apply 2° to obtain

$$U(f_n, g_n) \leq U(f_{n-1}, g_{n-1}) + A(f_{n-1}, g_{n-1})df_n + \langle B(f_{n-1}, g_{n-1}), dg_n\rangle.$$

Both sides are integrable, so

$$\mathbb{E}(U(f_n, g_n)|\mathcal{F}_{n-1}) \leq U(f_{n-1}, g_{n-1}) + A(f_{n-1}, g_{n-1})\mathbb{E}(df_n|\mathcal{F}_{n-1})$$
$$+ \langle B(f_{n-1}, g_{n-1}), \mathbb{E}(dg_n|\mathcal{F}_{n-1})\rangle.$$

However,

$$A(f_{n-1}, g_{n-1})\mathbb{E}(df_n|\mathcal{F}_{n-1}) + \langle B(f_{n-1}, g_{n-1}), \mathbb{E}(dg_n|\mathcal{F}_{n-1})\rangle$$
$$\leq A(f_{n-1}, g_{n-1})\mathbb{E}(df_n|\mathcal{F}_{n-1}) + |B(f_{n-1}, g_{n-1})||\mathbb{E}(dg_n|\mathcal{F}_{n-1})| \leq 0,$$

due to the assumption $\alpha|B| \leq -A$, the α-strong subordination and the inequality $\mathbb{E}(df_n|\mathcal{F}_{n-1}) \geq 0$, which follows from the submartingale property. This gives the claim. $\qquad \square$

Remark 4.1. Passing from f to $-f$ we get the analogous statement for supermartingales. Note that the only change in the conditions 1°–3° is that in the second property the inequality $\alpha|B| \leq -A$ must be replaced by $\alpha|B| \leq A$.

Remark 4.2. As in the martingale setting, if U is sufficiently regular (say, piecewise C^1), the typical approach while studying 2° is to look at the following, slightly stronger statement.

$2°'$ For any $x \in \mathbb{R}$, $h > 0$ and y, $k \in \mathcal{B}$ with $|k| \leq \alpha h$, the function $G = G_{x,y,h,k} :$ $\mathbb{R} \to \mathbb{R}$ given by $G(t) = U(x + th, y + tk)$ is nonincreasing. For any x, $h \in \mathbb{R}$ and y, $k \in \mathcal{B}$ with $|k| \leq |h|$, the function $G_{x,y,h,k}$ is concave.

This condition is verified directly, by straightforward analysis of the derivative of the function G.

Suppose that V is a given function, U is piecewise C^1 and satisfies $1°$, $2°'$ and $3°$: this will be our typical situation. Then the theorem above holds for a wider class of processes.

Definition 4.2. Fix a nonnegative number α. Let f be a submartingale and let g be an adapted sequence of integrable $\mathcal{B}$-valued random variables. We say that g is α-subordinate to f if there is a decomposition

$$f_n = d_n + a_n, \qquad g_n = e_n + b_n, \qquad n \geq 0,$$

such that

(i) a, b are adapted and satisfy $|b_n| \leq \alpha a_n$ for each $n \geq 0$,
(ii) d is a submartingale and e is α-strongly differentially subordinate to d.

One introduces α-subordination for supermartingales by replacing f with $-f$ in the definition above. Clearly, if g is α-strongly subordinate to a sub- or a supermartingale f, then g is α-subordinate to f (simply take $a = (0,0,\ldots)$ and $b = (0,0,\ldots)$). It is easy to see that the reverse implication is not true, since the inequality $|e_n + b_n| \leq |d_n + a_n|$ need not hold in general. The extension is particularly useful when $\alpha > 1$. The following pair (f, g) will be frequently used below. Suppose that d, e are real-valued martingales such that e is differentially subordinate to d and let a be a predictable nondecreasing process starting from 0. Take $df_{2n} = dd_n$, $dg_{2n} = de_n$ and $df_{2n+1} = a_{n+1}$, $dg_{2n+1} = \pm\alpha a_{n+1}$ for $n \geq 0$. Then g is α-subordinate to f, with respect to the filtration $(\mathcal{F}_0, \mathcal{F}_0, \mathcal{F}_1, \mathcal{F}_1, \mathcal{F}_2, \mathcal{F}_2, \ldots)$. Furthermore, when $\alpha > 1$, the α-strong differential subordination is violated at odd moves: the condition $|dg_{2n+1}| \leq |df_{2n+1}|$ fails to hold.

Theorem 4.2. Suppose that U is piecewise C^1 and satisfies $1°$, $2°'$ and $3°$. Then for any f, g as in Definition 4.2, we have

$$\mathbb{E}V(f_n, g_n) \leq 0, \qquad n = 0, 1, 2, \ldots.$$

Proof. As previously, we shall prove that $(U(f_n, g_n))_{n \geq 0}$ is a supermartingale. Let $n \geq 1$ be a fixed integer. Using the first part of $2°'$, we may write

$$U(f_n, g_n) = U(f_{n-1} + d_n + a_n, g_{n-1} + e_n + b_n) \leq U(f_{n-1} + d_n, g_{n-1} + e_n).$$

Now we repeat the reasoning from Theorem 4.2 (with A, B equal to the corresponding partial derivatives of U or their limits) to obtain

$$\mathbb{E}\big[U(f_{n-1} + d_n, g_{n-1} + e_n)|\mathcal{F}_{n-1}\big] \leq U(f_{n-1}, g_{n-1}).$$

This completes the proof. $\qquad \square$

Using the splicing argument, we obtain the following result, which can be regarded as a reverse to Theorem 4.2. Let $M^{\text{sub}}(x, y)$ denote the class of all simple pairs (f, g) such that f is a submartingale starting from x, g is a sequence starting from y and such that the following holds. For $n = 0, 1, 2, \ldots$,

$$df_{2n+1}, \; dg_{2n+1} \text{ are predictable with } |dg_{2n+1}| \leq \alpha df_{2n+1},$$

$$|dg_{2n+2}| \leq |df_{2n+2}|, \quad \mathbb{E}(df_{2n+2}|\mathcal{F}_{2n+1}) = \mathbb{E}(dg_{2n+2}|\mathcal{F}_{2n+1}) = 0.$$

In other words, the pairs (f, g) from $M(x, y)$ behave like martingales at even times, and move "to the right" along some line of slope belonging to $[-\alpha, \alpha]$ at odd times.

Theorem 4.3. *Let $\alpha \geq 0$. Suppose that $V : \mathbb{R} \times \mathbb{R} \to \mathbb{R}$ is a given Borel function such that*

$$\mathbb{E}V(f_n, g_n) \leq 0, \qquad n = 0, 1, 2, \ldots,$$

for any simple sequences f, g such that f is a submartingale and g is real valued and α-subordinate to f. Then there is $U : \mathbb{R} \times \mathbb{R} \to \mathbb{R}$ satisfying $1°$, $2°'$ and $3°$. Furthermore, the least such function is given by

$$U^0(x, y) = \sup\{\mathbb{E}V(f_n, g_n)\},$$

where the supremum is taken over all n and all pairs $(f, g) \in M^{\text{sub}}(x, y)$.

The proof is essentially the same as in the martingale setting. There is an analogous statement for supermartingales. We omit the further details.

We conclude this section by noting that Theorems 4.2 and 4.3 can be easily modified to the case when the sub- or supermartingale f is nonnegative.

4.2 Weak type estimates for general sub- or supermartingales

4.2.1 Formulation of the results

We begin with the following result. As usual, let $\mathcal{H}$ be a given separable Hilbert space.

Theorem 4.4. *Let $\alpha \geq 0$ be fixed. Suppose that f is a submartingale or a supermartingale and that g is a sequence of $\mathcal{H}$-valued integrable random variables which is α-subordinate to f. Then*

$$||g||_{1,\infty} \leq C_\alpha ||f||_1, \tag{4.4}$$

where

$$C_\alpha = \begin{cases} (\alpha + 1)\left[1 + (\alpha + 1)^{1/\alpha}\right] & \text{if } \alpha > 1, \\ 6 & \text{if } \alpha \leq 1. \end{cases}$$

The constant is the best possible, even if $\mathcal{H} = \mathbb{R}$.

It is important to note that Burkholder's method is not directly applicable here. Indeed, the inequality (4.4) is *not* equivalent to showing that for any $n \geq 0$ we have

$$\mathbb{P}(|g_n| \geq 1) \leq C_\alpha \mathbb{E}|f_n|.$$

This is because the sequence $(\mathbb{E}|f_n|)_{n \geq 0}$ is no longer nondecreasing. In fact, the latter inequality is not valid, which can be seen by considering the following easy example: $f_0 = g_0 \equiv -1$, $f_1 \equiv 0$, $g_1 \equiv -1$ and $df_n = dg_n \equiv 0$ for $n \geq 2$. Then f is a submartingale, g is α-subordinate to f (for any $\alpha \geq 0$) and, for any $n \geq 1$ we have $\mathbb{P}(|g_n| \geq 1) = 1$ and $\mathbb{E}|f_n| = 0$. Thus we need to reformulate the inequality (4.4) so that Burkholder's technique works. This is done by means of the following stronger inequality. Here and below, x^+ denotes the positive part of x: $x^+ = \max\{x, 0\}$.

Theorem 4.5. *Under the assumptions of Theorem 4.4, we have*

$$||g||_{1,\infty} \leq K_\alpha ||f^+||_1 - (C_\alpha - K_\alpha)\mathbb{E}f_0, \tag{4.5}$$

where

$$K_\alpha = \begin{cases} (\alpha + 1)^{1+1/\alpha} & \text{if } \alpha \geq 1, \\ 4 & \text{if } \alpha \leq 1. \end{cases}$$

The inequality is sharp.

Finally, we shall prove the following characterization of Hilbert spaces, related to the inequality (4.4).

Theorem 4.6. *Let $\mathcal{B}$ be a separate Banach space and let α be fixed nonnegative number. Suppose that for any pair (f, g) such that f is a submartingale and g is a sequence of $\mathcal{B}$-valued integrable random variables which is α-subordinate to f, we have*

$$\mathbb{P}(|g_n| \geq 1) \leq C_\alpha ||f||_1, \qquad n = 0, 1, 2, \ldots.$$

Then $\mathcal{B}$ is a Hilbert space.

See Theorem 3.28 above for the martingale version of this result.

4.2.2 Proof of Theorems 4.4 and 4.5

Proof of (4.5). In fact, we shall need a slight modification of Burkholder's technique, because of the existence of the term $\mathbb{E}f_0$ on the right. Namely, let $V : \mathbb{R} \times \mathcal{H} \to \mathbb{R}$ be given by $V(x, y) = 1_{\{|y| \geq 1\}} - K_\alpha x^+$. We will search for a special function U_α which satisfies the usual conditions 1° and 2° and the following modification of the initial condition:

3° $U_\alpha(x, y) \leq -(C_\alpha - K_\alpha)x$ for all $x \in \mathbb{R}$, $y \in \mathcal{H}$ such that $|y| \leq |x|$.

The existence of such U_α yields

$$\mathbb{E}V(f_n, g_n) \leq \mathbb{E}U_\alpha(f_n, g_n) \leq \mathbb{E}U_\alpha(f_0, g_0) \leq -(C_\alpha - K_\alpha)\mathbb{E}f_0,$$

for $n = 0, 1, 2, \ldots$. That is,

$$\mathbb{P}(|g_n| \geq 1) \leq K_\alpha \mathbb{E} f_n^+ - (C_\alpha - K_\alpha) \mathbb{E} f_0 \leq K_\alpha \| f^+ \|_1 - (C_\alpha - K_\alpha) \mathbb{E} f_0$$

and this implies (4.5) by the stopping time argument.

To introduce the needed special function, fix a nonnegative number α and consider the following subsets of $\mathbb{R} \times \mathcal{H}$. If $\alpha \geq 1$, then

$$D_1^\alpha = \{(x, y) : \alpha|x| + |y| \geq 1, \ x \leq 0\},$$
$$D_2^\alpha = \{(x, y) : |x| + |y| \geq 1, \ x \geq 0\},$$
$$D_3^\alpha = (\mathbb{R} \times \mathcal{H}) \setminus (D_1^\alpha \cup D_2^\alpha).$$

If $\alpha \in [0, 1)$, let $D_i^\alpha = D_i^1$ for $i = 1, 2, 3$. Now, if $\alpha \geq 1$, put

$$U_\alpha(x, y) = \begin{cases} 1 - K_\alpha x^+ & \text{if } (x, y) \in D_1^\alpha \cup D_2^\alpha, \\ 1 - (\alpha x - |y| + 1)(\alpha x + \alpha|y| + 1)^{1/\alpha} & \text{if } (x, y) \in D_3^\alpha \end{cases} \quad (4.6)$$

and, for $\alpha \in [0, 1)$, let $U_\alpha(x, y) = U_1(x, y)$. We shall verify the conditions $1°$, $2°'$ and $3°$. Clearly, we may restrict ourselves to the case $\alpha \geq 1$, since $V_\alpha = V_1$ and $U_\alpha = U_1$ for $\alpha \in [0, 1)$.

First we focus on the majorization property. Clearly, it suffices to establish this estimate on D_3^α, where it takes the form

$$1 - (\alpha x - |y| + 1)(\alpha x + \alpha|y| + 1)^{1/\alpha} \geq -K_\alpha x^+.$$

For a fixed x, the left-hand side increases as $|y|$ increases. Hence all we need is to show the estimate for $y = 0$: $1 - (\alpha x + 1)^{1+1/\alpha} \geq -K_\alpha x^+$. This is obvious for $x \leq 0$, since then $\alpha x + 1 \leq 1$. For $x \geq 0$ we use the fact that the function $j_1(x) = 1 - (\alpha x + 1)^{1+1/\alpha}$ is concave and lies above the linear $j_2(x) = -K_\alpha x$ on $[0, 1]$, because $j_1(0) = j_2(0)$ and $j_1(1) = j_2(1) + 1 > j_2(1)$.

We turn to $2°'$. Let $A_\alpha : \mathbb{R} \times \mathcal{H} \to \mathbb{R}$, $B_\alpha : \mathbb{R} \times \mathcal{H} \to \mathcal{H}$ be given by

$$A_\alpha(x, y) = \begin{cases} 0 & \text{if } (x, y) \in D_1^\alpha, \\ -K_\alpha & \text{if } (x, y) \in D_2^\alpha, \\ -(\alpha + 1)(\alpha x + \alpha|y| + 1)^{1/\alpha - 1}[\alpha x + 1 + (\alpha - 1)|y|] & \text{if } (x, y) \in D_3^\alpha \end{cases}$$

and

$$B_\alpha(x, y) = \begin{cases} 0 & \text{if } (x, y) \in D_1^\alpha \cup D_2^\alpha, \\ (\alpha + 1)(\alpha x + \alpha|y| + 1)^{1/\alpha - 1} y & \text{if } (x, y) \in D_3^\alpha. \end{cases}$$

The functions A_α, B_α coincide with the partial derivatives $U_{\alpha x}$, $U_{\alpha y}$ on the interiors of D_1^α, D_2^α and D_3^α. Observe that $A_\alpha(x, y) + \alpha|B_\alpha(x, y)| \leq 0$ for all x and y. This is clear for $(x, y) \in D_1^\alpha \cup D_2^\alpha$, while for remaining (x, y) we derive that

$$A_\alpha(x, y) + \alpha|B_\alpha(x, y)| = -(\alpha + 1)(\alpha x + \alpha|y| + 1)^{1/\alpha - 1}(\alpha x + 1 - |y|) \leq 0. \quad (4.7)$$

This proves the first half of condition $2^{\circ\prime}$. To establish the second half, fix $x \in \mathbb{R}$, $h > -x$, $y, k \in \mathcal{H}$ such that $|k| \leq |h|$. It suffices to show that the function $G(t) = U(x + th, y + tk)$ (defined for t such that $x + th \geq 0$) is concave. Note that G is linear on a large part of its domain; in fact, all we need is to check what happens on D_3^α. Precisely, fix $(x, y) \in D_3^\alpha$ and observe that

$$G''(0) = U_{\alpha xx}(x, y)h^2 + 2(U_{\alpha xy}(x, y)h, k) + (kU_{\alpha yy}(x, y), k) = I_1 + I_2,$$

where

$$
\begin{aligned}
I_1 &= (\alpha + 1)(\alpha x + \alpha|y| + 1)^{1/\alpha - 1}(|k|^2 - h^2) \leq 0, \\
I_2 &= (\alpha + 1)(1 - \alpha)|y|(\alpha x + \alpha|y| + 1)^{1/\alpha - 2}\left[h + (y \cdot k)/|y|\right]^2 \leq 0.
\end{aligned}
\tag{4.8}
$$

Furthermore,

$$G'(0) \leq U_{\alpha x}(x, y)h + |U_{\alpha y}(x, y)| \cdot |k| \leq h\left[U_{\alpha x}(x, y) + \alpha|U_{\alpha y}(x, y)|\right] \leq 0.$$

In addition, using $(y, k)/h \geq -|y|$ and the estimate $x + |y| \leq 1$ coming from the definition of D_3^α,

$$
\begin{aligned}
G'(0)/h &= -(\alpha + 1)(\alpha x + \alpha|y| + 1)^{1/\alpha - 1}[\alpha x + 1 + (\alpha - 1)|y| - (y \cdot k)/h] \\
&\geq -(\alpha + 1)(\alpha x + \alpha|y| + 1)^{1/\alpha} \geq -(\alpha + 1)^{1/\alpha + 1} = -K_\alpha.
\end{aligned}
$$

This shows that G is concave on $\mathbb{R}$. Finally, to check 3°, we use 2° and obtain that for $|y| \leq |x|$,

$$U_\alpha(x, y) \leq U_\alpha(0, 0) + U_{\alpha x}(0, 0)x + U_{\alpha y}(0, 0) \cdot y = -(\alpha + 1)x. \tag{4.9}$$

This completes the proof: the necessary integrability holds, since A_α and B_α are bounded. $\qquad\square$

Sharpness. It suffices to focus on the inequality (4.4). We shall construct appropriate examples on the probability space $([0, 1], \mathcal{B}(0, 1), |\cdot|)$. To gain some intuition about their structure, it is convenient to start with the case $\alpha \leq 1$, which is a bit easier.

First consider the following pair (f, g): let

$$
\begin{aligned}
f_0 &= -g_0 = -\frac{1}{6}[0, 1], \\
df_1 &= -dg_1 = -\frac{1}{3}\left[0, \frac{2}{3}\right] + \frac{2}{3}\left(\frac{2}{3}, 1\right], \\
df_2 &= dg_2 = \frac{1}{2}\left[0, \frac{1}{2}\right] - \frac{3}{2}\left(\frac{1}{2}, \frac{2}{3}\right] + \frac{3}{2}\left(\frac{2}{3}, \frac{3}{4}\right] - \frac{1}{2}\left(\frac{3}{4}, 1\right], \\
df_3 &= \left(\frac{1}{2}, \frac{2}{3}\right], \quad dg_3 \equiv 0, \\
df_n &= dg_n \equiv 0 \quad \text{for } n \geq 4
\end{aligned}
$$

(here, as usual, we identify an interval with its indicator function). It is easy to check that f is a submartingale and g is 0-subordinate to f (hence g is also α-subordinate to f for any $\alpha \geq 0$). In fact the moves in steps 1 and 2 are of martingale type in the sense that $\mathbb{E}df_1 = \mathbb{E}dg_1 = 0$ and $\mathbb{E}(df_2|f_1) = \mathbb{E}(dg_2|f_1) = 0$. Furthermore, we see that $|g_3| = |g_2| = 1$ almost surely and $f_3 = 2\left(\frac{2}{3}, \frac{3}{4}\right]$, so $||f_3||_1 = \frac{1}{6}$. Consequently, $\mathbb{P}(|g|^* \geq 1) = 6||f_3||_1$. However, we encounter here a problem, already mentioned above, which did not exist in the martingale setting. Namely, if f is a submartingale, then, in general, $||f||_1 \neq \lim_{n\to\infty} ||f_n||_1$. Unfortunately, this is also the case in our example: it is easy to see that $||f_2||_1 > ||f_3||_1$. Thus we need an additional modification of the pair (f, g) which will keep $||f_n||_1$ close to $1/6$ for all n. The idea is to split the set Ω into small parts and evolve the pair (f, g) according to the rules above on the first part, then on the second, and so on. To be more precise, let N be a large positive integer. Consider the pair $(\tilde{f}, \tilde{g})$ on $([0, 1], \mathcal{B}(0, 1), |\cdot|)$, defined as follows. Let $d\tilde{f}_0 = -d\tilde{g}_0 = -\frac{1}{6}[0, 1]$ and, for a fixed $n \in \{0, 1, 2, \ldots, N-1\}$ and $k = 1, 2, 3$,

$$d\tilde{f}_{3n+k}(x) = df_k(Nx - n), \quad d\tilde{g}_{3n+k}(x) = dg_k(Nx - n) \qquad \text{for } x \in (n/N, (n+1)/N]$$

and $d\tilde{f}_{3n+k}(x) = d\tilde{g}_{3n+k}(x) = 0$ for the remaining x. Now we have $|\tilde{g}_{3N}| = 1$ with probability 1 and, for any n and k as above,

$$||\tilde{f}_{3n+k}||_1 = \frac{n}{N}||f_3||_1 + \frac{1}{N}||f_k||_1 + \frac{N-n-1}{N} \cdot \frac{1}{6} \leq \frac{1}{6} + \frac{1}{N} \max_{0 \leq k \leq 3} ||f_k||_1.$$

Since N was arbitrary, the constant 6 is indeed the best possible.

Now we turn to the case $\alpha > 1$, in which the situation becomes more involved because the "basic" pair (f, g) has a more complicated structure. Let N be a large positive integer and put $\delta = 1/(2N)$. For $n = 1, 2, \ldots, N-1$, define

$$\ell_n = \frac{-1 + 2(n-1)\delta}{\alpha + 1}, \qquad r_n = (2n-1)\delta,$$

and put $\ell_N = 0$, $r_N = 2$. Consider a martingale f which evolves as follows:

(i) It starts from $-C_\alpha^{-1}$.

(ii) $f_1 \in \{\ell_1, r_1\}$.

(iii) If $n = 1, 2, \ldots, N-1$, then $f_{n+1} \in \{\ell_{n+1}, r_{n+1}\}$ on the set $\{f_n = r_n\}$.

(iv) On the set $\{f_n = \ell_n\}$, $f_n = f_{n+1} = \cdots = f_N$.

Furthermore, let $g_0 = -f_0$ and $dg_n = (-1)^n df_n$ for $n = 0, 1, 2, \ldots, N$. It is easy to verify that the random variable (f_N, g_N) satisfies $\alpha f_N + |g_N| = 1$ or $(f_N, g_N) = (2, 1)$. The final move (the only one which is not of martingale type) is the following. If $\alpha f_N \pm g_N = 1$, we take $df_{N+1} = -f_N$ and $dg_{N+1} = \pm \alpha df_{N+1}$; if $(f_n, g_N) = (2, 1)$, then $df_{N+1} = dg_{N+1} = 0$. It is evident that g is α-subordinate to f. However, it is worth mentioning here that the α-strong differential subordination *does not* hold, because of the final move of (f, g).

Let us now study the numbers $\mathbb{P}(|g|^* \geq 1)$, $||f||_1$ and $||f_{N+1}||_1$. The first of them equals 1, since $|g_{2N}| \equiv 1$. Furthermore,

$$||f||_1 \leq 2, \tag{4.10}$$

since $|f_n| \leq 2$ for any n. The next step is to note that $f_{N+1} \in \{0, 2\}$ and $f_{N+1} = 2$ if and only if the process f is nondecreasing. For convenience, put $r_0 = -C_\alpha^{-1}$; then we can write

$$\mathbb{P}(f_{N+1} = 2) = \prod_{n=1}^{N} \frac{r_{n-1} - \ell_n}{r_n - \ell_n}$$

$$= \frac{r_0 - \ell_1}{r_1 - \ell_1} \cdot \frac{r_{N-1} - \ell_N}{r_N - \ell_N} \prod_{n=2}^{N-1} \frac{r_{n-1} - \ell_n}{r_n - \ell_n}$$

$$= \frac{r_0 + (\alpha + 1)^{-1}}{2\delta + (\alpha + 1)^{-1}} \cdot \frac{1 - \delta}{2} \prod_{n=2}^{N-1} \left(1 - \frac{2\delta(\alpha + 1)}{1 + \alpha\delta(2n - 1)}\right)$$

$$\leq \frac{(1 + r_0(\alpha + 1))(1 - \delta)}{2(1 + 2\delta(\alpha + 1))} \exp\left(-2\delta(\alpha + 1) \sum_{n=2}^{N-1} (1 + \alpha\delta(2n - 1))^{-1}\right)$$

$$\leq \frac{(1 + r_0(\alpha + 1))(1 - \delta)}{2(1 + 2\delta(\alpha + 1))} \left(\frac{1 + (2N - 1)\delta\alpha}{1 + 3\delta\alpha}\right)^{-(\alpha+1)/\alpha}.$$

Here in the first inequality we have used the elementary bound $1 - x \leq e^{-x}$ and in the second one we have exploited the fact that

$$2\delta \sum_{n=2}^{N-1} (1 + \alpha\delta(2n - 1))^{-1} \geq \int_{3\delta}^{(2N-1)\delta} (1 + \alpha x)^{-1} dx = \frac{1}{\alpha} \log \frac{1 + (2N - 1)\delta\alpha}{1 + 3\delta\alpha}.$$

Now recall that $\delta = (2N)^{-1}$, so letting $N \to \infty$ yields

$$\mathbb{P}(f_{N+1} = 2) \to \frac{(\alpha + 1)r_0 + 1}{2}(1 + \alpha)^{-(\alpha+1)/\alpha} = (2C_\alpha)^{-1}$$

and, in consequence, $||f_{N+1}||_1 \to C_\alpha^{-1}$. This would complete the proof, but, as in the case $\alpha \leq 1$, we have a problem because of the inequality $||f||_1 > ||f_{N+1}||_1$. This is dealt with exactly in the same manner as for $\alpha \leq 1$. Namely, we split the probability space into a large number of small parts and copy the above pair into each part on a separate period of time. Finally, using (4.10), we are able to ensure that the first norm of f is arbitrarily close to C_α^{-1}. $\square$

On the search of the suitable majorant. We shall only study the case $\alpha = 1$, the other values of α can be treated likewise. The first question which comes to one's mind is how to discover the right version of (4.4) (that is, how to deduce that (4.5) is the right inequality to prove). As observed above, the inequality $\mathbb{P}(|g_n| \geq 1) \leq$

$\beta \mathbb{E}|f_n|$ is not a good choice. Nonetheless, let $V(x, y) = 1_{\{|y| \geq 1\}} - \beta|x|$ for x, $y \in \mathbb{R}$ and consider the function $U^0 : \mathbb{R} \times \mathbb{R} \to \mathbb{R}$ given by

$$U^0(x, y) = \sup\{\mathbb{E}V(f_n, g_n)\},$$

where the supremum is taken over all n and all $(f, g) \in M^{\text{sub}}(x, y)$. The key observation is that $U^0 = U^1$, where

$$U^1(x, y) = \sup\{\mathbb{P}(|g_n| \geq 1) - \beta\mathbb{E}f_n^+\},$$

with the supremum over the same parameters as previously. The bound $U^0 \leq U^1$ comes from the trivial estimate $f_n^+ \leq |f_n|$. To get the reverse, pick any f, g, n as above and consider $\bar{f}$, $\bar{g}$ given by $\bar{f}_k = f_k$ and $\bar{g}_k = g_k$ for $k \leq n$ and $d\bar{f}_{n+1} = f_n^-$, $d\bar{f}_{n+2} = d\bar{f}_{n+3} = \cdots = d\bar{g}_{n+1} = d\bar{g}_{n+2} = \cdots \equiv 0$. Then $\bar{f}$ is a submartingale, $\bar{g}$ is 1-subordinate to $\bar{f}$ and

$$\mathbb{P}(|g_n| \geq 1) - \beta\mathbb{E}f_n^+ = \mathbb{P}(|\bar{g}_{n+1}| \geq 1) - \beta\mathbb{E}|\bar{f}_{n+1}| \leq U^0(x, y).$$

Consequently, we have $U^1 \leq U^0$, since f, g and n were arbitrary. This explains why the nonnegative part of f appears in (4.5). Now, using the standard arguments we can show that U^0 satisfies the conditions $1°$ and $2°$. However, there is no chance for the initial condition to hold, which can be seen, for example, by considering the deterministic pair $(-1, 1)$. All we can hope for is the estimate $U(x, y) \leq U(x, x)$ for $|y| \leq |x|$ and hence, all we can get using Burkholder's method is the inequality

$$\mathbb{P}(|g_n| \geq 1) \leq \beta\mathbb{E}f_n^+ + \mathbb{E}U^0(f_0, f_0).$$

This leads directly to the estimate (4.5).

Now let us collect some information on U^0. First, $U^0(x, y) \leq 1 - \beta x^+$ for all x, y. Moreover, we have equality for $|x| + |y| \geq 1$. This is trivial for $|y| \geq 1$, and for $|y| < 1$ it can be seen by using the following example, which already appeared while studying the weak type (1,1) inequality for martingales. Namely, take (f, g) such that $(f_0, g_0) \equiv (x, y)$, df_1 is a centered random variable taking values $1 \pm |y|$ and $|dg_1| = |df_1|$, so that $|g_1| \equiv 1$. Finally, inside the square $|x| + |y| < 1$, we *guess* that the special function is linear along the line segments of slope -1. This gives the formula (4.6). $\square$

4.2.3 Proof of Theorem 4.6

We shall only present here the proof in the case $\alpha \in [0, 1]$. The remaining case $\alpha > 1$ can be dealt with similarly, but the calculations are a bit more complicated.

The key tool in studying our problem is the special function $U_B^0 : \mathbb{R} \times B \to \mathbb{R}$, defined by

$$U_B^0(x, y) = \inf\{\mathbb{E}f_\infty^+\}.$$

Here the infimum is taken over all $(f, g) \in M^{\text{sub}}(x, y)$ satisfying $\mathbb{P}(|g_\infty| \geq 1) = 1$. First, let us show an auxiliary fact.

Lemma 4.1. *If* $|y| = 1/6$, *then* $U_{\mathcal{B}}^0(-1/6, |y|) \geq 1/3$.

Proof. Suppose, on the contrary, that there is a pair $(f, g) \in M^{\mathrm{sub}}(-1/6, y)$ with $\mathbb{P}(|g_n| \geq 1) = 1$ and $\mathbb{E}f_n^+ < 1/6$ for some n. For a given small positive ε, consider a pair $(\overline{f}, \overline{g}) = (f + \varepsilon, g - y'\varepsilon)$, which of course belongs to $M^{\mathrm{sub}}(-1/6 + \varepsilon, y - y'\varepsilon)$. Possibly, the condition $\mathbb{P}(|\overline{g}_n| \geq 1) = 1$ does not hold any more: to ensure this, we modify the pair by adding the differences $d\overline{f}_{n+1}$, $d\overline{g}_{n+1}$ satisfying $\mathbb{P}(|\overline{g}_{n+1}| \geq 1) = 1$ and $\mathbb{E}(d\overline{f}_{n+1}|\mathcal{F}_n) = 0$, $\mathbb{E}(d\overline{g}_{n+1}|\mathcal{F}_n) = 0$ (so α-subordination is not violated). It is easy to see that when we pick sufficiently small ε, then we may do this so that the inequality $\mathbb{E}\overline{f}_{n+1}^+ < 1/6$ still holds. Now we use the portioning argument: for any $\delta > 0$ there is a pair $(F, G) \in M^{\mathrm{sub}}(-1/6 + \varepsilon, y - y'\varepsilon)$ for which $\mathbb{P}(|G_\infty| \geq 1) = 1$ and $||F||_1 = ||F^+||_1 \leq \max\{-\mathbb{E}\overline{f}_0, \mathbb{E}\overline{f}_{n+1}^+\} + \delta$. Consequently, if δ is sufficiently small, we get a pair (F, G) such that F is a submartingale, G is α-subordinate to F, $\mathbb{P}(|G_\infty| \geq 1) = 1$ and $||F||_1 < 1/6$. We get a contradiction: (4.4) does not hold. This proves the claim. $\qquad\square$

We are ready for the proof of Theorem 4.6. By using the splicing argument, we can show that the function $-U_{\mathcal{B}}^0$ satisfies the condition 2°. We shall prove that

$$U_{\mathcal{B}}^0(x, y) = U_\alpha(x, y) = \begin{cases} x^+ & \text{if } |x| + |y| \geq 1, \\ \frac{1}{4}((1+x)^2 - |y|^2) & \text{if } |x| + |y| < 1. \end{cases} \qquad (4.11)$$

Step 1. This is standard when $|x| + |y| \geq 1$ (by considering appropriate pairs (f, g)).

Step 2. If $|x| + |y| < 1$ and $\mathcal{B} = \mathbb{R}$, this can be easily seen by modifying the example which appears in the proof of the sharpness. It can also be established using Theorem 4.3: see the search which leads to (3.6) above.

Step 3. When $|x| + |y| < 1$ and $\mathcal{B}$ is general, note that $U_{\mathcal{B}}^0(x, y) \leq U_{\mathbb{R}}^0(x, |y|)$ for any $x \in \mathbb{R}$ and $y \in \mathcal{B}$. This follows immediately from the fact that when $y \neq 0$, then each pair $(f, g) \in M^{\mathrm{sub}}(x, |y|)$ gives rise to a pair $(f, gy') \in M^{\mathrm{sub}}(x, y)$; the same holds for $y = 0$, with y' replaced by any vector of norm one. On the other hand, by Lemma 4.1, we have $U_{\mathcal{B}}^0(-1/6, y) \geq U_{\mathbb{R}}^0(-1/6, |y|)$. Next, fix $y \in \mathcal{B}$ of norm one and consider the line segment

$$\{(t, ty) : t \in [-1/2, 1/2]\} \subset \mathbb{R} \times \mathcal{B}.$$

The function $U_{\mathcal{B}}^0$ is convex along this segment; furthermore, $U_{\mathcal{B}}^0(t, ty) = U_{\mathbb{R}}^0(t, t|y|)$ for $t = \pm 1/2$ (see Step 1) and $t = -1/6$ (see Lemma 4.1 and the beginning of this step). However, $t \mapsto U_{\mathbb{R}}^0(t, t|y|)$, $t \in [-1/2, 1/2]$, is linear: thus (4.11) holds on every line segment of the above form.

Step 4. Now, suppose that $|x| + |y| < 1$ and $|x| \neq |y|$. Then we consider the line segment

$$\{(x + t, y + y't) : |x + t| + |y + y't| \leq 1\} \subset \mathbb{R} \times \mathcal{B}$$

and note there are three values of t such that $U_\mathcal{B}^0(x+t, y+y't) = U_\mathbb{R}^0(x+t, |y+y't|)$: two corresponding to the endpoints of the segment and one inside, coming from the previous step. Since $t \mapsto U_\mathbb{R}^0(x+t, |y+y't|)$ is linear, we get that $U_\mathcal{B}^0(x,y) = U_\mathbb{R}^0(x, |y|)$; this completes the proof of (4.11).

Now take any $a, b \in \mathcal{B}$ and note that by 2°′, $2U_\mathcal{B}^0(0, a) \geq U_\mathcal{B}^0(-|b|, a-b) + U_\mathcal{B}^0(|b|, a+b)$. When a and b are sufficiently small, this is equivalent to

$$|a+b|^2 + |a-b|^2 \leq 2|a|^2 + 2|b|^2$$

and, by homogeneity, extends to all $a, b \in \mathcal{B}$. Putting $a+b$, $a-b$ in the place of a and b, we get the reverse estimate. Consequently, the parallelogram identity is satisfied and, by the famous result of Jordan and von Neumann [105], $\mathcal{B}$ is a Hilbert space.

4.3 Weak type inequalities for nonnegative sub- and supermartingales

4.3.1 Formulation of the results

Now let us turn to the case when the dominating process is nonnegative. We have the following.

Theorem 4.7. *Let α be fixed nonnegative number. Assume that f is a nonnegative submartingale and g is a sequence of $\mathcal{H}$-valued integrable random variables which is α-subordinate to f. Then*

$$||g||_{1,\infty} \leq (\alpha+2)||f||_1. \tag{4.12}$$

The constant $\alpha + 2$ is the best possible, even if $\mathcal{H} = \mathbb{R}$.

In the supermartingale setting, the behavior of the constants is completely different: there are two expressions, depending on whether α is big or small.

Theorem 4.8. *Let $\alpha \geq 0$ be fixed and suppose that f is a nonnegative supermartingale and g is a sequence of $\mathcal{H}$-valued integrable random variables which is α-subordinate to f. Then for $0 < p \leq 1$,*

$$||g||_{p,\infty} \leq C_{\alpha,\infty}||f||_p, \tag{4.13}$$

where

$$C_{\alpha,\infty} = \begin{cases} 2 & \text{if } \alpha \leq 1, \\ \alpha+1 & \text{if } \alpha > 1. \end{cases}$$

The constant is the best possible, even if $\mathcal{H} = \mathbb{R}$. For $p > 1$ the inequality does not hold in general with any finite constant.

4.3.2 Proof of Theorem 4.7

Proof of (4.12). Here Burkholder's technique is directly applicable. It suffices to show that for any f and g as in the statement and any nonnegative integer n we have

$$\mathbb{P}(|g_n| \geq 1) \leq (\alpha + 2)\mathbb{E}f_n.$$

Fix $\alpha \geq 0$ and let $U, V : [0, \infty) \times \mathcal{H} \to \mathbb{R}$ be given by

$$U(x, y) = \begin{cases} (|y| - (\alpha + 1)x)(x + |y|)^{1/(\alpha+1)} & \text{if } x + |y| < 1, \\ 1 - (\alpha + 2)x & \text{if } x + |y| \geq 1 \end{cases} \quad (4.14)$$

and $V(x, y) = 1_{\{|y| \geq 1\}} - (\alpha + 2)x$. We will show that U satisfies the properties $1°$, $2°'$ and $3°$. To show the majorization, we may restrict ourselves to x, y satisfying $x + |y| < 1$. Substituting $x + |y| = r^{\alpha+1} \in [0, 1)$, we have

$$V(x, y) - U(x, y) = -r^{\alpha+2} - (1 - r)(\alpha + 2)x < 0.$$

To establish $2°'$, we first note that

$$U_x(x, y) + \alpha|U_y(x, y)| = \begin{cases} -(\alpha + 2)(x + |y|)^{-\alpha/(\alpha+1)}x & \text{if } x + |y| < 1, \\ -(\alpha + 2) & \text{if } x + |y| > 1 \end{cases}$$

is nonpositive: this implies the first half of $2°'$. To show the second half, let $G = G_{x,y,h,k}$ be defined as usual. A direct computation gives $G''(0) = 0$ if $x + |y| > 1$ and

$$G''(0) = I + II, \quad (4.15)$$

where

$$I = -\frac{\alpha + 2}{\alpha + 1}(x + |y|)^{-\alpha/(\alpha+1)}(h^2 - |k|^2),$$

$$II = -\frac{\alpha(\alpha + 2)}{(\alpha + 1)^2}(x + |y|)^{-(2\alpha+1)/(\alpha+1)}|y|(h - y' \cdot k)^2,$$

so $G''(0) \leq 0$ if $|k| \leq |h|$. Finally, for $x > 0$, $x + |y| = 1$ and $|k| \leq h$,

$$G'(0-) = -(\alpha + 2)h + \frac{\alpha + 2}{\alpha + 1}(|y|h + y \cdot k) \geq -(\alpha + 2)h = G'(0+),$$

so $2°'$ holds. To prove the initial condition, use $2°'$ to obtain

$$U(x, y) \leq U(0, 0) + U_x(0+, 0)x + U_y(0+, 0) \cdot y = 0.$$

Thus, it suffices to apply Theorem 4.2: the necessary integrability is satisfied, since the partial derivatives of U are bounded on $(0, \infty) \times \mathbb{R} \setminus \{(x, y) : x + |y| = 1\}$. $\square$

Remark 4.3. For $x + |y| < 1$ we have

$$U(x, y) \leq 1 - (\alpha + 2)x, \tag{4.16}$$

which can be seen, for example, by checking that for fixed x, $U(x, y)$ increases as $|y|$ increases. This inequality will be needed later.

Sharpness. Let $\alpha \geq 0$ be fixed and suppose $\beta = \beta(\alpha)$ has the property that

$$\mathbb{P}(|g_n| \geq 1) \leq \beta \mathbb{E} f_n \tag{4.17}$$

for any f, g such that f is a nonnegative submartingale and g is a real-valued process which is α-subordinate to f. Let $U^0 : [0, \infty) \times \mathbb{R} \to \mathbb{R}$ be given by

$$U^0(x, y) = \sup\{\mathbb{P}(|g_n| \geq 1) - \beta \mathbb{E} f_n\}, \tag{4.18}$$

where the supremum is taken over all n and all pairs $(f, g) \in M^{\mathrm{sub}}(x, y)$. We shall exploit the properties $1°$, $2°'$ and $3°$ to obtain $\beta \geq \alpha + 2$. The reasoning we are going to present now will be frequently repeated during the study of other estimates.

Introduce the notation $A(y) = U^0(0, y)$ and $B(x) = U^0(x, 0)$ for all $x \geq 0$ and $y \in \mathbb{R}$. Let N be a given positive integer and let $\delta = (N(\alpha + 1))^{-1}$. We have

$$A(0) = B(0) \leq 0, \qquad A(1) \geq 1 \qquad \text{and} \qquad B(1) \geq 1 - \beta. \tag{4.19}$$

Indeed: the first inequality comes from $3°$, while the second is an immediate consequence of $1°$. Finally, to prove the third estimate, we look at the pair (f, g) starting from $(1, 0)$ and such that $df_1 = dg_1$ takes values ± 1, each with probability $1/2$; then $|g_1| \equiv 1$ and $\mathbb{E} f_1 = 1$.

Using the first part of $2°'$ and then the second, we get

$$
\begin{aligned}
A(y) &\geq U^0(\delta, y + \alpha\delta) \\
&\geq \frac{\delta}{y + (\alpha + 1)\delta} B(y + (\alpha + 1)\delta) + \frac{y + \alpha\delta}{y + (\alpha + 1)\delta} A(y + (\alpha + 1)\delta).
\end{aligned} \tag{4.20}
$$

Furthermore, again by $2°$,

$$B(y) \geq \frac{(\alpha + 1)\delta}{2y + (\alpha + 1)\delta} A(y) + \frac{2y}{2y + (\alpha + 1)\delta} U^0\left(y + \frac{(\alpha + 1)\delta}{2}, \frac{(\alpha + 1)\delta}{2}\right)$$

and

$$
\begin{aligned}
U^0\left(y + \frac{(\alpha + 1)\delta}{2}, \frac{(\alpha + 1)\delta}{2}\right) &\geq \frac{2y + (\alpha + 1)\delta}{2y + 2(\alpha + 1)\delta} B(y + (\alpha + 1)\delta) \\
&\quad + \frac{(\alpha + 1)\delta}{2y + 2(\alpha + 1)\delta} A(y + (\alpha + 1)\delta).
\end{aligned}
$$

Consequently, we obtain

$$B(y) \geq \frac{(\alpha+1)\delta}{2y+(\alpha+1)\delta}B(y) + \frac{y}{y+(\alpha+1)\delta}B(y+(\alpha+1)\delta)$$
$$+ \frac{y(\alpha+1)\delta}{(2y+(\alpha+1)\delta)(y+(\alpha+1)\delta)}A(y+(\alpha+1)\delta).$$

Multiply both sides by $(\alpha+1)^{-1}$, add it to (4.20) and substitute $y = k(\alpha+1)\delta$. After some manipulations we arrive at the inequality

$$(\alpha+1)\big[A((k+1)(\alpha+1)\delta)-A(k(\alpha+1)\delta)\big]$$
$$+ B((k+1)(\alpha+1)\delta) - B(k(\alpha+1)\delta) \tag{4.21}$$
$$\leq \frac{A((k+1)(\alpha+1)\delta) - A(k(\alpha+1)\delta)}{2k+1}.$$

Writing this for $k = 0, 1, 2, \ldots, N-1$ and summing the obtained estimates gives

$$(\alpha+1)(A(1) - A(0)) + B(1) - B(0) \leq \sum_{k=0}^{N-1} \frac{A((k+1)(\alpha+1)\delta) - A(k(\alpha+1)\delta)}{2k+1},$$

which, by (4.19), implies

$$\alpha + 2 - \beta \leq \sum_{k=0}^{N-1} \frac{A((k+1)(\alpha+1)\delta) - A(k(\alpha+1)\delta)}{2k+1}. \tag{4.22}$$

To complete the proof it suffices to show that the right-hand side tends to 0 as N goes to infinity. To do this, we use (4.20) and the estimates $A(y) \leq 1 - y$, $B(x) \geq -x$, which follow directly from the definition of U^0. We obtain

$$A((k+1)(\alpha+1)\delta) - A(k(\alpha+1)\delta)$$
$$= \frac{k(\alpha+1)+\alpha}{(k+1)(\alpha+1)}A((k+1)(\alpha+1)\delta) - A(k(\alpha+1)\delta) + \frac{A((k+1)(\alpha+1)\delta)}{(k+1)(\alpha+1)}$$
$$\leq \frac{-B((k+1)(\alpha+1)\delta) + A((k+1)(\alpha+1)\delta)}{(k+1)(\alpha+1)} \leq \frac{1}{(k+1)(\alpha+1)}.$$

Therefore the right-hand side of (4.22) is of order $O(1/N)$ as $N \to \infty$ and the claim follows. $\square$

On the search of the suitable majorant. As usual, we restrict our attention to the case $\mathcal{H} = \mathbb{R}$. Let $\beta > 0$ be a constant in the weak type inequality (to be determined) and let us start with the function U^0, given by (4.18). Clearly, we have $U^0(x, y) \leq 1 - \beta x$, furthermore, the reverse inequality holds if $x + |y| \geq 1$ (consider standard examples: see, e.g., the weak-type $(1, 1)$ inequality for martingales). Thus all we need is to find the special function when $x + |y| < 1$. This will be accomplished by imposing three assumptions.

We start with the property

(A1) U is continuous on $[0, \infty) \times \mathbb{R}$, of class C^1 on the set

$$\{(x, y) \in [0, \infty) \times \mathbb{R} : x + |y| < 1\}$$

and satisfies $U(0,0) = 0$.

By the symmetry condition $U(x, y) = U(x, -y)$, which can be assumed with no loss of generality, we may restrict ourselves to $y \geq 0$. Let $A(y) = U(0, y)$ and $B(x) = U(x, 0)$ for all x, $y \geq 0$. We impose the following "structural" condition on U:

(A2) For any x, $y > 0$ with $x + y < 1$,

$$U(x, y) = \frac{y}{x + y} A(x + y) + \frac{x}{x + y} B(x + y).$$

In other words, in the first quadrant the function U is linear along the line segments of slope -1.

The final assumption is the following. We know that for any $y \geq 0$, the function $G_y(t) = U(t, y + \alpha t)$, $t \geq 0$, is nonincreasing. It is natural to expect the extremal behavior $G'_y(0+) = 0$, at least for some y. Otherwise, we would replace U with the function $\overline{U}(x, y) = U(x, y) + ay$ for some small $a > 0$, which also would have all the required properties (and would give a better constant). But which y should satisfy $G'_y(0+) = 0$? A little thought leads to the conjecture: all $y \in [0, 1)$. This implies the assumption

(A3) For any $y \in [0, 1)$,
$$U_x(0+, y) + \alpha U_y(0+, y) = 0.$$

Conditions (A1), (A2) and (A3) yield the special function studied above. To see this, note that by (A2),

$$\frac{U(x, y) - A(x + y)}{x} = \frac{B(x + y) - A(x + y)}{x + y},$$

so letting $x \to 0$ gives

$$U_x(0+, y) - U_y(0+, y) = \frac{B(y) - A(y)}{y},$$

or, by (A3),

$$-(\alpha + 1)A'(y) = \frac{B(y) - A(y)}{y}. \tag{4.23}$$

Next, we have $U_y(x, 0) = 0$ by the symmetry of U, which implies, by (A2),

$$B'(x) = \frac{B(x) - A(x)}{x}.$$

Combining this with (4.23) gives $B'(x) = -(\alpha+1)A'(x)$, or $B(x) = c-(\alpha+1)A(x)$ for some constant c. However, by (A1), $A(0) = B(0) = 0$, so $c = 0$; thus, coming back to (4.23), we get

$$-(\alpha + 1)A'(y) = -\frac{(\alpha + 2)A(y)}{y}.$$

Consequently, $A(y) = \bar{c}y^{(\alpha+2)/(\alpha+1)}$ and $B(x) = -\bar{c}(\alpha + 1)x^{(\alpha+2)/(\alpha+1)}$ for some constant $\bar{c}$ and all $x \in [0, 1]$, $y \in [0, 1]$. However, as we have observed at the beginning, $A(1) = 1$ and $B(1) = 1 - \beta$. This implies $\bar{c} = 1$ and $\beta = \alpha + 2$. It suffices to use (A2) to obtain the desired candidate for U. $\qquad\square$

4.3.3 Proof of Theorem 4.8

Proof of (4.13). It suffices to prove the theorem for $\alpha \geq 1$. Since f is nonnegative, we have $||f||_p = ||f_0||_p$ and the inequality can be rewritten in the form

$$\mathbb{P}(|g_n| \geq 1) \leq 2^p \mathbb{E}f_0^p.$$

Let $V, U : [0, \infty) \times \mathcal{H} \to \mathbb{R}$ be given by $V(x, y) = 1_{\{|y|\geq 1\}}$ and

$$U(x, y) = U_\alpha(-x, y), \qquad (4.24)$$

where U_α is given by (4.6). We will verify the conditions $1°$, $2°$ and $3°$. The first two of them have been already established in the proof of (4.4). To check $3°$, suppose that $|y| \leq x$. If $x \leq 1/(\alpha+1)$, then $U(x, y) \leq (\alpha+1)x \leq ((\alpha+1)x)^p$; if $x > 1/(\alpha+1)$, then $U(x, y) = 1$ which also does not exceed $((\alpha+1)x)^p$. Since the necessary integrability is satisfied, (4.13) follows. $\qquad\square$

Sharpness. Suppose that $p \leq 1$ and $\mathcal{H} = \mathbb{R}$. First, if $\alpha \leq 1$, we consider a centered random variable ξ taking values in the set $\{-1/2, 3/2\}$ and put $(f_0, g_0) \equiv (1/2, 1/2)$, $(f_1, g_1) = (f_2, g_2) = \cdots = (1/2 + \xi, 1/2 - \xi)$. Then f is a nonnegative martingale, g is α-subordinate to f (for any $\alpha \geq 0$), $||f||_p = 1/2$ and $|g_1| = 1$ almost surely. If $\alpha > 1$, then let $(f_0, g_0) \equiv ((\alpha+1)^{-1}, (\alpha+1)^{-1})$ and $(f_1, g_1) \equiv (0, 1)$. One easily checks that f, g have the necessary properties and the two sides of (4.13) are equal.

On the other hand, suppose that $p > 1$ and let $\delta > 0$. On the probability space $([0, 1], \mathcal{B}([0, 1]), \mathbb{P})$, let ξ be a centered random variable taking the values $-\delta$ and $1 - \delta$. Consider the sequence $(f_n, g_n)_{n=0}^2$, given by $f_0 = g_0 = \delta$, $f_1 = g_1 = \delta + \xi$, $f_2 \equiv 0$, $g_2 = 1_{\{\xi=1-\delta\}}$. Then f is a nonnegative supermartingale with $||f||_p^p = ||f_1||_p^p = \delta$ and $\mathbb{P}(|g_2| \geq 1) = \delta$. Now we use the "portioning argument": fix a positive integer N and let $\bar{f}, \bar{g}$ be given as follows: let $(\bar{f}_0, \bar{g}_0) \equiv (\delta, \delta)$ and, for $n = 0, 1, 2, \ldots, N - 1$, $k = 1, 2$,

$$\bar{f}_{2n+k}(\omega) = f_k(N\omega - n), \quad \bar{g}_{2n+k}(\omega) = g_k(N\omega - n) \quad \text{for } \omega \in (n/N, (n+1)/N].$$

We easily see that f is a nonnegative supermartingale and g is α-subordinate to f for any $\alpha \geq 0$. We have, for any $n = 0, 1, 2, \ldots, N-1$ and $k = 1, 2$,

$$||\bar{f}_{2n+k}||_p^p = \frac{1}{N}||f_k||_p^p + \frac{N-n-1}{N}\delta^p \leq \frac{\delta}{N} + \delta^p.$$

On the other hand, $||\bar{g}||_{p,\infty}^p \geq \mathbb{P}(\bar{g}_{2N} = 1) = \delta$. Since N and δ were arbitrary, we see that the ratio $||\bar{g}||_{p,\infty}/||\bar{f}||_p$ can be made as large as we wish. Thus the weak type inequality does not hold with any finite constant in the case when $p > 1$. $\quad\square$

On the search of the suitable majorant. We shall only focus on the case $\alpha \leq 1$ and leave the argumentation for $\alpha > 1$ to the reader. Assume that $\mathcal{H} = \mathbb{R}$ and write the formula

$$U^0(x, y) = \sup\{\mathbb{E}V(f_n, g_n)\} = \sup\{\mathbb{P}(|g_n| \geq 1)\},$$

where supremum is taken over all n and all appropriate pairs (f, g). Clearly $U^0(x, y) \leq 1$ and, using standard examples, we see that $U^0(x, y) = 1$ for $x+|y| \geq 1$. How do we find U^0 on the set $\{(x, y) : x + |y| < 1\}$? We have $U(0, y) = 0$ for $|y| < 1$, since there are no nontrivial nonnegative supermartingales starting from 0. If $x > 0$, then it is not difficult to see that if (f, g) starts from (x, y), the best move is to jump either to $(0, y - x)$ or to $(\frac{1+x-y}{2}, \frac{1-x+y}{2})$, with probabilities determined by the condition $\mathbb{E}df_1 = 0$. Thus, (f_1, g_1) lands on one of the lines $x = 0$ or $x + y = 1$, on which we know how to proceed. It is easy to see that such an example gives precisely (4.24). $\quad\square$

4.4   Moment estimates for submartingales

4.4.1   Formulation of the results

We start with a negative fact.

Theorem 4.9. *Let $\alpha \geq 0$ and $0 < p < \infty$ be fixed. Then there is no finite constant β_p such that for any submartingale f and any real-valued g which is α-subordinate to f we have*

$$||g||_p \leq \beta_p||f||_p.$$

However, if the dominating process is assumed to be nonnegative, the moment inequality does hold for p between 1 and infinity. Precisely, we have the following statement.

Theorem 4.10. *Suppose that $\alpha \geq 0$ is fixed. Let f be a nonnegative submartingale and suppose g is an $\mathcal{H}$-valued process which is α-subordinate to f. Then*

$$||g||_p \leq (p_\alpha^* - 1)||f||_p, \tag{4.25}$$

where $p_\alpha^* = \max\{(\alpha + 1)p, p/(p-1)\}$. The inequality is sharp, even if $\mathcal{H} = \mathbb{R}$.

4.4.2 Lack of moment inequalities in the general case

This is clear for $0 < p \leq 1$, since then there are no moment estimates even in the martingale setting (see Chapter 3). Suppose then that $1 < p < \infty$ and consider a Markov process (f, g) taking values in $\mathbb{R} \times \mathbb{R}$, given on the probability space $([0, 1], \mathcal{B}([0, 1]), |\cdot|)$ by the following conditions. Let $K > 0$ be a number, to be specified later.

(i) The process starts from $(0, 0)$.

(ii) The state $(0, 0)$ leads to $(-(p+1)/2, (p+1)/2)$ or to $((p+1)/2, -(p+1)/2)$, each event has probability $1/2$.

(iii) The state $(-(p+1)/2, (p+1)/2)$ leads to $(-1, p)$ or to $(-p, 1)$, each event has probability $1/2$.

(iv) The state $(-x, px)$ $(0 < x < K)$ leads to $(0, (p-1)x)$ or to $(-2x, (p+1)x)$, each event has probability $1/2$.

(v) The state $(-x, px)$ $(x \geq K)$ leads to $(0, px)$ with probability 1.

(vi) The state $(-2x, (p+1)x)$ $(x > 0)$ leads to $(-(1+1/p)x, (p+1)x)$ with probability 1.

(vii) All the remaining states are absorbing.

It is easy to see that f, g are simple: they terminate after a number of steps (say, after L moves). Furthermore, f is a submartingale and g is α-subordinate to f for any $\alpha \geq 0$. We have that f has finite p-th norm (which depends on K) and we have $\mathbb{P}(f_L = (p+1)/2) = 1/2$, $\mathbb{P}(f_L = -p) = 1/4$ and $\mathbb{P}(f_L = 0) = 1/4$. Moreover, if $k = 0, 1, 2, \ldots$ satisfies $p(1+1/p)^k < K$, then

$$\mathbb{P}(g_L = (p-1)(1+1/p)^k) = 2^{-k-3}.$$

Consequently,

$$\|g\|_p \geq \frac{1}{8}(p-1)^p \sum_{k=0}^{k_0} \left[\frac{1}{2}\left(1+\frac{1}{p}\right)^p\right]^k,$$

where k_0 is the largest integer k satisfying $p(1+1/p)^k < K$.

Now we proceed as follows. Pick any positive number M. Then there is K such that $\|g_L\|_p > M\|f_L\|_p$. Next we repeat the "portioning argument" used in the proof of the weak type inequalities above. Namely, we divide the probability space $([0, 1], \mathcal{B}([0, 1]), |\cdot|)$ into N parts and copy the above process into each part on a separate period of time. Let $(\bar{f}_n, \bar{g}_n)_{n=0}^{LN}$ be the process we obtain in this manner. Then $\|\bar{g}_{LN}\|_p = \|g\|_p > M\|f_L\|_p$ and, for any $m = Ln + \ell$, where $n = 0, 1, 2, \ldots, N-1$ and $\ell = 0, 1, 2, \ldots, L$,

$$\|\bar{f}_n\|_p \leq \frac{n}{N}\|f_L\|_p + \frac{1}{N}\|f\|_p \leq \|f_L\|_p + \|f\|_p/N.$$

Consequently, if N is sufficiently large, the ratio $\|\bar{g}\|_p/\|\bar{f}\|_p$ exceeds M; since M was arbitrary, the claim follows.

4.4.3 Proof of (4.25)

With no loss of generality, we may assume that $||f||_p$ is finite. Then so is $||df_n||_p$ for each n, and hence also $||g_n||_p < \infty$, due to differential subordination. Let $V(x, y) = |y|^p - (p_\alpha^* - 1)^p x^p$ for $x \geq 0$, $y \in \mathcal{H}$. We consider the cases $1 < p < (\alpha + 2)/(\alpha + 1)$, $(\alpha + 2)/(\alpha + 1) \leq p < 2$ and $p \geq 2$ separately.

The case $1 < p < (\alpha + 2)/(\alpha + 1)$. We use the integration argument. Let $u : [0, \infty) \times \mathcal{H} \to \mathbb{R}$ be given by (4.14) and $k(t) = p^{3-p}(1 - p(\alpha + 1)/(\alpha + 2))t^{p-1}$. Then

$$U(x, y) = \int_0^\infty k(t)u(x/t, y/t)\mathrm{d}t$$

admits the following explicit formula:

$$U(x, y) = p^{2-p}(|y| - (p-1)^{-1}x)(x + |y|)^{p-1}. \tag{4.26}$$

This is precisely Burkholder's function corresponding to the moment inequality for martingales; consequently, $1°$ and $3°$ hold. It suffices to apply Fubini's theorem:

$$\mathbb{E}V(f_n, g_n) \leq \mathbb{E}U(f_n, g_n) \leq \mathbb{E}U(f_0, g_0) \leq 0$$

and we are done. To see that Fubini's theorem applies, note that $|u(x, y)| \leq c(|x| + |y|)$ for some absolute constant c, and thus

$$\int_0^\infty k(t)|u(x/t, y/t)|\mathrm{d}t < \tilde{c}(|x|^p + |y|^p)$$

for another universal constant $\tilde{c}$.

The case $(\alpha + 2)/(\alpha + 1) \leq p < 2$. This is the most elaborate part. Assume first that p, α satisfy

$$p(\alpha^2 - 1) \leq \alpha^2 + 2\alpha - 1. \tag{4.27}$$

Let $U : \mathbb{R} \times \mathcal{H} \to \mathbb{R}$ be defined by the formula

$$U(x, y) = p(1 - 1/p_\alpha^*)^{p-1}(|y| - (p_\alpha^* - 1)x)(x + |y|)^{p-1}.$$

To show the majorization $U \geq V$, we note that both sides are homogeneous of order p; consequently, it suffices to prove the inequality for $x + |y| = 1$. It is equivalent to

$$F(x) := p(1 - 1/p_\alpha^*)^{p-1}(1 - p_\alpha^* x) - (1 - x)^p + (p_\alpha^* - 1)^p x^p \geq 0, \qquad x \in [0, 1].$$

It suffices to verify the following easy facts: $F(1/p_\alpha^*) = F'(1/p_\alpha^*) = 0$, $F(1) \geq 0$ and there is $p_0 \in (1/p_\alpha^*, 1)$ such that F is convex on $(0, p_0)$ and concave on $(p_0, 1)$. To prove $2°$, we derive that

$$\frac{U_x(x, y) + \alpha|U_y(x, y)|}{c_p(x + |y|)^{p-2}} = -(p(\alpha + 1) - 1)x - \alpha|y| + \alpha|(-p(\alpha + 1) + \alpha + 2)x + |y||,$$

where $c_p = p(1 - 1/p_\alpha^*)^{p-1}$. This expression is nonpositive, since

$$-(p(\alpha + 1) - 1)x - \alpha|y| + \alpha\big((-p(\alpha + 1) + \alpha + 2)x + |y|\big) = -(p - 1)(\alpha + 1)^2 x$$

and

$$-(p(\alpha + 1) - 1)x - \alpha|y| - \alpha\big((-p(\alpha + 1) + \alpha + 2)x + |y|\big)$$
$$= [p(\alpha^2 - 1) - \alpha^2 - 2\alpha + 1]x - 2\alpha|y| \leq 0. \tag{4.28}$$

Here we have used the assumption (4.27) on p and α. The next step is to establish the concavity of the function $G = G_{x,y,h,k}$, given by the usual formula. We have

$$G''(0) = I + II + III, \tag{4.29}$$

where

$$I = -c_p p^2 (\alpha + 1)(x + |y|)^{p-2}(h^2 - |k|^2),$$
$$II = -pc_p(x + |y|)^{p-3}\big[(p(\alpha + 1) - (\alpha + 2))x + (\alpha + 1)|y|\big](h + y' \cdot k)^2,$$
$$III = -c_p\left(p(\alpha + 1) - \frac{p}{p-1}\right)\frac{(x + |y|)^{p-1}}{|y|}(|k|^2 - y' \cdot k)^2.$$

Thus, $G''(0) \leq 0$ when $|k| \leq |h|$ and $2°$ is valid. Finally, the condition $3°$ is evident. By Burkholder's method, (4.25) follows, since we have the necessary integrability.

Now we turn to the case when

$$p(\alpha^2 - 1) > \alpha^2 + 2\alpha - 1. \tag{4.30}$$

In particular, this implies $\alpha > 1$ and $p > (\alpha + 2)/(\alpha + 1)$ (or $\alpha(p-1) + p - 2 > 0$). We start with an auxiliary fact.

Lemma 4.2. *We have*

$$\alpha \leq p_\alpha^* - 1 \quad and \quad \frac{1}{\alpha(p-1) + p - 2} \leq p_\alpha^* - 1 = p(\alpha + 1) - 1. \tag{4.31}$$

Proof. The first inequality is equivalent to $(p - 1)(\alpha + 1) \geq 0$. To show the second estimate, it suffices to do this for the least possible p, since as p increases, so do the right-hand side and the denominator of the left-hand side. By (4.30), the smallest value is $p = (\alpha^2 + 2\alpha - 1)/(\alpha^2 - 1)$. If we plug this into the estimate, we obtain, after some manipulations,

$$\frac{\alpha(\alpha + 1)^2}{(\alpha - 1)^2} \geq 1,$$

since $\alpha > 1$. $\square$

The inequalities in (4.31) give

$$\beta(\alpha, p) := \frac{\alpha}{(\alpha(p-1) + p - 2)^{p-1}} - (p_\alpha^* - 1)^p \leq 0.$$

We are ready to introduce the special function $U : \mathbb{R} \times \mathcal{H} \to \mathbb{R}$. When $|y| \geq (p_\alpha^* - 1)x$, define

$$U(x, y) = p(1 - 1/p_\alpha^*)^{p-1}(|y| - (p_\alpha^* - 1)x)(x + |y|)^{p-1}.$$

If $x/(\alpha(p-1) + p - 2) \leq |y| < (p_\alpha^* - 1)x$, we set

$$U(x, y) = |y|^p - (p_\alpha^* - 1)^p x^p.$$

Finally, for the remaining values of x, y, put

$$U(x, y) = [(p-1)(\alpha + 1)]^{2-p} \left(|y| - \frac{x}{p-1} \right) (x + |y|)^{p-1} + \beta(\alpha, p)x^p.$$

The next step is to verify 1°, 2° and 3°. The majorization is established in much the same manner as before, using the homogeneity of U and V. To show that $U_x(x, y) + \alpha|U_y(x, y)| \leq 0$, first we focus on the situation when $|y| \geq (p_\alpha^* - 1)x$. We repeat the argumentation presented in the previous case: see (4.28). Since

$$[p(\alpha^2 - 1) - \alpha^2 - 2\alpha + 1]x - 2\alpha|y| \leq [-p(\alpha + 1)^2 - \alpha^2 + 1]x \leq 0,$$

the bound holds. If $x/(\alpha(p-1) + p - 2) \leq |y| < (p_\alpha^* - 1)x$, then

$$U_x(x, y) + \alpha|U_y(x, y)| = p\alpha|y|^{p-1} - p(p_\alpha^* - 1)^p x^{p-1}$$
$$\leq -p(p-1)(p_\alpha^* - 1)^{p-1}(\alpha + 1)x^{p-1} \leq 0.$$

Finally, when $|y| < x/(\alpha(p-1) + p - 2)$, we get, omitting the term $\beta(\alpha, p)x^p$,

$$U_x(x, y) + \alpha|U_y(x, y)| \leq \frac{p}{p-1} \left[\frac{(p-1)(\alpha+1)}{x + |y|} \right]^{2-p} [(p - 2 + (p-1)\alpha)|y| - x],$$

which is nonpositive. To show that $G''_{x,y,h,k}(0) \leq 0$, we may assume that $|y| < (p_\alpha^* - 1)x$, since for large $|y|$ we may repeat the reasoning from the previous case: see (4.29). If $x/(\alpha(p-1) + p - 2) \leq |y| < (p_\alpha^* - 1)x$, then

$$G''(0) = I + II + III, \tag{4.32}$$

where

$$I = -p(2-p)|y|^{p-2}(y' \cdot k)^2,$$
$$II = -p((p-1)(p_\alpha^* - 1)^p x^{p-2} - |y|^{p-2})|k|^2,$$
$$III = -p(p-1)(p_\alpha^* - 1)^p x^{p-2}(h^2 - |k|^2).$$

All the expressions are nonpositive: to see that $II \leq 0$, use the second inequality in (4.31) to get

$$(p-1)(p_\alpha^* - 1)^p x^{p-2} - |y|^{p-2} \geq x^{p-2} [(p-1)(p_\alpha^* - 1)^p - (\alpha(p-1) + p - 2)^{2-p}]$$
$$\geq x^{p-2}(p_\alpha^* - 1)^{p-1}(p-1) \left[p_\alpha^* - 1 - \alpha - \frac{p-2}{p-1} \right]$$

and note that the expression in the square bracket is nonnegative. Finally, if $|y| < x/(\alpha(p-1)+p-2)$, then we omit $\beta(\alpha,p)$ and obtain

$$G''(0) \leq [(p-1)(\alpha+1)]^{2-p}[I+II], \tag{4.33}$$

where

$$I = p(2-p)(x+|y|)^{p-3}|y|(h\operatorname{sgn} x + y' \cdot k)^2,$$
$$II = p(x+|y|)^{p-2}(h^2 - |k|^2).$$

It remains to check 3°. This is easy: if $|y| \leq x$, then

$$U(x,y) \leq x^p(1 - (p_\alpha^* - 1)^p) \leq 0.$$

This proves (4.25), since we have the necessary integrability.

The case $p \geq 2$. This time the special function $U : [0,\infty) \times \mathbb{R} \to \mathbb{R}$ is given by

$$U(x,y) = \begin{cases} p(1-1/p_\alpha^*)^{p-1}(|y|-(p_\alpha^*-1)x)(x+|y|)^{p-1} & \text{if } |y| > (p_\alpha^*-1)x, \\ |y|^p - (p_\alpha^*-1)^p x^p & \text{if } |y| \leq (p_\alpha^*-1)x. \end{cases}$$

This function is of class C^1 and satisfies the majorization $U \geq V$, by a similar reasoning as above. Furthermore, the calculations presented previously show that property 2° is valid on the set $|y| > (p_\alpha^* - 1)x$. To show that 2° holds for $|y| < (p_\alpha^* - 1)x$, we derive that

$$U_x(x,y) + \alpha|U_y(x,y)| = -p(p(\alpha+1)-1)^p x^{p-1} + p\alpha|y|^{p-1}$$
$$\leq -p(p-1)(\alpha+1)(p(\alpha+1)-1)^{p-1}x^{p-1} \leq 0$$

and, for $G = G_{x,y,h,k}$ given by the usual formula,

$$G''(0) = I + II + III, \tag{4.34}$$

where

$$I = -p(p-1)|y|^{p-2}(h^2 - |k|^2),$$
$$II = -p(p-2)|y|^{p-2}(|k|^2 - (y' \cdot k)^2),$$
$$III = -p(p-1)h^2\big[(p(\alpha+1)-1)^p x^{p-2} - |y|^{p-2}\big].$$

All the terms are nonpositive. Finally, 3° is obvious: if $|y| \leq x$, then $U(x,y) \leq (1 - (p_\alpha^* - 1)^p)x^p \leq 0$. This completes the proof. $\qquad\square$

Sharpness. Let $\alpha \geq 0$ and $p > 1$ be fixed. If $p \leq (\alpha+2)/(\alpha+1)$, then $p_\alpha^* - 1 = p^* - 1$ and this is optimal even in the moment estimate for nonnegative martingales and their differential subordinates. Next, suppose that we have $p > (\alpha + 2)/(\alpha + 1)$ and assume that the moment inequality holds with some constant $\beta > 0$. Define

$$U^0(x, y) = \sup\{\mathbb{E}|g_n|^p - \beta^p \mathbb{E} f_n^p\},$$

where the supremum is taken over all n and all appropriate f and g. The function U^0 is homogeneous of order p and, by Theorem 4.3, the function $w : [0, 1] \to \mathbb{R}$ given by $w(x) = U^0(x, 1 - x)$ is concave. Furthermore, the function $x \mapsto U^0(x, 1 + \alpha x) = (x(\alpha + 1) + 1)^p w\left(\frac{x}{x(\alpha+1)+1}\right)$, is nonincreasing, so

$$(\alpha + 1)pw(0) + w'(0+) \leq 0.$$

On the other hand, using the majorization and concavity of w,

$$(1 - 1/p_\alpha^*)^p - \beta^p/(p_\alpha^*)^p \leq w(1/p_\alpha^*) \leq w(0) + w'(0+)/p_\alpha^* \tag{4.35}$$
$$= ((\alpha + 1)pw(0) + w'(0+))/p_\alpha^* \leq 0.$$

That is, $\beta \geq p_\alpha^* - 1$ and we are done. $\square$

On the search of the suitable majorant. We concentrate on the case $\mathcal{H} = \mathbb{R}$ and $p > (\alpha + 2)/(\alpha + 1)$. Clearly, it suffices to determine $w(x) = U(x, 1 - x)$, $x \in [0, 1]$; one gets the formula for U on the whole $[0, \infty) \times \mathbb{R}$ using homogeneity and symmetry. The above considerations suggest that the optimal constant should be equal to $p_\alpha^* - 1$. Consequently, let $v(x) = (1 - x)^p - (p_\alpha^* - 1)^p x^p$. If our hypothesis about the optimal constant is true, then equality must hold throughout (4.35). This implies that w is linear on $[0, p_\alpha^*]$ and, since $w \geq v$ and $w(1/p_\alpha^*) = v(1/p_\alpha^*)$, we conclude that

$$w(x) = p(1 - 1/p_\alpha^*)^{p-1}(1 - p_\alpha^* x) \qquad \text{for } x \in [0, p_\alpha^*].$$

This gives

$$U_p(x, y) = p(1 - 1/p_\alpha^*)^{p-1}(|y| - (p_\alpha^* - 1)x)(x + |y|)^{p-1} \tag{4.36}$$

on the set $\{(x, y) : |y| \geq (p_\alpha^* - 1)x$. A natural idea is to *assume* that the above formula holds on the whole set $[0, \infty) \times \mathbb{R}$. However, this works only for those p and α that satisfy (4.27): otherwise, $U_x + \alpha|U_y| \leq 0$ does not hold. To overcome this problem, one considers another choice for U on the set $\{(x, y) : |y| < (p_\alpha^* - 1)x\}$: namely, $U(x, y) = V(x, y)$ there. This turns out to work only for $p \geq 2$: for smaller p the inequality $G''(0) \leq 0$ is not valid. To be more precise, G fails to be concave when $|y|$ is small in comparison with x; on the other hand, for such x and y, Burkholder's function (4.26) works. Therefore, if p, α satisfy (4.30), we are led to the following function U: we put (4.36) for $|y| \geq (p_\alpha^* - 1)x$, $U_p(x, y) = |y|^p - (p_\alpha^* - 1)x^p$ for $(p_\alpha^* - 1)x > |y| \geq \beta x$ and

$$U(x, y) = -c_1 x^p + c_2(|y| - (p - 1)^{-1} x)(x + |y|)^{p-1},$$

for some positive c_1, c_2 and β to be found. These are obtained by verifying $1°$, $2°$ and the assumption that U is of class C^1. $\square$

4.5 Moment estimates for supermartingales

4.5.1 Formulation of the result

It follows from Theorem 4.9 that there is no hope for moment inequalities for general supermartingales and their α-subordinates. However, under the additional assumption that the dominating supermartingale is nonnegative, we can show such an estimate. Let

$$
\beta_p = \begin{cases} 2\left(\frac{2-p}{2-2p}\right)^{1/p} & \text{if } p \in (-\infty, 0) \cup (0, 1), \\ 2e^{1/2} & \text{if } p = 0. \end{cases}
$$

Theorem 4.11. *Let $\alpha \in [0,1]$ be fixed. Assume that f is a nonnegative super-martingale and g is an $\mathcal{H}$-valued process which is α-subordinate to f. Then for $-\infty < p < 1$ we have*

$$
\|g\|_p \le \beta_p \|f\|_p \tag{4.37}
$$

and the constant is the best possible. It is already the best possible when f is assumed to be a nonnegative martingale and g is its ± 1 transform. For $p \ge 1$, the inequality (4.37) does not hold in general with any finite constant.

4.5.2 Proof of Theorem 4.11

Proof of (4.37). We shall focus on the case $0 < p < 1$; for $p \le 0$ the argumentation is similar, see [137] for details. Let f be a nonnegative supermartingale and g be α-subordinate to f. We have $\|f\|_p = \|f_0\|_p$ for $p \le 1$ and hence it suffices to show that

$$
\|g_n\|_p \le \beta_p \|f_0\|_p, \qquad n = 0, 1, 2, \ldots. \tag{4.38}
$$

Let $U_p, V_p : [0, \infty) \times \mathcal{H} \to \mathbb{R}$ be given by

$$
U_p(x, y) = (|y| + (1-p)^{-1}x)(x + |y|)^{p-1}, \tag{4.39}
$$

and $V_p(x, y) = |y|^p$. We will prove the conditions $1°$, $2°$ and $3°$. To prove the majorization, calculate the partial derivative

$$
\frac{\partial U_p}{\partial x}(x, y) = p(1-p)^{-1}(x + |y|)^{p-2}(x + (2-p)|y|) > 0,
$$

and since $U_p(0+, y) = V_p(0, y)$, $1°$ follows. To show $2°$ and $3°$, we use the identity

$$
U_p(x, y) = \int_0^\infty k(t) u(x/t, y/t) \mathrm{d}t, \tag{4.40}
$$

where u is given by (4.24) and $k(t) = p(2-p)t^{p-1}/2$. The use of Fubini's theorem is permitted and we get

$$
\mathbb{E}U_p(f_n, g_n) \le \mathbb{E}U_p(f_0, g_0) \le \beta_p^p \mathbb{E}f_0^p,
$$

which is the claim. $\qquad\square$

Sharpness. Let $p \in (0,1)$ be fixed. Of course, it suffices to show the optimality of the constant for $\alpha = 1$. Define

$$U^0(x,y) = \sup\{\mathbb{E}|g_n|^p\},$$

where the supremum is taken over all n, all nonnegative martingales f and all g which are ± 1-transforms of f. We have that $U^0(0,y) = |y|^p$, directly from the definition; furthermore, U^0 is homogeneous of order p and diagonally concave. Consequently, for any $\delta > 0$,

$$
\begin{aligned}
U^0(1,0) &\geq \frac{1}{1+\delta}U^0(1+\delta, -\delta) + \frac{\delta}{1+\delta}U^0(0,1) \\
&= \frac{1}{1+\delta}U^0(1+\delta, -\delta) + \frac{\delta}{1+\delta} \\
&\geq \frac{1-\delta}{1+\delta}U^0(1+2\delta, 0) + \frac{\delta}{1+\delta}U^0(0, 1+2\delta) + \frac{\delta}{1+\delta} \\
&= \frac{1-\delta}{1+\delta}(1+2\delta)^p U^0(1,0) + \frac{\delta(1+2\delta)^p}{1+\delta} + \frac{\delta}{1+\delta}.
\end{aligned}
$$

Subtract $U^0(1,0)$ from both sides, divide by δ and let δ go to 0. We get $U^0(1,0) \geq (1-p)^{-1}$. Using the diagonal concavity again, we obtain

$$U^0(1/2, 1/2) \geq \frac{1}{2}U^0(1,0) + \frac{1}{2}U^0(0,1) \geq \frac{2-p}{2-2p}.$$

In other words, for any $\varepsilon > 0$ there is a pair (f,g) starting from $(1/2, 1/2)$ such that f is a nonnegative martingale, g is its ± 1 transform and

$$\mathbb{E}|g_n|^p \geq \frac{2-p}{2-2p} - \varepsilon = (\beta_p^p - 2^p\varepsilon)\mathbb{E}f_0^p = (\beta_p^p - 2^p\varepsilon)\|f\|_p^p.$$

This shows that β_p is indeed the best possible in (4.37). $\square$

On the search of the suitable majorant. We write the definition of U^0:

$$U^0(x,y) = \sup\{\mathbb{E}|g_n|^p\},$$

where the supremum is taken over all n, all nonnegative supermartingales f and all g which are α-subordinate to f. It is quite easy to derive U^0 directly from the definition. A little thought and experimentation lead to the conjecture that one has to consider only those pairs in which the process f is a martingale. Having observed this, the following natural example arises. Fix $(x,y) \in (0,\infty) \times \mathbb{R}$, $\delta > 0$ and consider a Markov martingale (f,g) such that

(i) It starts from (x,y).

(ii) For any n we have $dg_n \equiv df_n$ or $dg_n \equiv -df_n$.

(iii) Any state in $(0, \infty) \times (0, \infty)$ leads to the half-line $\{(x, 0) : x > 0\}$ or to $\{(0, y) : y > 0\}$.

(iv) Any state in $(0, \infty) \times (-\infty, 0)$ leads to the half-line $\{(x, 0) : x > 0\}$ or to $\{(0, y) : y < 0\}$.

(v) Any state of the form $(x, 0)$ leads to $(0, x)$ or to $(x + \delta, -\delta)$.

(vi) All states lying on the y-axis are absorbing.

It is not difficult to see that if one computes $\mathbb{E}|g_n|^p$ and lets $\delta \to 0$, one obtains the formula (4.39). Note that this reasoning is an alternative proof of the optimality of the constant β_p. $\quad\square$

Moment inequalities for $p \geq 1$. We will prove that $\beta_p = \infty$ for $p \geq 1$. This is true for $p = 1$ since then the norm inequality fails to hold even in the case f is a nonnegative martingale and g is its ± 1 transform. Assume $p > 1$ and consider the following processes f, g on the interval $[0, 1]$ with Lebesgue measure. For $n = 0, 1, 2, \ldots$ set

$$f_{2n} = 2^{n/(p-1)}[0, 2^{-np/(p-1)}], \quad f_{2n+1} = \frac{1}{2}f_{2n}, \quad dg_n = (-1)^n df_n.$$

It is easy to check that f is a nonnegative supermartingale and $\|f_{2n}\|_p = 1$, $\|f_{2n+1}\|_p = \frac{1}{2}$ for any n, which gives $\|f\|_p = 1$. Note that

$$g_{2n+2} = g_{2n} + f_{2n+2} - 2f_{2n+1} + f_{2n} = g_{2n} + f_{2n+2},$$

which implies

$$g_{2n} = \sum_{k=0}^{n} f_{2k} = \sum_{k=0}^{n} 2^{k/(p-1)}[0, 2^{-kp/(p-1)}]$$

$$\geq \sum_{k=0}^{n} 2^{k/(p-1)}[2^{-(k+1)p/(p-1)}, 2^{-kp/(p-1)}]$$

and

$$\mathbb{E}g_{2n}^p \geq \sum_{k=0}^{n}(1 - 2^{-p/(p-1)}) = (n+1)(1 - 2^{-p/(p-1)}).$$

This shows $\|g\|_p = \infty$ and proves the claim. $\quad\square$

4.6   Logarithmic estimates

4.6.1   Formulation of the results

The next step of our analysis is to study logarithmic inequalities. Here the situation is much more complicated and the calculations turn out to be much more elaborate. Throughout, α is a fixed positive number which belongs to $[0, 1]$.

First, we need the following easy fact, which we state here without proof.

Lemma 4.3. (i) *There is a unique $K_0 = K_0(\alpha) \in (1, \infty)$ such that*

$$K_0 + \log(K_0 - 1) = \alpha + 1 + \log\left(\frac{2\alpha + 1}{\alpha + 1}\right). \tag{4.41}$$

(ii) *If $K \in (1, K_0)$, then there is a unique $c = c(K, \alpha) \in (\frac{\alpha+1}{2\alpha+1}, (K-1)^{-1})$ satisfying*

$$\alpha + \frac{\alpha + 1}{2\alpha + 1} \cdot \frac{1}{c} - \log(c(K-1)) - K = 0. \tag{4.42}$$

For $K \geq K_0$, let

$$c = c(K, \alpha) = \left[\exp(\alpha + 1 - K)\frac{\alpha + 1}{(2\alpha + 1)(K - 1)}\right]^{1/2}. \tag{4.43}$$

The main result of this section is the following.

Theorem 4.12. *Assume that f is a nonnegative submartingale and g is a sequence of integrable real-valued random variables which is α-subordinate to f. Then for all $K > 1$ we have*

$$\|g\|_1 \leq K \sup_n \mathbb{E}f_n \log^+ f_n + L(K, \alpha), \tag{4.44}$$

where

$$L(K, \alpha) = \begin{cases} \dfrac{c}{\alpha + 2} + \dfrac{(\alpha + 1)^2}{c(2\alpha + 1)^2(\alpha + 2)} + \dfrac{\alpha(2\alpha^2 + 5\alpha + 3)}{(\alpha + 2)(2\alpha + 1)} & \text{if } K < K_0, \\[2mm] \dfrac{(\alpha + 1)(2\alpha^2 + 3\alpha + 2)}{(2\alpha + 1)(\alpha + 2)} & \text{if } K \geq K_0. \end{cases}$$

The constant $L(K, \alpha)$ is the best possible. Furthermore, for $K \leq 1$ the inequality does not hold in general with any universal $L(K, \alpha) < \infty$.

Therefore, as in Theorem 3.9, there are two different expressions for the constants $L(K, \alpha)$ depending on whether K is small or large. Let us stress here that the above result is valid only for $\alpha \in [0, 1]$. We do not know the optimal constants in the case $\alpha > 1$.

Comparing (3.88) and (4.44), we see a slight difference: in the second estimate we have the positive part of the logarithm. The special function used in the proof of Theorem 4.12 will enable us to show the following Log^+ version of Theorem 3.9.

Theorem 4.13. *Let f, g be two real-valued martingales such that g is differentially subordinate to f. Then for $K > 1$,*

$$\|g\|_1 \leq K \sup_n \mathbb{E}|f_n| \log^+ |f_n| + L(K, 0). \tag{4.45}$$

The constant $L(K, 0)$ is the best possible. Furthermore, for $K \leq 1$ the inequality does not hold in general with any universal $L(K) < \infty$.

4.6.2 Proof of Theorem 4.12

Proof of (4.44). Let $V_K : [0, \infty) \times \mathbb{R} \to \mathbb{R}$ be given by

$$V_K(x, y) = |y| - Kx \log^+ x. \tag{4.46}$$

Now we will introduce the special functions corresponding to the inequality (4.44). Assume first that $1 < K < K_0(\alpha)$. Consider the following subsets of $[0, \infty) \times \mathbb{R}$:

$$D_1 = \{(x, y) : x + |y| \leq c - (2\alpha + 1)^{-1}\},$$
$$D_2 = \{(x, y) : x + |y| > c - (2\alpha + 1)^{-1}, \; x \leq \alpha/(2\alpha + 1)\},$$
$$D_3 = \{(x, y) : -x + |y| \geq c - 1, \; \alpha/(2\alpha + 1) < x \leq 1\},$$
$$D_4 = \{(x, y) : -x + |y| < c - 1, \; c - (2\alpha + 1)^{-1} \leq x + |y| \leq c + 1\},$$
$$D_5 = \{(x, y) : c + 1 < x + |y| \leq K/(K - 1)\},$$
$$D_6 = \{(x, y) : K/(K - 1) - x < |y| \leq x/(K - 1)\},$$
$$D_0 = [0, \infty) \times \mathbb{R} \setminus (D_1 \cup D_2 \cup \cdots \cup D_6)$$

(see Figure 4.1). Note that D_5 is nonempty, due to Lemma 4.3 (ii).

Let

$$p = p_{K,\alpha} = \frac{\alpha + 1}{c(\alpha + 2)} \left(c - \frac{1}{2\alpha + 1} \right)^{\alpha/(\alpha + 1)},$$

$$\lambda = \frac{\alpha + 1}{c(2\alpha + 1)} \exp\left(-1 + \frac{2\alpha + 1}{\alpha + 1} c \right),$$

and introduce $U = U_{K,\alpha} : [0, \infty) \times \mathbb{R} \to \mathbb{R}$ by

$$U(x, y) =$$

$$\begin{cases} p_{K,\alpha}(x + |y|)^{1/(\alpha+1)}(-(\alpha + 1)x + |y|) + L(K, \alpha) & \text{on } D_1, \\[2mm] -\alpha x + |y| + \alpha + \lambda \exp\left[-\frac{2\alpha+1}{\alpha+1}\left(x + |y| - \frac{\alpha}{2\alpha+1} \right) \right]\left(x + \frac{1}{2\alpha+1} \right) & \text{on } D_2, \\[2mm] -\alpha x + |y| + \alpha + \lambda \exp\left[-\frac{2\alpha+1}{\alpha+1}\left(-x + |y| + \frac{\alpha}{2\alpha+1} \right) \right](1 - x) & \text{on } D_3, \\[2mm] \frac{|y|^2 - x^2}{2c} + \left(\frac{1}{c} - \log(c(K - 1)) - K \right)(x - 1) + \frac{c}{2} + \frac{1}{2c} & \text{on } D_4, \\[2mm] |y| - (x - 1)[\log(x + |y| - 1) + K + \log(K - 1)] & \text{on } D_5, \\[2mm] K|y| - x - Kx \log[(K - 1)(x + |y|)/K] & \text{on } D_6, \\[2mm] |y| - Kx \log x & \text{on } D_0. \end{cases}$$

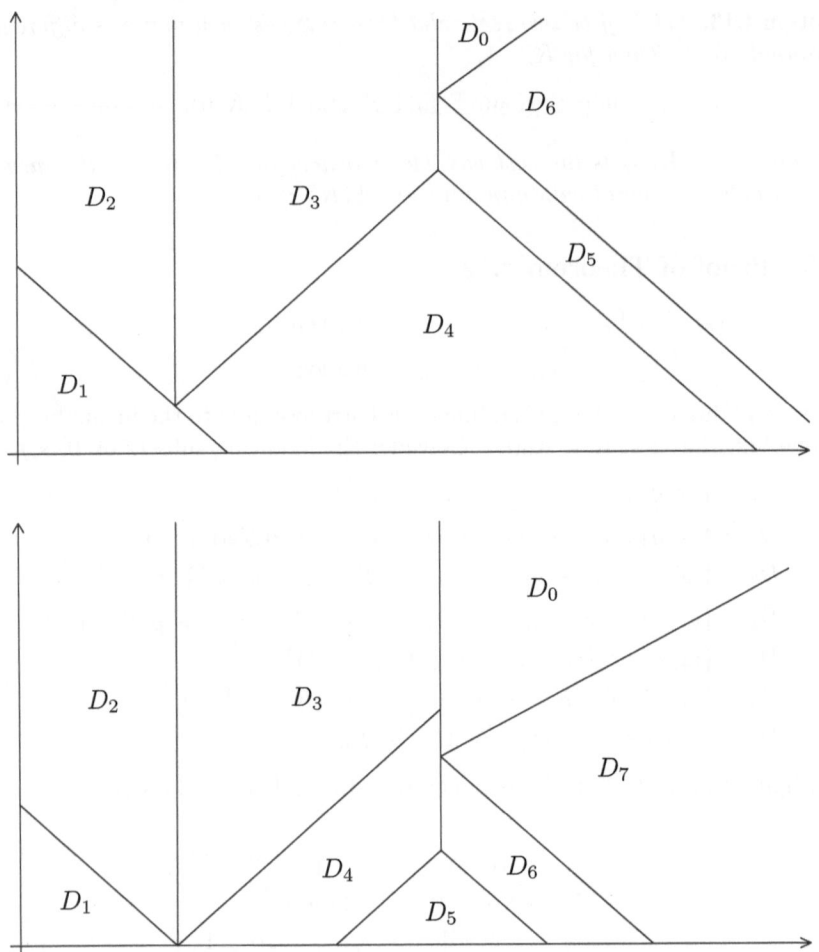

Figure 4.1: The sets D_i, intersected with $\mathbb{R}^2_+$, in case $1 < K < K_0$ (upper picture) and $K \geq K_0$ (lower picture).

Now assume that $K \geq K_0(\alpha)$ and consider the following subsets of $[0, \infty) \times \mathbb{R}$:

$$D_1 = \{(x, y) : x + |y| \leq \alpha/(2\alpha + 1)\},$$
$$D_2 = \{(x, y) : x + |y| > \alpha/(2\alpha + 1),\ x \leq \alpha/(2\alpha + 1)\},$$
$$D_3 = \{(x, y) : -x + |y| \geq \alpha/(2\alpha + 1),\ \alpha/(2\alpha + 1) < x \leq 1\},$$
$$D_4 = \{(x, y) : c - 1 \leq -x + |y| < \alpha/(2\alpha + 1),\ x \leq 1\},$$
$$D_5 = \{(x, y) : |x - 1| + |y| < c\},$$
$$D_6 = \{(x, y) : c + 1 \leq x + |y| \leq K/(K - 1)\},$$

$$D_7 = \{(x,y) : K/(K-1) - x < |y| \le x/(K-1)\},$$
$$D_0 = [0,\infty) \times \mathbb{R} \setminus (D_1 \cup D_2 \cup \cdots \cup D_7)$$

(see Figure 4.1). Set

$$p = p_{K,\alpha} = \alpha^{\alpha/(\alpha+1)}(2\alpha+1)^{1/(\alpha+1)}(\alpha+2)^{-1}$$

(with the convention $0^0 = 1$) and let $U = U_{K,\alpha} : [0,\infty) \times \mathbb{R} \to \mathbb{R}$ be given by

$$U(x,y) = \begin{cases} p_{K,\alpha}(x+|y|)^{1/(\alpha+1)}(-(\alpha+1)x + |y|) + L(K,\alpha) & \text{on } D_1, \\ -\alpha x + |y| + \alpha + \exp\left[-\frac{2\alpha+1}{\alpha+1}\left(x + |y| - \frac{\alpha}{2\alpha+1}\right)\right]\left(x + \frac{1}{2\alpha+1}\right) & \text{on } D_2, \\ -\alpha x + |y| + \alpha + \exp\left[-\frac{2\alpha+1}{\alpha+1}\left(-x + |y| + \frac{\alpha}{2\alpha+1}\right)\right](1-x) & \text{on } D_3, \\ -(1-x)\log\left[\frac{2\alpha+1}{\alpha+1}(1 - x + |y|)\right] + (\alpha+1)(1-x) + |y| & \text{on } D_4, \\ \frac{|y|^2 - x^2}{2c} + \left(\frac{1}{c} - \log(c(K-1)) - K\right)(x-1) + \frac{c}{2} + \frac{1}{2c} & \text{on } D_5, \\ |y| - (x-1)[\log(x + |y| - 1) + K + \log(K-1)] & \text{on } D_6, \\ K|y| - x - Kx\log[(K-1)(x+|y|)/K] & \text{on } D_7, \\ |y| - Kx\log x & \text{on } D_0. \end{cases}$$

Now we shall verify the conditions 1°, 2°′ and 3°. To show that $U(x,y) \ge V(x,y)$, we may assume that $y \ge 0$, due to identity $U(x,y) = U(x,-y)$ on $[0,\infty) \times \mathbb{R}$. The majorization follows from the three properties below:

(i) $\lim_{y\to\infty}(U(x,y) - V(x,y)) \ge 0$,

(ii) $\lim_{y\to\infty}(U_y(x,y) - V_y(x,y)) = 0$,

(iii) the function $y \mapsto U(x,y)$ is convex and $y \mapsto V(x,y)$ is linear on $[0,\infty)$.

Now we turn to 2°′. It can be easily verified that if $\alpha \in [0,1]$ and $K > 1$ are fixed, then U is continuous on $[0,\infty) \times \mathbb{R}$. Furthermore, U_y is continuous on $(0,\infty) \times \mathbb{R}$ and U_x is continuous on $(0,\infty) \times \mathbb{R} \setminus \{(x,y) : x = 1, |y| \ge c\}$. Indeed, one easily checks the continuity of U and its partial derivatives in the interiors of the sets D_i and hence one only needs to see whether the functions agree at the common boundaries. We omit the tedious calculations. The condition 2°′ is studied in the following.

Lemma 4.4. *For any $y \in \mathbb{R}$ and $|\gamma| \le 1$, the function $t \mapsto U(t, y + \gamma t)$ is concave. Furthermore, if $|\gamma| \le \alpha$, then this function is nonincreasing.*

Proof. Fix x, y, $h > 0$ and $k \in \mathbb{R}$ such that $|k| \le h$. Introduce the function $G = G_{x,y,h,k}$ defined by the usual formula. We shall show that

$$G''(0) \le 0 \text{ if } (x,y) \text{ lies in the interior of one of the sets } D_i, \tag{4.47}$$

$$G'_{1,y,h,k}(0-) \ge G'_{1,y,h,k}(0+) \text{ for any } y > c \tag{4.48}$$

and
$$G'_{0,y,h,k}(0+) \le 0 \qquad \text{if } |k| \le \alpha h. \tag{4.49}$$

This clearly will yield the claim. We shall only present the detailed proof in the case $K \ge K_0$; the case $K < K_0$ can be treated similarly.

We start with (4.47). If $(x,y) \in D_1^\circ$, this has been already shown: see, e.g., the proof of the weak type inequality for nonnegative submartingales. If (x,y) lies in D_2°, then

$$G''(0) = \frac{2\alpha+1}{\alpha+1} \exp\left[-\frac{2\alpha+1}{\alpha+1}\left(x+|y|-\frac{\alpha}{2\alpha+1}\right)\right]$$
$$\times (h+k)\left\{\left[\frac{2\alpha+1}{\alpha+1}\left(x+\frac{1}{2\alpha+1}\right)-2\right]h + \frac{2\alpha+1}{\alpha+1}\left(x+\frac{1}{2\alpha+1}\right)k\right\} \le 0.$$

The latter inequality holds because

$$|k| \le h, \quad \frac{2\alpha+1}{\alpha+1}\left(x+\frac{1}{2\alpha+1}\right)-2 \le -1 \text{ and } \frac{2\alpha+1}{\alpha+1}\left(x+\frac{1}{2\alpha+1}\right) \le 1.$$

If $(x,y) \in D_3^\circ$, then we derive that

$$G''(0) = \frac{2\alpha+1}{\alpha+1} \exp\left[-\frac{2\alpha+1}{\alpha+1}\left(-x+|y|+\frac{\alpha}{2\alpha+1}\right)\right]$$
$$\times (h-k)\left\{\left[\frac{2\alpha+1}{\alpha+1}(1-x)-2\right]h - \frac{2\alpha+1}{\alpha+1}(1-x)k\right\} \le 0,$$

which follows from

$$|k| \le h, \quad \frac{2\alpha+1}{\alpha+1}(1-x)-2 \le -1 \quad \text{and} \quad \frac{2\alpha+1}{\alpha+1}(1-x) \le 1.$$

For $(x,y) \in D_4^\circ$ we have

$$G''(0) = \frac{-h+k}{1-x+|y|}\left[\left(2-\frac{1-x}{1-x+|y|}\right)h + \frac{1-x}{1-x+|y|}k\right] \le 0,$$

which follows from

$$|k| \le h, \quad 2-\frac{1-x}{1-x+|y|} \ge 1 \quad \text{and} \quad \frac{1-x}{1-x+|y|} \le 1.$$

If $(x,y) \in D_5^\circ$, then $G''(0) = c^{-1}(k^2-h^2) \le 0$. If $(x,y) \in D_6^\circ$, one easily checks that
$$G''(0) = (x+y-1)^{-2}(h+k)[(-x-2y+1)h + (x-1)k]$$
is nonpositive, as $|k| \le h$ and $-x-2y+1 \le 1-x < 0$. If $(x,y) \in D_7^\circ$, then
$$G''(0) = K(x+y)^{-2}(h+k)[h(-x-2y)+xk] \le 0,$$
since $|k| \le h$ and $-x-2y \le -x < 0$. Finally, on D_8°, $G''(0) = -Kx^{-1}h^2 \le 0$.

Now we will prove (4.48). Since U_y is continuous, it suffices to show that for $y > c$ we have

$$U_x(1-, y) \geq U_x(1+, y). \tag{4.50}$$

However,

$$U_x(1-, y) = \begin{cases} \log[\frac{2\alpha+1}{\alpha+1}y] - \alpha - 1 & \text{if } y \in (c, \alpha/(2\alpha+1)), \\ -\alpha - \exp[-\frac{2\alpha+1}{\alpha+1}y + 1] & \text{if } y \geq \alpha/(2\alpha+1) \end{cases}$$

is nondecreasing as a function of y, while

$$U_x(1+, y) = \begin{cases} -\log|y| - K - \log(K-1) & \text{if } y \in (c, (K-1)^{-1}), \\ -K & \text{if } y \geq (K-1)^{-1} \end{cases}$$

is nonincreasing. Consequently, it suffices to check (4.50) for $y = c$, and an easy computation shows that for this choice of y both sides of the estimate are equal.

Finally, we see that (4.49) can be rewritten as $U_x(0+, y) + \alpha|U_y(0+, y)| \leq 0$, and we easily check that in fact we have equality here. □

The initial condition reads

3° If $|y| \leq x$, then $U(x, y) \leq L(K, \alpha)$.

To show it, we use the previous lemma: the function $t \mapsto U(xt, yt)$ is concave and since $U_x(0+, 0) = U_y(0+, 0) = 0$, it is nonincreasing. Hence $U(x, y) \leq U(0, 0) = L(K, \alpha)$. Thus, by Burkholder's method, (4.44) follows; the necessary integrability can be verified by a bit lengthy but straightforward computations (for analogous argumentation, see the proof of the logarithmic estimate in the martingale setting). □

Sharpness, $K \geq K_0$. Let $\varepsilon > 0$. We will show below that there exist a submartingale f taking values in $[0, 1]$ and g, which is α-subordinate to f, such that $||g||_1 \geq L(K, \alpha) - \varepsilon$. For such a pair, we have $||g||_1 - K \sup_n \mathbb{E}f_n \log^+ f_n = ||g||_1 \geq L(K, \alpha) - \varepsilon$ and hence the constant $L(K, \alpha)$ in (4.44) cannot be replaced by a smaller one. See Theorem 4.16 below. □

Sharpness of (4.44), $1 < K < K_0$. This is quite involved. Suppose that the inequality (4.44) holds with some universal constant $\beta(K, \alpha)$ and let

$$U^0(x, y) = \sup\{\mathbb{E}|g_n| - K\mathbb{E}f_n \log^+ f_n\},$$

where the supremum is taken over all (f, g) starting from (x, y) such that f is a nonnegative submartingale and y satisfies $|dg_n| \leq |df_n|$, $|\mathbb{E}(|dg_n|\mathcal{F}_{n-1})| \leq \alpha\mathbb{E}(df_n|\mathcal{F}_{n-1})$ for all $n \geq 1$. For $x \geq 0$, $y \in \mathbb{R}$, let $A(y) = U^0(0, y)$, $B(x) = U^0(x, 0)$ and $C(y) = U^0(\alpha/(2\alpha+1), y - \alpha/(2\alpha+1))$. Clearly, A is even. We split the remaining part of the proof into a few intermediate steps.

Step 1. We start with the observation that for any $x \geq 0$ and $y \in \mathbb{R}$,

$$U^0(x, y + \delta) \leq U^0(x, y) + \delta. \tag{4.51}$$

To see this, take (f, g) as in the definition of $U^0(x, y + \delta)$, and observe that by the triangle inequality,

$$\mathbb{E}\left[|g_n| - K f_n \log^+ f_n\right] \leq \mathbb{E}\left[|(g_n - \delta)| - K f_n \log^+ f_n\right] + \delta \leq U^0(x, y) + \delta$$

and it suffices to take the supremum over (f, g).

Step 2. We will establish the bound

$$B\left(K/(K - 1)\right) \geq -K/(K - 1). \tag{4.52}$$

To this end we introduce the function $W : [1, \infty) \times \mathbb{R} \to \mathbb{R}$ by the formula

$$W(x, y) = \inf_{\lambda > 1/x} [U^0(\lambda x, \lambda y)/\lambda + K x \log \lambda].$$

This function enjoys $1°$, $2°$ (with obvious restriction to the domain of W) and is finite. Indeed, $1°$ is a consequence of the fact that for any $x \geq 1$ and $\lambda > 1/x$,

$$U^0(\lambda x, \lambda y)/\lambda + K x \log \lambda \geq |y| - K \lambda x \log(\lambda x) + K x \log \lambda = V(x, y).$$

To prove $2°$, one shows that for any $(x, y) \in (1, \infty) \times \mathbb{R}$ and any $\varepsilon > 0$ there is $\delta > 0$ such that if $a \in (0, \delta)$ and $x - a \geq 1$, then

$$W(x, y) \geq (W(x + a, y \pm a) + W(x - a, y \mp a))/2 - \varepsilon.$$

Furthermore, W has the following homogeneity-type property: for any $x \geq 1$, $y \in \mathbb{R}$ and $\mu \geq 1/x$,

$$W(\mu x, \mu y) = \mu W(x, y) - K \mu x \log \mu. \tag{4.53}$$

By properties $1°$ and $2°$, we have, for $x = K/(K - 1)$,

$$\begin{aligned}
W(x, 0) &\geq \frac{K\delta}{x + K\delta} W(1, x - 1) + \frac{x}{x + K\delta} W(x + \delta, -\delta) \\
&\geq \frac{K\delta}{x + K\delta}(x - 1) + \frac{x}{x + K\delta} W(x + \delta, -\delta)
\end{aligned} \tag{4.54}$$

and

$$\begin{aligned}
W(x + \delta, -\delta) &\geq \frac{K\delta}{x + 2\delta} W\left(\frac{(K - 1)(x + 2\delta)}{K}, \frac{x + 2\delta}{K}\right) \\
&\quad + \frac{x + 2\delta - K\delta}{x + 2\delta} W(x + 2\delta, 0) \\
&\geq \frac{K\delta}{x + 2\delta}\left[\frac{x + 2\delta}{K} - (K - 1)(x + 2\delta)\log\left(\frac{(K - 1)(x + 2\delta)}{K}\right)\right] \\
&\quad + \frac{x + 2\delta - K\delta}{x + 2\delta} W(x + 2\delta, 0)
\end{aligned}$$

$$= \frac{K\delta}{x+2\delta}\left[\frac{x+2\delta}{K} - (K-1)(x+2\delta)\log\left(\frac{(K-1)(x+2\delta)}{K}\right)\right]$$

$$+ \frac{x+2\delta - K\delta}{x+2\delta}\left[\frac{x+2\delta}{x}W(x,0) - K(x+2\delta)\log\left(1+\frac{2\delta}{x}\right)\right],$$

where in the last passage we have exploited (4.53). Insert this into (4.54), subtract $W(x,0)$ from both sides, divide throughout by δ and let $\delta \to 0$. We obtain $W(x,0) \geq -K/(K-1)$. The final observation is that $W(x,0) \leq U^0(x,0)$, so (4.52) follows.

Step 3. Now we will prove that

$$B(c+1) \geq -c\alpha - (\alpha+1)/(2\alpha+1). \tag{4.55}$$

To this end, note that by the properties $1°$ and $2°$, for any $x \geq 1$,

$$B(x) \geq \frac{\delta}{x-1+\delta}U(1,x-1) + \frac{x-1}{x-1+\delta}U(x+\delta,-\delta)$$

$$\geq \frac{\delta(x-1)}{(x-1+\delta)} + (x-1)\frac{U(x+\delta,-\delta)}{x-1+\delta}$$

and

$$U(x+\delta,-\delta) \geq \frac{\delta}{x-1+2\delta}U(1,-(x-1+2\delta)) + \frac{x-1+\delta}{x-1+2\delta}B(x+2\delta)$$

$$\geq \delta + (x-1)B\frac{x+2\delta}{x-1+\delta}.$$

Combining these two estimates we obtain, after some manipulations,

$$\frac{B(x)}{x-1} \geq \frac{B(x+2\delta)}{x+2\delta-1} + \frac{2\delta}{x-1} - \frac{\delta^2 x}{(x-1+\delta)(x-1)},$$

so, by induction,

$$\frac{B(x)}{x-1} \geq \frac{B(x+2N\delta)}{x+2N\delta-1} + \sum_{k=0}^{N-1}\left[\frac{2\delta}{x+2k\delta-1} - \frac{\delta^2(x+2k\delta)}{(x+(2k+1)\delta-1)(x+2k\delta-1)}\right].$$

Now set $x = c+1$, $\delta = (\frac{K}{K-1} - c - 1)/(2N)$ and let $N \to \infty$ to obtain

$$B(c+1) \geq c(K-1)B(K/(K-1)) - c\log(c(K-1)),$$

as the sum converges to the integral $\int_{c+1}^{K/(K-1)}(x-1)^{-1}dx = \log(c(K-1))$. Now (4.55) follows from (4.52) and (4.42).

Step 4. The next step is to establish the bound

$$B(c - (2\alpha+1)^{-1}) \geq \frac{\alpha+1}{c(2\alpha+1)+\alpha}A(c-(2\alpha+1)^{-1})$$

$$+ \frac{c(2\alpha+1)-1}{c(2\alpha+1)+\alpha}\left[-c\alpha + \frac{\alpha(\alpha+1)}{2\alpha+1} + \frac{(\alpha+1)^2}{c(2\alpha+1)^2}\right]. \tag{4.56}$$

We proceed as previously. Using the concavity of $t \mapsto U^0(t, c - (2\alpha + 1)^{-1} - t)$ and $t \mapsto U^0(t, c + 1 - t)$ we can bound $B(c - (2\alpha + 1)^{-1})$ from below by a convex combination of $A(c - (2\alpha + 1)^{-1})$, $B(c + 1)$ and $U^0(1, c)$. It suffices to use (4.55) and $U^0(1, c) \geq V(1, c) = c$ to obtain the desired inequality.

   Step 5. Now we will deal with the estimate

$$A(c - (2\alpha + 1)^{-1}) \geq c + \frac{2\alpha^2 + 2\alpha - 1}{2\alpha + 1} - \frac{2\alpha(\alpha + 1)}{c(2\alpha + 1)^2} + \frac{B(c - (2\alpha + 1)^{-1})}{c(2\alpha + 1)}. \quad (4.57)$$

This is the most elaborate part. We will show the inequality only for $\alpha > 0$; the case $\alpha = 0$ can be treated similarly. Use 2° to obtain, for any $y \geq c - (2\alpha + 1)^{-1}$ and $0 < \delta < \alpha$,

$$
\begin{aligned}
A(y) &\geq U^0(\delta, y + \alpha\delta) \geq (2\alpha + 1)\delta C(y + (\alpha + 1)\delta)/\alpha \\
&\quad + (\alpha - (2\alpha + 1)\delta)A(y + (\alpha + 1)\delta)/\alpha.
\end{aligned} \quad (4.58)
$$

Similarly, using (iv) and (i), one gets

$$
\begin{aligned}
C(y + (\alpha + 1)\delta) &\geq (2\alpha + 1)\delta y/(2 + (2\alpha + 1)\delta) + \delta/2 \\
&\quad + \frac{(2\alpha + 1)(\alpha + 1)\delta}{\alpha(2 + (2\alpha + 1)\delta)} A(y) + \frac{2\alpha - (2\alpha + 1)(\alpha + 1)\delta}{\alpha(2 + (2\alpha + 1)\delta)} C(y).
\end{aligned} \quad (4.59)
$$

Multiply both sides of (4.58) by

$$\lambda = \left(2\alpha + 3 + \sqrt{(2\alpha + 1)^2 - 4\delta(2\alpha + 1)}\right)/(4 + 2\delta(2\alpha + 1)).$$

and add it to (4.59) to obtain, after some manipulations,

$$
\begin{aligned}
(A(y) - y)\gamma_1 - (C(y) - y)\gamma_2 &\geq r[(A(y + (\alpha + 1)\delta) - (y + (\alpha + 1)\delta))\gamma_1 \\
&\quad - (C(y + (\alpha + 1)\delta) - (y + (\alpha + 1)\delta))\gamma_2] \quad (4.60) \\
&\quad + (\lambda - 1)(\alpha + 1)\delta + \delta/2,
\end{aligned}
$$

where

$$\gamma_1 = \lambda - \delta \frac{(\alpha + 1)(2\alpha + 1)}{\alpha(2 + \delta(2\alpha + 1))}, \quad \gamma_2 = \frac{2\alpha - (\alpha + 1)(2\alpha + 1)\delta}{\alpha(2 + \delta(2\alpha + 1))}$$

and

$$r = \frac{(2 + \delta(2\alpha + 1))(\alpha - \lambda(2\alpha + 1)\delta)}{2\alpha - (\alpha + 1)(2\alpha + 1)\delta} = 1 - \delta \frac{(2\alpha + 1)(2\lambda - (2\alpha + 1))}{2\alpha} + o(\delta).$$

By induction, (4.60) gives, for any integer $N \geq 1$,

$$
\begin{aligned}
(A(y) - y)\gamma_1 - (C(y) - y)\gamma_2 &\geq r^N[(A(y + N(\alpha + 1)\delta) - (y + N(\alpha + 1)\delta))\gamma_1 \\
&\quad - (C(y + N(\alpha + 1)\delta) - (y + N(\alpha + 1)\delta))\gamma_2] \\
&\quad + [(\lambda - 1)(\alpha + 1)\delta + \delta/2](r^N - 1)/(r - 1).
\end{aligned}
$$

Now fix $z > y$, set $\delta = (z-y)/N$ (here N is sufficiently large, so that $\delta < \alpha/(2\alpha+1)$) and let $N \to \infty$. Then $\gamma_1 \to \alpha + 1$, $\gamma_2 \to 1$ and $r^N \to \exp[(y-z)\frac{2\alpha+1}{2\alpha(\alpha+1)}]$, so we obtain, after some computations,

$$(A(y) - y)(\alpha + 1) - (C(y) - y) \geq \exp\left[(y-z)\frac{2\alpha+1}{2\alpha(\alpha+1)}\right] [(A(z) - z)(\alpha + 1)$$
$$- (C(z) - z) - \alpha(2\alpha(\alpha+1)+1)/(2\alpha+1)]$$
$$+ \alpha(2\alpha(\alpha+1)+1)/(2\alpha+1).$$

Now let $z \to \infty$. Since $A(z) = U^0(0,z) \geq V(0,z) = z$ and, by (4.51), $C(z) \leq C(0) + z$, we obtain

$$(A(y) - y)(\alpha + 1) - (C(y) - y) \geq \alpha(2\alpha(\alpha+1)+1)/(2\alpha+1). \tag{4.61}$$

Take $y = c - (2\alpha + 1)^{-1}$ and use the following consequence of $2°$:

$$C(c - (2\alpha+1)^{-1}) \geq \frac{\alpha B(c - (2\alpha + 1)^{-1})}{c(2\alpha + 1) - 1} + \frac{c(2\alpha + 1) - (\alpha + 1)}{c(2\alpha + 1) - 1} A(c - (2\alpha+1)^{-1}).$$

As a result, we obtain (4.57).

Step 6. The last inequality we need is

$$(\alpha + 2)U^0(0,0) \geq (\alpha + 1)A(c - (2\alpha + 1)^{-1}) + B(c - (2\alpha + 1)^{-1}). \tag{4.62}$$

Fix a positive integer N and let $\delta = (c - (2\alpha+1)^{-1})/N$, $k < N$. Arguing as above, one can establish the inequalities

$$A(k(\alpha + 1)\delta) \geq \frac{k(\alpha + 1) + \alpha}{(k + 1)(\alpha + 1)} A((k + 1)(\alpha + 1)\delta) + \frac{B((k + 1)(\alpha + 1)\delta)}{(k + 1)(\alpha + 1)} \tag{4.63}$$

and

$$B(k(\alpha + 1)\delta) \geq \frac{2k(\alpha + 1) + \alpha}{(2k + 1)(k + 1)(\alpha + 1)} A((k + 1)(\alpha + 1)\delta)$$
$$+ \left[\frac{1}{(2k + 1)(k + 1)(\alpha + 1)} + \frac{k}{k + 1}\right] B((k + 1)(\alpha + 1)\delta). \tag{4.64}$$

Multiply (4.63) throughout by $\alpha + 1 - (2k + 1)^{-1}$ and add it to (4.64). After some manipulations, one obtains

$$(\alpha + 1)[A(k(\alpha + 1)\delta) - A((k + 1)(\alpha + 1)\delta)] + B(k(\alpha + 1)\delta) - B((k + 1)(\alpha + 1)\delta)$$
$$\geq (A(k(\alpha + 1)\delta) - A((k + 1)(\alpha + 1)\delta))/(2k + 1) \geq (\alpha + 1)\delta/(2k + 1),$$

where the latter inequality follows from (4.51). Write the above estimate for $k = 0, 1, 2, \ldots, N - 1$ and add the obtained inequalities to get

$$(\alpha + 1)A(0) + B(0) \geq (\alpha + 1)A(c - (2\alpha + 1)^{-1}) + B(c - (2\alpha + 1)^{-1})$$
$$+ (\alpha + 1)\delta \sum_{k=0}^{N-1} (2k + 1)^{-1}.$$

It suffices to let $N \to \infty$ to obtain (4.62); the last term in the estimate above tends to 0, as it is of order $N^{-1} \log N$.

Step 7. Combine (4.56) and (4.57) to get

$$A(c - (2\alpha + 1)^{-1}) \geq c - (2\alpha + 1)^{-1} + \alpha + (\alpha + 1)/(c(2\alpha + 1)^2),$$
$$B(c - (2\alpha + 1)^{-1}) \geq \alpha + 1 - c\alpha.$$

Plugging this into (4.62) yields $U^0(0,0) \geq L(K, \alpha)$. Now use 3° with $x = y = 0$ to complete the proof. $\qquad\qquad\qquad\qquad\qquad\qquad\qquad\qquad\qquad\qquad\qquad\qquad\qquad\square$

On the search of a suitable majorant. Some arguments will be similar to those used in the analogous search in the martingale setting: see Subsection 3.6.2 in Chapter 3 above. However, the reasoning will be much more complicated. We shall focus on the case $1 < K < K_0$ and $\alpha \in (0, 1]$. We let $V(x, y) = |y| - Kx \log^+ x$ and consider the function

$$U^0(x, y) = \sup\{\mathbb{E}V(f_n, g_n)\},$$

where the supremum is taken over appropriate parameters. Clearly, it suffices to find the special function on the first quadrant $[0, \infty) \times [0, \infty)$.

Step 1. Let us try to gain some understanding of what the set $\{U^0 = V\}$ should look like. As in the martingale setting, it is natural to conjecture that it is of the form $\{(x, y) : |y| \geq \gamma(x)\}$ for some curve γ. However, it readily turns out that it is not exactly the case. Namely, because the function $x \mapsto x \log^+ x$ is not differentiable at $x = 1$, it can be seen that if $|y|$ is sufficiently large, then $U^0(x, y) = V(x, y)$. Loosely speaking, it is not possible to increase $\|g\|_1$ with keeping $\mathbb{E}f \log^+ f$ small at the same time. This gives rise to the following assumption.

(A1)   There is $c = c(K, \alpha)$ such that $U(x, y) = V(x, y)$ for $|y| \geq c$. Furthermore, there is $\gamma : (1, \infty) \to \mathbb{R}$ such that for $x > 1$ we have $U(x, y) = V(x, y)$ if and only if $|y| \geq \gamma(x)$.

Step 2. *Further assumptions.* As usual, we shall impose a kind of a regularity condition on U. However, since V is not of class C^1, because of the line $x = 1$, it seems plausible to assume that

(A2)   The function U is continuous on $[0, \infty) \times \mathbb{R}$. Furthermore, it is of class C^1 on the set $(0, \infty) \times \mathbb{R} \setminus [\{1\} \times ((-\infty, -c] \cup [c, \infty))]$.

Now it is time for structural assumptions. After many experiments and calculations, one conjectures that the set $[0, \infty) \times [0, \infty)$ should be divided into regions $D_0 - D_6$ which look similarly to those on the upper part of Figure 4.1. Here we do not specify any parameters of these domains: we only take the "peak" of D_4 to be equal to $(1, c)$. Next, some further experimentation leads to

(A3) U is linear along line segments of slope -1 contained in D_1, D_2, D_5 and D_6, and linear along line segments of slope 1 contained in D_3 and D_4.

Furthermore, as in the case of the weak type estimate for nonnegative submartingales, we impose the condition

(A4) $U_x(0+, y) + \alpha U_y(0, y) = 0$ for $y \geq 0$.

We have that

$$\partial D_2 \cap \partial D_3 = \{b\} \times [c + b - 1, \infty)$$

for some $b \in (0, 1)$. In order to determine b, we need additional information on U: the above conditions (A1)–(A4) do not help. Namely, one requires additional smoothness of U at the boundary $\partial D_2 \cap \partial D_3$. To state the assumption, we need the function $G : [-b, 1 - b] \to \mathbb{R}$, given by $G(s) = U(b + s, y - s)$ ($y > c + b - 1$ is fixed). Note that according to (A3), G is linear on $[-b, 0]$ for any such y.

(A5)   For any $y > c + b - 1$, $G''(0) = 0$.

Step 3. Determining of γ and U on $D_5 \cup D_6$. This is similar to the analogous calculation in the martingale setting: in fact, the formula for γ, and the formula for U on D_6 turn out to be the same.

Step 4. Determining of U on $D_2 \cup D_3$. Let $A(y) = U(0, y)$ and $C(y) = U(b, y - b)$, $y \in \mathbb{R}$. The assumption (A3) gives

$$U(x, y) = \frac{x}{b} C(x + y) + \frac{b - x}{b} A(x + y), \tag{4.65}$$

for $(x, y) \in D_2$, and

$$\begin{aligned}
U(x, y) &= \frac{x - b}{1 - b} U(1, -x + y + 1) + \frac{1 - x}{1 - b} C(y - x + 2b) \\
&= \frac{x - b}{1 - b} (-x + y + 1) + \frac{1 - x}{1 - b} C(y - x + 2b)
\end{aligned} \tag{4.66}$$

for $(x, y) \in D_3$. Here we have used the first part of (A1). Now, (A5) combined with (4.66) yields the differential equation

$$C''(y + b) = \frac{1 - C'(y + b)}{1 - b}, \qquad y > c + b - 1,$$

from which it follows that

$$C(y + b) = y + K_1 \exp\left(-\frac{y}{1 - b}\right) + K_2, \qquad y \geq c + b - 1,$$

for some constants K_1 and K_2. Next, by (A2) we have $U_x(b-, y) = U_x(b+, y)$ for $y \geq c + b - 1$, which gives

$$2C'(y + b) = \frac{A(y + b) - C(y + b)}{b} + \frac{1 - b + y - C(y + b)}{1 - b},$$

or

$$A(y + b) = K_1 \frac{1 - 2b}{1 - b} \exp\left(-\frac{y}{1 - b}\right) + y + b + \frac{K_2}{1 - b}.$$

Finally, (A4) and (4.65) imply the differential equation

$$\frac{C(y) - A(y)}{b} + (\alpha + 1)A'(y) = 0,$$

and plugging the above formulas for A and C gives

$$b = \frac{\alpha}{2\alpha + 1} \qquad \text{and} \qquad K_2 = \frac{\alpha(\alpha + 1)}{2\alpha + 1}.$$

Next, by (A2), we get the connection between K_1 and c, by means of the equality $U_x(1-, c) = U_x(1+, c)$. By the formula for U on D_5, which we have already obtained in the previous step, we get

$$K_1 = (-\alpha + \log(c(K - 1)) + K) \cdot \frac{\alpha + 1}{2\alpha + 1} \exp\left(-1 + \frac{2\alpha + 1}{\alpha + 1}c\right).$$

Step 5. Determining of U on D_1. Arguing as in the proof of the weak type estimate, we compute that U is given on D_1 by

$$U(x, y) = a_1(x + |y|)^{1/(\alpha+1)}(-(\alpha + 1)x + |y|) + a_2,$$

where a_1, a_2 are some constants. Using (A2) and the formula for U on D_2, we get

$$a_1 = \frac{2\alpha + 1 - K - \log(c(K - 1))}{\alpha + 2}\left(c - \frac{1}{2\alpha + 1}\right)^{-1/(\alpha+1)}$$

and

$$a_2 = \left[-\alpha + \log(c(K - 1)) + K\right]\left(\frac{1}{2\alpha + 1} + \frac{1}{\alpha + 2}\left(c - \frac{1}{2\alpha + 1}\right)\right)$$
$$+ \frac{1}{\alpha + 2}\left(c - \frac{1}{2\alpha + 1}\right) - \frac{1}{2\alpha + 1}.$$

Step 6. Determining of U on D_4 and the value of c. The formula for U on D_4 is obtained by the arguments used previously (for example, we can derive the differential equation for $U(\cdot, 0)$ on $(c - (2\alpha + 1)^{-1}, c)$). The only thing which needs to be explained is how we get the equation (4.42), since, so far, the expressions for U on $D_1 \cup D_2 \cup D_3 \cup D_4$ depend on c. This is simple: by (A2) and (A3),

$$U_x\left(0, c - \frac{1}{2\alpha + 1}\right) - U_y\left(0, c - \frac{1}{2\alpha + 1}\right) = U_x\left(c + \frac{\alpha}{2\alpha + 1}, \frac{\alpha + 1}{2\alpha + 1}\right)$$
$$+ U_x\left(c + \frac{\alpha}{2\alpha + 1}, \frac{\alpha + 1}{2\alpha + 1}\right),$$

which follows from the fact that the points $(0, c - \frac{1}{2\alpha+1})$ and $(c + \frac{\alpha}{2\alpha+1}, -\frac{\alpha+1}{2\alpha+1})$ lie on the line segment of slope -1, contained in $\overline{D_1} \cup -\overline{D_5}$. The left-hand side and the right-hand side of the equality above can be computed by the formulas for U on D_1 and D_5, which we have already provided. We obtain

$$K + \log(c(K-1)) - 2\alpha - 1 = 1 - (\log c + K + \log(K-1)) - \left(c - \frac{\alpha+1}{2\alpha+1}\right) \cdot \frac{2}{c},$$

which, after simple manipulations, becomes precisely (4.42). This completes the search. □

4.6.3 Proof of Theorem 4.13

Proof of (4.45). We simply extend one of the special functions used in the proof of (4.44) to a function on the whole $\mathbb{R} \times \mathbb{R}$. Specifically, let $V : \mathbb{R} \times \mathbb{R} \to \mathbb{R}$ be given by $V(x, y) = V_K(|x|, y) = |y| - K|x| \log^+ |x|$. The corresponding special function $U : \mathbb{R} \times \mathbb{R} \to \mathbb{R}$ is given by $U(x, y) = U_{K,0}(|x|, y)$, where $U_{K,0}$ was defined above. Then the function U satisfies 1° and the appropriate versions of the conditions 2° and 3°. This follows immediately from the analysis presented in the previous subsection and the fact that $(U_{K,0})_x(0+, y) = 0$ for all y. Thus, (4.45) follows. □

Sharpness. This can be shown in a similar manner as previously. We omit the details and refer the interested reader to [149]. □

4.7 Inequalities for bounded sub- and supermartingales

4.7.1 Formulation of the results

Theorem 4.14. *Let $\alpha \in [0, 1]$. Suppose f is a submartingale satisfying $\|f\|_\infty \leq 1$ and g is an adapted sequence of real-valued random variables which is α-subordinate to f. Then for $\lambda > 0$ we have the sharp inequality*

$$\mathbb{P}(|g|^* \geq \lambda) \leq U_\lambda(-1, 1), \tag{4.67}$$

where U_λ is given by (4.72), (4.73) or (4.75), depending on whether $\lambda \in (0, 2]$, $\lambda \in (2, 4)$ or $\lambda \geq 4$. In particular, for $\lambda \geq 4$,

$$\mathbb{P}(|g|^* \geq \lambda) \leq \gamma e^{-\lambda/(2\alpha+2)}, \tag{4.68}$$

where

$$\gamma - \frac{1+\alpha}{2\alpha+4}\left(\alpha + 1 + 2^{-\frac{\alpha+2}{\alpha+1}}\right) \exp\left(\frac{2}{\alpha+1}\right).$$

We turn to the Φ-inequalities for α-subordinates of bounded submartingales. To be consistent with the literature, we present the results in the case when the dominating processes f are nonnegative, but an easy argument can be used to transfer these to general bounded f. See Remark 4.4 below.

Let Φ be a nondecreasing convex function on $[0, \infty)$, which is twice differentiable on $(0, \infty)$ and such that Φ' is convex on $(0, \infty)$ and $\Phi(0) = \Phi'(0+) = 0$.

Theorem 4.15. *Let α be a fixed nonnegative number. Assume that f is a nonnegative submartingale bounded from above by 1 and g is an adapted sequence of $\mathcal{H}$-valued random variables which is α-subordinate to f. Then for Φ as above,*

$$\sup_n \mathbb{E}\Phi\left(\frac{|g_n|}{\alpha + 1}\right) \leq \frac{1 + \alpha}{2 + \alpha} \int_0^\infty \Phi(t)e^{-t}\,dt. \tag{4.69}$$

The inequality is sharp, even if $\mathcal{H} = \mathbb{R}$.

For Φ as above, but Φ' is concave on $(0, \infty)$, the optimal constants are not known. However, we will establish a partial result in this direction: the following sharp L^1 inequality.

Theorem 4.16. *Let α be a fixed nonnegative number. Assume that f is a nonnegative submartingale bounded from above by 1 and let g be an adapted sequence of real-valued random variables which is α-subordinate to f. Then*

$$||g||_1 \leq \frac{(\alpha + 1)(2\alpha^2 + 3\alpha + 2)}{(2\alpha + 1)(\alpha + 2)} \tag{4.70}$$

and the bound is the best possible.

4.7.2   Proof of Theorem 4.14

Proof of (4.67). The special function corresponding to our problem admits three different formulas, depending on the value of λ. Let S denote the strip $[-1, 1] \times \mathbb{R}$ and consider the following subsets of S. For $0 < \lambda \leq 2$,

$$A_\lambda = \{(x, y) \in S : |y| \geq x + \lambda - 1\},$$
$$B_\lambda = \{(x, y) \in S : 1 - x \leq |y| < x + \lambda - 1\},$$
$$C_\lambda = \{(x, y) \in S : |y| < 1 - x \text{ and } |y| < x + \lambda - 1\}.$$

For $\lambda \in (2, 4)$, define

$$A_\lambda = \{(x, y) \in S : |y| \geq \alpha x + \lambda - \alpha\},$$
$$B_\lambda = \{(x, y) \in S : \alpha x + \lambda - \alpha > |y| \geq x - 1 + \lambda\},$$
$$C_\lambda = \{(x, y) \in S : x - 1 + \lambda > |y| \geq 1 - x\},$$
$$D_\lambda = \{(x, y) \in S : 1 - x > |y| \geq -x - 3 + \lambda \text{ and } |y| < x - 1 + \lambda\},$$
$$E_\lambda = \{(x, y) \in S : -x - 3 + \lambda > |y|\}.$$

Finally, for $\lambda \geq 4$, let

$$A_\lambda = \{(x, y) \in S : |y| \geq \alpha x + \lambda - \alpha\},$$

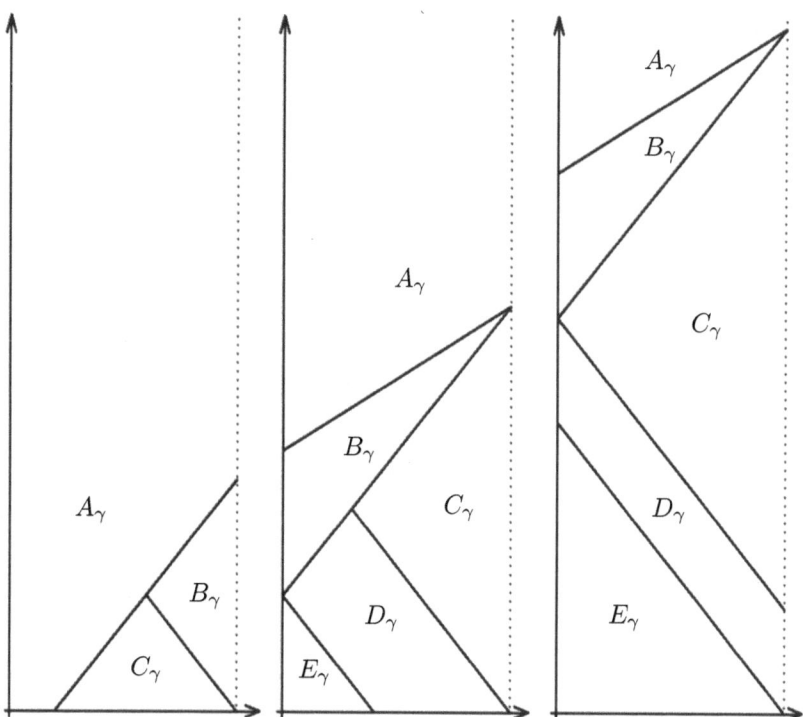

Figure 4.2: The sets $A_\lambda - E_\lambda$, intersected with $\mathbb{R}_+^2$, in the case $\lambda \leq 2$ (left picture), $2 < \lambda < 4$ (middle picture) and $\lambda \geq 4$ (right picture).

$$B_\lambda = \{(x,y) \in S : \alpha x + \lambda - \alpha > |y| \geq x - 1 + \lambda\},$$
$$C_\lambda = \{(x,y) \in S : x - 1 + \lambda > |y| \geq -x - 3 + \lambda\},$$
$$D_\lambda = \{(x,y) \in S : -x - 3 + \lambda > |y| \geq 1 - x\},$$
$$E_\lambda = \{(x,y) \in S : 1 - x > |y|\}.$$

Let $V(x,y) = 1_{\{|y| \geq \lambda\}}$ for $(x,y) \in S$. Let $H : S \times (-1, \infty) \to \mathbb{R}$ be the function given by

$$H(x,y,z) = \frac{1}{\alpha+2}\left[1 + \frac{(x+1+|y|)^{1/(\alpha+1)}((\alpha+1)(x+1) - |y|)}{(1+z)^{(\alpha+2)/(\alpha+1)}}\right]. \quad (4.71)$$

Now let us define the special functions $U_\lambda : S \to \mathbb{R}$. For $0 < \lambda \leq 2$, let

$$U_\lambda(x,y) = \begin{cases} 1 & \text{if } (x,y) \in A_\lambda, \\ \frac{2-2x}{1+\lambda-x-|y|} & \text{if } (x,y) \in B_\lambda, \\ 1 - \frac{(\lambda-1+x-|y|)(\lambda-1+x+|y|)}{\lambda^2} & \text{if } (x,y) \in C_\lambda. \end{cases} \quad (4.72)$$

For $2 < \lambda < 4$, set

$$
U_\lambda(x,y) = \begin{cases}
1 & \text{if } (x,y) \in A_\lambda, \\
1 - (\alpha(x-1) - |y| + \lambda) \cdot \frac{2\lambda - 4}{\lambda^2} & \text{if } (x,y) \in B_\lambda, \\
\frac{2-2x}{1+\lambda-x-|y|} - \frac{2(1-x)(1-\alpha)(\lambda-2)}{\lambda^2} & \text{if } (x,y) \in C_\lambda, \\
\frac{2(1-x)}{\lambda}\left[1 - \frac{(1-\alpha)(\lambda-2)}{\lambda}\right] - \frac{(1-x)^2 - |y|^2}{\lambda^2} & \text{if } (x,y) \in D_\lambda, \\
a_\lambda H(x,y,\lambda-3) + b_\lambda & \text{if } (x,y) \in E_\lambda,
\end{cases}
\tag{4.73}
$$

where

$$
a_\lambda = -\frac{2(1+\alpha)(\lambda-2)^2}{\lambda^2}, \quad b_\lambda = 1 - \frac{4(\lambda-2)(1-\alpha)}{\lambda^2}.
\tag{4.74}
$$

For $\lambda \geq 4$, set

$$
U_\lambda(x,y) = \begin{cases}
1 & \text{if } (x,y) \in A_\lambda, \\
1 - \frac{\alpha(x-1)-|y|+\lambda}{4} & \text{if } (x,y) \in B_\lambda, \\
\frac{2-2x}{1+\lambda-x-|y|} - \frac{(1-x)(1-\alpha)}{4} & \text{if } (x,y) \in C_\lambda, \\
\frac{(1-x)(1+\alpha)}{4} \exp\left(\frac{3+x+|y|-\lambda}{2(\alpha+1)}\right) & \text{if } (x,y) \in D_\lambda, \\
a_\lambda H(x,y,1) + b_\lambda & \text{if } (x,y) \in E_\lambda,
\end{cases}
\tag{4.75}
$$

where

$$
a_\lambda = -b_\lambda = -\frac{(1+\alpha)}{2} \exp\left(\frac{4-\lambda}{2\alpha+2}\right).
\tag{4.76}
$$

Now let us study the properties of U_λ. For the sake of clarity, we have decided to group these into two lemmas.

Lemma 4.5. *For $\lambda > 2$, let ϕ_λ, ψ_λ denote the partial derivatives of U_λ with respect to x, y on the interiors of A_λ, B_λ, C_λ, D_λ, E_λ, extended continuously to the whole of these sets. The following statements hold.*

(i) *The functions U_λ, $\lambda > 2$, are continuous on $S \setminus \{(1,\pm\lambda)\}$.*
(ii) *Let*

$$
S_\lambda = \{(x,y) \in [-1,1] \times \mathbb{R} : |y| \neq \alpha x + \lambda - \alpha \text{ and } |y| \neq x + \lambda - 1\}.
$$

Then

$$
\phi_\lambda, \ \psi_\lambda, \ \lambda > 2, \ \text{are continuous on } S_\lambda.
\tag{4.77}
$$

(iii) *For any $(x,y) \in S$, the function $\lambda \mapsto U_\lambda(x,y)$, $\lambda > 0$, is left-continuous.*
(iv) *For any $\lambda > 2$ we have the inequality*

$$
\phi_\lambda \leq -\alpha|\psi_\lambda|.
\tag{4.78}
$$

(v) *For $\lambda > 2$ and any $(x,y) \in S$ we have $V(x,y) \leq U_\lambda(x,y) \leq 1$.*

Proof. We start by computing the derivatives. Let $y' = y/|y|$ stand for the sign of y, with $y' = 0$ if $y = 0$. For $\lambda \in (2, 4)$ we have

$$\phi_\lambda(x, y) = \begin{cases} 0 & \text{if } (x, y) \in A_\lambda, \\ -\frac{(2\lambda-4)\alpha}{\lambda^2} & \text{if } (x, y) \in B_\lambda, \\ -\frac{2\lambda-2|y|}{(1+\lambda-x-|y|)^2} + \frac{(2\lambda-4)(1-\alpha)}{\lambda^2} & \text{if } (x, y) \in C_\lambda, \\ -\frac{2}{\lambda}\left[1 - \frac{(1-\alpha)(\lambda-2)}{\lambda}\right] + \frac{2(1-x)}{\lambda^2} & \text{if } (x, y) \in D_\lambda, \\ -c_\lambda(x + |y| + 1)^{-\alpha/(\alpha+1)}(x + 1 + \frac{\alpha}{\alpha+1}|y|) & \text{if } (x, y) \in E_\lambda, \end{cases}$$

$$\psi_\lambda(x, y) = \begin{cases} 0 & \text{if } (x, y) \in A_\lambda, \\ \frac{2\lambda-4}{\lambda^2}y' & \text{if } (x, y) \in B_\lambda, \\ \frac{2-2x}{(1+\lambda-x-|y|)^2}y' & \text{if } (x, y) \in C_\lambda, \\ \frac{2y}{\lambda^2} & \text{if } (x, y) \in D_\lambda, \\ c_\lambda(x + |y| + 1)^{-\alpha/(\alpha+1)}\frac{y}{1+\alpha} & \text{if } (x, y) \in E_\lambda, \end{cases}$$

where

$$c_\lambda = 2(1+\alpha)(\lambda - 2)^{\alpha/(\alpha+1)}\lambda^{-2}.$$

Finally, for $\lambda \geq 4$, we have

$$\phi_\lambda(x, y) = \begin{cases} 0 & \text{if } (x, y) \in A_\lambda, \\ -\frac{\alpha}{4} & \text{if } (x, y) \in B_\lambda, \\ -\frac{2\lambda-2|y|}{(1+\lambda-x-|y|)^2} + \frac{1-\alpha}{4} & \text{if } (x, y) \in C_\lambda, \\ -\frac{x+1+2\alpha}{8}\exp\left(\frac{x+|y|+3-\lambda}{2(\alpha+1)}\right) & \text{if } (x, y) \in D_\lambda, \\ -c_\lambda(x + |y| + 1)^{-\alpha/(\alpha+1)}(x + 1 + \frac{\alpha}{\alpha+1}|y|) & \text{if } (x, y) \in E_\lambda, \end{cases}$$

$$\psi_\lambda(x, y) = \begin{cases} 0 & \text{if } (x, y) \in A_\lambda, \\ \frac{1}{4}y' & \text{if } (x, y) \in B_\lambda, \\ \frac{2-2x}{(1+\lambda-x-|y|)^2}y' & \text{if } (x, y) \in C_\lambda, \\ \frac{(1-x)}{8}\exp\left(\frac{x+|y|+3-\lambda}{2(\alpha+1)}\right)y' & \text{if } (x, y) \in D_\lambda, \\ c_\lambda(x + |y| + 1)^{-\alpha/(\alpha+1)}\frac{y}{1+\alpha} & \text{if } (x, y) \in E_\lambda, \end{cases}$$

where

$$c_\lambda = (1+\alpha)2^{-(2\alpha+3)/(\alpha+1)}\exp\left(\frac{4-\lambda}{2(\alpha+1)}\right).$$

Now the properties (i), (ii), (iii) follow by straightforward computation. To prove (iv), note first that for any $\lambda > 2$ the condition (4.78) is clearly satisfied on the sets A_λ and B_λ. Suppose $(x, y) \in C_\lambda$. Then $\lambda - |y| \in [0, 4]$, $1 - x \leq \min\{\lambda - |y|, 4 - \lambda + |y|\}$ and (4.78) takes the form

$$-2(\lambda - |y|) + \frac{2\lambda - 4}{\lambda^2}(1 - \alpha)(1 - x + \lambda - |y|)^2 + 2\alpha(1 - x) \leq 0,$$

or

$$-2(\lambda - |y|) + \frac{1-\alpha}{4} \cdot (1 - x + \lambda - |y|)^2 + 2\alpha(1 - x) \le 0, \qquad (4.79)$$

depending on whether $\lambda < 4$ or $\lambda \ge 4$. As $(2\lambda - 4)/\lambda^2 \le \frac{1}{4}$, it suffices to show (4.79). If $\lambda - |y| \le 2$, then, as $1 - x \le \lambda - |y|$, the left-hand side does not exceed

$$-2(\lambda - |y|) + (1-\alpha)(\lambda - |y|)^2 + 2\alpha(\lambda - |y|) = (\lambda - |y|)(-2 + (1-\alpha)(\lambda - |y|) + 2\alpha)$$
$$\le (\lambda - |y|)(-2 + 2(1-\alpha) + 2\alpha) = 0.$$

Similarly, if $\lambda - |y| \in (2, 4]$, then we use the bound $1 - x \le 4 - \lambda + |y|$ and conclude that the left-hand side of (4.79) is not greater than

$$-2(\lambda - |y|) + 4(1 - \alpha) + 2\alpha(4 - \lambda + |y|) = -2(\lambda - |y| - 2)(1 + \alpha) \le 0$$

and we are done with the case $(x, y) \in C_\lambda$.

Assume that $(x, y) \in D_\lambda$. For $\lambda \in (2, 4)$, the inequality (4.78) is equivalent to

$$-\frac{2}{\lambda}\left[1 - \frac{(1-\alpha)(\lambda - 2)}{\lambda}\right] + \frac{2 - 2x}{\lambda^2} \le -\frac{2\alpha|y|}{\lambda^2},$$

or, after some simplifications, $\alpha|y| + 1 - x \le 2 + \alpha\lambda - 2\alpha$. It is easy to check that $\alpha|y| + 1 - x$ attains its maximum for $x = -1$ and $|y| = \lambda - 2$ and then we have equality. If $(x, y) \in D_\lambda$ and $\lambda \ge 4$, then (4.78) takes the form $-(2\alpha + 1 + x) \le -\alpha(1 - x)$, or $(x + 1)(\alpha + 1) \ge 0$. Finally, on the set E_λ, the inequality (4.78) is obvious.

   (v) By (4.78), we have $\phi_\lambda \le 0$, so $U_\lambda(x, y) \ge U_\lambda(1, y) = \chi_{\{|y| \ge \lambda\}}$. Furthermore, as $U_\lambda(x, y) = 1$ for $|y| \ge \lambda$ and $\psi_\lambda(x, y)y' \ge 0$ on S_λ, the second estimate follows. $\square$

Lemma 4.6. *Let x, h, y, k be fixed real numbers, satisfying x, $x + h \in [-1, 1]$ and $|k| \le |h|$. Then for any $\lambda > 2$ and $\alpha \in [0, 1)$,*

$$U_\lambda(x + h, y + k) \le U_\lambda(x, y) + \phi_\lambda(x, y)h + \psi_\lambda(x, y)k. \qquad (4.80)$$

Proof. It suffices to show that the function

$$G(t) = G_{x,y,h,k}(t) = U_\lambda(x + th, y + tk),$$

defined on the set $\{t : |x + th| \le 1\}$, satisfies $G''(0) \le 0$ if the derivative exists, and $G'(0-) \ge G'(0+)$ otherwise. Clearly, we may assume that $h \ge 0$, changing the signs of both h and k if necessary. Due to the symmetry of U_λ, it suffices to consider $y \ge 0$ only.

   We start from the observation that $G''(0) = 0$ on the interior of A_λ and $G'_+(0) \le G'_-(0)$ for $(x, y) \in A_\lambda \cap \overline{B}_\lambda$. The latter inequality holds since $U_\lambda \equiv 1$ on A_λ and $U_\lambda \le 1$ on B_λ. For the remaining inequalities, we consider the cases $\lambda \in (2, 4)$, $\lambda \ge 4$ separately.

The case $\lambda \in (2,4)$. The inequality $G''(0) \leq 0$ is clear for (x,y) lying in the interior of B_λ. On C_λ, we have

$$G''(0) = -\frac{4(h+k)(h(\lambda-y)-k(1-x))}{(1-x-y+\lambda)^3} \leq 0, \qquad (4.81)$$

which follows from $|k| \leq h$ and the fact that $\lambda - y \geq 1 - x$. For (x,y) in the interior of D_λ,

$$G''(0) = \frac{-h^2+k^2}{\lambda^2} \leq 0,$$

as $|k| \leq h$. Finally, on E_λ, the concavity follows from the fact that the function H has this property: see the proof of the weak type estimate (4.12).

It remains to check the inequalities for one-sided derivatives. By Lemma 4.5 (ii), the points (x,y) for which G is not differentiable at 0 do not belong to S_λ. Since we excluded the set $A_\lambda \cap \overline{B}_\lambda$, they lie on the line $y = x - 1 + \lambda$. For such points (x,y), the left derivative equals

$$G'_-(0) = -\frac{2\lambda-4}{\lambda^2}(\alpha h - k),$$

while the right one is given by

$$G'_+(0) = \frac{-h+k}{2(\lambda-y)} + \frac{(2\lambda-4)(1-\alpha)h}{\lambda^2},$$

or

$$G'_+(0) = -\frac{2h}{\lambda}\left[1 - \frac{(1-\alpha)(\lambda-2)}{\lambda}\right] + \frac{2(1-x)h+2yk}{\lambda^2},$$

depending on whether $y \geq 1-x$ or $y < 1-x$. In the first case, the inequality $G'_+(0) \leq G'_-(0)$ reduces to

$$(h-k)\left(\frac{1}{2(\lambda-y)} - \frac{2(\lambda-2)}{\lambda^2}\right) \geq 0,$$

while in the remaining one

$$\frac{2}{\lambda^2}(h-k)(y-(\lambda-2)) \geq 0.$$

Both inequalities follow from the estimate $\lambda - y \leq 2$ and the condition $|k| \leq h$.

The case $\lambda \geq 4$. On the set B_λ the concavity is clear. For C_λ, (4.81) holds. If (x,y) lies in the interior of D_λ, then

$$G''(0) = \frac{1}{8}\exp\left(\frac{3+x+y-\lambda}{2(\alpha+1)}\right)\left[\frac{1-x}{2(\alpha+1)}\cdot(-h^2+k^2) - \left(2 - \frac{1-x}{\alpha+1}\right)(h^2+hk)\right] \leq 0,$$

since $|k| \leq h$ and $(1-x)/(\alpha+1) \leq 2$. The concavity on E_λ is a consequence of the appropriate property of H. It remains to check the inequality for one-sided derivatives. By Lemma 4.5 (ii), we may assume $y = x + \lambda - 1$, and the inequality $G'_+(0) \leq G'_-(0)$ reads

$$\frac{1}{2}(h - k)\left(\frac{1}{\lambda - y} - \frac{1}{2}\right) \geq 0,$$

which is obvious, as $\lambda - y \leq 2$.   □

Now we are ready to establish the inequality (4.67). We have checked that the functions U_λ, V_λ satisfy 1° and 2°. Furthermore, we have that $U_y \geq 0$ for $y \geq 0$ and $U_x \leq 0$ on S. Consequently, we have the following version of the initial condition:

3° $U_\lambda(x, y) \leq U_\lambda(-1, 1)$ if $|y| \leq |x|$.

This establishes (4.67).   □

4.7.3   Sharpness of (4.67)

We will deal only with the case $\lambda \geq 4$, since the remaining cases can be studied in a similar manner. Let

$$U^0(x, y) = \sup \mathbb{P}(|g_n| \geq \lambda),$$

where the supremum is taken over all n and all appropriate f and g. Let $A(y) = U^0(-1, y)$ and $B(x) = U^0(x, 0)$ for all $x \in [-1, 1]$ and $y \in \mathbb{R}$. Note that $U^0(1, y) = 1_{\{|y| \geq \lambda\}}$, directly from the definition.

Step 1. First we will show that

$$A(\lambda - 2) \geq \frac{\alpha + 1}{2}. \tag{4.82}$$

This follows immediately from 2°, if $\alpha = 1$: $A(\lambda - 2) \geq U^0(1, \lambda) = 1$. If $\alpha < 1$, then, for $\delta < (1 - \alpha)/(1 + \alpha)$,

$$A(\lambda - 2) \geq U^0(-1 + \delta, \lambda - 2 + \alpha\delta)$$

$$\geq \left(1 - \frac{\delta}{2}\right) A(\lambda - 2 + (\alpha + 1)\delta) + \frac{\delta}{2} U^0(1, \lambda - 3 + (\alpha + 1)\delta)$$

$$= \left(1 - \frac{\delta}{2}\right) A(\lambda - 2 + (\alpha + 1)\delta)$$

$$\geq \left(1 - \frac{\delta}{2}\right) U^0\left(-1 + \frac{\alpha + 1}{1 - \alpha}\delta, \lambda - 2 + (\alpha + 1)\delta + \alpha\frac{\alpha + 1}{1 - \alpha}\delta\right)$$

$$\geq \left(1 - \frac{\delta}{2}\right)\left[\frac{(\alpha + 1)\delta}{2(1 - \alpha)} U^0(1, \lambda) + \left(1 - \frac{(\alpha + 1)\delta}{2(1 - \alpha)}\right) A(\lambda - 2)\right]$$

$$= \left(1 - \frac{\delta}{2}\right)\left[\frac{(\alpha + 1)\delta}{2(1 - \alpha)} + \left(1 - \frac{(\alpha + 1)\delta}{2(1 - \alpha)}\right) A(\lambda - 2)\right].$$

Subtract $A(\lambda - 2)$ from both sides, divide throughout by δ and let $\delta \to 0$ to obtain (4.82).

Step 2. Now we will prove that

$$A(2) \geq \frac{\alpha+1}{2} \exp\left(-\frac{\lambda-4}{2\alpha+2}\right). \tag{4.83}$$

To do this, note that for any $y \in [2, \lambda - 2]$ and $\delta < 1$,

$$A(y) \geq \left(1 - \frac{\delta}{2}\right) A(y + (\alpha+1)\delta)$$

(see the above chain of equalities and inequalities for the case $y = \lambda$). Let N be a positive integer and put $\delta = (\lambda - 4)/(N(\alpha+1))$: here N must be so large that $\delta < 1$. The above inequality, by induction, gives

$$A(2) \geq \left(1 - \frac{\delta}{2}\right)^N A(\lambda - 2)$$

and letting $N \to \infty$ yields (4.83).

Step 3. The next move is to establish the bound

$$B(y - 1) \geq (\alpha+1)(A(2) - A(y)), \qquad |y| \leq 2. \tag{4.84}$$

To do this, we repeat the arguments leading to (4.21); however, we replace $B(y)$ by $B(y - 1)$ there, and obtain the bound

$$(\alpha+1)\left[A(y+(\alpha+1)\delta)-A(y)\right]+B(y+(\alpha+1)\delta)-B(y) \leq \delta\frac{A(y+(\alpha+1)\delta)-A(y)}{2y+(\alpha+1)\delta}.$$

Now, if N is a (large) positive integer and we let $\delta = (2 - y)/((\alpha+1)N)$, then writing this inequality for y, $y + (\alpha+1)\delta$, $y + 2(\alpha+1)\delta$, ..., $y + (N-1)(\alpha+1)\delta$ and summing the obtained estimates, we get

$$(\alpha+1)(A(y) - A(2)) + B(y - 1) - B(1)$$

$$\geq \delta \sum_{k=0}^{N-1} \frac{A(y+(k+1)(\alpha+1)\delta) - A(y+k(\alpha+1)\delta)}{2y+(\alpha+1)\delta}$$

We have $B(1) = 0$ and letting $N \to \infty$ we easily check that the sum on the right-hand side converges to 0. This yields (4.84).

Step 4. This is the final step. Let $y \in [1, 2]$. By $2°$, we have

$$A(y) \geq U^0(-1 + \delta, y + \alpha\delta) \geq \frac{\delta}{y}B(y + (\alpha+1)\delta - 1) + \left(1 - \frac{\delta}{y}\right)A(y + (\alpha+1)\delta),$$

which, combined with (4.84), gives

$$A(y) \geq \frac{\delta}{y(\alpha+2)}A(2) + \left(1 - \frac{(\alpha+2)\delta}{y}\right)A(y + (\alpha+1)\delta).$$

This can be rewritten in the equivalent form

$$A(y) - \frac{\alpha+1}{\alpha+2}A(2) \geq \left(1 - \frac{(\alpha+2)\delta}{y}\right)\left(A(y + (\alpha+1)\delta) - \frac{\alpha+1}{\alpha+2}A(2)\right).$$

Assume that $\delta = (N(\alpha+1))^{-1}$, where N is a large positive integer, and apply this inequality to $y = 1$, $y = 1 + (\alpha+1)\delta$, $\ldots$, $y = 1 + (N-1)(\alpha+1)\delta$. Multiplying the obtained estimates gives

$$A(1) - \frac{\alpha+1}{\alpha+2}A(2) \geq \prod_{k=0}^{N-1}\left(1 - \frac{(\alpha+2)\delta}{1 + k(\alpha+1)\delta}\right) \cdot \left(A(2) - \frac{\alpha+1}{\alpha+2}A(2)\right).$$

Now if we let $N \to \infty$, a straightforward analysis gives that the product above converges to $2^{-(\alpha+2)/(\alpha+1)}$. Consequently,

$$A(1) \geq \frac{\alpha+1}{\alpha+2}A(2) + 2^{-(\alpha+2)/(\alpha+1)} \cdot \frac{A(2)}{\alpha+2}.$$

Note that after applying (4.83), the right-hand side above coincides with the right-hand side of (4.68); thus the claim follows, directly from the definition of A and U^0.

On the search of the suitable majorant. This is more or less analogous to the search in the martingale case. By experimentation, one splits the strip $[-1, 1] \times \mathbb{R}$ into regions A_λ, B_λ, $\ldots$ and conjectures linearity of the special function along the lines of slope -1 or 1. We omit the details: however, see the papers by Hammack [87] and the author [127]. The examples constructed there can be used to derive U^0 directly.   □

4.7.4   Proof of Theorem 4.15

Proof of (4.69). Let $S = [0, 1] \times \mathcal{H}$ and consider $V : S \to \mathbb{R}$ given by $V(x, y) = \Phi\left(\frac{|y|}{\alpha+1}\right)$. The special function $U : S \to \mathbb{R}$, corresponding to the Φ-inequality, is defined by the following formula. If $x + |y| \leq 1$, then

$$U(x, y) = \left[\frac{\alpha+1}{\alpha+2} + \frac{1}{\alpha+2}(|y| - (\alpha+1)x)(x + |y|)^{1/(\alpha+1)}\right]\Psi(1),$$

while for $x + |y| > 1$,

$$U(x, y) = (1 - x)\Psi\left(\frac{x + |y| + \alpha}{\alpha+1}\right) + x\Phi\left(\frac{x + |y| + \alpha}{\alpha+1} - 1\right).$$

Here, for $t \geq 1$,

$$\Psi(t) = e^t \int_t^\infty \Phi(s - 1)e^{-s}ds.$$

Lemma 4.7. *Let* $F(t) = \Psi(t) - \Phi(t-1)$, $t \geq 1$. *Then* $F'(1+) = F(1) = \Psi'(1+) = \Psi(1) > 0$, $F(t) > 0$, $F'(t) > 0$, $F''(t) \geq 0$ *and* $F(t) \leq tF'(t)$ *for* $t > 1$.

Proof. The first assertion is evident. The next three inequalities follow from the fact that Φ, Φ' Φ'' are increasing, applied to the identities

$$\Psi'(t) = e^t \int_t^\infty \Phi'(s-1)ds, \qquad \Psi''(t) = e^t \int_t^\infty \Phi''(s-1)ds.$$

Finally, we observe that $\lim_{t\downarrow 1} tF'(t) - F(t) = F'(1+) - F(1) = 0$ and $(tF'(t) - F(t))' = tF''(t) \geq 0$: this establishes the last inequality. $\square$

Lemma 4.8. *The function* U *is of class* C^1 *in the interior of* S.

Proof. We compute that

$$U_x(x,y) = \begin{cases} -\Psi(1)\left(x + \frac{2\alpha+\alpha^2}{(\alpha+1)(\alpha+2)}|y|\right)(x+|y|)^{-\alpha/(\alpha+1)} & \text{if } x+|y| < 1, \\ -\frac{\alpha}{\alpha+1}\Psi'\left(\frac{x+|y|+\alpha}{\alpha+1}\right) - \frac{x}{\alpha+1}F'\left(\frac{x+|y|+\alpha}{\alpha+1}\right) & \text{if } x+|y| > 1 \end{cases}$$

and

$$U_y(x,y) = \begin{cases} \frac{1}{\alpha+1}\Psi(1)(x+|y|)^{-\alpha/(\alpha+1)}y & \text{if } x+|y| < 1, \\ \left[\frac{1}{\alpha+1}\Psi'\left(\frac{x+|y|+\alpha}{\alpha+1}\right) - \frac{x}{\alpha+1}F'\left(\frac{x+|y|+\alpha}{\alpha+1}\right)\right]y' & \text{if } x+|y| > 1. \end{cases}$$

It remains to note that the partial derivatives match at the boundary $\{(x,y) \in S : x+|y| = 1\}$. Furthermore, observe that U_x and U_y can be extended to continuous functions on the whole strip: denote the extensions by A and B, respectively. $\square$

Now let us verify the conditions $1°$, $2°$ and $3°$. Since $U_x < 0$ (see the formula above), we have

$$U(x,y) \geq U(1,y) = \Phi\left(\frac{|y|}{\alpha+1}\right) = V(x,y).$$

To show $2°$, observe that

$$A(x,y) + \alpha|B(x,y)| = \begin{cases} -\Psi(1)x(x+|y|)^{-\alpha/(\alpha+1)} & \text{if } x+|y| \leq 1, \\ -xF'\left(\frac{x+|y|+\alpha}{\alpha+1}\right) & \text{if } x+|y| > 1 \end{cases}$$

is nonpositive. Now, fix $x \in [0,1]$, $h \geq 0$ and y, $k \in \mathcal{H}$ and consider the function $G - G_{x,y,h,k}$ defined as usually. Then G is of class C^1, satisfies $G''(0) \leq 0$ if $x+|y| < 1$ (see (4.15)) and, for $x+|y| > 1$,

$$G''(0) = I + II + III + IV,$$

where

$$I = -\frac{1}{\alpha+1}F'\left(\frac{x+|y|+\alpha}{1+\alpha}\right)(h^2 - |k|^2),$$

$$II = -\frac{\alpha}{(\alpha+1)^2}F'\left(\frac{x+|y|+\alpha}{1+\alpha}\right)(h+y'\cdot k)^2,$$

$$III = -\frac{x}{(1+\alpha)^2}F''\left(\frac{x+|y|+\alpha}{1+\alpha}\right)(h+y'\cdot k)^2,$$

$$IV = -\frac{1}{(1+\alpha)|y|}\left[(x+|y|)F'\left(\frac{x+|y|+\alpha}{1+\alpha}\right) - F\left(\frac{x+|y|+\alpha}{1+\alpha}\right)\right](|k|^2 - (y'\cdot k)^2).$$

Thus, $G''(0) \leq 0$ if $|k| \leq |h|$; to see that the expression in the square brackets in IV is nonnegative, use the last inequality from Lemma 4.7:

$$F\left(\frac{x+|y|+\alpha}{1+\alpha}\right) \leq \frac{x+|y|+\alpha}{1+\alpha}F'\left(\frac{x+|y|+\alpha}{1+\alpha}\right) \leq (x+|y|)F'\left(\frac{x+|y|+\alpha}{1+\alpha}\right).$$

Consequently, $2°$ follows. Applying this condition yields $3°$:

$$U(x,y) \leq U(0,0) + A(0,0)x + B(0,0)y \leq U(0,0) = \frac{1+\alpha}{2+\alpha}\int_0^\infty \Phi(t)e^{-t}dt.$$

The necessary integrability holds and therefore (4.69) is established. $\qquad\square$

Sharpness. Let $U^0 : S \to \mathbb{R}$ be given by

$$U^0(x,y) = \sup \mathbb{E}\Phi\left(\frac{|g_n|}{1+\alpha}\right),$$

where the supremum is taken over all n and all appropriate processes f, g. Denote $A(y) = U^0(0,y)$ and $B(x) = U^0(x,0)$ for all $x \in [0,1]$ and $y \in \mathbb{R}$.

Step 1. First we will establish the bound

$$A(1) \geq \int_0^\infty e^{-s}\Phi(s)ds. \qquad (4.85)$$

To do this, suppose that $y > 1$ and $\delta < 1$. We have, by $2°$,

$$A(y) \geq U^0(\delta, y + \alpha\delta) \geq (1-\delta)A(y + (\alpha+1)\delta) + \delta U^0(1, y + (\alpha+1)\delta - 1)$$

$$= (1-\delta)A(y + (\alpha+1)\delta) + \delta\Phi\left(\frac{y + (\alpha+1)\delta - 1}{1+\alpha}\right),$$

since there are no nontrivial submartingales bounded by 1 starting from 1. Consequently, by induction,

$$A(1) \geq (1-\delta)^N A(1 + N(\alpha+1)\delta) + \delta\sum_{k=1}^N (1-\delta)^{k-1}\Phi(k\delta).$$

Therefore, if we set $\delta = y/N$ and let $N \to \infty$, we get

$$A(1) \geq e^{-y}A(1 + (\alpha + 1)y) + \int_0^y e^{-s}\Phi(s)ds \geq \int_0^y e^{-s}\Phi(s)ds.$$

Letting $y \to \infty$, we obtain (4.85).

Step 2. Now we will prove that

$$A(0) \geq \frac{\alpha + 1}{\alpha + 2}A(1), \tag{4.86}$$

which, together with (4.85) and the definition of A and U^0 will give the claim. We apply the inequality which appears right below (4.21), let $N \to \infty$ and obtain $(\alpha + 1)A(1) + B(1) - (\alpha + 2)A(0) \leq 0$. Since $B(1) = U^0(1, 0) = \Phi(0) = 0$, the inequality (4.86) follows.   □

On the search of the suitable majorant. Define $U^0 : [0, 1] \times \mathbb{R} \to \mathbb{R}$ by the formula

$$U^0(x, y) = \sup\{\mathbb{E}V(|g_n|)\},$$

where the supremum is taken over all n and all $(f, g) \in M^{\mathrm{sub}}(x, y)$ such that f takes values in $[0, 1]$. Of course, we have $U^0(1, y) = V(|y|)$, since there is only one submartingale bounded by 1 and starting from 1. Some experimentation lead to the following assumptions (as usual, we switch from U^0 to U):

(A1) U is continuous on the strip $[0, 1] \times \mathbb{R}$ and of class C^1 in its interior.
(A2)   When restricted to $[0, 1] \times [0, \infty)$, U is linear along the line segments of slope -1.
(A3) $U_x(0+, y) + \alpha U_y(0, y) = 0$.

These three conditions lead to the special function used above. We shall not present the calculations since they are completely analogous to those used in the previous estimates.   □

Remark 4.4. To obtain a related estimate in the case when f takes values in $[-1, 1]$, use the special function $\overline{U} : [-1, 1] \times \mathcal{H} \to \mathbb{R}$ given by

$$\overline{U}(x, y) = U\left(\frac{x + 1}{2}, \frac{y}{2}\right),$$

where U is defined above. Obviously, the function $\overline{U}$ satisfies the majorization

$$\overline{U}(x, y) \geq \Phi(|y|/2)$$

and enjoys 2°. Furthermore, 3° takes the form

$$\overline{U}(x, y) \leq \frac{\alpha + 1}{\alpha + 2} + \frac{2^{-(\alpha+2)/(\alpha+1)}}{\alpha + 2} \int_0^\infty \Phi(t)e^{-t}dt.$$

provided $|y| \leq |x|$: this is due to the fact that the right-hand side equals

$$\sup_{|y| \leq |x|} U\left(\frac{x+1}{2}, \frac{y}{2}\right).$$

Thus, the function $\overline{U}$ leads to the estimate

$$\sup_{n \geq 0} \mathbb{E}\Phi\left(\frac{|g_n|}{2(\alpha+1)}\right) \leq \frac{\alpha+1}{\alpha+2} + \frac{2^{-(\alpha+2)/(\alpha+1)}}{\alpha+2} \int_0^\infty \Phi(t)e^{-t}dt,$$

if g is α-subordinate to a submartingale f taking values in $[-1, 1]$.

Similar reasoning can be applied to transfer the results from $[0, 1]$-valued case to $[-1, 1]$-valued case, or in the reverse direction.

4.7.5   Proof of Theorem 4.16

Proof of (4.70). Let $V : [0, 1] \times \mathbb{R} \to \mathbb{R}$ be given by $V(x, y) = |y|$. To define the corresponding U, consider the following subsets of $[0, 1] \times \mathbb{R}$:

$$D_1 = \left\{(x, y) : x \leq \frac{\alpha}{2\alpha+1}, \quad x + |y| > \frac{\alpha}{2\alpha+1}\right\},$$

$$D_2 = \left\{(x, y) : x \geq \frac{\alpha}{2\alpha+1}, \quad -x + |y| > -\frac{\alpha}{2\alpha+1}\right\},$$

$$D_3 = \left\{(x, y) : x \geq \frac{\alpha}{2\alpha+1}, \quad -x + |y| \leq -\frac{\alpha}{2\alpha+1}\right\},$$

$$D_4 = \left\{(x, y) : x \leq \frac{\alpha}{2\alpha+1}, \quad x + |y| \leq \frac{\alpha}{2\alpha+1}\right\}.$$

Let $H : \mathbb{R}^2 \to \mathbb{R}$ be defined by

$$H(x, y) = (|x| + |y|)^{1/(\alpha+1)}((\alpha+1)|x| - |y|)$$

and, finally, introduce $U : [0, 1] \times \mathbb{R} \to \mathbb{R}$ by

$$U(x, y) = -\alpha x + |y| + \alpha + \exp\left[-\frac{2\alpha+1}{\alpha+1}\left(x + |y| - \frac{\alpha}{2\alpha+1}\right)\right]\left(x + \frac{1}{2\alpha+1}\right)$$

if $(x, y) \in D_1$,

$$U(x, y) = -\alpha x + |y| + \alpha + \exp\left[-\frac{2\alpha+1}{\alpha+1}\left(-x + |y| + \frac{\alpha}{2\alpha+1}\right)\right](1 - x)$$

if $(x, y) \in D_2$,

$$U(x, y) = -(1 - x)\log\left[\frac{2\alpha+1}{\alpha+1}(1 - x + |y|)\right] + (\alpha+1)(1 - x) + |y|$$

if $(x, y) \in D_3$ and

$$U(x, y) = -\frac{\alpha^2}{(2\alpha + 1)(\alpha + 2)}\left[1 + \left(\frac{2\alpha + 1}{\alpha}\right)^{(\alpha+2)/(\alpha+1)} H(x, y)\right] + \frac{2\alpha^2}{2\alpha + 1} + 1$$

if $(x, y) \in D_4$.

It suffices to verify the conditions 1°, 2°, 3° and check the necessary integrability (which is easy, since f, and hence also each g_n, are bounded). The first and the third condition are relatively simple, and the concavity property was in fact verified in the proof of the logarithmic inequality (4.44): there are many similarities between the above U and the special function used there. We omit the details. □

Sharpness of (4.70). If $\alpha = 0$, then we take the constant processes $f = g \equiv (1, 1, 1, \ldots)$: then the two sides of (4.70) are equal. Thus, from now on, $\alpha > 0$. Let

$$U^0(x, y) = \sup\{\mathbb{E}|g_n|\},$$

where the supremum is taken over all n and appropriate f, g. Let $A(y) = U^0(0, y)$, $B(x) = U^0(x, 0)$ and $C(y) = U^0(\alpha/(2\alpha + 1), y - \alpha/(2\alpha + 1))$.

Now we split the reasoning into three steps.

Step 1. First we observe that, by the diagonal concavity of U^0 and the fact that $U^0(1, y) = |y|$ for all y,

$$B\left(\frac{\alpha}{2\alpha + 1}\right) \geq \frac{\alpha + 1}{2\alpha + 1}A\left(\frac{\alpha}{2\alpha + 1}\right) + \frac{\alpha}{2\alpha + 1}U^0\left(0, -\frac{\alpha + 1}{2\alpha + 1}\right)$$

$$= \frac{\alpha + 1}{2\alpha + 1}A\left(\frac{\alpha}{2\alpha + 1}\right) + \frac{\alpha(\alpha + 1)}{(2\alpha + 1)^2}.$$

(4.87)

Step 2. Next we observe that (4.61) holds; indeed, all the arguments leading to it are valid, since $U^0(1, y) = |y|$ for all y. Applying this inequality for $y = \alpha/(2\alpha+1)$ yields

$$\left[A\left(\frac{\alpha}{2\alpha + 1}\right) - \frac{\alpha}{2\alpha + 1}\right](\alpha + 1) - C\left(\frac{\alpha}{2\alpha + 1}\right) + \frac{\alpha}{2\alpha + 1} \geq \frac{2\alpha^3 + 2\alpha^2 + \alpha}{2\alpha + 1},$$

which can be rewritten as

$$(\alpha + 1)A\left(\frac{\alpha}{2\alpha + 1}\right) - B\left(\frac{\alpha}{2\alpha + 1}\right) \geq \alpha(\alpha + 1),$$

(4.88)

since $B\left(\frac{\alpha}{2\alpha+1}\right) = C\left(\frac{\alpha}{2\alpha+1}\right)$. Combining this with (4.87) gives

$$A\left(\frac{\alpha}{2\alpha + 1}\right) \geq \frac{2\alpha^2 + 2\alpha + 1}{2\alpha + 1}$$

(4.89)

and plugging this into (4.87) yields

$$B\left(\frac{\alpha}{2\alpha+1}\right) \geq \frac{(\alpha+1)^2}{2\alpha+1} \tag{4.90}$$

Step 3. Let N be a fixed positive integer and let $\delta = \alpha/((2\alpha+1)N)$. Note that we may use the inequality (4.21): the reasoning which gives this estimate is still working. So, write this bound for any $k = 0, 1, 2, \ldots, N-1$ and sum the obtained estimates to get

$$(\alpha+1)\left[A\left(\frac{\alpha}{2\alpha+1}\right) - A(0)\right] + B\left(\frac{\alpha}{2\alpha+1}\right) - B(0)$$
$$\leq \sum_{k=0}^{N-1} \frac{A((k+1)(\alpha+1)\delta) - A(k(\alpha+1)\delta)}{2k+1}. \tag{4.91}$$

Note that for any y and d we have

$$|A(y+d) - A(y)| \leq |d|, \tag{4.92}$$

in view of triangle inequality. Indeed, if f, g are as in the definition of $U^0(0,y)$, then $\mathbb{E}|g_n| \leq \mathbb{E}|g_n + d| + |d| \leq U^0(0, y+d) + |d|$, so taking the supremum over f, g and n gives $A(y) - A(y+d) \leq |d|$; replacing d by $-d$ and y by $y+d$ gives $A(y+d) - A(y) \leq |d|$ and (4.92) follows. Consequently, if N goes to ∞, then the right-hand side of (4.91) converges to 0 and we get

$$(\alpha+2)A(0) \geq (\alpha+1)A\left(\frac{\alpha}{2\alpha+1}\right) + B\left(\frac{\alpha}{2\alpha+1}\right),$$

since $A(0) = B(0)$. Finally, applying (4.89) and (4.90) gives

$$A(0) \geq \frac{(\alpha+1)(2\alpha^2 + 3\alpha + 2)}{(2\alpha+1)(\alpha+2)},$$

which yields the claim. Indeed, there is a pair (f,g) starting from $(0,0)$ such that f is a nonnegative submartingale bounded by 1 and g is α-subordinate to f such that $\|g\|_1$ is arbitrarily close to the right-hand side of (4.70). □

On the search of the suitable majorant. Essentially, one needs to repeat the argumentation from Step 4 in the analogous search corresponding to the logarithmic inequality (4.44); then one gets the formula for the special function on $D_1 \cup D_2$. To deal with the set D_4, use the argumentation from the search corresponding to the inequality (4.12). Finally, the formula on set D_3 is obtained by exploiting the assumption

(A) When restricted to $D_3 \cap \{y \geq 0\}$, U is linear along the line segments of slope 1.

We omit the details.   □

4.8 On the optimal control of nonnegative submartingales

4.8.1 Formulation of the result

Throughout, the function $U : [0, \infty) \times \mathbb{R} \to \mathbb{R}$ is given by (4.98) below.

Theorem 4.17. *Let f be a nonnegative submartingale starting from $x \geq 0$ and let g be a sequence starting from $y \in \mathbb{R}$ such that*

$$|dg_n| \leq |df_n| \quad and \quad |\mathbb{E}(dg_n|\mathcal{F}_{n-1})| \leq \mathbb{E}(df_n|\mathcal{F}_{n-1}) \qquad for\ all\ \ n \geq 1. \quad (4.93)$$

Then

$$\mathbb{P}(g^* \geq 0) \leq ||f||_1 + U(x, y) \quad (4.94)$$

and the inequality is sharp.

This leads to the following result on the optimal control of submartingales. Let $L : [0, \infty) \times \mathbb{R} \times [0, 1] \to \mathbb{R}$ be given by (4.113) below.

Theorem 4.18. *Let f be a nonnegative submartingale starting from $x \geq 0$ and let g start from $y \in \mathbb{R}$. Suppose that (4.93) holds. If g satisfies the one-sided estimate*

$$\mathbb{P}(g^* \geq \beta) \geq t, \quad (4.95)$$

where $t \in [0, 1]$ is a fixed number, then

$$||f||_1 \geq L(x, y - \beta, t) \quad (4.96)$$

and the bound is the best possible. In particular, if g is strongly differentially subordinate to f, then we have a sharp inequality

$$||f||_1 \geq L(x, x - \beta, t). \quad (4.97)$$

Finally, Theorem 4.17 leads to the following one-sided weak type inequality.

Theorem 4.19. *Assume that f is a nonnegative submartingale and g is strongly differentially subordinate to f. Then for any $\lambda > 0$ we have*

$$\lambda \mathbb{P}(g^* \geq \lambda) \leq \frac{8}{3}||f||_1$$

and the constant $8/3$ is the best possible.

4.8.2   Proof of Theorem 4.17

Proof of (4.94). Let f, g be as in the statement. By the stopping time argument, it suffices to establish the bound

$$\mathbb{P}(g_n \geq 0) - \mathbb{E}f_n \leq U(x, y), \qquad n = 0, 1, 2, \ldots.$$

Let $V : [0, \infty) \times \mathbb{R} \to \mathbb{R}$ be given by $V(x, y) = 1_{\{y \geq 0\}} - x$. To introduce the corresponding special function, consider first the following subsets of $[0, \infty) \times \mathbb{R}$:

$$D_0 = \{(x, y) : x + y \geq 0\},$$
$$D_1 = \{(x, y) : (x - 8)/3 \leq y < -x\},$$
$$D_2 = \{(x, y) : x - 4 < y < (x - 8)/3\},$$
$$D_3 = ([0, \infty) \times \mathbb{R}) \smallsetminus (D_0 \cup D_1 \cup D_2).$$

Let $U : [0, \infty) \times \mathbb{R} \to \mathbb{R}$ be given by

$$U(x, y) = \begin{cases} 1 - x & \text{if } (x, y) \in D_0, \\ \frac{1}{16}(3x + 3y + 8)^{1/3}(-5x + 3y + 8) & \text{if } (x, y) \in D_1, \\ -\frac{1}{4}x(6x - 6y - 16)^{1/3} & \text{if } (x, y) \in D_2, \\ \frac{2x}{x-y} - x & \text{if } (x, y) \in D_3. \end{cases} \qquad (4.98)$$

In addition, we introduce $A, B : [0, \infty) \times \mathbb{R} \to \mathbb{R}$ by

$$A(x, y) = \begin{cases} -1 & \text{if } (x, y) \in D_0, \\ \frac{1}{4}(3x + 3y + 8)^{-2/3}(-5x - 3y - 8) & \text{if } (x, y) \in D_1, \\ \frac{1}{2}(6x - 6y - 16)^{-2/3}(-4x + 3y + 8) & \text{if } (x, y) \in D_2, \\ -\frac{2y}{(x-y)^2} - 1 & \text{if } (x, y) \in D_3 \end{cases}$$

and

$$B(x, y) = \begin{cases} 0 & \text{if } (x, y) \in D_0, \\ \frac{1}{4}(3x + 3y + 8)^{-2/3}(x + 3y + 8) & \text{if } (x, y) \in D_1, \\ \frac{1}{2}x(6x - 6y - 16)^{-2/3} & \text{if } (x, y) \in D_2, \\ \frac{2x}{(x-y)^2} & \text{if } (x, y) \in D_3. \end{cases}$$

It is straightforward to check that U is continuous and of class C^1 on $E = \{(x, y) : x > 0, x + y \neq 0\}$; in addition, $A = U_x$ and $B = U_y$ on this set. Let us verify that U and V satisfy the conditions $1°$, $2°'$ (property $3°$ is not required here, due to the form of the problem). The majorization is easy: for a fixed x, the function $y \mapsto U(x, y)$ is nondecreasing (since $B \geq 0$), $\lim_{y \to -\infty} U(x, y) = -x$ and $U(x, y) = 1 - x$ for sufficiently large y. Let us turn to $2°'$. Fix $x \geq 0$, $y \in \mathbb{R}$, $a \in [-1, 1]$ and consider the function $G = G_{x,y,a} : [-x, \infty) \to \mathbb{R}$ given by $G(t) =$

$U(x+t, y+at)$. We must prove that G is concave and nonincreasing. Observe that $G'_{x,y,a}(-x+) \leq 0$: this follows directly from the estimate $A(0, y) + |B(0, y)| \leq 0$ for all y (which is straightforward to check). Consequently, $2^{\circ\prime}$ will be established if we show the concavity of G, which, as we already know, reduces to verifying the inequalities $G''(0) \leq 0$ for $(x, y) \in D_i^\circ$ and $G'(0-) \geq G'(0+)$ for $(x, y) \notin E$. If (x, y) lies in D_0°, then $G''(0) = 0$. For $(x, y) \in D_1^\circ$, a little calculation yields

$$G''(0) = \frac{1}{4}(3x + 3y + 8)^{-5/3}(a+1)\big[(a-3)(-x + 3y + 8) + 8(a-1)x\big],$$

which is nonpositive: this follows from $|a| \leq 1$ and $-x + 3y + 8 \geq 0$, coming from the definition of D_1. If $(x, y) \in D_2^\circ$, then

$$G''(0) = 2(1-a)(6x - 6y - 16)^{-5/3}\big[-x(a+1) + (-x + 3y + 8)\big] \leq 0,$$

since $-x + 3y + 8 \leq 0$, by the definition of D_2. Finally, if (x, y) belongs to D_3°, we have

$$G''(0) = 4(1-a)(x-y)^{-3}\big[-(a+1)x + (x+y)\big] \leq 0,$$

because $x + y \leq 0$. It remains to check the inequality for one-sided derivatives. The condition $x > 0$, $(x, y) \notin E$ is equivalent to $x > 0$ and $x + y = 0$. If $(x, y) \in \partial D_0 \cap \partial D_1$ (so $y = -x \in [-2, 0)$), then after some straightforward computations,

$$G'(0-) = \frac{1}{8}[(a+1)y + 4a - 4] \geq \frac{1}{4}(a+1) - 1 \geq -1 = G'(0+).$$

On the other hand, if $(x, y) \in \partial D_3 \cap \partial D_0$ and $x > 2$, then

$$G'(0-) = \frac{a+1}{8x} - 1 \geq -1 = G'(0+)$$

and we are done. Thus, all that remains is to verify the required integrability; but this follows immediately from the easy estimates

$$|U(x, y)| \leq 1 + x, \quad |A(x, y)| \leq 1, \quad |B(x, y)| \leq 1,$$

valid for all $x \geq 0$ and $y \in \mathbb{R}$. Thus, Burkholder's method guarantees that (4.94) holds true. □

Sharpness. Let $\delta > 0$ be a fixed small number, to be specified later. Consider a Markov family (f_n, g_n) on $[0, \infty) \times \mathbb{R}$, with the transitions described as follows:

(i) The states $\{(x, y) : y \geq 0\}$ and $\{(0, y) : y \leq -8/3\}$ are absorbing.
(ii) For $-x \leq y < 0$, the state (x, y) leads to $(x + y, 0)$ or to $(x - y, 2y)$, with probabilities $1/2$.
(iii) The state $(x, y) \subset D_1$, $x > 0$, leads to $(0, y+x)$ or to $(\frac{3x+3y}{4} + 2 + \delta, \frac{x+y}{4} - 2 - \delta)$, with probabilities p_1 and $1-p_1$, where $p_1 = (-x+3y+8+4\delta)/(3x+3y+8+4\delta)$.
(iv) The state $(x, y) \in D_2$, $x > 0$, leads to $(0, y-x)$ or to $(\frac{3x-3y}{2} - 4, \frac{x-y}{2} - 4)$, with probabilities p_2 and $1 - p_2$, where $p_2 = (x - 3y - 8)/(3x - 3y - 8)$.

(v) The state $(x, y) \in D_3$, $x > 0$, leads to $(0, y - x)$ or to $(\frac{x-y}{2}, \frac{y-x}{2})$, with probabilities p_3 and $1 - p_3$, where $p_3 = -(x + y)/(x - y)$.

(vi) The state $(0, y)$, $y \in (-8/3, 0)$, leads to $(0 + 2\delta, y + 2\delta)$.

Let $x \geq 0$ and $y \in \mathbb{R}$. It is not difficult to check that under the probability measure $\mathbb{P}_{x,y} = \mathbb{P}(\cdot | (f_0, g_0) = (x, y))$, the sequence f is a nonnegative submartingale and g satisfies $dg_n = \pm df_n$ for $n \geq 1$ (so (4.93) holds). In fact, the steps described in (i)–(v) are martingale moves in the sense that

$$\mathbb{E}_{x,y}((f_{n+1}, g_{n+1}) | (f_n, g_n) = (x', y')) = (x', y'),$$

provided the conditioning event has nonzero probability and (x', y') belongs to one of the sets from (i)–(v). Set

$$P^\delta(x, y) = \mathbb{P}_{x,y}(g^* \geq 0), \quad M^\delta(x, y) = \lim_{n \to \infty} \mathbb{E}_{x,y} f_n. \tag{4.99}$$

Usually we will omit the upper index and write P, M instead of P^δ, M^δ, but it should be kept in mind that these functions do depend on δ. We will prove that if this parameter is sufficiently small, then

$$P(x, y) - M(x, y) \text{ is arbitrarily close to } U(x, y). \tag{4.100}$$

This will clearly yield the claim. It is convenient to split the remaining part of the proof into a few steps.

1°. *The case $y \geq 0$.* Here (4.100) is trivial: $P(x, y) = 1$, $M(x, y) = x$ for all δ, and $U(x, y) = 1 - x$.

2°. *The case $x = 0$, $y \leq -8/3$.* Again, (4.100) is obvious: $P(x, y) = M(x, y) = 0$ for all δ, and $U(x, y) = 0$.

3°. *The case $-x \leq y < 0$.* For any $\delta > 0$ and $n \geq 1$, it is easy to see that $\mathbb{P}_{x,y}(g_n \geq 0) = 1 - 2^{-n}$ and $\mathbb{E}_{x,y} f_n = x$. Letting $n \to \infty$, we get $P(x, y) = 1$ and $M(x, y) = x$, which yields (4.100), since $U(x, y) = 1 - x$.

4°. *The case $\{(x, y) : y = x/3 - 8/3, x \in (0, 2)\} \cup \{(x, y) : x = 0, y \in (-8/3, 0)\}$.* This is the most technical part. Let us first deal with

$$A(x) := P(x, x/3 - 8/3), \quad \text{and} \quad B(x) := P(0, 4x/3 - 8/3)$$

for $0 < x < 2$. We will prove that

$$\lim_{\delta \to 0} A(x) = \frac{2}{3} \left(\frac{x}{2}\right)^{1/3} + \frac{1}{3} \left(\frac{x}{2}\right)^{4/3}, \quad \lim_{\delta \to 0} B(x) = \frac{4}{3} \left(\frac{x}{2}\right)^{1/3} - \frac{1}{3} \left(\frac{x}{2}\right)^{4/3}. \tag{4.101}$$

This will be done by showing that

$$\lim_{\delta \to 0} [A(x) + B(x)] = 2(x/2)^{1/3}, \quad \lim_{\delta \to 0} [2A(x) - B(x)] = (x/2)^{4/3}. \tag{4.102}$$

To get the first statement above, note that by (iii) and the Markov property, we have

$$P\left(x, \frac{x}{3} - \frac{8}{3}\right) = \frac{x}{x+\delta}P\left(x+\delta, \frac{x}{3} - \frac{8}{3} - \delta\right) + \frac{\delta}{x+\delta}P\left(0, \frac{4}{3}x - \frac{8}{3}\right),$$

which can be rewritten in the form

$$A(x) = \frac{x}{x+\delta}P\left(x+\delta, \frac{x}{3} - \frac{8}{3} - \delta\right) + \frac{\delta}{x+\delta}B(x). \tag{4.103}$$

Similarly, using (iv) and the Markov property, we get

$$P\left(x+\delta, \frac{x}{3} - \frac{8}{3} - \delta\right) = \frac{x+\delta}{x+3\delta}A(x+3\delta) + \frac{2\delta}{x+3\delta}P\left(0, -\frac{2}{3}x - \frac{8}{3} - 2\delta\right)$$

$$= \frac{x+\delta}{x+3\delta}A(x+3\delta),$$

where in the last passage we have used 2°. Plugging this into (4.103) yields

$$A(x) = \frac{x}{x+3\delta}A(x+3\delta) + \frac{\delta}{x+\delta}B(x). \tag{4.104}$$

Analogous argumentation, with the use of (iii), (vi) and the Markov property, leads to the equation

$$B(x) = \frac{2\delta}{x+3\delta}A(x+3\delta) + \frac{x+\delta}{x+3\delta}B(x+3\delta). \tag{4.105}$$

Adding (4.104) to (4.105) gives

$$A(x) + \frac{x}{x+\delta}B(x) = \frac{x+2\delta}{x+3\delta}\left[A(x+3\delta) + \frac{x+3\delta}{x+4\delta}B(x+3\delta)\right] + c(x,\delta)\cdot\delta^2, \tag{4.106}$$

where

$$c(x,\delta) = -\frac{2B(x+3\delta)}{(x+3\delta)(x+4\delta)},$$

which can be bounded in absolute value by $2/x^2$. Now we use (4.106) several times: if N is the largest integer such that $x + 3N\delta < 2$, then

$$A(x) + \frac{x}{x+\delta}B(x) = \frac{x+2\delta}{x+3\delta} \cdot \frac{x+5\delta}{x+6\delta} \cdots \frac{x+3N\delta - \delta}{x+3N\delta} \tag{4.107}$$

$$\times \left[A(x+3N\delta) + \frac{x+3N\delta}{x+3N\delta+\delta}B(x+3N\delta)\right] + cN\delta^2,$$

where $|c| < 2/x^2$. Now we will study the limit behavior of the terms on the right as $\delta \to 0$. First, note that for any $k = 1, 2, \ldots, N$,

$$\frac{x+3k\delta - \delta}{x+3k\delta} = \exp\left(-\frac{\delta}{x+3k\delta}\right) \cdot \exp(d(k)\delta^2),$$

where $|d(k)| \leq 1/x^2$; consequently,

$$\prod_{k=1}^{N} \frac{x + 3k\delta - \delta}{x + 3k\delta} = \exp\left(-\sum_{k=1}^{N} \frac{\delta}{x + 3k\delta}\right) \exp(\tilde{d}N\delta^2), \qquad (4.108)$$

for some $\tilde{d}$ satisfying $|\tilde{d}| \leq 1/x^2$. Since $N = O(1/\delta)$ for small δ, we conclude that the product in (4.108) converges to $\exp(-\frac{1}{3}\int_x^2 t^{-1}dt) = (x/2)^{1/3}$ as $\delta \to 0$.

The next step is to show that the expression in the square brackets in (4.107) converges to 2 as $\delta \to 0$. First observe that

$$B(x + 3N\delta) = P\left(0, 4x/3 + 4N\delta - \frac{8}{3}\right) = P\left(2\delta, 4x/3 + 4N\delta - \frac{8}{3} + 2\delta\right) = 1,$$

where in the first passage we have used the definition of B, in the second we have exploited (vi), and the latter is a consequence of 1° and 3°. To show that $A(x + 3N\delta)$ converges to 1, use (4.103), with x replaced by $x + 3N\delta$, to get

$$A(x + 3N\delta) = \frac{x + 3N\delta}{x + 3N\delta + 3\delta} P\left(x + 3N\delta + \delta, \frac{x + 3N\delta}{3} - \frac{8}{3} - \delta\right) + \frac{\delta}{x + 3N\delta + \delta}.$$

Note that, by the definition of N, the point under P lies in D_3, and arbitrarily close to the line $y = -x$, if δ is sufficiently small. Thus, by (v) and 2°,

$$P\left(x + 3N\delta + \delta, \frac{x + 3N\delta}{3} - \frac{8}{3} - \delta\right)$$

can be made arbitrary close to 1, provided δ is small enough. Summarizing, letting $\delta \to 0$ in (4.107) yields the first limit in (4.102). To get the second one, multiply both sides of (4.105) by $1/2$, subtract it from (4.104) and proceed as previously.

Next we show that $C(x) := M(x, x/3 - 8/3)$, $D(x) := M(0, 4x/3 - 8/3)$, $x \in (0, 2)$, satisfy

$$\lim_{\delta \to 0} C(x) = \frac{2}{3}\left(\frac{x}{2}\right)^{1/3} + \frac{4}{3}\left(\frac{x}{2}\right)^{4/3}, \qquad \lim_{\delta \to 0} D(x) = \frac{4}{3}\left(\frac{x}{2}\right)^{1/3} - \frac{4}{3}\left(\frac{x}{2}\right)^{4/3}. \quad (4.109)$$

We proceed in a similar manner: arguing as before, C and D satisfy the same system of equations as A and B, that is, (4.104) and (4.105). The only difference in the further considerations is that $C(x + 3N\delta) \to 2$ and $D(x + 3N\delta) \to 0$ as $\delta \to 0$.

The final step is to combine (4.101) and (4.109). We get

$$\lim_{\delta \to 0} \left[A(x) - C(x)\right] = U\left(x, x/3 - 8/3\right), \qquad \lim_{\delta \to 0} \left[B(x) - D(x)\right] = U\left(0, 4x/3 - 8/3\right),$$

as desired.

5°. *The remaining (x, y)'s.* Now (4.100) is easily deduced from the previous cases using the Markov property: the point (x, y) leads, in at most two steps, to the state for which we have already calculated the (limiting) values of P and M. For example, if $(x, y) \in D_3$, we have

$$P(x, y) = p_3 P(0, y - x) + (1 - p_3)P\left(\frac{x - y}{2}, \frac{y - x}{2}\right) = 1 - p_3 = \frac{2x}{x - y},$$

$$M(x, y) = p_3 M(0, y - x) + (1 - p_3)M\left(\frac{x - y}{2}, \frac{y - x}{2}\right) = x,$$

in view of 1° and 3°. Since $U(x, y) = P(x, y) - M(x, y)$, (4.100) follows. The other states are checked similarly. The proof of the sharpness is complete.

We conclude this section with an observation which follows immediately from the above considerations. It will be needed later in the proof of Theorem 4.18.

Remark 4.5. The function $Q : [0, \infty) \times \mathbb{R} \to [0, 1]$, given by

$$Q(x, y) = \lim_{\delta \to 0} P(x, y),$$

is continuous.

4.8.3 Proof of Theorem 4.18

We start with the following auxiliary fact.

Lemma 4.9.

(i) *Suppose that*

$$y < -x < 0 \quad and \quad \left(\frac{4x}{x - 3y}\right)^{1/3} \frac{2x - 2y}{x - 3y} < t < 1. \tag{4.110}$$

Then there is a unique positive number $C_0 = C_0(x, y, t)$ satisfying $C_0 \leq 8/(x - 3y)$ (equivalently, $(C_0 x, C_0 y) \in D_1$) and

$$\frac{1}{16}C_0^2(x + y)(5x - 3y) + C_0(x + y) + 4 = t(3C_0(x + y) + 8)^{2/3}. \tag{4.111}$$

(ii) *Suppose that*

$$y < -x < 0 \quad and \quad \frac{2x}{x - y} < t \leq \left(\frac{4x}{x - 3y}\right)^{1/3} \frac{2x - 2y}{x - 3y}. \tag{4.112}$$

Then there is a unique positive number $C_1 = C_1(x, y, t)$ such that $8/(x-3y) \leq C_1 \leq 4/(x - y)$ (equivalently, $(C_1 x, C_1 y) \in D_2$) and

$$C_1^2 x(x - y) = 2t(6C_1(x - y) - 16)^{2/3}.$$

Proof. We will only prove (i), since the second part can be established essentially in the same manner. Let

$$F(C) = \frac{1}{16}C^2(x+y)(5x-3y) + C(x+y) + 4 - t(3C(x+y)+8)^{2/3}.$$

It can be readily verified that F' is convex and satisfies $F'(0+) = (x+y)(1-t) < 0$: thus F is either decreasing on $(0, 8/(x-3y))$, or decreasing on $(0, x_0)$ and increasing on $(x_0, 8/(x-3y)$ for some x_0 from $(0, 8/(x-3y))$. It suffices to note that $F(0) = 4 - 4t > 0$ and

$$F\left(\frac{8}{x-3y}\right) = 4\left(\frac{4x}{x-3y}\right)^{2/3}\left[\left(\frac{4x}{x-3y}\right)^{1/3}\frac{2x-2y}{x-3y} - t\right] < 0.$$

The claim is proved.   □

Let $L : [0, \infty) \times \mathbb{R} \times [0, 1] \to \mathbb{R}$ be given by

$$L(x, y, t) = \begin{cases} (x-y)/2 & \text{if } t = 1, \\ \frac{t}{C_0} - (3C_0(x+y)+8)^{1/3}\left(\frac{-5x+3y}{16} + \frac{1}{2C_0}\right) & \text{if (4.110) holds,} \\ \frac{t}{C_1} + \frac{C_1 x}{4}\left(\frac{x(x-y)}{2t}\right)^{1/2} & \text{if (4.112) holds,} \\ x & \text{if } t \le \frac{2x}{x-y}. \end{cases}$$
$$(4.113)$$

Proof of (4.96) *and* (4.97). Clearly, it suffices to prove the inequality for $\beta = 0$, replacing (f, g) by $(f, g - \beta)$, if necessary. Let $C > 0$ be an arbitrary constant. Application of (4.94) to the sequences Cf, Cg yields $\mathbb{P}(g^* \ge 0) - C\|f\|_1 \le U(Cx, Cy)$, which, by (4.95), leads to the bound

$$\|f\|_1 \ge \frac{t - U(Cx, Cy)}{C}. \tag{4.114}$$

If one maximizes the right-hand side over C, one gets precisely $L(x, y, t)$. This follows from a straightforward but lengthy analysis of the derivative with the aid of the previous lemma. We omit the details. To get (4.97), note that for fixed x and t, the function $L(x, \cdot, t)$ is nonincreasing. This follows immediately from (4.114) and the fact that $U(x, \cdot)$ is nondecreasing, which we have already exploited.   □

Sharpness of (4.96). As previously, we may restrict ourselves to $\beta = 0$. Fix $x \ge 0$, $y \in \mathbb{R}$ and $t \in [0, 1]$. If $x + y \ge 0$, then the examples studied in 1° and 3° in the preceding section give equality in (4.96). Hence we may and do assume that $x + y < 0$. Consider three cases.

(i) Suppose that $t \le 2x/(x - y)$. Take $C > 0$ such that $(Cx, Cy) \in D_3$ and take the Markov pair (f, g), with $(f_0, g_0) = (Cx, Cy)$, from the previous section. Then the pair $(f/C, g/C)$ gives equality in (4.96) (see 5°).

(ii) Let $2x(x - y) < t < 1$ and take $0 < \varepsilon < 1 - t$. Recall the function Q defined in Remark 4.5. First we will show that

$$Q(Cx, Cy) = t + \varepsilon \qquad \text{for some } C = C(\varepsilon, t) > 0. \tag{4.115}$$

Indeed, for large C we have $(Cx, Cy) \in D_3$, so $P(Cx, Cy) = 2x/(x - y)$ regardless of the value of δ (see 5°), and, in consequence, $Q(Cx, Cy) = 2x/(x - y)$. Similarly, $P(0, 0) = 1$ for any δ, so $Q(0, 0) = 1$. Thus (4.115) follows from Remark 4.5. Another observation, to be needed at the end of the proof, is that

$$\liminf_{\varepsilon \to 0} C(\varepsilon, t) > 0. \tag{4.116}$$

Otherwise, we would have a contradiction with (4.115), Remark 4.5 and the equality $Q(0, 0) = 1$.

Now fix $\delta > 0$ and consider the Markov pair (f, g), starting from (Cx, Cy), studied in the previous section. If δ is taken sufficiently small, then the following two conditions are satisfied: first, by (4.100), we have $\mathbb{P}(g^* \geq 0) - ||f||_1 \geq U(Cx, Cy) - \varepsilon$; second, by the definition of Q, $\mathbb{P}(g^* \geq 0) = P(Cx, Cy) \in (t, t+2\varepsilon)$. In other words, for this choice of δ, the pair $(f/C, g/C)$ starts from (x, y), satisfies (4.93), we have $\mathbb{P}((g/C)^* \geq 0) \geq t$ and

$$||f/C||_1 \leq (t - U(Cx, Cy) + 3\varepsilon)/C \leq L(x, y, t) + 3\varepsilon/C.$$

To get the claim, it suffices to note that ε was arbitrary and that (4.116) holds.

(iii) Finally, assume that $t = 1$. Then the following Markov pair (f, g), starting from (x, y), gives equality in (4.96):

- For $y \geq 0$, the state $(0, y)$ is absorbing.
- if $x \neq 0$, then (x, y) leads to $(0, y+x)$ and $(2x, y - x)$, with probabilities $1/2$.
- if $y < 0$, the state $(0, y)$ leads to $(-y/2, y/2)$.

The analysis is similar to the one presented in the case 3° in the previous section. The details are left to the reader. □

We turn to the proof of the weak type inequality from Theorem 4.19.

4.8.4   Proof of Theorem 4.19

Fix $\lambda > 0$ and apply Theorem 4.17 to the martingales $8f/(3\lambda)$ and $8g/(3\lambda) - 8/3$, conditionally on $\mathcal{F}_0$. Taking expectation of both sides, we get

$$\mathbb{P}(g^* \geq \lambda) = \mathbb{P}\left(\left(\frac{8g}{3\lambda} - \frac{8}{3}\right)^* \geq 0\right) \leq \left\|\frac{8f}{3\lambda}\right\|_1 + \mathbb{E}U\left(\frac{8f_0}{3\lambda}, \frac{8g_0}{3\lambda} - \frac{8}{3}\right).$$

Therefore, the theorem will be established if we show that for all points $(x, y) \in [0, \infty) \times \mathbb{R}$ satisfying $|y| \leq |x|$ we have

$$U\left(x, y - \frac{8}{3}\right) \leq 0$$

and that the equality holds for at least one such point. This is straightforward: as already mentioned above, the function $y \mapsto U(x, y)$ is nondecreasing, so

$$U\left(x, y - \frac{8}{3}\right) \leq U\left(x, x - \frac{8}{3}\right) \leq U\left(0, -\frac{8}{3}\right) = 0,$$

because for any y, the function $t \mapsto U(t, y + t)$, $t \geq 0$, is nonincreasing.

4.9 Escape inequalities

As in the martingale setting, the weak type $(1, 1)$ inequalities above lead to interesting escape estimates. For any $\mathcal{H}$-valued sequence $x = (x_n)$, recall the counting function $C_\varepsilon(x)$.

4.9.1 Formulation of the results.

We start with a theorem for general sub- and supermartingales.

Theorem 4.20. *Assume that f is a sub- or supermartingale and g is an adapted sequence of integrable, $\mathcal{H}$-valued random variables which is α-subordinate to f. Then for any positive integer j we have*

$$\mathbb{P}(C_\varepsilon(g) \geq j) \leq \frac{C_\alpha \|f\|_1}{\varepsilon j^{1/2}}.$$

If $\alpha \in [0, 1]$, then both the constant C_α and the exponent $1/2$ are the best possible.

Under additional assumption on the sign of the dominating process, we have the following.

Theorem 4.21. *Assume that f is a nonnegative submartingale and g is an adapted sequence of integrable, $\mathcal{H}$-valued random variables which is α-subordinate to f. Then for any positive integer j we have*

$$\mathbb{P}(C_\varepsilon(g) \geq j) \leq \frac{(\alpha + 2)\|f\|_1}{\varepsilon j^{1/2}}.$$

If $\alpha \in [0, 1]$, then both the constant $\alpha + 2$ and the exponent $1/2$ are the best possible.

Theorem 4.22. *Assume that f is a nonnegative supermartingale and g is an adapted sequence of integrable, $\mathcal{H}$-valued random variables which is α-subordinate to f. Then for any positive integer j we have*

$$\mathbb{P}(C_\varepsilon(g) \geq j) \leq \frac{2\|f\|_1}{\varepsilon j^{1/2}}.$$

If $\alpha \in [0,1]$, then both the constant 2 and the exponent $1/2$ are the best possible.

4.9.2 Proof

The reasoning leading to the announced inequalities is the same as in the martingale setting, using the random variable $S(g, \tau)$ and the corresponding sequence h taking values in the enlarged Hilbert space $\ell^2(\mathcal{H})$. To see that the constants C_α, $\alpha + 2$ and 2 are the best possible, simply write the inequalities for $\varepsilon = j = 1$ and note that for any $\delta > 0$ we have $\mathbb{P}(|g|^* \geq 1 + \delta) \leq \mathbb{P}(C_\varepsilon(g) \geq 1)$. Thus the sharpness follows from the optimality of the constants in the corresponding weak type inequalities. To show that the exponent $1/2$ is the best possible, consider a symmetric random walk f over integers, starting from 1 and stopped at 0. Then $\|f\|_1 = 1$ and

$$\mathbb{P}(C_1(f) \geq j) = \mathbb{P}(f_1 > 0, f_2 > 0, \ldots, f_j > 0) = O(j^{-1/2})$$

as $j \to \infty$. This proves the claim. $\qquad\square$

4.10 Notes and comments

Section 4.1. The notion of strong subordination appeared first in the special case $\alpha = 1$ in the paper [34] of Burkholder, motivated by related domination of Itô processes (see [33]). Then it was extended to an arbitrary $\alpha \geq 0$ by Choi [45] (see also [47]). The concept of α-subordination considered above in the discrete-time setting seems to be new; however, its continuous-time version was used by Wang in [200] (for $\alpha = 1$) and by the author in [132] (for general α).

Section 4.2. Theorems 4.4 and 4.5 in the case $\alpha \in [0, 1]$ are due to Hammack [88]. The result for $\alpha > 1$ is new, but a corresponding result for continuous-time processes was obtained by the author in [150].

Section 4.3. Theorems 4.7 and 4.8 for $\alpha = 1$ first appeared in Burkholder's paper [33] and concerned Itô processes. The more general statement (still with $\alpha = 1$, but this time dealing with discrete-time submartingales and their 1-strong subordinates) can be found in [34]. Theorem 4.7 under strong α-subordination, $\alpha \geq 0$, was established by Choi in [47] (the special function was used by Choi [49] in the proof of a related result concerning nonnegative subharmonic functions).

Section 4.4. Theorem 4.10 was proved by Burkholder in [34] in the particular case $\alpha = 1$; a related estimate for Itô processes can be found in [33]. The result for

$\alpha \in [0, 1]$ is due to Choi [45] (see also Choi [46] for a related result for Itô processes and Choi [48] for a version concerning nonnegative subharmonic functions; both papers deal with the case $0 \leq \alpha \leq 1$). The above extension to the general values of $\alpha \geq 0$ is new.

Section 4.5. The results presented in that section are taken from the author's paper [137].

Section 4.6. The logarithmic estimates presented above were established by the author in [149].

Section 4.7. Theorem 4.14 was established by Hammack [87] in the particular case $\alpha = 1$. The above statement for $\alpha \in [0, 1]$ is due to the author [127]. It is worth to mention here that both papers contain the proof of a slightly stronger fact. Namely, it is shown that for a fixed $\alpha \in [0, 1]$ and $\lambda \geq 0$, the above function U_λ satisfies

$$U_\lambda(x, y) = \sup\{\mathbb{P}(|g|^* \geq \lambda)\},$$

where the supremum is taken over all f, g such that f is a submartingale satisfying $f_0 \equiv x$ and $||f||_\infty \leq 1$, while g starts from y and

$$|dg_n| \leq |df_n|, \quad |\mathbb{E}(dg_n|\mathcal{F}_{n-1})| \leq \alpha \mathbb{E}(df_n|\mathcal{F}_{n-1}), \qquad n = 1, 2, \ldots.$$

Theorem 4.15 was proved by Burkholder in [34] in the case when $\alpha = 1$. The inequality (4.69) for $\alpha \in [0, 1)$ appears in the paper [107] by Kim and Kim, but its sharpness is not established there. The above version of the theorem, for general α and with the proof of the optimality of the constant, seems to be new. Theorem 4.16 comes from the author's paper [129].

Section 4.8. The contents of this section are taken from [140].

Section 4.9. The escape inequalities studied above were proved by Burkholder in [34] (for $\alpha = 1$, but the results carry over to general α essentially with no changes in the reasoning).

Chapter 5

Inequalities in Continuous Time

It is natural to ask whether the inequalities studied in the previous chapters can be transferred to continuous-time processes. The answer is affirmative. First let us extend the notions of differential subordination and strong differential subordination.

5.1 Dominations in the continuous-time case

Let $(\Omega, \mathcal{F}, \mathbb{P})$ be a complete probability space s filtered by a nondecreasing right-continuous family $(\mathcal{F}_t)_{t \geq 0}$ of sub-σ-fields of $\mathcal{F}$. Assume that $\mathcal{F}_0$ contains all the events of probability 0. Suppose that $X = (X_t)_{t \geq 0}$, $Y = (Y_t)_{t \geq 0}$ are adapted semimartingales taking values in a certain separable Hilbert space. Assume also that these processes have right-continuous paths with left limits. Let $H = (H_t)_{t \geq 0}$ be an adapted predictable real-valued process. Then the Itô integral of H with respect to X, that is,

$$Y_t = \int_0^t H_s \mathrm{d}X_s, \qquad t \geq 0,$$

denoted by $Y = H \cdot X$, is precisely the continuous-time analogue of a transform. Furthermore, the condition of being ± 1-transform corresponds here to the assumption $H_s \in \{-1, 1\}$ for all $s \geq 0$. Similarly, if one considers H taking values in the interval $[0, 1]$, then one obtains the continuous version of a nonsymmetric transform.

Next we turn to the extension of differential subordination. Note that in discrete time this domination can be rewritten in the following form:

$$S_n^2(f) - S_n^2(g) \text{ is nondecreasing and nonnegative as a function of } n, \qquad (5.1)$$

where $S_n(f) = \left(\sum_{k=0}^n |df_k|^2\right)^{1/2}$ is the nth term of the square function of f. This formulation suggests how to extend the domination to the continuous time: simply replace the square function by its continuous counterpart, the square bracket. To recall this notion, let us state here the following classical result.

Lemma 5.1. *Let X be a cadlag semimartingale taking values in $\mathcal{H}$. Then there is a family of nonnegative functions $[X, X] = ([X, X]_t)_{t \geq 0}$ such that $[X, X]_0 = |X_0|^2$, the function $[X, X]_t$ is $\mathcal{F}_t$-measurable for all $t \geq 0$, the function $t \mapsto [X, X]_t(\omega)$ is right-continuous and nondecreasing on $[0, \infty)$ for all $\omega \in \Omega$, and $[X, X]_t$ is the limit in probability of the sums $Q_{n,t}$ as $n \to \infty$ for all $t \geq 0$, where*

$$Q_{n,t} = |X_0|^2 + \sum_{k=0}^{2^n - 1} |X_{t(k+1)2^{-n}} - X_{tk2^{-n}}|^2.$$

This was established by Doléans [69] for real martingales, but the proof extends to Hilbert-space-valued martingales with no essential changes. Motivated by (5.1), we give the following definition.

Definition 5.1. We say that a semimartingale Y is differentially subordinate to a semimartingale X, if the process $([X, X]_t - [Y, Y]_t)_{t \geq 0}$ is nondecreasing and nonnegative as a function of t.

Remark 5.1. Let us repeat that this definition is consistent with that in the discrete-time setting. If we treat the martingales $f = (f_n)_{n \geq 0}$ and $g = (g_n)_{n \geq 0}$ as as continuous-time processes (setting $X_t = f_{\lfloor t \rfloor}$ and $Y_t = g_{\lfloor t \rfloor}$, $t \geq 0$), then the above condition is equivalent to saying that $|df_n|^2 - |dg_n|^2 \geq 0$ for all n.

As an example, suppose that X is a semimartingale, H is a predictable process taking values in $[-1, 1]$ and let $Y = H \cdot X$. Then Y is differentially subordinate to X, which follows immediately from the equality $[X, X]_t - [Y, Y]_t = \int_0^t |H_s|^2 d[X, X]_s$ for all $t \geq 0$.

We will now extend the notion of α-strong subordination to the continuous-time setting. Suppose that X is a real-valued semimartingale and let

$$X_t = X_0 + M_t + A_t, \qquad t \geq 0, \tag{5.2}$$

be Doob-Meyer decomposition of X: M is a local martingale, A is a finite variation process and both processes start from 0. In general, the decomposition is not unique; however, if X is assumed to be a sub- or supermartingale, then there is unique *predictable* process A in (5.2). In what follows, whenever the dominating process X is a sub- or supermartingale, we will consider this particular Doob-Meyer decomposition only. If X takes values in a separable Hilbert space (say, ℓ^2), then we may give meaning to (5.2), simply by using the decomposition for each coordinate separately. Let Y be a semimartingale taking values in $\mathcal{H}$ and let

$$Y_t = Y_0 + N_t + B_t, \qquad t \geq 0, \tag{5.3}$$

be its Doob-Meyer decomposition. In the statement below, $|A_t|$ denotes the total variation of A on the interval $[0, t]$.

Definition 5.2. Let α be a fixed nonnegative number and let X, Y be two semi-martingales as above. Let X be a general semimartingale. We say that Y is α-strongly subordinate to X if Y is differentially subordinate to X and there are decompositions (5.2) and (5.3) such that the process $(\alpha|A_t| - |B_t|)_{t \geq 0}$ is nondecreasing as a function of t.

Remark 5.2. This generalizes α-subordination from the discrete time. Indeed, suppose that $g = (g_n)_{n \geq 0}$ is α-subordinate to $f = (f_n)_{n \geq 0}$, and treat both sequences as continuous-time processes by taking $X_t = f_{\lfloor t \rfloor}$ and $Y_t = g_{\lfloor t \rfloor}$ for $t \geq 0$. Then Y is α-subordinate to X, which can be seen by considering Doob-Meyer decompositions

$$X_t = f_0 + \sum_{k=1}^{\lfloor t \rfloor} \left[df_k - \mathbb{E}(df_k|\mathcal{F}_{k-1}) \right] + \sum_{k=1}^{\lfloor t \rfloor} \mathbb{E}(df_k|\mathcal{F}_{k-1}),$$

$$Y_t = g_0 + \sum_{k=1}^{\lfloor t \rfloor} \left[dg_k - \mathbb{E}(dg_k|\mathcal{F}_{k-1}) \right] + \sum_{k=1}^{\lfloor t \rfloor} \mathbb{E}(dg_k|\mathcal{F}_{k-1}).$$

Now, let us establish a useful lemma. Recall that for any semimartingale X there is a unique continuous local martingale part X^c of X, which satisfies for all $t \geq 0$ the equality

$$[X, X]_t = [X^c, X^c]_t + \sum_{0 \leq s \leq t} |\Delta X_s|^2,$$

where $\Delta X_s = X_s - X_{s-}$ denotes the jump of X at time s (we set $X_{0-} = 0$). In fact, $[X^c, X^c]_t$ equals $[X, X]_t^c$, the pathwise continuous part of $[X, X]_t$. A little thought leads to the following easy statement.

Lemma 5.2. *Suppose that X and Y are semimartingales. Then Y is differentially subordinate to X if and only if Y^c is differentially subordinate to X^c and $|\Delta Y_s| \leq |\Delta X_s|$ for all $s \geq 0$.*

The purpose of this chapter is to extend the inequalities from the previous two chapters to this new, continuous-time setting. Let us introduce the strong and weak norms of semimartingales. First, if X, Y are sub-, super- or martingales and $1 \leq p \leq \infty$, we set

$$||X||_p = \sup_{t \geq 0} ||X_t||_p, \quad ||X||_{p,\infty} = \sup_{t \geq 0} ||X_t||_{p,\infty} \quad \text{and} \quad ||X||_{\log} - \sup_{t \geq 0} \mathbb{E}|X_t| \log |X_t|.$$

For general semimartingales, we need to modify these and put

$$|||X|||_p = \sup_{\tau \in \mathcal{T}} ||X_\tau||_p, \quad |||X|||_{p,\infty} = \sup_{\tau \in \mathcal{T}} ||X_\tau||_{p,\infty}$$

and

$$|||X|||_{\log} = \sup_{\tau \in \mathcal{T}} \mathbb{E}|X_\tau| \log |X_\tau|$$

where $\mathcal{T}$ is the class of all bounded stopping times relative to $(\mathcal{F}_t)_{t \geq 0}$.

Lemma 5.3.

(i) *Suppose that X a local martingale such that $|||X|||_p < \infty$ for some $1 < p < \infty$. Then X is a martingale and we have $|||X|||_p = ||X||_p$.*

(ii) *Suppose that X a local martingale satisfying $|||X|||_{\log} < \infty$. Then X is a martingale.*

Proof. (i) Let $(\tau_n)_{n\geq 1}$ be a localizing sequence for X. By Doob's maximal L^p inequality,

$$||X^*_{\tau_n}||_p \leq \frac{p}{p-1} \sup_{t\geq 0} ||X_{\tau_n \wedge t}||_p \leq \frac{p}{p-1}|||X|||_p$$

for any $n \geq 1$. Consequently, by Fatou's lemma, X^* is in L^1 and the martingale property of X follows from Lebesgue's dominated convergence theorem. It remains to note that $||X||_p \leq |||X|||_p$ is obvious, while the reverse inequality follows from Doob's optional sampling theorem.

(ii) This can be shown in the same manner, but Doob's inequality is replaced by the logarithmic bound

$$||X^*_{\tau_n}||_1 \leq K \sup_{t\geq 0} \mathbb{E}|X_{\tau_n \wedge t}| \log |X_{\tau_n \wedge t}| + L \leq K|||X|||_{\log} + L.$$

This completes the proof. $\square$

5.2 Inequalities for stochastic integrals and Bichteler's theorem

Now we shall present an approximation procedure, which enables to transfer the inequalities presented in the previous two chapters to analogous statements for stochastic integrals. Assume for a moment that the integrator X takes values in $\mathbb{R}$. Let $\mathbf{Z}$ denote the collection of all processes $Z = (Z_t)_{t\geq 0}$ of the form

$$Z_t = a_0 X(0) + \sum_{k=1}^{n} a_k (X(\tau_k \wedge t) - X(\tau_{k-1} \wedge t)),$$

where n is a positive integer, the coefficients a_k belong to $[-1, 1]$ and $(\tau_k)_{k=0}^n$ is a nondecreasing family of simple stopping times with $\tau_0 \equiv 0$ (here we have used the notation $X(t) = X_t$). If X is a martingale (or submartingale, or nonnegative supermartingale, etc.), then $(X(\tau_k))_{k=0}^n$ inherits this property, by means of Doob's optional sampling theorem.

The following statement is a direct consequence of Proposition 4.1 of Bichteler [11].

Theorem 5.1. *Suppose that X is a semimartingale and let Y be the integral, with respect to X, of a predictable process H bounded in absolute value by 1.*

(i) *If $1 < p < \infty$ and $||X||_p < \infty$, then there is a sequence $(Z^j)_{j \geq 1}$ in $\mathbf{Z}$ such that*

$$\lim_{j \to \infty} ||Z^j - Y||_p = 0.$$

(ii) *If $1 \leq p < \infty$ and $||X||_p < \infty$, then there is a sequence $(Z^j)_{j \geq 1}$ in $\mathbf{Z}$ such that*

$$\lim_{j \to \infty} |Z^j - Y|^* = 0 \qquad \textit{almost surely.}$$

This theorem enables extensions of the previous results to the case of stochastic integrals, both in the scalar- and vector-valued setting. We shall illustrate this on the following two examples of inequalities.

Theorem 5.2. *Suppose that X is an $\mathcal{H}$-valued martingale and let Y be the integral, with respect to X, of a predictable process H bounded in absolute value by 1. Then*

$$||Y||_p \leq (p^* - 1)||X||_p, \qquad 1 < p < \infty,$$

and

$$||Y||_{1,\infty} \leq 2||X||_1.$$

Both estimates are sharp.

Proof. By Lebesgue's monotone convergence theorem we may assume that $\mathcal{H} = \mathbb{R}^d$ for some positive integer d. Let us first establish the moment bound for a fixed $1 < p < \infty$. Obviously, we may restrict ourselves to those X which are bounded in L^p. For any $1 \leq i \leq d$, the pair (X^i, Y^i) satisfies the conditions of Theorem 5.1 and hence, for any $\varepsilon > 0$, there exist a nondecreasing family $\tau^i = (\tau^i_k)^{n_i}_{k=0}$ of simple stopping times such that $\tau^i_0 \equiv 0$ and a sequence $(a^i_k)^{n_i}_{k=0}$ for which the corresponding process $Z^i \in \mathbf{Z}$ satisfies

$$||Z^i - Y^i||_p \leq \varepsilon. \tag{5.4}$$

With no loss of generality, we may assume that the sequence of simple stopping times is common for all $i = 1, 2, \ldots, d$, using $0 \equiv \tau_0 \leq \tau_1 \leq \tau_2 \leq \cdots \leq \tau_N$ obtained by interlacing the sequences τ^i. Introduce the processes

$$f = (X(0), X(\tau_1), X(\tau_2), \ldots, X(\tau_N), X(\tau_N), \ldots),$$
$$g = (Z(0), Z(\tau_1), Z(\tau_2), \ldots, Z(\tau_N), Z(\tau_N), \ldots), \tag{5.5}$$

where $Z = (Z^1, Z^2, \ldots, Z^d)$. Then g is differentially subordinate to f: indeed,

$$|g_0|^2 = \sum_{k=1}^{d} |Z^k(0)|^2 = \sum_{k=1}^{d} (a^k_0)^2 |X^k(0)|^2 \leq \sum_{k=1}^{d} |X^k(0)|^2 = |f_0|^2$$

and, similarly, for $n \geq 1$,

$$|dg_n|^2 = \sum_{k=1}^{d} |Z^k(\tau_n) - Z^k(\tau_{n-1})|^2$$

$$= \sum_{k=1}^{d} (a_n^k)^2 |X^k(\tau_n) - X^k(\tau_{n-1})|^2 \leq \sum_{k=1}^{d} |X^k(\tau_n) - X^k(\tau_{n-1})|^2 = |df_n|^2.$$

Consequently,

$$||Z||_p = ||g||_p \leq (p^* - 1)||f||_p \leq (p^* - 1)||X||_p$$

and it suffices to let $\varepsilon \to 0$ and use (5.4) to obtain the moment estimate.

The proof of the weak type estimate is slightly different. We may assume that $||X||_1 < \infty$. Introduce $Z = (Z^1, Z^2, \ldots, Z^d)$ as previously, but this time (5.4) is replaced by

$$\mathbb{P}(|Z^i - Y^i|^* \geq \varepsilon) \leq \varepsilon, \tag{5.6}$$

where ε is a small positive number. Next, let $(\tau_k)_{k=0}^{N}$ be as previously, consider $\tau = \inf\{t : |Z_t| \geq 1\}$ and define f, g as in (5.5), but with τ_k replaced by $\tau_k \wedge \tau$, $k = 0, 1, 2, \ldots, N$. The differential subordination of g to f is preserved, so

$$\mathbb{P}(|Z|^* \geq 1) = \mathbb{P}(\tau \leq \tau_N) = \mathbb{P}(|Z(\tau_N \wedge \tau)| \geq 1) = \mathbb{P}(|g_N| \geq 1) \leq 2||f_N||_1 \leq 2||X||_1.$$

It remains to note that by (5.6) we have

$$\mathbb{P}(|Y|^* \geq 1 - N\varepsilon) \geq \mathbb{P}(|Z|^* \geq 1) - N\varepsilon,$$

so letting $\varepsilon \to 0$ yields the claim. $\qquad\square$

The other estimates can be obtained in a similar manner. We omit the straightforward details. In the next section we shall present an alternative tool to establish results of this type: but this requires some further work with corresponding special functions.

5.3 Burkholder's method in continuous time

As we shall see, the special functions used in the discrete-time case can also be exploited in the continuous setting. To gain some intuition on how this will be done, let us look at a typical problem. Let $V : \mathcal{H} \times \mathcal{H} \to \mathbb{R}$ be a given Borel function and suppose that we want to show that for all bounded stopping times τ and all local martingales X, Y such that Y is differentially subordinate to X, we have

$$\mathbb{E}V(X_\tau, Y_\tau) \leq 0. \tag{5.7}$$

Let U be the corresponding special function coming from the discrete time. Then the usual approach works, if only we find an argument which will assure that

$$\mathbb{E}U(X_\tau, Y_\tau) \leq \mathbb{E}U(X_0, Y_0) \tag{5.8}$$

for all X, Y and τ as above. Note that in contrast with the discrete-time case, here we no longer have the "step-by-step" procedure, which was the heart of the matter: it reduced the estimate $\mathbb{E}U(f_n, g_n) \leq \mathbb{E}U(f_0, g_0)$ to the problem of proving the "one-step" inequality $\mathbb{E}U(f_1, g_1) \leq \mathbb{E}U(f_0, g_0)$. To overcome this difficulty, we will show (5.8) directly, using Itô's formula and a convolution argument (needed to improve the smoothness of U). A similar approach will also be successful while studying sub- or supermartingale inequalities under α-strong subordination.

5.3.1 Local martingale inequalities under differential subordination

This subsection is devoted to the study of results which imply various versions of (5.8), under the assumption that X, Y are local martingales and Y is differentially subordinate to X. Throughout, we assume that $\mathcal{H} = \ell^2$. Consider the following property, which a given function $U : \mathcal{H} \times \mathcal{H} \to \mathbb{R}$ might have:

$$U(x, y) = U((0, x_1, x_2, \ldots), (0, y_1, y_2, \ldots)), \tag{5.9}$$

where $x = (x_1, x_2, \ldots)$, $y = (y_1, y_2, \ldots)$.

The first step in our analysis is the following fact.

Lemma 5.4. *Let $U : \mathcal{H} \times \mathcal{H} \to \mathbb{R}$ be a continuous function satisfying (5.9) such that U is bounded on bounded sets, of class C^1 on $\mathcal{H} \times \mathcal{H} \setminus \{(x, y) : |x||y| = 0\}$ and of class C^2 on D_i, $i \geq 1$, where (D_i) is a sequence of open connected sets such that the union of the closures of D_i is $\mathcal{H} \times \mathcal{H}$. Suppose that for each $i \geq 1$ there is a measurable $c_i : D_i \to [0, \infty)$ satisfying the following condition: for any $(x, y) \in D_i$ with $|x||y| \neq 0$ and any $h, k \in \mathcal{H}$,*

$$(hU_{xx}(x, y), h) + 2(hU_{xy}(x, y), k) + (kU_{yy}(x, y), k) \leq -c_i(x, y)(|h|^2 - |k|^2). \tag{5.10}$$

Assume further that for each $n \geq 1$ there exists M_n such that

$$\sup c_i(x, y) \leq M_n < \infty, \tag{5.11}$$

where the supremum is taken over all $(x, y) \in D_i$ such that $1/n^2 \leq |x|^2 + |y|^2 \leq n^2$ and then over all $i \geq 1$. Let X, Y be two bounded martingales with bounded quadratic variations. If Y is differentially subordinate to X, then for all $0 \leq s \leq t$,

$$\mathbb{E}U(X_t, Y_t) \leq \mathbb{E}U(X_s, Y_s). \tag{5.12}$$

Remark 5.3. Before we turn to the proof, let us make here an important observation. It is clear that the key assumption above is the inequality (5.10); all the

other conditions are rather technical. This assumption is closely related to the
concavity of the function $G(t) = U(x + th, y + tk)$, $t \in \mathbb{R}$, where x, y, h, $k \in \mathcal{H}$,
$|k| \leq |h|$. In fact, (5.10) is slightly stronger, since it imposes an upper bound for
the second derivative of G also when $|k| > |h|$.

Proof of Lemma 5.4. It suffices to show the claim for $s = 0$. By the boundedness
condition, there is an integer L such that

$$\|X\|_\infty + \|Y\|_\infty + [X, X]_\infty + [Y, Y]_\infty \leq L - 2.$$

Therefore, given $\varepsilon > 0$, there is an integer $d = d(\varepsilon)$ satisfying

$$\sum_{m \geq d} ([X^m, X^m]_\infty + [Y^m, Y^m]_\infty) \leq \varepsilon/L. \tag{5.13}$$

Now, let $a > 2/L$ and

$$X^{(d)} = (a, X^1, X^2, \ldots, X^{d-1}, 0, 0, \ldots), \quad Y^{(d)} = (a, Y^1, Y^2, \ldots, Y^{d-1}, 0, 0, \ldots)$$

and $Z^{(d)} = (X^{(d)}, Y^{(d)})$. Let $g : \mathbb{R}^d \times \mathbb{R}^d \to [0, \infty)$ be a C^∞ function, supported
on a unit ball of $\mathbb{R}^d \times \mathbb{R}^d$ and satisfying $\int_{\mathbb{R}^d \times \mathbb{R}^d} g = 1$. For a given integer $\ell > L$,
define $U^\ell : \mathbb{R}^d \times \mathbb{R}^d \to \mathbb{R}$ by the convolution

$$U^\ell(x, y) = \int_{\mathbb{R}^d \times \mathbb{R}^d} U(x + u/\ell, y + v/\ell) g(u, v) \mathrm{d}u \mathrm{d}v.$$

Here by $U(x + u/\ell, y + v/\ell)$ we mean $U((x + u/\ell, 0, 0, \ldots), (y + v/\ell, 0, 0, \ldots))$. The
function U^ℓ is of class C^∞, so we may apply Itô's formula to obtain

$$U^\ell(Z_t^{(d)}) = I_0 + I_1 + I_2 + \frac{1}{2} I_3, \tag{5.14}$$

where

$$I_0 = U^\ell(Z_0^{(d)}),$$

$$I_1 = \int_{0+}^t U_x^\ell(Z_{s-}^{(d)}) \mathrm{d}X_s^{(d)} + \int_{0+}^t U_y^\ell(Z_{s-}^{(d)}) \mathrm{d}Y_s^{(d)},$$

$$I_2 = \sum_{m,n=1}^{d-1} \left[\int_{0+}^t U_{x_m x_n}^\ell(Z_{s-}^{(d)}) \mathrm{d}[X^{mc}, X^{nc}]_s \right. \tag{5.15}$$

$$\left. + 2 \int_{0+}^t U_{x_m y_n}^\ell(Z_{s-}^{(d)}) \mathrm{d}[X^{mc}, Y^{nc}]_s + \int_{0+}^t U_{y_m y_n}^\ell(Z_{s-}^{(d)}) \mathrm{d}[Y^{mc}, Y^{nc}]_s \right],$$

$$I_3 = \sum_{0 < s \leq t} [U^\ell(Z_s^{(d)}) - U^\ell(Z_{s-}^{(d)}) - \nabla U^\ell(Z_{s-}^{(d)}) \cdot \Delta Z_s^{(d)}].$$

We will deal with each of the terms I_1, I_2 and I_3 separately. Since U is of class C^1
on $\mathcal{H} \times \mathcal{H} \setminus \{|x| |y| = 0\}$ and the martingales $X^{(d)}$, $Y^{(d)}$ are bounded and bounded

away from 0, both stochastic integrals in I_1 are martingales and, consequently, $\mathbb{E}I_1 = 0$. To handle I_2, integrate by parts to get, for $|x| \geq a$, $|y| \geq a$,

$$U_{xx}^\ell(x,y) = \int_{\mathbb{R}^d \times \mathbb{R}^d} U_{xx}(x+u/\ell, y+v/\ell)g(u,v)dudv$$

and, similarly,

$$U_{xy}^\ell(x,y) = \int_{\mathbb{R}^d \times \mathbb{R}^d} U_{xy}(x+u/\ell, y+v/\ell)g(u,v)dudv,$$

$$U_{yy}^\ell(x,y) = \int_{\mathbb{R}^d \times \mathbb{R}^d} U_{yy}(x+u/\ell, y+v/\ell)g(u,v)dudv.$$

Thus, by (5.10), if $|x| \geq a$, $|y| \geq a$, then

$$(hU_{xx}^\ell(x,y), h) + 2(hU_{xy}^\ell(x,y), k) + (kU_{yy}^\ell(x,y), k) \leq -c(x,y)(|h|^2 - |k|^2), \quad (5.16)$$

with

$$c(x,y) = \sum_{i \geq 1} \int_{\mathbb{R}^d \times \mathbb{R}^d} c_i(x+u/\ell, y+v/\ell)g(u,v)dudv$$

(we extend c_i to the whole $\mathcal{H} \times \mathcal{H}$ setting $c_i(x,y) = 0$ for $(x,y) \notin D_i$). Let $0 \leq s_0 < s_1 \leq t$. For any $j \geq 0$, let $(\eta_i^j)_{1 \leq i \leq i_j}$ be a sequence of nondecreasing finite stopping times with $\eta_0^j = s_0$, $\eta_{i_j}^j = s_1$ such that $\lim_{j \to \infty} \max_{1 \leq i \leq i_j - 1} |\eta_{i+1}^j - \eta_i^j| = 0$. Keeping j fixed, we apply, for each $i = 0, 1, 2, \ldots, i_j$, the inequality (5.16) to $x = X_{s_0-}^{(d)}$, $y = Y_{s_0-}^{(d)}$ and $h = h_i^j = X_{\eta_{i+1}^j}^{(d)c} - X_{\eta_i^j}^{(d)c}$, $k = k_i^j = Y_{\eta_{i+1}^j}^{(d)c} - Y_{\eta_i^j}^{(d)c}$. Summing the obtained $i_j + 1$ inequalities and letting $j \to \infty$ yields

$$\sum_{m=1}^d \sum_{n=1}^d \Big[U_{x_m x_n}^\ell(Z_{s_0-}^{(d)})[(X^{(d)c})^m, (X^{(d)c})^n]_{s_0}^{s_1} + 2U_{x_m y_n}^\ell(Z_{s_0-}^{(d)})[(X^{(d)c})^m, (Y^{(d)c})^n]_{s_0}^{s_1}$$

$$+ U_{y_m y_n}^\ell(Z_{s_0-}^{(d)})[(Y^{(d)c})^m, (Y^{(d)c})^n]_{s_0}^{s_1} \Big]$$

$$\leq -c(Z_{s_0-}^{(d)}) \sum_{m=1}^d \big([(X^{(d)c})^m, (X^{(d)c})^m]_{s_0}^{s_1} - [(Y^{(d)c})^m, (Y^{(d)c})^m]_{s_0}^{s_1} \big)$$

$$= -c(Z_{s_0-}^{(d)}) \sum_{m=1}^{d-1} \big([X^{mc}, X^{mc}]_{s_0}^{s_1} - [Y^{mc}, Y^{mc}]_{s_0}^{s_1} \big)$$

where we have used the notation $[S,T]_{s_0}^{s_1} = [S,T]_{s_1} - [S,T]_{s_0}$. By approximation, this yields

$$I_2 \leq -\sum_{m=1}^d \int_{0+}^t c(Z_{s-}^{(d)})d([X^{mc}, X^{mc}]_s - [Y^{mc}, Y^{mc}]_s). \quad (5.17)$$

By differential subordination,

$$
d\left[\sum_{m=1}^{d-1}\left([X^{mc}, X^{mc}]_s - [Y^{mc}, (Y^{mc}]_s)\right)\right]
$$

$$
= d\left[[X^c, X^c]_s - [Y^c, Y^c]_s - \sum_{m \geq d}\left([X^{mc}, X^{mc}]_s - [Y^{mc}, Y^{mc}]_s\right)\right]
$$

$$
\geq -d\sum_{m \geq d}\left([X^{mc}, X^{mc}]_s + [Y^{mc}, Y^{mc}]_s\right)
$$

Furthermore, $c(Z_{s-}^{(d)}) \leq M_\ell$ for all s. Indeed, we have $\sqrt{2}a \leq |Z_{s-}^{(d)}| \leq \ell - 2$, so, for any (u, v) from the unit ball of $\mathbb{R}^d \times \mathbb{R}^d$,

$$
1/\ell \leq |Z_{s-}^{(d)} + (u/\ell, v/\ell)| \leq \ell
$$

and it suffices to use (5.11). Therefore,

$$
I_2 \leq M_\ell \sum_{m \geq d}\left([X^{mc}, X^{mc}]_0^t + [Y^{mc}, Y^{mc}]_0^t\right) \leq \varepsilon,
$$

where in the latter passage we have exploited (5.13). It remains to analyze I_3. Let $x, y, h, k \in \mathcal{H}$ satisfy $|x| \geq 1/\ell$, $|y| \geq 1/\ell$ and let $G = G_{x,y,h,k} : \mathbb{R} \to \mathbb{R}$ be given by $G(t) = U^\ell(x+th, y+tk)$. By (5.16) and mean value theorem, there is $t_0 \in (0, 1)$ such that

$$
U^\ell(x + h, y + k) - U^\ell(x, y) - \nabla U^\ell(x, y) \cdot (h, k)
$$
$$
= G(1) - G(0) - G'(0) = G''(t_0) \leq -c(x + t_0 h, y + t_0 k)(|h|^2 - |k|^2).
$$

Apply this to $(x, y) = Z_{s-}^{(d)}$, $(h, k) = \Delta Z_s^{(d)}$ to obtain

$$
U^\ell(Z_s^{(d)}) - U^\ell(Z_{s-}^{(d)}) - \nabla U^\ell(Z_{s-}^{(d)}) \cdot \Delta Z_s^{(d)}
$$
$$
\leq -c(Z_{s-}^{(d)} + t_0 \Delta Z_s^{(d)})(|\Delta X_s^{(d)}|^2 - |\Delta Y_s^{(d)}|^2).
$$

Using differential subordination,

$$
|\Delta X_s^{(d)}|^2 - |\Delta Y_s^{(d)}|^2 = |\Delta X_s|^2 - |\Delta Y_s|^2 - \sum_{m \geq d}(|\Delta X_s^m|^2 - |\Delta Y_s^m|^2)
$$
$$
\geq -\sum_{m \geq d}(|\Delta X_s^m|^2 + |\Delta Y_s^m|^2).
$$

Furthermore, by the definition of $Z^{(d)}$,

$$
\sqrt{2}a \leq |Z_{s-}^{(d)} + t_0 \Delta Z_s^{(d)}| \leq (1 - t_0)|Z_{s-}^{(d)}| + t_0|Z_s^{(d)}| \leq \ell - 2,
$$

so $c(Z^{(d)}_{s-} + t_0 \Delta Z^{(d)}_s) \le M_\ell$. Putting all these facts together yields

$$I_3 \le M_\ell \sum_{0 < s \le t} \sum_{m \ge d} (|\Delta X^m_s|^2 + |\Delta Y^m_s|^2) \le \varepsilon,$$

where in the last inequality we have used (5.13).

Now, plug the above estimates for I_i into (5.14) to get

$$\mathbb{E}U^\ell(X^{(d)}_t, Y^{(d)}_t) \le \mathbb{E}U^\ell(X^{(d)}_0, Y^{(d)}_0) + 2\varepsilon.$$

Letting $\ell \to \infty$ and using the continuity of U and boundedness of X and Y, we get, by Lebesgue's dominated convergence theorem,

$$\mathbb{E}U(X^{(d)}_t, Y^{(d)}_t) \le \mathbb{E}U(X^{(d)}_0, Y^{(d)}_0) + 2\varepsilon$$

(here we treat $X^{(d)}_s$, $Y^{(d)}_s$ as elements of $\mathcal{H}$, using the embedding $(X^{(d)}_s, 0, 0, \ldots)$ and similarly for $Y^{(d)}_s$). Now let $d \to \infty$ and $\varepsilon \to 0$ to get

$$\mathbb{E}U((a, X_t), (a, Y_t)) \le \mathbb{E}U((a, X_0), (a, Y_0)),$$

and, finally, take $a \to 0$ to get the claim, in view of (5.9). $\qquad\square$

Lemma 5.4, though powerful, does not work if the local martingales have possibly unbounded jumps and we need to strengthen this result.

Theorem 5.3. *Suppose that U satisfies the conditions of Lemma 5.4 and, in addition, that U_x, U_y are bounded on any set of the form $1/n^2 \le |x|^2 + |y|^2 \le n^2$. Let X, Y be local martingales such that Y is differentially subordinate to X. Then there is a nondecreasing sequence $(T_n)_{n \ge 1}$ going to ∞ such that, for any $a > 0$ and $0 \le s \le t$,*

$$\mathbb{E}\big[U((a, X_{T_n \wedge t}), (a, Y_{T_n \wedge t})) - U((a, X_0), (a, Y_0)) \big| \mathcal{F}_0\big] \le 0.$$

Since we do not impose any extra integrability assumptions on X, Y, the left-hand side may happen to be $-\infty$ on a set of positive measure.

Proof of Theorem 5.3. The argumentation is similar to that of the previous lemma. Let n be a fixed positive integer and introduce the stopping time

$$T_n = \inf\{t \ge 0 : |X_t| + |Y_t| + [X, X]_t + [Y, Y]_t \ge n\}.$$

For any fixed $\varepsilon > 0$, there is a positive integer d (depending on n and ε) for which

$$\sum_{m \ge d} ([X^m, X^m]_{T_n-} + [Y^m, Y^m]_{T_n-}) < \varepsilon/(n+2)$$

with probability 1. For a given $a > 0$, put

$$X^{(d)} = (a, X^1, X^2, \ldots, X^{d-1}, 0, 0, \ldots) \quad \text{and} \quad Y^{(d)} = (a, Y^1, Y^2, \ldots, Y^{d-1}, 0, 0, \ldots)$$

and $Z^{(d)} = (X^{(d)}, Y^{(d)})$. Now let ℓ be a fixed positive integer, $g : \mathbb{R}^d \times \mathbb{R}^d \to [0, \infty)$ be a C^∞ function, supported on the unit ball and satisfying $\int_{\mathbb{R}^d \times \mathbb{R}^d} g = 1$ and define $U^\ell : \mathbb{R}^d \times \mathbb{R}^d \to \mathbb{R}$ by the convolution

$$U^\ell(x, y) = \int_{\mathbb{R}^d \times \mathbb{R}^d} U\big((x + u/\ell, 0, 0, \ldots), (y + v/\ell, 0, 0, \ldots)\big) \mathrm{d}u \mathrm{d}v.$$

Let $(\tau_k)_{k \geq 1}$ be a nondecreasing sequence of stopping times which simultaneously localizes $X^{(d)}$, $Y^{(d)}$ and the stochastic integrals $U_x(Z^{(d)}) \cdot X^{(d)}$, $U_y(Z^{(d)}) \cdot Y^{(d)}$. The processes $(X^{(d)}_{T_n \wedge \tau_k \wedge t})_{t \geq 0}$ and $(Y^{(d)}_{T_n \wedge \tau_k \wedge t})_{t \geq 0}$ are martingales, but we cannot use the previous lemma, since they do not have the required boundedness. Therefore, we need some extra effort: let $t > 0$ and apply Itô's formula on the set $\{T_n > 0\}$. We get

$$U^\ell(Z^{(d)}_{T_n \wedge \tau_k \wedge t}) = I_0 + I_1 + \frac{1}{2} I_2 + I_3 + I_4, \tag{5.18}$$

where

$$I_0 = U^\ell(Z^{(d)}_{T_n \wedge \tau_k \wedge r}),$$

$$I_1 = \int_{0+}^{T_n \wedge \tau_k \wedge t} U^\ell_x(Z^{(d)}_{s-}) \mathrm{d}X^{(d)}_s + \int_{0+}^{T_n \wedge \tau_k \wedge t} U^\ell_y(Z^{(d)}_{s-}) \mathrm{d}Y^{(d)}_s,$$

$$I_2 = \sum_{m,n=1}^{d-1} \Big[\int_{0+}^{T_n \wedge \tau_k \wedge t} U^\ell_{x_m x_n}(Z^{(d)}_{s-}) \mathrm{d}[X^m, X^n]^c_s$$

$$+ 2 \int_{0+}^{T_n \wedge \tau_k \wedge t} U^\ell_{x_m y_n}(Z^{(d)}_{s-}) \mathrm{d}[X^m, Y^n]^c_s + \int_{0+}^{T_n \wedge \tau_k \wedge t} U^\ell_{y_m y_n}(Z^{(d)}_{s-}) \mathrm{d}[Y^m, Y^n]^c_s \Big],$$

$$I_3 = \sum_{0 < s < T_n \wedge \tau_k \wedge t} \big[U^\ell(Z^{(d)}_s) - U^\ell(Z^{(d)}_{s-}) - \nabla U^\ell(Z^{(d)}_{s-}) \cdot \Delta Z^{(d)}_s \big],$$

$$I_4 = U^\ell(Z^{(d)}_{T_n \wedge \tau_k \wedge t}) - U^\ell(Z^{(d)}_{T_n \wedge \tau_k \wedge t-}) - \nabla U^\ell(Z^{(d)}_{T_n \wedge \tau_k \wedge t-}) \cdot \Delta Z^{(d)}_{T_n \wedge \tau_k \wedge t}.$$

Now, $I_1 = I_t(t)$ is a martingale, since $(\tau_k)_{k \geq 1}$ is its localizing sequence. Furthermore, $I_2 \leq \varepsilon$ and $I_3 \leq \varepsilon$: this is proved exactly in the same manner as in the previous lemma. Plugging these facts into (5.18), subtracting $U^\ell(Z^{(d)}_{T_n \wedge \tau_k \wedge t})$ from both sides and taking the conditional expectation gives

$$\mathbb{E}\Big[U^\ell(Z^{(d)}_{T_n \wedge \tau_k \wedge t-}) + \nabla U^\ell(Z^{(d)}_{T_n \wedge \tau_k \wedge t-}) \cdot \Delta Z^{(d)}_{T_n \wedge \tau_k \wedge t} \big| \mathcal{F}_0 \Big] \leq U^\ell(Z^{(d)}_0) + 2\varepsilon$$

almost surely on the set $\{T_n > 0\}$. Letting $k \to \infty$ and using Lebesgue's dominated convergence theorem we obtain

$$\mathbb{E}\Big[U^\ell(Z^{(d)}_{T_n \wedge t-}) + \nabla U^\ell(Z^{(d)}_{T_n \wedge t-}) \cdot \Delta Z^{(d)}_{T_n \wedge t} \big| \mathcal{F}_0 \Big] \leq U^\ell(Z^{(d)}_0) + 2\varepsilon$$

almost surely on $\{T_n > 0\}$. Next we let $\ell \to \infty$, then $d \to \infty$, and finally $\varepsilon \to 0$ to get, again by Lebesgue's dominated convergence theorem,

$$\mathbb{E}\Big[U((a, X_{T_n \wedge t-}), (a, Y_{T_n \wedge t-}))$$
$$+ \nabla U((a, X_{T_n \wedge t-}), (a, Y_{T_n \wedge t-})) \cdot \Delta((a, X_{T_n \wedge t}), (a, Y_{T_n \wedge t})) | \mathcal{F}_0\Big] \qquad (5.19)$$
$$\leq U((a, X_0), (a, Y_0))$$

on $\{T_n > 0\}$. Now, by (5.10), for any $x, y, h, k \in \mathcal{H}$ with $|k| \leq |h|$, the function $G = G_{x,y,h,k} : \mathbb{R} \to \mathbb{R}$, given by $G(t) = U(x + t, y + h)$, is concave. This implies $U(x + h, y + k) \leq U(x, y) + U_x(x, y) \cdot h + U_y(x, y) \cdot k$. Using this fact for $x = (a, X_{T_n \wedge t-})$, $y = (a, Y_{T_n \wedge t-})$, $h = \Delta(a, X_{T_n \wedge t})$ and $k = \Delta(a, X_{T_n \wedge t})$, and combining this inequality with (5.19) yields

$$\mathbb{E}\big[U((a, X_{T_n \wedge t}), (a, Y_{T_n \wedge t}))1_{\{T_n > 0\}} | \mathcal{F}_0\big] \leq U((a, X_0), (a, Y_0))1_{\{T_n > 0\}}.$$

This is precisely the claim, since

$$U((a, X_{T_n \wedge t}), (a, Y_{T_n \wedge t})) - U((a, X_0), (a, Y_0))$$
$$= \big[U((a, X_{T_n \wedge t}), (a, Y_{T_n \wedge t})) - U((a, X_0), (a, Y_0))\big]1_{\{T_n > 0\}}.$$

The proof is complete. $\square$

The above theorem concerns functions defined on the whole $\mathcal{H} \times \mathcal{H}$, but the reasoning can be easily transferred to more general situations. Suppose that D is a closed subset of $\mathcal{H} \times \mathcal{H}$ satisfying the following two conditions. First, if $x, y, \overline{x}, \overline{y}$ are vectors from $\mathcal{H}$ such that $|\overline{x}| = |x|$, $|\overline{y}| = |y|$ and $(x, y) \in D$, then $(\overline{x}, \overline{y}) \in D$. Second, for any $0 < r < 1$, there is $\varepsilon > 0$ such that rD_ε, the ε-neighborhood of rD, is contained in D.

Theorem 5.4. *Suppose that D is as above and contains the range of (X, Y): $\mathbb{P}((X, Y)_t \in D$ for all $t \geq 0) = 1$. Assume that $U : D \to \mathbb{R}$ satisfies the conditions from the previous theorem, restricted to D. Then there is a nondecreasing sequence $(T_n)_{n \geq 1}$ going to ∞ with the following property. For any $0 < r < 1$ there is $a(r) > 0$ such that*

$$\mathbb{E}\big[U((a, rX_{T_n \wedge t}), (a, rY_{T_n \wedge t})) - U((a, rX_0), (a, rY_0)) | \mathcal{F}_0\big] \leq 0$$

for any $t \geq 0$ and $a \in (0, a(r))$.

The proof goes along the same lines. The only change is that instead of the truncated process $X^{(d)} = (a, X^1, X^2, \ldots, X^{d-1}, 0, 0, \ldots)$, one has to consider $X^{(d,r)} = (a, rX^1, rX^2, \ldots, rX^{d-1}, 0, 0, \ldots)$ and a similar $Y^{(d,r)}$.

In some situations, the special function U is not of class C^1, but is piecewise C^1: see, e.g., the special function U_1 given by (3.6). Then Theorems 5.3 and 5.4 cannot be applied. Here is a modification which enables us to handle such problems.

Theorem 5.5. *Suppose that D is as above. Assume that a function $U : \mathcal{H} \times \mathcal{H} \to \mathbb{R}$ is bounded on bounded sets, of class C^1 on $D \setminus \{|x||y| = 0\}$ and of class C^2 on the sets D_i, where the union of the closures of D_i is equal to D. Assume that for any $(x, y) \in D_i$, $|x||y| \neq 0$ and $h, k \in \mathcal{H}$,*

$$(hU_{xx}(x, y), h) + 2(hU_{xy}(x, y), k) + (kU_{yy}(x, y), k) \leq -c_i(x, y)(|h|^2 - |k|^2), \quad (5.20)$$

where $c_i : D_i \to [0, \infty)$ is measurable and such that for all $0 < r < 1$,

$$\sup_{i} \sup_{(x,y) \in D_i \cap rD} c_i(x, y) \leq M_r.$$

Here $M_r < M_s < \infty$ for $0 < r < s < 1$. Assume further that

$$U(x + h, y + k) \leq U(x, y) + U_x(x, y) \cdot h + U_y(x, y) \cdot k \quad (5.21)$$

for $(x, y) \in D$ with $|x||y| \neq 0$ and $h, k \in \mathcal{H}$ such that $|k| \leq |h|$. Let X and Y be two local martingales such that Y is differentially subordinate to X and let $T = \inf\{t \geq 0 : (X_t, Y_t) \notin D\}$. Then there is a nondecreasing sequence $(T_n)_{n \geq 1}$ going to ∞ with the following property. For any $0 < r < 1$ there is $a(r) > 0$ such that

$$\mathbb{E}\big[U((a, rX_{T \wedge T_n \wedge t}), (a, rY_{T \wedge T_n \wedge t})) - U((a, rX_0), (a, rY_0))|\mathcal{F}_0\big] \leq 0 \quad (5.22)$$

for all $t \geq 0$ and $a \in (0, a(r))$.

This theorem can be proved in the same manner as Theorems 5.3 and 5.4: we use the processes $X^{(d,r)}$, $Y^{(d,r)}$ and repeat the reasoning with T_n replaced by $T \wedge T_n$. The details are left to the reader.

Before we proceed, let us formulate here a very important observation.

Remark 5.4. Sometimes we will need to establish the inequality in the case when one or both processes X, Y take values in $\mathbb{R}$. Then the above results, Lemma 5.4, Theorems 5.3, 5.4 and 5.5, are not directly applicable. Namely, we use the fact that the Hilbert space $\mathcal{H}$ has at least two dimensions when we replace the processes X, Y by (a, X) and (a, Y), in order to bound them away from 0 (where we do not control the derivative of U). However, if we do control the derivative, the replacement is not necessary and $\mathcal{H}$ may be taken to be equal to $\mathbb{R}$. That is to say, suppose that the function U, given on $\mathbb{R} \times \mathbb{R}$, is of class C^1 on the whole domain (or on a given subset D, depending on which theorem we focus). Then the above results remain valid and the assertions simplify to

$$\mathbb{E}\big[U(X_{T_n \wedge t}, Y_{T_n \wedge t}) - U(X_0, Y_0)|\mathcal{F}_0\big] \leq 0$$

or, respectively,

$$\mathbb{E}\big[U(rX_{T_n \wedge t}, rY_{T_n \wedge t}) - U(rX_0, rY_0)|\mathcal{F}_0\big] \leq 0,$$
$$\mathbb{E}\big[U(rX_{T \wedge T_n \wedge t}, rY_{T \wedge T_n \wedge t}) - U(rX_0, rY_0)|\mathcal{F}_0\big] \leq 0,$$

for some nondecreasing sequence (T_n) going to ∞. Similarly, if U is given on $\mathbb{R} \times \mathcal{H}$ and is of class C^1 on $\mathbb{R} \times \mathcal{H} \smallsetminus \{(x, y) : y = 0\}$ (or, on $D \smallsetminus \{(x, y) : y = 0\}$, respectively), then the above results are valid and we have

$$\mathbb{E}\big[U(X_{T_n \wedge t}, (a, Y_{T_n \wedge t})) - U(X_0, (a, Y_0))\big|\mathcal{F}_0\big] \leq 0,$$

with analogous statements for versions of Theorems 5.3, 5.4 and 5.5.

5.3.2 Inequalities for sub- and supermartingales under α-strong subordination

When the dominating process X is a sub- or supermartingale the situation is a bit simpler, since we immediately get the reduction to the finite-dimensional case. Indeed, if Y is $\mathcal{H}$-valued and α-subordinate to X, then for any $d = 1, 2, \ldots$ the truncated process $Y^{(d)} = (Y^1, Y^2, \ldots, Y^d, 0, 0, \ldots)$ is also α-subordinate to X, but takes values in $\mathbb{R}^d$. This makes the computations a bit less complex, but the general idea is the same as in the martingale setting. In fact, all the theorems from the previous subsection are easily transferred. We will only show the changes which need to be implemented to obtain a submartingale version of Lemma 5.4. Then it will be clear how to modify the remaining theorems to obtain appropriate extensions which deal with sub- and supermartingales.

Lemma 5.5. *Let $U : \mathbb{R} \times \mathcal{H} \to \mathbb{R}$ be a continuous function such that U is bounded on bounded sets, of class C^1 on $\mathbb{R} \times \mathcal{H} \smallsetminus \{(x, y) : y = 0\}$ and of class C^2 on D_i, $i \geq 1$, where (D_i) is a sequence of open connected sets such that the union of the closures of D_i is $\mathbb{R} \times \mathcal{H}$. Suppose that for each $i \geq 1$ there is a measurable function $c_i : D_i \to [0, \infty)$ satisfying the following condition: for any $(x, y) \in D_i$ and $h \in \mathbb{R}$, $k \in \mathcal{H}$,*

$$U_{xx}(x, y)h^2 + 2(hU_{xy}(x, y), k) + (kU_{yy}(x, y), k) \leq -c_i(x, y)(h^2 - |k|^2). \quad (5.23)$$

Assume further that there is $\alpha \geq 0$ such that for all $x \in \mathbb{R}$, $y \in \mathcal{H}$ we have

$$U_x(x, y) + \alpha|U_y(x, y)| \leq 0. \quad (5.24)$$

Moreover, suppose that for each $n \geq 1$ there exists M_n such that

$$\sup c_i(x, y) \leq M_n < \infty, \quad (5.25)$$

where the supremum is taken over all $(x, y) \in D_i$ such that $1/n^2 \leq x^2 + |y|^2 \leq n^2$ and then over all $i \geq 1$. Let X be a bounded submartingale and let Y be a bounded $\mathcal{H}$-valued semimartingale. Assume that both processes have bounded quadratic variations and Y is α-subordinate to X. Then for all $0 \leq s \leq t$,

$$\mathbb{E}U(X_t, Y_t) \leq \mathbb{E}U(X_s, Y_s). \quad (5.26)$$

Proof (sketch). Clearly, we may take $s = 0$. By the above argument and Lebesgue's dominated convergence theorem, it suffices to show the assertion for $\mathcal{H} = \mathbb{R}^d$. Replacing Y by (a, Y) (and, at the end, letting $a \to 0$), we may and do assume that Y is bounded away from 0. As in the proof of Lemma 5.4, we approximate U by C^∞ functions $U^\ell : \mathbb{R} \times \mathbb{R}^d \to \mathbb{R}$, which inherit the properties (5.23) and (5.24) in view of the integration by parts. Let $X = X_0 + M + A$ be the unique Doob-Meyer decomposition with A predictable, and let $Y = Y_0 + N + B$ be the Doob-Meyer decomposition coming from α-subordination. An application of Itô's formula yields

$$U^\ell(X_t, Y_t) = I_0 + I_1 + I_2 + I_3 + \frac{1}{2} I_4,$$

where, this time,

$$I_0 = U^\ell(X_0, Y_0),$$

$$I_1 = \int_{0+}^t U_x^\ell(X_{s-}, Y_{s-}) \mathrm{d}M_s + \int_{0+}^t U_y^\ell(X_{s-}, Y_{s-}) \mathrm{d}N_s,$$

$$I_2 = \int_{0+}^t U_x^\ell(X_{s-}, Y_{s-}) \mathrm{d}A_s + \int_{0+}^t U_y^\ell(X_{s-}, Y_{s-}) \mathrm{d}B_s,$$

$$I_3 = \int_{0+}^t U_{xx}^\ell(X_{s-}, Y_{s-}) \mathrm{d}[X^c, X^c]_s + 2 \sum_{n=1}^d \int_{0+}^t U_{xy_n}^\ell(X_{s-}, Y_{s-}) \mathrm{d}[X^c, Y^{nc}]_s$$

$$+ \sum_{m,n=1}^d \int_{0+}^t U_{y_m y_n}^\ell(X_{s-}, Y_{s-}) \mathrm{d}[Y^{mc}, Y^{nc}]_s,$$

$$I_4 = \sum_{0 < s \leq t} \left[U^\ell(X_s, Y_s) - U^\ell(X_{s-}, Y_{s-}) - \nabla U^\ell(X_{s-}, Y_{s-}) \cdot \Delta(X_s, Y_s) \right].$$

The difference in comparison with (5.15) is that the term I_1 appearing there has been split into I_1 and I_2 above. Note that the terms I_1, I_3 and I_4 can be dealt with in exactly the same manner as in the proof of Lemma 5.4. To handle I_2, we make use of (5.24) and α-subordination to obtain

$$I_2 \leq \int_{0+}^t U_x^\ell(X_{s-}, Y_{s-}) \mathrm{d}A_s + \int_{0+}^t |U_y^\ell(X_{s-}, Y_{s-})| \mathrm{d}|B_s|$$

$$\leq \int_{0+}^t U_x^\ell(X_{s-}, Y_{s-}) \mathrm{d}A_s + \int_{0+}^t |U_y^\ell(X_{s-}, Y_{s-})| \mathrm{d}(\alpha A_s) \qquad (5.27)$$

$$= \int_{0+}^t \left[U_x^\ell(X_{s-}, Y_{s-}) + \alpha |U_y^\ell(X_{s-}, Y_{s-})| \right] \mathrm{d}A_s \leq 0.$$

This completes the proof.   □

Therefore, we see that, in comparison with the martingale setting there are two essential changes in the formulation. The first one is the technical assumption that U is differentiable at the y-axis: this is due to the fact that X is real valued, see Remark 5.4 above. However, this condition can be omitted if X is assumed to be nonnegative, by replacing X with $X + a$ for some $a > 0$; in such a case, it suffices to assume that $U : [0, \infty) \times \mathbb{R} \to \mathbb{R}$ is continuous and of class C^1 on $(0, \infty) \times \mathbb{R}$ (and of class C^2 on the sets D_i, and so on). The second additional assumption is the condition (5.24). Note that we had a similar situation in the discrete-time setting, where we also added this assumption to make Burkholder's method for submartingales work.

In the same manner, adding (5.24), one obtains the submartingale versions of Theorems 5.3, 5.4 and 5.5. The formulations are clear and the details are left to the reader.

We conclude this subsection by noting that if the dominating process X is assumed to be a supermartingale, then (5.24) should be replaced by

$$-U_x(x, y) + \alpha |U_y(x, y)| \leq 0.$$

This can be seen by modifying (5.27) accordingly.

5.4 Inequalities for continuous-time martingales

Theorem 5.6. *Let $U_1, U_\infty : \mathcal{H} \times \mathcal{H} \to \mathbb{R}$ be the functions given by (3.6) and (3.12), respectively. Let X be a martingale and let Y be a local martingale such that Y is differentially subordinate to X.*

(i) *For any $t \geq 0$ we have*
$$\mathbb{E}U_1(X_t, Y_t) \leq 0.$$

(ii) *If X is square-integrable, then for any $t \geq 0$,*
$$\mathbb{E}U_\infty(X_t, Y_t) \leq 0. \tag{5.28}$$

Proof. Let $D = \{(x, y) \in \mathcal{H} \times \mathcal{H} : |x| + |y| \leq 1\}$ and introduce the stopping time $T = \inf\{s \geq 0 : (X_s, Y_s) \notin D\}$.

(i) Observe that
$$\mathbb{E}U_1(X_t, Y_t) \leq \mathbb{E}U_1(X_{T \wedge t}, Y_{T \wedge t}). \tag{5.29}$$

To see this, we use the inequality $U_1(x, y) \leq 1 - 2|x|$, established in the proof of (3.4). By the conditional Jensen inequality, we get

$$\mathbb{E}U_1(X_t, Y_t)1_{\{T < t\}} \leq \mathbb{E}(1 - 2|X_t|)1_{\{T < t\}}$$
$$= \mathbb{E}\left[1_{\{T < t\}}\mathbb{E}\left(1 - 2|X_t| \big| \mathcal{F}_{T \wedge t}\right)\right]$$
$$\leq \mathbb{E}\left[1_{\{T < t\}}\left(1 - 2|X_{T \wedge t}|\right)\right] = \mathbb{E}U_1(X_{T \wedge t}, Y_{T \wedge t})1_{\{T < t\}}$$

and it suffices to add the trivial equality

$$\mathbb{E}U_1(X_t, Y_t)1_{\{T \geq t\}} = \mathbb{E}U_1(X_{T \wedge t}, Y_{T \wedge t})1_{\{T \geq t\}}.$$

Now let us apply Theorem 5.5 to the stopped martingales $(X_{t \wedge s})_{s \geq 0}$ and $(Y_{t \wedge s})_{s \geq 0}$, with D given above and $D_1 = D^{\circ}$. It is obvious that the function U_1 satisfies (5.20) with $c_1 = 1$ on D_1. Furthermore, the condition (5.21) has been verified in the proof of (3.4). Thus, (5.22) holds for some nondecreasing sequence $(T_n)_{n \geq 1}$. Now let $a \to 1$ and $r \to 1$. Since $|U(x,y)| \leq A|x| + B$ for some absolute constants $A, B > 0$, we may apply Lebesgue's dominated convergence theorem to get that for any $n \geq 1$,

$$\mathbb{E}U_1(X_{T \wedge T_n \wedge t}, Y_{T \wedge T_n \wedge t}) \leq \mathbb{E}U_1(X_0, Y_0) \leq 0.$$

Here we have used the fact that $U_1(x,y) \leq 0$ if $|y| \leq |x|$. Since

$$U(X_{T \wedge T_n \wedge t}, Y_{T \wedge T_n \wedge t}) \geq -1 - 2|X_{T \wedge t}|,$$

the use of Fatou's lemma is permitted and we obtain

$$\mathbb{E}U_1(X_{T \wedge t}, Y_{T \wedge t}) \leq 0.$$

It remains to combine this with (5.29).

(ii) The argumentation is similar, but first we show that Y is in L^2. To do this, let τ be any bounded stopping time and apply Theorem 5.3 to the function $U(x,y) = |y|^2 - |x|^2$ and processes X, $Y^{\tau} = (Y_{\tau \wedge t})_{t \geq 0}$. We obtain

$$\mathbb{E}|Y_{T_n \wedge \tau \wedge t}|^2 \leq \mathbb{E}|X_{T_n \wedge t}|^2 \leq ||X||_2^2, \qquad t \geq 0,$$

for some nondecreasing sequence $(T_n)_{n \geq 1}$ converging to ∞. Now let $n \to \infty$ and $t \to \infty$ and use Fatou's lemma to get $\mathbb{E}|Y_\tau|^2 \leq ||X||_2^2$. Thus $|||Y|||_2 < \infty$ and hence Y is a martingale, by virtue of Lemma 5.3. A similar reasoning, conditionally on $\mathcal{F}_{T \wedge t}$, shows that

$$\mathbb{E}(|Y_t|^2 - |X_t|^2) \leq \mathbb{E}(|Y_{T \wedge t}|^2 - |X_{T \wedge t}|^2), \tag{5.30}$$

where T was introduced at the beginning of the proof. The next move is to show that for any $t \geq 0$,

$$\mathbb{E}U_\infty(X_t, Y_t) \leq \mathbb{E}U_\infty(X_{T \wedge t}, Y_{T \wedge t}). \tag{5.31}$$

We have $U_\infty(x,y) \geq (|y| - 1)^2 - |x|^2$ for all x, $y \in \mathcal{H}$ (see the proof of Lemma 3.1). Therefore, by Jensen's inequality and (5.30),

$$\begin{aligned}
\mathbb{E}U_\infty(X_t, Y_t)1_{\{T < t\}} &\leq \mathbb{E}\big[(|Y_t| - 1)^2 - |X_t|^2\big]1_{\{T < t\}} \\
&= \mathbb{E}\big[|Y_t|^2 - |X_t|^2 - 2|Y_t| + 1\big]1_{\{T < t\}} \\
&\leq \mathbb{E}\big[|Y_{T \wedge t}|^2 - |X_{T \wedge t}|^2 - 2|Y_{T \wedge t}| + 1\big]1_{\{T < t\}} \\
&= \mathbb{E}U_\infty(X_{T \wedge t}, Y_{T \wedge t})1_{\{T < t\}}
\end{aligned}$$

and it suffices to add the trivial estimate

$$\mathbb{E}U_\infty(X_{T \wedge t}, Y_{T \wedge t})1_{\{T \geq t\}} = \mathbb{E}U_\infty(X_t, Y_t)1_{\{T \geq t\}}$$

to get (5.31). Apply Theorem 5.5 to the stopped martingales $(X_{T \wedge t})_{t \geq 0}$ and $(Y_{T \wedge t})_{t \geq 0}$ and D, D_1 as above. The conditions (5.20) and (5.21) hold (for the second of them, see Lemma 3.1). Thus (5.22) is valid: we let $a \to 0$, $r \to 1$ and $n \to \infty$ and use Lebesgue's dominated convergence theorem. Here the majorant is given by $(|Y|^*)^2 + (|X|^*)^2 + 1$, which is integrable by Doob's maximal inequality. As the result, we obtain

$$\mathbb{E}U_\infty(X_{T \wedge t}, Y_{T \wedge t}) \leq \mathbb{E}U_\infty(X_0, Y_0) \leq 0$$

and the use of (5.31) completes the proof. $\qquad\square$

Thus, the inequalities for basis functions U_1 and U_∞ are valid. By Fubini's theorem and localization, all the inequalities of Chapter 3 which were established using the integration method are also valid in the continuous-time setting. Let us gather these results in the three theorems below.

Theorem 5.7. *Suppose that X, Y are $\mathcal{H}$-valued local martingales such that Y is differentially subordinate to X. Then*

$$
\begin{aligned}
|||Y|||_p &\leq (p^* - 1)|||X|||_p, & 1 < p < \infty, \\
|||Y|||_p &\leq C_{p,q}|||X|||_q, & 1 \leq p < q < \infty, \\
|||Y|||_{q,\infty} &\leq c_{p,q}|||X|||_p, & 0 \leq p, q < \infty, \\
|||Y|||_1 &\leq K|||X|||_{\log} + L(K)
\end{aligned}
\tag{5.32}
$$

and all the inequalities are sharp.

Proof. For example, to establish the first inequality, we may assume that $|||X|||_p < \infty$ (otherwise the inequality is trivial). Thus X is a martingale, by Lemma 5.3. Let $T_n = \inf\{t \geq 0 : |X_t| + |Y_t| \geq n\}$ and let τ be a bounded stopping time. We have, by Lemma 5.2,

$$|Y_{T_n \wedge \tau \wedge t}| \leq |Y_0| + n + |\Delta Y_{T_n \wedge \tau \wedge t}| \leq |X_0| + n + |\Delta X_{T_n \wedge t}|,$$

so $Y_{T_n \wedge \tau \wedge t} \in L^p$. Thus we may apply the integration method to the processes X and $(Y_{T_n \wedge \tau \wedge t})_{t \geq 0}$ and we get

$$||Y_{T_n \wedge \tau \wedge t}||_p \leq (p^* - 1)||X||_p.$$

It suffices to let $t \to \infty$ and $n \to \infty$ and apply Fatou's lemma to obtain $||Y_\tau||_p \leq (p^*-1)||X||_p$. This is what we need, since τ was arbitrary. The remaining estimates are proved in a similar manner. $\qquad\square$

Analogously, one shows the following result for nonnegative local martingales.

Theorem 5.8. *Assume that X, Y are local martingales such that Y is differentially subordinate to X. Then*

(i) *If $X \geq 0$ and Y is $\mathcal{H}$-valued, then*

$$
\begin{aligned}
|||Y|||_p &\leq C_{p+,p}|||X|||_p, & 1 < p < \infty, \\
|||Y|||_{p,\infty} &\leq c_{p,p}^+|||X|||_p, & 1 \leq p < \infty.
\end{aligned}
$$

(ii) *If X is $\mathcal{H}$-valued and $Y \geq 0$, then*

$$|||Y|||_p \leq C_{p,p+}|||X|||_p.$$

All the inequalities are sharp.

Now we turn to the estimates in the case when the dominating process X is bounded. Let Φ be an increasing convex function on $[0, \infty)$, such that $\Phi(0) = 0$, $\Phi'(0+) = 0$ and Φ is twice differentiable on $(0, \infty)$.

Theorem 5.9. *Assume that X, Y are $\mathcal{H}$-valued local martingales such that Y is differentially subordinate to X and $|||X|||_\infty \leq 1$.*

(i) *We have, for any $\lambda > 0$,*

$$\mathbb{P}(|Y|^* \geq \lambda) \leq P(\lambda),$$

where $P(\lambda)$ is given in (3.102).

(ii) *If Φ' is convex, then for any bounded stopping time τ,*

$$\sup_t \mathbb{E}\Phi(|Y_{\tau \wedge t}|) \leq \frac{1}{2} \int_0^\infty \Phi(t)e^{-t}dt.$$

(iii) *If Φ' is concave, then for any bounded stopping time τ,*

$$\sup_t \mathbb{E}\Phi(|Y_{\tau \wedge t}|) \leq \Phi(1).$$

All the estimates above are sharp.

Proof. (i) For $\lambda \leq 2$ we repeat the argumentation from the discrete-time case. Hence, let $\lambda > 2$ be fixed. It suffices to show that for any $t \geq 0$,

$$\mathbb{P}(|Y_t| \geq 1) \leq P(\lambda).$$

Here the situation is a bit more complicated, as the function U_λ defined by (3.106) is *not* of class C^1 on $(-1, 1) \times (-\lambda, \lambda)$, so the argument with the stopping time $T = \inf\{t \geq 0 : |Y_t| \geq \lambda\}$ and Theorem 5.5 is not applicable. Thus we need some extra effort. The key observation is that the function $\overline{U}_\lambda : \{(x, y) \in \mathcal{H} \times \mathcal{H} : |x| \leq 1\} \to \mathbb{R}$, given by

$$\overline{U}_\lambda(x, y) = \begin{cases} U_\lambda(x, y) & \text{on } D_0 \cup D_1 \cup D_2 \cup D_3, \\ 1 - (\lambda^2 - 1 - |y|^2 + |x|^2)(4(\lambda - 1))^{-1} & \text{on } D_4 \end{cases}$$

is an alternative special function leading to (3.102) for $\lambda > 2$. To see this, note that $\overline{U}_\lambda \geq U_\lambda$, so $\overline{U}_\lambda$ has the majorization property. Furthermore, since $\overline{U}_\lambda$ is of class C^1 on $D_3 \cup D_4$ and the dependence on x and y on the set D_4 is only through

$|y|^2 - |x|^2$, the function has the concavity property as well. Introduce the stopping time $T = \inf\{t : (X_t, Y_t) \in D_3 \cup D_4\}$. The next step is to show that

$$\mathbb{P}(|Y_t| \geq \lambda) \leq \mathbb{E}\overline{U}_\lambda(X_{T \wedge t}, Y_{T \wedge t}). \qquad (5.33)$$

To establish this bound note that $\{|Y_t| \geq 1\} \subseteq \{T \leq t\}$, so by Chebyshev's inequality,

$$\mathbb{P}(|Y_t| \geq 1) \leq \mathbb{E}\left[1 - \frac{\lambda^2 - 1 - |Y_t|^2 + |X_t|^2}{4(\lambda - 1)}\right]1_{\{T \leq t\}}.$$

Now plug (5.30) to obtain (5.33). The final step is to apply Theorem 5.5 to the function $\overline{U}_\lambda$ and the set $D_0 \cup D_1 \cup D_2$. We get

$$\mathbb{E}\overline{U}_\lambda(X_{T \wedge t}, Y_{T \wedge t}) \leq P(\lambda)$$

which is the claim.

(ii) Follows from the integration argument.

(iii) Evident from simplicity of the special function.   □

The inequalities for "nonsymmetric differentially subordinate martingales" studied in Section 3.9 also extend to the continuous-time setting. Recall the constants $K_{p,\infty}$, $L^{ns}(K)$ and c_p given there.

Theorem 5.10. *Assume that X, Y are local martingales such that $-X/2 + Y$ is differentially subordinate to $X/2$.*

(i) *If X and Y are $\mathcal{H}$-valued, then*

$$|||Y|||_{p,\infty} \leq K_{p,\infty}|||X|||_p, \qquad 1 \leq p < \infty$$

and

$$|||Y|||_1 \leq K \sup_t \mathbb{E}|X_t| \log |X_t| + L^{ns}(K).$$

Both inequalities are sharp.

(ii) *If X and Y are real-valued, then*

$$|||Y|||_p \leq c_p|||X|||_p$$

and the constant c_p is the best possible.

Proof. (i) By Theorem 5.6, we have that if X is a martingale, then we have $\mathbb{E}U_1(X_t/2, -X_t/2 + Y_t) \leq 0$ and if, in addition, $X \in L^2$, then we also have $\mathbb{E}U_\infty(X_t/2, -X_t/2 + Y_t) \leq 0$. These two basic inequalities combined with the integration argument lead to the announced estimates: see the discrete-time setting.

(ii) We will exploit Theorem 5.3. Using an additional trick, we will not deal with the explicit formula for Choi's special function; it will be enough for us that such a function exists. Let $\mathcal{M}(x, y)$ denote the class of all pairs (f, g) of simple martingales starting from (x, y) satisfying the following condition: there is a deterministic sequence $\theta = (\theta_n)_{n \geq 0}$ with terms in $[0, 1]$ such that for $n \geq 1$, $dg_n = \theta_n df_n$. Introduce the function $V : \mathbb{R} \times \mathbb{R} \to \mathbb{R}$ by $V_p(x, y) = |y|^p - c_p^p |x|^p$. Then the following lemma is clear (see Chapter 2).

Lemma 5.6. *Let $U_p : \mathbb{R}^2 \to \mathbb{R}^2$ be given by*

$$U_p(x, y) = \sup \mathbb{E} V_p(f_n, g_n),$$

where the supremum is taken over all n and all pairs $(f, g) \in \mathcal{M}(x, y)$. Then $U_p \geq V_p$ and U_p is concave along any line of slope belonging to $[0, 1]$. Furthermore, U_p is homogeneous of order p.

The next auxiliary fact is the following.

Lemma 5.7. *Let U_p be as in the previous lemma. Then for any $x \in \mathbb{R}$ and $1 < p < \infty$, the function $U_p(x, \cdot) : t \mapsto U_p(x, t)$ is convex.*

Proof. Let $y_1, y_2 \in \mathbb{R}$ and $\lambda \in (0, 1)$. For any $(f, g) \in \mathcal{M}(x, 0)$, we have, by the convexity of the function $t \mapsto |t|^p$,

$$\mathbb{E} V_p(f_n, \lambda y_1 + (1 - \lambda)y_2 + g_n) \leq \lambda \mathbb{E} V_p(f_n, y_1 + g_n) + (1 - \lambda)\mathbb{E} V_p(f_n, y_2 + g_n)$$
$$\leq \lambda \mathbb{E} U_p(x, y_1) + (1 - \lambda)\mathbb{E} U_p(x, y_2),$$

since $(f, y_i + g) \in \mathcal{M}(x, y_i)$ for $i = 1$, 2. It suffices to take the supremum over all n and $(f, g) \in \mathcal{M}(x, 0)$ to complete the proof. $\square$

Now we come back to the proof of the moment inequality. We cannot verify (5.10) (or its nonsymmetric version), since we do not know whether U_p is sufficiently smooth. To overcome this problem, let $\varepsilon > 0$ and $g : \mathbb{R}^2 \to [0, \infty)$ be a C^∞ function satisfying $\int_{\mathbb{R}^2} g = 1$, supported on the ball of radius 1 and center $(0, 0)$. Define $\overline{U}_p, \overline{V}_p : \mathbb{R}^2 \to \mathbb{R}$ by the convolutions

$$\overline{U}_p(x, y) = \int_{\mathbb{R}^2} U_p(x + \varepsilon u, y + \varepsilon v)g(u, v)dudv$$

and

$$\overline{V}_p(x, y) = \int_{\mathbb{R}^2} V_p(x + \varepsilon u, y + \varepsilon v)g(u, v)dudv.$$

Since $U_p \geq V_p$, we have that $\overline{U}_p$ majorizes $\overline{V}_p$. The function $\overline{U}_p$ is of class C^∞ and inherits the convexity and concavity properties described in the two lemmas above. In particular,

$$\overline{U}_{pxx} + 2\overline{U}_{pxy} + \overline{U}_{pyy} \leq 0, \quad \overline{U}_{pxx} \leq 0, \quad -\overline{U}_{pyy} \leq 0. \tag{5.34}$$

We will prove that $U : \mathbb{R} \times \mathbb{R} \to \mathbb{R}$ given by $U(x, y) = \overline{U}_p(2x, x+y)$ satisfies (5.10) with

$$c(x, y) = -2(\overline{U}_{pxx}(2x, x+y) + \overline{U}_{pxy}(2x, x+y)),$$

which is nonnegative due to (5.34). To do this, note that the first two inequalities in (5.34) can be rewritten in the form

$$\pm(2\overline{U}_{pxy} + U_{pyy}) \leq -\left(2\overline{U}_{pxx} + 2\overline{U}_{pxy} + \overline{U}_{pyy}\right).$$

Therefore,

$$\begin{aligned}
&U_{xx}(x, y)h^2 + 2U_{xy}(x, y)hk + U_{yy}(x, y)k^2 \\
&= \left[4\overline{U}_{pxx}(2x, x+y) + 4\overline{U}_{pxy}(2x, x+y) + \overline{U}_{pyy}(2x, x+y)\right]h^2 \\
&\quad + 2\left[2\overline{U}_{pxy}(2x, x+y) + \overline{U}_{pyy}(2x, x+y)\right]hk + \overline{U}_{pyy}(2x, x+y)k^2 \\
&\leq \left[4\overline{U}_{pxx}(2x, x+y) + 4\overline{U}_{pxy}(2x, x+y) + \overline{U}_{pyy}(2x, x+y)\right]h^2 \\
&\quad - \left[2\overline{U}_{pxx}(2x, x+y) + 2\overline{U}_{pxy}(2x, x+y) + \overline{U}_{pyy}(2x, x+y)\right](h^2 + k^2) \\
&\quad + \overline{U}_{pyy}(2x, x+y)k^2 \\
&= -c(x, y)(h^2 - k^2).
\end{aligned}$$

The remaining part of the proof is straightforward. We may assume that $|||X|||_p < \infty$, which implies that $|||Y|||_p < \infty$ (by Burkholder's estimate for differentially subordinate martingales). We have that $|U_p(x, y)| \leq \kappa(|x|^p + |y|^p)$ for all x, y and some absolute constant κ: indeed, $U_p(x, y) \geq |y|^p - c_p^p|x|^p$ and U_p can be bounded from above by Burkholder's function $(x, y) \mapsto \alpha_p(|y| - (p^* - 1)|x|)(|x| + |y|)^{p-1}$, directly from the definition of U_p. Thus, a similar bound is valid for $\overline{U}_p$ and, in turn, for U:

$$|U(x, y)| \leq \bar{\kappa}(|x|^p + |y|^p) \tag{5.35}$$

Consequently, by Theorem 5.3 and Remark 5.4,

$$\mathbb{E}\left[U(X_{T_n \wedge t}/2, -X_{T_n \wedge t}/2 + Y_{T_n \wedge t}) - U(X_0/2, -X_0/2 + Y_0)|\mathcal{F}_0\right] \leq 0, \qquad t \geq 0,$$

for some nondecreasing sequence (T_n) going to ∞. Now we use (5.35) with the condition $X, Y \in L^p$ to obtain

$$\mathbb{E}U(X_t/2, -X_t/2 + Y_t) \leq \mathbb{E}U(X_0/2, -X_0/2 + Y_0),$$

by Lebesgue's dominated convergence theorem (the majorant is $\bar{\kappa}((|X|^*)^p + (|Y|^*)^p)$, which is integrable thanks to Doob's inequality). That is,

$$\mathbb{E}\overline{U}_p(X_t, Y_t) \leq \mathbb{E}\overline{U}_p(X_0, Y_0),$$

and, consequently, $\mathbb{E}\overline{V}_p(X_t, Y_t) \leq \mathbb{E}\overline{U}_p(X_0, Y_0)$. Now we let $\varepsilon \to 0$ (ε was the parameter used in the convolution) to get

$$\mathbb{E}V_p(X_t, Y_t) \leq \mathbb{E}U_p(X_0, Y_0) \leq 0,$$

which yields the claim. $\qquad\qquad\square$

Finally, we turn to the results concerning optimal control of martingales.

Theorem 5.11. *Assume that X, Y are real-valued martingales such that $X_0 \equiv x$, Y is differentially subordinate to X and $\mathbb{P}(\sup_t Y_t \geq 0) \geq t$. Then*

$$|||X|||_p \geq L_p(x, |x|, t), \qquad 1 \leq p < \infty,$$

and the inequality is sharp.

Theorem 5.12. *Assume that X, Y are real-valued local martingales such that Y is differentially subordinate to X. Then*

$$\mathbb{P}(\sup_t Y_t \geq 1) \leq K_{p,\infty}|||X|||_p, \qquad 1 \leq p < \infty,$$

and the constant is the best possible.

Proof. As in the discrete-time case, the key ingredient in the proof of both these estimates is the inequality

$$\mathbb{P}(Y^* \geq 0) \leq |||X|||_p^p + U_p(x, y), \tag{5.36}$$

where X, Y are real-valued local martingales starting from x and y, respectively, such that the process $([X, X]_t - [Y, Y]_t)_{t \geq 0}$ is nondecreasing as a function of t (this is not differential subordination: this process may take negative values). We consider two cases $p > 1$ and $p = 1$ separately.

The case $p > 1$. In fact we will focus on the case $1 < p < 2$; for $p \geq 2$ the reasoning is similar. With no loss of generality, we take $|||X|||_p < \infty$: this implies that $|||Y|||_p$ is also finite. We will apply Theorem 5.5 combined with Remark 5.4. Let $V_p(x, y) = 1_{\{y \geq 0\}} - |x|^p$ and let U_p be the function coming from the discrete-time setting. Finally, set $D = \mathbb{R} \times (-\infty, 0)$. Then U_p satisfies the assumptions of Theorem 5.5: the inequality (5.20) can be extracted from the inequalities (3.159)–(3.162), while (5.21) was established directly in the discrete-time case. The remaining properties are easy to verify. Consequently, we have

$$\mathbb{E}U_p(rX_{T \wedge T_n \wedge t}, rY_{T \wedge T_n \wedge t}) \leq \mathbb{E}U_p(rX_0, rY_0) = U_p(rx, ry)$$

and letting $r \to 1$, $n \to \infty$ we get

$$\mathbb{E}V_p(X_{T \wedge t}, Y_{T \wedge t}) \leq \mathbb{E}U_p(X_{T \wedge t}, Y_{T \wedge t}) \leq U_p(x, y),$$

by Lebesgue's dominated convergence theorem. This yields (5.36) by the stopping time argument.

The case $p = 1$. We may assume that $|||X|||_1 < \infty$ and, by localizing, that X is a martingale. We cannot use "the real-valued version" of Theorem 5.5, since U_1, given by (3.142), is not differentiable on the set $\{(0, y) : y < -2\}$. On the other hand, we know that U_1 is diagonally concave and, as one easily checks, the function $U_1(x, \cdot)$ is convex on $(-\infty, 0]$ for any $x \in \mathbb{R}$. Introduce the function

$$\overline{U}_1(x, y) = \int_{\mathbb{R}^2} U_1(x + \varepsilon u, y + \varepsilon v) g(u, v) du dv, \qquad (x, y) \in \mathbb{R}^2,$$

where g and ε are as usual (assume also that $\varepsilon < -y$). Then $\overline{U}_1$ is also diagonally concave and $\overline{U}_1(x, \cdot)$ is convex on $(-\infty, -\varepsilon]$ for any $x \in \mathbb{R}$. This gives

$$\pm 2\overline{U}_{1xy} \leq -\overline{U}_{1xx} - \overline{U}_{1yy} \qquad \text{on } \mathbb{R}^2$$

and

$$\overline{U}_{1yy} \geq 0 \qquad \text{on } \mathbb{R} \times (-\infty, -\varepsilon]. \tag{5.37}$$

Thus, for any x, y, h and $k \in \mathbb{R}$,

$$\overline{U}_{1xx}(x, y)h^2 + 2\overline{U}_{1xy}(x, y)hk + \overline{U}_{1yy}(x, y)k^2$$
$$\leq \overline{U}_{1xx}(x, y)h^2 - \frac{1}{2}\Delta\overline{U}_1(x, y)(h^2 + k^2) + \overline{U}_{1yy}(x, y)k^2$$
$$= -(\overline{U}_{1yy}(x, y) - \overline{U}_{1xx}(x, y))(h^2 - k^2)$$
$$= -c(x, y)(h^2 - k^2).$$

Now, if $(x, y) \in D = \mathbb{R} \times (-\infty, -\varepsilon)$, then $c(x, y) \geq 0$ in virtue of (5.37). Thus, $\overline{U}_1$ satisfies the assumptions of Theorem 5.5 (with $\mathcal{H} = \mathbb{R}$) and hence

$$\mathbb{E}\overline{U}_1(X_{T \wedge T_n \wedge t}, Y_{T \wedge T_n \wedge t}) \leq \overline{U}(x, y), \tag{5.38}$$

where T is the exit time of D and (T_n) is a certain nondecreasing sequence going to ∞. Since $U(x, y) \geq 1 - |x|$, $\overline{U}_1$ also satisfies an inequality of this type: $\overline{U}_1(x, y) \geq -\kappa(1 + |x|)$ for some universal constant κ. By (5.38),

$$\mathbb{E}\left[\kappa + \kappa|X_{T \wedge T_n \wedge t}| + \overline{U}_1(X_{T \wedge T_n \wedge t}, Y_{T \wedge T_n \wedge t})\right] \leq \overline{U}(x, y) + \kappa + \kappa\mathbb{E}|X_{T \wedge T_n \wedge t}|$$
$$\leq \overline{U}(x, y) + \kappa + \kappa\mathbb{E}|X_{T \wedge t}|.$$

where in the second passage we have used Doob's optional sampling theorem. If we let $n \to \infty$ and then $\varepsilon \to 0$, we obtain, by Fatou's lemma,

$$\mathbb{E}U_1(X_{T \wedge t}, Y_{T \wedge t}) \leq U_1(x, y).$$

Hence $\mathbb{E}V_1(X_{T \wedge t}, Y_{T \wedge t}) \leq U_1(x, y)$, which yields (5.36) by the stopping time argument. $\qquad \square$

5.5 Inequalities for continuous-time sub- and supermartingales

Now we will focus on the inequalities where the dominating process is a sub- or a supermartingale.

5.5.1 Weak type inequalities for local submartingales

We start from the generalization of Theorems 4.4 and 4.5. For any $\alpha \geq 0$, let C_α, K_α be the constants defined there.

Theorem 5.13. *Let $\alpha \geq 0$. Assume that X is a local submartingale and Y is α-subordinate to X. Then*

$$\mathbb{P}(|Y|^* \geq 1) \leq K_\alpha |||X^+|||_1 - (C_\alpha - K_\alpha)\mathbb{E}X_0 \tag{5.39}$$

and

$$\mathbb{P}(|Y|^* \geq 1) \leq C_\alpha |||X|||_1. \tag{5.40}$$

Both inequalities are sharp.

Proof. Sharpness is clear, since the constant C_α is already optimal in the discrete-time case. Thus we will be done if we establish the first estimate, since $|||X^+|||_1 \leq |||X|||_1$ and $-\mathbb{E}X_0 \leq ||X_0||_1 \leq |||X|||_1$. By localization, we may and do assume that X is a submartingale. Recall the function U_α, given by (4.6), and some of its properties. This function is of class C^1 on the set

$$D = \left\{ (x, y) \in \mathbb{R} \times \mathcal{H} : \frac{|y| - 1}{\alpha \vee 1} < x < 1 - |y| \right\}$$

and of class C^2 on $D \setminus \{(x, y) : y = 0\}$. Furthermore, it satisfies (5.20) and (5.21) on D (see (4.8)) as well as the inequality $U_{\alpha x} + \alpha |U_{\alpha y}| \leq 0$, due to (4.7). Consequently, applying the submartingale version of Theorem 5.5 yields

$$\mathbb{E}\left[U_\alpha(rX_{T \wedge T_n \wedge t}, rY_{T \wedge T_n \wedge t}) - U_\alpha(rX_0, rY_0)\right] \leq 0,$$

for some nondecreasing sequence $(T_n)_{n \geq 0}$ going to ∞.

By (4.9) we have $U_\alpha(rX_0, rY_0) \leq -r(\alpha + 1)X_0$, whence

$$\mathbb{E}U_\alpha(rX_{T \wedge T_n \wedge t}, rY_{T \wedge T_n \wedge t}) \leq -r(\alpha + 1)\mathbb{E}X_0 = -r(C_\alpha - K_\alpha)\mathbb{E}X_0.$$

Now we let n go to ∞ and r go to 1. Observe that $U_\alpha(rX_{T \wedge T_n \wedge t}, rY_{T \wedge T_n \wedge t}) \geq -1 - K_\alpha |rX_{T \wedge t}|$, so by Fatou's lemma,

$$\mathbb{E}U_\alpha(X_{T \wedge t}, Y_{T \wedge t}) \leq -(C_\alpha - K_\alpha)\mathbb{E}X_0.$$

Thus, using the majorization property of U_α, we get

$$\mathbb{P}(Y_{T \wedge t} \geq 1) \leq K_\alpha \mathbb{E}X_{T \wedge t}^+ - (C_\alpha - K_\alpha)\mathbb{E}X_0 \leq K_\alpha |||X|||_1 - (C_\alpha - K_\alpha)\mathbb{E}X_0,$$

which yields the claim by the stopping time argument. $\qquad \square$

5.5.2 Weak type and moment inequalities for nonnegative local submartingales

Theorem 5.14. *Let* $\alpha \geq 0$. *Assume that* X *is a nonnegative local submartingale and* Y *is* α-*subordinate to* X. *Then*

$$\mathbb{P}(|Y|^* \geq 1) \leq (\alpha + 2)|||X|||_1 \tag{5.41}$$

and

$$|||Y|||_p \leq (p_\alpha^* - 1)|||X|||_p, \qquad 1 < p < \infty. \tag{5.42}$$

Both inequalities are sharp.

Proof. It suffices to prove inequalities (5.41) and (5.42); their sharpness has been already established in the discrete-time setting. First we focus on the weak type estimate. Let U be the special function given by (4.14) and let T be the first exit time from the set $D = \{(x, y) : 0 \leq x \leq 1 - |y|\}$. By (4.16) and Doob's optional sampling theorem,

$$\begin{aligned}
\mathbb{E}U(X_t, Y_t)1_{\{T<t\}} &\leq \mathbb{E}(1 - (\alpha + 2)X_t)1_{\{T<t\}} \\
&\leq \mathbb{E}(1 - (\alpha + 2)X_{T\wedge t})1_{\{T<t\}} \\
&= \mathbb{E}U(X_{T\wedge t}, Y_{T\wedge t})1_{\{T<t\}}
\end{aligned}$$

and adding this to the trivial equality $\mathbb{E}U(X_t, Y_t)1_{\{T\geq t\}} = \mathbb{E}U(X_{T\wedge t}, Y_{T\wedge t})1_{\{T\geq t\}}$, we obtain

$$\mathbb{E}U(X_t, Y_t) \leq \mathbb{E}U(X_{T\wedge t}, Y_{T\wedge t}). \tag{5.43}$$

Now we shall use the submartingale version of Theorem 5.5, with U and D as above. The assumptions are satisfied: the inequality (5.10) follows from (4.15) and (5.21) was established in the discrete-time setting. Consequently, we obtain

$$\mathbb{E}U(rX_{T\wedge T_n\wedge t}, (a, rY_{T\wedge T_n\wedge t})) \leq \mathbb{E}U(rX_0, (a, rY_0))$$

for some nondecreasing sequence $(T_n)_{n\geq 1}$ going to ∞. We have that

$$|U(rX_0, (a, rY_0))| \leq 1 + (\alpha + 2)X_0$$

and

$$|U(rX_{T\wedge T_n\wedge t}, (a, rY_{T\wedge T_n\wedge t}))| \leq 2 + (\alpha + 2)X_{T\wedge t},$$

so letting $n \to \infty$, $a \to 0$ and $r \to 1$ yields, by Lebesgue's dominated convergence theorem,

$$\mathbb{E}U(X_{T\wedge t}, Y_{T\wedge t}) \leq \mathbb{E}U(X_0, Y_0) \leq 0.$$

Combining this with (5.43) gives

$$\mathbb{E}U(X_t, Y_t) \leq 0, \tag{5.44}$$

which implies the weak type estimate by the stopping time argument.

We turn to the moment estimate. We may and do assume that $|||X|||_p < \infty$. Let N be a positive integer and let $\tau = \inf\{t : |Y_t| \geq N\}$. Then the stopped process $\bar{Y} = (Y_{\tau \wedge t})_{t \geq 0}$ is α-subordinate to X and satisfies $|||Y|||_p < \infty$, because

$$|Y_{\tau \wedge t}| \leq |Y_{\tau \wedge t-}| + |\Delta Y_{\tau \wedge t}| \leq N + |\Delta X_{\tau \wedge t}|.$$

Now, if $p < (\alpha + 2)/(\alpha + 1)$, we exploit the integration argument and (5.44), obtaining

$$\mathbb{E}U(X_t, \bar{Y}_t) \leq 0,$$

where U is given by (4.26). By the majorization property, this implies

$$||Y_{\tau \wedge t}||_p \leq (p_\alpha^* - 1)|||X|||_p$$

and it suffices to let $t \to \infty$ and $N \to \infty$. For the remaining values of p we make use of the submartingale version of Theorem 5.5 and obtain

$$\mathbb{E}U(a + X_{T_n \wedge t}, Y_{T_n \wedge \tau \wedge t}) \leq \mathbb{E}U(a + X_0, Y_0) \leq 0,$$

where $(T_n)_{n \geq 1}$ is a certain nondecreasing sequence of stopping times (here U is the corresponding special function). It suffices to use the inequality $|U(x,y)| \leq \bar{c}(x^p + |y|^p)$, valid for all $x \geq 0$, $y \in \mathcal{H}$ and some absolute constant $\bar{c}$. Then Lebesgue's dominated convergence theorem yields

$$\mathbb{E}U(X_t, Y_{\tau \wedge t}) \leq 0,$$

which, as previously, immediately yields the claim. $\qquad\qquad\qquad\square$

5.5.3   Weak type and moment inequalities for nonnegative local supermartingales

Theorem 5.15. *Let $\alpha \geq 0$. Assume that X is a nonnegative local supermartingale and Y is α-subordinate to X. Then*

$$\mathbb{P}(|Y|^* \geq 1) \leq 2^p|||X|||_p^p, \qquad 0 < p \leq 1, \qquad (5.45)$$

and, for β_p as in Theorem 4.11

$$|||Y|||_p \leq \beta_p|||X|||_p, \qquad p < 1. \qquad (5.46)$$

Both inequalities are sharp.

Proof. This is established exactly in the same manner as the previous theorem. We leave the details to the reader. $\qquad\qquad\qquad\square$

5.5.4 Logarithmic estimates for nonnegative local submartingales

Theorem 5.16. *Let $\alpha \in [0,1]$. Assume that X is a nonnegative local submartingale and Y a real-valued semimartingale which is α-subordinate to X. Then for $K > 1$,*

$$|||Y|||_1 \le K|||X|||_{\log^+} + L(K,\alpha). \tag{5.47}$$

The constant is the best possible.

Proof. We may and do assume that $|||X|||_{\log^+} < \infty$. For a fixed α and K, let U be the corresponding function considered in the discrete-time case. This function is concave along the lines of slope belonging to the interval $[-1, 1]$. Furthermore, when $y \in \mathbb{R}$ and $|\gamma| \le \alpha$, then the function $t \mapsto U(t, y + \alpha t)$ is nonincreasing; finally, for any $x \ge 0$ the function $U(x, \cdot)$ is convex. These properties are inherited by the function $\overline{U} : [\varepsilon, \infty) \times \mathbb{R} \to \mathbb{R}$, given by

$$\overline{U}(x, y) = \int_{\mathbb{R}^2} U(x - \varepsilon u, y - \varepsilon v) g(u, v) du dv,$$

where $\varepsilon > 0$ and $g : \mathbb{R}^2 \to \mathbb{R}$ are as usual. Now, arguing as in the proof of Theorem 5.11, we show that $\overline{U}$ satisfies $1°$ (with $\overline{V}$ obtained from $V(x,y) = |y| - Kx \log^+ x$ by the appropriate convolution) and $2°$. Consequently, by the submartingale version of Theorem 5.3, if $a > 2\varepsilon$ and $t \ge 0$,

$$\mathbb{E}\overline{U}(a + X_{T_n \wedge t}, Y_{T_n \wedge t}) \le \mathbb{E}\overline{U}(a + X_0, Y_0)$$

for some nondecreasing sequence (T_n) going to ∞. This implies

$$\mathbb{E}\overline{V}(a + X_{T_n \wedge t}, Y_{T_n \wedge t}) \le \mathbb{E}\overline{U}(a + X_0, Y_0) \le L(K, \alpha).$$

Here in the latter passage we have used the inequality $U(x, y) \le L(K, \alpha)$ for $|y| \le |x|$ and the fact that $|a + x + u\varepsilon| \ge |y + v\varepsilon|$ for any (u, v) belonging to the unit ball of $\mathbb{R}^2$. Obviously, we have

$$|\overline{V}(x, y)| \le |y| + K(x + \varepsilon) \log^+(x + \varepsilon) \le \kappa(x \log^+ x + 1)$$

for some absolute constant κ (not depending on ε). Therefore, by Lebesgue's dominated convergence theorem, if we let $N \to \infty$, $\varepsilon \to 0$ and then $a \to 0$, we obtain

$$\mathbb{E}|Y_t| \le K\mathbb{E}X_t \log^+ X_t + L(K, \alpha) \le K|||X|||_{\log^+} + L(K, \alpha).$$

Now if pick a bounded stopping time τ and replace Y with $(Y_{\tau \wedge t})_{t \ge 0}$, we get, taking sufficiently large t,

$$\mathbb{E}|Y_\tau| \le K|||X|||_{\log^+} + L(K, \alpha).$$

This yields the claim, since τ was arbitrary. $\qquad\square$

5.5.5   Inequalities for bounded local submartingales

Finally, we formulate the results in the case when the dominating process is a continuous-time submartingale bounded by 1. The reasoning is similar to that presented above and we leave the details to the interested reader.

Theorem 5.17. *Let $\alpha \in [0,1]$. Assume that X is a local submartingale satisfying $|||X|||_\infty \le 1$ and Y is an adapted process which is α-subordinate to X. Then for $\lambda > 0$ we have the sharp inequality*

$$\mathbb{P}(|Y|^* \ge \lambda) \le U_\lambda(-1, 1), \tag{5.48}$$

where U_λ is given by (4.72), (4.73) or (4.75), depending on whether $\lambda \in (0,2]$, $\lambda \in (2,4)$ or $\lambda \ge 4$. In particular, for $\lambda \ge 4$,

$$\mathbb{P}(|Y|^* \ge \lambda) \le \gamma e^{-\lambda/(2\alpha+2)}, \tag{5.49}$$

where

$$\gamma = \frac{1+\alpha}{2\alpha+4}\left(\alpha+1+2^{-\frac{\alpha+2}{\alpha+1}}\right)\exp\left(\frac{2}{\alpha+1}\right).$$

Let Φ be a nondecreasing convex function on $[0,\infty)$, which is twice differentiable on $(0,\infty)$ and such that Φ' is convex on $(0,\infty)$ and $\Phi(0) = \Phi'(0+) = 0$.

Theorem 5.18. *Let $\alpha \ge 0$. Assume that X is a nonnegative submartingale bounded from above by 1 and let Y be an $\mathcal{H}$-valued process which is α-subordinate to X. Then for Φ as above and any bounded stopping time τ,*

$$\mathbb{E}\Phi\left(\frac{|Y_\tau|}{\alpha+1}\right) \le \frac{1+\alpha}{2+\alpha}\int_0^\infty \Phi(t)e^{-t}dt.$$

The inequality is sharp, even if $\mathcal{H} = \mathbb{R}$.

Theorem 5.19. *Let $\alpha \ge 0$ be a fixed number. Assume that X is a nonnegative submartingale bounded from above by 1 and let Y be a real-valued process which is α-subordinate to X. Then*

$$\|Y\|_1 \le \frac{(\alpha+1)(2\alpha^2+3\alpha+2)}{(2\alpha+1)(\alpha+2)}.$$

The constant on the right is the best possible.

5.6 Inequalities for smooth functions on Euclidean domains

The estimates for continuous-time processes studied above yield interesting bounds for smooth functions defined on subdomains of $\mathbb{R}^n$.

5.6.1 Inequalities for harmonic functions

Let D be an open connected set of points $x = (x_1, x_2, \ldots, x_n) \in \mathbb{R}^n$, where n is a fixed positive integer. Let $\mathcal{H}$ be a separable Hilbert space. Suppose that $u : D \to \mathcal{H}$ is harmonic: the partial derivatives $u_k = \frac{\partial u}{\partial x_k}$ and $u_{jk} = \frac{\partial^2 u}{\partial x_j \partial x_k}$ exist and are continuous, and $\Delta u = \sum_{k=1}^n u_{kk} = 0$, the zero vector of $\mathcal{H}$. Then $u_k(x) \in \mathcal{H}$ and we put

$$|\nabla u(x)| = \left(\sum_{k=1}^n |u_k(x)|^2 \right)^{1/2}.$$

Fix a point $\xi \in D$ and let D_0 be a bounded subdomain of D satisfying $\xi \in D_0 \subset D_0 \cup \partial D_0 \subset D$. Denote by $\mu_{D_0}^\xi$ the harmonic measure on ∂D_0 with respect to ξ. If $1 \le p < \infty$, let

$$\|u\|_p = \sup \left(\int_{\partial D_0} |u(x)|^p \mu_{D_0}^\xi(dx) \right)^{1/p}$$

be the strong pth norm of u: here the supremum is taken over all D_0 as above. The norm $\|u\|_\infty$ is defined by the essential supremum of u, as usual. Similarly, let

$$\|u\|_{p,\infty} = \sup_{D_0} \sup_{\lambda > 0} \lambda \left[\mu_{D_0}^\xi(\{x \in \partial D_0 : |u(x)| \ge \lambda\}) \right]^{1/p}$$

denote the corresponding weak pth norm of u. Now let $v : D \to \mathcal{H}$ be another harmonic function.

Definition 5.3. We say that v is *differentially subordinate* to u if, for all $x \in D$, $|\nabla v(x)| \le |\nabla u(x)|$.

For example, if $n = 2$, D is the unit disc of $\mathbb{R}^2$ and u, v are conjugate harmonic functions on D, then v is differentially subordinate to u. This follows immediately from the Cauchy-Riemann equations.

The key fact which connects the differential subordination of harmonic functions with that of martingales is the following. Suppose that $\xi \in D$, u and v are as above and that v is differentially subordinate to u. Assume, in addition, that $|v(\xi)| \le |u(\xi)|$. If $B = (B_t)_{t \ge 0}$ is a Brownian motion in $\mathbb{R}^n$, started at ξ and stopped at the exit time of D, then $X = (u(B_t))_{t \ge 0}$ and $Y = (v(B_t))_{t \ge 0}$ are continuous-time martingales such that Y is differentially subordinate to X. This follows immediately from the identity

$$[X, X]_t - [Y, Y]_t = |u(\xi)|^2 - |v(\xi)|^2 + \int_{0+}^t |\nabla u(B_s)|^2 - |\nabla v(B_s)|^2 \, ds.$$

Moreover, we have $||X||_p = ||u||_p$, $||X||_{p,\infty} = ||u||_{p,\infty}$ and similarly for Y and v. Consequently, the inequalities studied in Section 5.3 above immediately give corresponding results for harmonic functions under differential subordination. However, it should be stressed here that in general, these new estimates may, but need not remain sharp.

For example, we have the following result, concerning the strong type (p, p) and weak type $(1, 1)$ estimate.

Theorem 5.20. *Suppose that u, v are harmonic functions on D such that v is differentially subordinate to u and $|v(\xi)| \leq |u(\xi)|$. Then*

$$||v||_p \leq (p^* - 1)||u||_p, \qquad 1 < p < \infty.$$

It is not known whether the first inequality is sharp (except for the trivial case $p = 2$). Furthermore,

$$||v||_{1,\infty} \leq 2||u||_1$$

and the constant 2 is the best possible, even if $n = 1$ and $\mathcal{H} = \mathbb{R}$.

Proof. In view of the remarks preceding the theorem, it suffices to establish the sharpness of the weak type estimate. Consider $D = (-1, 3)$, $\xi = 0$ and $u(x) = 1 + x$, $v(x) = 1 - x$. Then the assumptions on u and v are satisfied. Furthermore, if $-1 < a < 0 < b < 3$ and $D_0 = (a, b)$, then the harmonic measure on ∂D_0 with respect to 0 is given by

$$\mu_{D_0}^0(\{a\}) = b/(b - a), \qquad \mu_{D_0}^0(\{b\}) = -a/(b - a).$$

Consequently, $||u||_1 = 1$ and, when $\lambda < 2$,

$$\limsup_{\lambda \uparrow 2} \lambda \mu(|v| \geq \lambda) = \lim_{\lambda \uparrow 2} \lambda = 2.$$

This completes the proof. $\square$

5.6.2 Inequalities for smooth functions

All that was said in the previous subsection can be easily extended to a much wider setting. Suppose that u, $v : D \to \mathcal{H}$ are continuous functions with continuous first- and second-order partial derivatives. In particular, they need not be harmonic.

Definition 5.4. Let α be a fixed nonnegative number. We say that v is α-subordinate to u, if v is differentially subordinate to u and for any $x \in D$ we have

$$|\Delta v(x)| \leq \alpha |\Delta u(x)|. \tag{5.50}$$

Assuming that u is sub- or superharmonic (and hence real-valued), we may transfer the results of Section 5.4 above to this new setting. This is again based on the observation that composition of such functions with Brownian motion leads to continuous-time processes for which the α-subordination is satisfied. Indeed, the Doob-Meyer decomposition we consider is given by

$$X_t = u(\xi) + \int_{0+}^{t} \nabla u(B_s)\mathrm{d}B_s + \frac{1}{2}\int_{0+}^{t} \Delta u(B_s)\mathrm{d}s,$$

$$Y_t = v(\xi) + \int_{0+}^{t} \nabla v(B_s)\mathrm{d}B_s + \frac{1}{2}\int_{0+}^{t} \Delta v(B_s)\mathrm{d}s$$

and (5.50) implies that the process

$$\left(\frac{\alpha}{2}\int_{0+}^{t} |\Delta u(B_s)|\mathrm{d}s - \frac{1}{2}\int_{0+}^{t} |\Delta v(B_s)|\mathrm{d}s\right)_{t\geq 0}$$

is nonnegative and nondecreasing. We omit the further details. See also Chapter 6 for related results under an additional orthogonality property.

5.7   Notes and comments

Section 5.1. The notion of differential subordination in the continuous-time case is due to Bañuelos and Wang [8] and Wang [200]. The extension of the strong differential subordination in the particular case $\alpha = 1$ can also be found in [200]; for the general case, see [132].

Section 5.2. The results described in this part are taken from [200].

Section 5.3. The weak type estimates (for $1 \leq p \leq 2$) as well as the moment and tail inequalities were established by Wang in [200]. The weak type inequality for $p \geq 2$, in the real-valued case, was proved by Suh [189]. The above-simplified approach, based on the integration argument, is entirely new (but see [151]).

Section 5.4. The results presented there are new, except for Theorems 5.13 and 5.16 (see [150] and [149], respectively).

Section 5.5. Differential subordination of harmonic functions appeared first in Burkholder's paper [29] and was studied in many subsequent works: see Bañuelos and Wang [8], Burkholder [32], [34], [36], Choi [48], [49], [50], Hammack [88], Janakiraman [103], Suh [189], and the author [125], [126], [130], [133], [156] and [161].

5.7 Notes and comments

Chapter 6

Inequalities for Orthogonal Semimartingales

This chapter can be seen as a continuation of the previous one: we shall study the estimates for continuous-time semimartingales under an additional orthogonality assumption. However, we have decided to gather the results in a separate chapter. This is due to their close connection to the classical inequalities for conjugate harmonic functions on the unit disc of the complex plane.

6.1 Orthogonality and modification of Burkholder's method

Throughout this chapter we assume that $\mathcal{H} = \ell^2$. This can be done with no loss of generality and will be very convenient for our purposes.

Definition 6.1. Let X, Y be two semimartingales taking values in $\mathcal{H}$. We say that X and Y are orthogonal, if for any nonnegative integers i, j the process $[X^i, Y^j]$ is constant (here X^i, Y^j denote the ith and jth coordinate of X and Y, respectively).

For example, if B^1 and B^2 are independent Brownian motions, then they are orthogonal (we treat them as ℓ^2-valued martingales using the embedding $(B_t^1, 0, 0, \ldots)$, $(B_t^2, 0, 0, \ldots)$, $t \geq 0$). Another very important example is the following. Let B be a d-dimensional Brownian motion and let H, K be $\mathbb{R}^d$-valued predictable processes satisfying the condition $H_s \cdot K_s = 0$ for all $s > 0$ (here $\cdot$ is the scalar product in $\mathbb{R}^d$). Then the processes

$$X_t = X_0 + \int_{0+}^t H_s \mathrm{d}B_s, \qquad Y_t = Y_0 + \int_{0|}^t K_s \mathrm{d}B_s$$

are orthogonal: $[X, Y]_t = X_0 Y_0 + \int_{0+}^t H_s \cdot K_s \mathrm{d}s = X_0 Y_0$. This example can be easily modified to give vector-valued orthogonal processes.

Let us make here an important observation.

Remark 6.1. If Y is differentially subordinate to X and both processes are orthogonal, then Y has continuous paths. This follows immediately from $|\Delta Y_s| \leq |\Delta X_s|$ (see Lemma 5.2) and the equality $\Delta X_s^i \cdot \Delta Y_s^j = 0$ for all i, j, which is a direct consequence of orthogonality.

In this chapter we will work with continuous-time semimartingales under the assumptions of (strong) differential subordination and orthogonality. Burkholder's method can be transfered to this more restrictive setting as follows. Consider the "orthogonal version" of Lemma 5.4.

Theorem 6.1. *Let U be a continuous function on $\mathcal{H} \times \mathcal{H}$ satisfying (5.9), bounded on bounded sets and C^1 on $\mathcal{H} \times \mathcal{H} \setminus (\{|x| \neq 0\} \cup \{|y| \neq 0\})$, whose first-order derivative is bounded on bounded sets not containing 0, the origin of $\mathcal{H} \times \mathcal{H}$. Moreover, assume that U is C^2 on D_i, $i \geq 1$, where D_i is a sequence of open connected sets, such that the union of the closures of the D_i is $\mathcal{H} \times \mathcal{H}$. Suppose that for each $i \geq 1$, there exists a nonnegative measurable function c_i defined on D_i such that for $(x,y) \in D_i$ with $|x||y| \neq 0$,*

$$(hU_{xx}(x,y),h) + (kU_{yy}(x,y),k) \leq -c_i(x,y)(|h|^2 - |k|^2) \tag{6.1}$$

for all $h, k \in \mathcal{H}$. Assume that U satisfies

$$U(x+h,y) - U(x,y) - U_x(x,y) \cdot h \leq 0 \tag{6.2}$$

for all x, y, $h \in \mathcal{H}$ such that $|x||y| \neq 0$. Assume further that there exists a nondecreasing sequence $(M_n)_{n \geq 1}$ such that

$$\sup c_i(x,y) < M_n < \infty, \tag{6.3}$$

where the supremum is taken over all $(x,y) \in D_i$ such that $1/n^2 \leq |x|^2 + |y|^2 \leq n^2$ and all $i > 1$. Let X and Y be bounded $\mathcal{H}$-valued orthogonal martingales with bounded quadratic variations such that Y is differentially subordinate to X. Then for any $0 \leq s \leq t$ we have

$$\mathbb{E}U(X_t, Y_t) \leq \mathbb{E}U(X_s, Y_s). \tag{6.4}$$

The proof goes along the same lines. Note that the condition (5.10) from Lemma 5.4 is replaced above by the two inequalities (6.1) and (6.2). The reason for this change comes from a slightly different form of the terms I_2 and I_3 defined in (5.15). Due to the orthogonality, we have $[X^m, Y^n] = 0$ for all m, n, so

$$I_2 = \sum_{m,n=1}^{d-1} \left[\int_{0+}^{t} U_{x_m x_n}^{\ell}(Z_{s-}^{(d)}) \mathrm{d}[X^{mc}, X^{nc}]_s + \int_{0+}^{t} U_{y_m y_n}^{\ell}(Z_{s-}^{(d)}) \mathrm{d}[Y^{mc}, Y^{nc}]_s \right].$$

Second, Y has continuous paths, and so

$$I_3 = \sum_{0 < s \leq t} [U^{\ell}(X_s^{(d)}, Y_s^{(d)}) - U^{\ell}(X_{s-}^{(d)}, Y_s^{(d)}) - U_x^{\ell}(X_{s-}^{(d)}, Y_s^{(d)}) \cdot \Delta X_s^{(d)}].$$

It is clear that (6.1) and (6.2) control the above two terms; all the remaining arguments are essentially the same as those used in the proof of Lemma 5.4.

Theorems 5.3, 5.4 and 5.5 can be analogously transferred to the orthogonal setting. The further extensions to the case when the dominating process X is a sub- or supermartingale are also easy to obtain. We omit the straightforward details.

6.2 Weak type inequalities for martingales

6.2.1 Formulation of the result

As usual, we start with the weak type estimates.

Theorem 6.2. *Assume that the real-valued local martingales X and Y are orthogonal. Then for $1 \leq p \leq 2$ we have*

$$|||Y|||_{p,\infty} \leq K_p |||X|||_p, \tag{6.5}$$

where

$$K_p^p = \frac{1}{\Gamma(p+1)} \cdot \frac{\pi^{p-1}}{2^{p-1}} \cdot \frac{1 + \frac{1}{3^2} + \frac{1}{5^2} + \frac{1}{7^2} + \cdots}{1 - \frac{1}{3^{p+1}} + \frac{1}{5^{p+1}} - \frac{1}{7^{p+1}} + \cdots}. \tag{6.6}$$

The constant is the best possible.

Before we turn to the proof, let us make here some important observations. First, the above theorem concerns real-valued processes. We do not know the best constants in the case when the martingales are Hilbert-space-valued and it seems that these values are different (see the moment estimates in the next section). Secondly, we do not know the best constant in the case $p > 2$, even in the real-valued case.

6.2.2 Proof of Theorem 6.2

Let $H = \{(\alpha, \beta) : \beta > 0\}$ denote the upper half-plane and define $\mathcal{W} : H \to \mathbb{R}$ as the Poisson integral

$$\mathcal{W}(\alpha, \beta) = \frac{2^p}{\pi^{p+1}} \int_{-\infty}^{\infty} \frac{\beta |\log |t||^p}{(\alpha - t)^2 + \beta^2} \, d\iota.$$

Clearly, $\mathcal{W}$ is harmonic on H. Furthermore, we have

$$\lim_{(\alpha, \beta) \to (t, 0)} \mathcal{W}(\alpha, \beta) = \left| \frac{2}{\pi} \log |t| \right|^p$$

for $t \neq 0$. Consider the conformal map ϕ on $S = \{(x, y) : |y| < 1\}$, defined by

$$\phi(x, y) = \phi(z) = ie^{\pi z/2} = \left(-e^{\pi x/2} \sin(\pi y/2), e^{\pi x/2} \cos(\pi y/2) \right).$$

It is easy to check that ϕ maps S onto H. Introduce $V, U : \mathbb{R} \times \mathbb{R} \to \mathbb{R}$ by
$V(x, y) = 1_{\{|y| \geq 1\}} - K_p^p |x|^p$ and

$$U(x, y) = \begin{cases} 1 - K_p^p |x|^p & \text{if } |y| \geq 1, \\ 1 - K_p^p W(\phi(x, y)) & \text{if } |y| < 1. \end{cases} \tag{6.7}$$

Note that when $|y| \leq 1$, the substitution $s = e^{-\pi x/2} t$ yields

$$U(x, y) = 1 - \frac{K_p^p}{\pi} \int_{\mathbb{R}} \frac{\cos(\pi y/2) \left| \frac{2}{\pi} \log |s| + x \right|^p}{\left(\sin(\pi y/2) + s \right)^2 + \cos^2 \left(\frac{\pi y}{2} \right)} \, ds. \tag{6.8}$$

Let us study the key properties of U.

Lemma 6.1.

(i) *The function U satisfies*

$$U(x, y) = U(x, -y) = U(-x, y) \qquad \text{on } \mathbb{R}^2.$$

(ii) *For any $(x, y) \in \mathbb{R}^2$ with $|y| < 1$ we have $U_{xx}(x, y) \leq 0$ and $U_{yy}(x, y) \geq 0$.*
(iii) *If $(x, y) \in (0, \infty) \times (0, 1)$, then $U_{xxx}(x, y) \geq 0$.*
(iv) *For any $x, y \in \mathbb{R}$ such that $|y| \leq |x|$ we have $U(x, y) \leq 0$.*
(v) *We have*
$$U(x, y) \geq 1_{\{|y| \geq 1\}} - K_p^p |x|^p \quad \text{for all } x, y \in \mathbb{R}. \tag{6.9}$$

Proof. (i) Substitute $s := -s$ and $s := 1/s$ in (6.8).

(ii) Since U is harmonic in the strip, it suffices to deal with the first estimate. Using Fubini's theorem we verify that

$$U_{xx}(x, y) = -\frac{p(p-1)K_p^p}{\pi} \int_{\mathbb{R}} \frac{\cos(\frac{\pi}{2} y) \left| \frac{2}{\pi} \log |s| + x \right|^{p-2}}{(\sin(\frac{\pi}{2} y) + s)^2 + \cos^2(\frac{\pi}{2} y)} \, ds$$

and it is evident that the expression on the right is nonpositive.

(iii) We have

$$U_x(x, y) = c \int_{\mathbb{R}} \frac{\cos(\frac{\pi}{2} y) \left| \frac{2}{\pi} \log |s| + x \right|^{p-2} \left(\frac{2}{\pi} \log |s| + x \right)}{(s - \sin(\frac{\pi}{2} y))^2 + \cos^2(\frac{\pi}{2} y)} \, ds,$$

where $c = -pK_p^p/\pi$. Therefore, for $\varepsilon \in (0, x)$ we have

$$2U_x(x, y) - U_x(x - \varepsilon, y) - U_x(x + \varepsilon, y) = c \int_{\mathbb{R}} \frac{f_{x,\varepsilon} \left(\frac{2}{\pi} \log |s| \right) \cos(\frac{\pi}{2} y)}{(s - \sin(\frac{\pi}{2} y))^2 + \cos^2(\frac{\pi}{2} y)} \, ds = I,$$

where

$$f_{x,\varepsilon}(h) = 2|x + h|^{p-2}(x + h) - |x - \varepsilon + h|^{p-2}(x - \varepsilon + h) - |x + \varepsilon + h|^{p-2}(x + \varepsilon + h).$$

The expression I, after splitting it into integrals over the nonpositive and nonnegative half-line, and substitution $s = \pm e^r$, can be rewritten in the form

$$I = c \int_{-\infty}^{\infty} f_{x,\varepsilon}\left(\frac{2}{\pi}r\right) g^y(r) dr,$$

where

$$g^y(r) = \frac{\cos(\frac{\pi}{2}y)e^r}{(e^r - \sin(\frac{\pi}{2}y))^2 + \cos^2(\frac{\pi}{2}y)} + \frac{\cos(\frac{\pi}{2}y)e^r}{(e^r + \sin(\frac{\pi}{2}y))^2 + \cos^2(\frac{\pi}{2}y)}.$$

Observe that $f_{x,\varepsilon}(h) \geq 0$ for $h \geq -x$, due to the concavity of the function $t \mapsto t^{p-1}$ on $(0, \infty)$. Moreover, note that we have $f_{x,\varepsilon}(-x + h) = -f_{x,\varepsilon}(-x - h)$ for all h. Finally, g^y is even and, for $r > 0$,

$$(g^y)'(r) = \frac{\cos(\frac{\pi}{2}y)e^r(1 - e^r)}{[(e^r - \sin(\frac{\pi}{2}y))^2 + \cos^2(\frac{\pi}{2}y)]^2} + \frac{\cos(\frac{\pi}{2}y)e^r(1 - e^r)}{[(e^r + \sin(\frac{\pi}{2}y))^2 + \cos^2(\frac{\pi}{2}y)]^2} \leq 0.$$

This implies $I \leq 0$ and, since $\varepsilon \in (0, x)$ was arbitrary, the function $U_x(\cdot, y) : x \mapsto U_x(x, y)$ is convex on $(0, \infty)$.

(iv) The inequality is clear if $|y| \geq 1$, so we may assume that $|y| < 1$. By the symmetry of U (see (i)), it suffices to deal with positive x and y only. First we show that

$$U_{xy}(x, y) \leq 0 \qquad \text{for } x \geq 0, \, y \in (0, 1). \tag{6.10}$$

Since U is harmonic on the strip $\mathbb{R} \times [-1, 1]$, so is U_x and hence we have $U_{xyy}(x, y) = -U_{xxx}(x, y) \leq 0$ for $x \geq 0$ and $y \in (0, 1)$. Since $U_y(x, 0) = 0$, which is a consequence of (i), we see that $U_{xy}(x, 0) = 0$ and therefore (6.10) follows.

Let $0 \leq y \leq x \leq 1$ and consider the function $\Phi(t) = U(tx, ty)$, $t \in [-1, 1]$. Then Φ is even and, by (ii) and (6.10),

$$\Phi''(t) = x^2 U_{xx}(tx, ty) + 2xy U_{xy}(tx, ty) + y^2 U_{yy}(tx, ty)$$
$$\leq x^2 \Delta U(tx, ty) + 2xy U_{xy}(tx, ty) \leq 0$$

for $t \in (-1, 1)$. This implies

$$U(x, y) = \Phi(1) \leq \Phi(0) = U(0, 0) = 1 - K_p^p \mathcal{W}(0, 1)$$
$$= 1 - K_p^p \cdot \frac{2^{p+1}}{\pi^{p+1}} \int_0^\infty \frac{|\log t|^p}{t^2 + 1} dt$$
$$= 1 - K_p^p \cdot \frac{2^{p+1}}{\pi^{p+1}} \int_{-\infty}^\infty \frac{|s|^p e^s}{e^{2s} + 1} ds$$
$$= 1 - K_p^p \cdot \frac{2^{p+2}}{\pi^{p+1}} \int_0^\infty s^p e^{-s} \sum_{k=0}^\infty (-e^{-2s})^k ds$$
$$= 1 - K_p^p \cdot \frac{2^{p+2}}{\pi^{p+1}} \Gamma(p+1) \sum_{k=0}^\infty \frac{(-1)^k}{(2k+1)^{p+1}}$$
$$= 1 - 1 = 0,$$

where in the latter passage we have exploited the well-known identity

$$\frac{\pi^2}{8} = 1 + \frac{1}{3^2} + \frac{1}{5^2} + \frac{1}{7^2} + \cdots .$$

(v) It suffices to deal with $|y| < 1$. First, one can easily verify that for all $x, y \in \mathbb{R}$ we have

$$|x + y|^p + |x - y|^p \le 2|x|^p + 2|y|^p. \tag{6.11}$$

Consequently, using (i),

$$
\begin{aligned}
2U(x, y) + 2K_p^p |x|^p &= U(x, y) + U(-x, y) + 2K_p^p |x|^p \\
&= 2 - \frac{K_p^p}{\pi} \int_{\mathbb{R}} \frac{\cos(\frac{\pi y}{2})}{\left(\sin(\frac{\pi y}{2}) + s\right)^2 + \cos^2(\frac{\pi y}{2})} \\
&\qquad \times \left[\left|\frac{2}{\pi} \log |s| + x\right|^p + \left|\frac{2}{\pi} \log |s| - x\right|^p - 2|x|^p \right] ds \\
&\ge 2 - \frac{2K_p^p}{\pi} \int_{\mathbb{R}} \frac{\cos(\frac{\pi y}{2}) \left|\frac{2}{\pi} \log |s|\right|^p}{\left(\sin(\frac{\pi y}{2}) + s\right)^2 + \cos^2(\frac{\pi y}{2})} ds \\
&= 2U(0, y) \ge 2U(0, 0) = 0.
\end{aligned}
$$

Here in the latter inequality we used the fact that $y \mapsto U(0, y)$ is even and convex on $[-1, 1]$: see (i) and (ii). □

Proof of (6.5). By localizing, we may assume that X and Y are martingales. Furthermore, we may restrict ourselves to $X \in L^p$, since otherwise the inequality is obvious. Let $\eta > 0$ and set $D = [-\eta, \eta] \times [-1, 1]$, $D_1 = D^\circ$ and $T = \inf\{t : (X_t, Y_t) \notin D\}$. Apply Burkholder's method ("orthogonal" version of Theorem 5.5) to X, Y and the function U. This function is of class C^∞ on D_1. Since $U(\cdot, y)$ is concave for any fixed $|y| < 1$ (see part (ii) of Lemma 6.1), we have

$$U(x + h, y) \le U(x, y) + U_x(x, y)h.$$

For any $(x, y) \in D_1$ and $h, k \in \mathbb{R}$,

$$U_{xx}(x, y)h^2 + U_{yy}(x, y)k^2 = -U_{yy}(x, y)(h^2 - k^2)$$

and $U_{yy}(x, y)$ is nonnegative and bounded on any rD, $r \in (0, 1)$. Thus there is a sequence $(T_n)_{n \ge 1}$ such that for all $r \in (0, 1)$,

$$\mathbb{E}U(rX_{T \wedge T_n \wedge t}, rY_{T \wedge T_n \wedge t}) \le \mathbb{E}U(rX_0, rY_0) \le 0.$$

We have, by (6.9),

$$U(rX_{T \wedge T_n \wedge t}, rY_{T \wedge T_n \wedge t}) + K_p^p |X_{T \wedge T_n \wedge t}|^p \ge 0,$$

so letting $r \to 1$ and $n \to \infty$ we get, by Fatou's lemma,

$$\mathbb{E}U(X_{T \wedge t}, Y_{T \wedge t}) \le 0.$$

It suffices to apply (6.9) to get the claim. □

Sharpness. Let $B = ((B_t^1, B_t^2))_{t \geq 0}$ be a standard two-dimensional Brownian motion starting from 0 and let $T = \inf\{t : |B_t^2| = 1\}$. Let $p' > p$. Then $T \in L^{p'/2}$, since it is even exponentially integrable and, by the Burkholder-Davis-Gundy inequality, $B \in L^{p'}$. Now, apply Itô's formula to B stopped at T and the function U. We get

$$\mathbb{E}U(B_{T \wedge t}) = \mathbb{E}U(B_0) = 0$$

(the left-hand side is integrable, by (6.9)). By Doob's maximal inequality, $|B|^* \in L^{p'}$, so, using Lebesgue's dominated convergence theorem, we let $t \to \infty$ to obtain $\mathbb{E}U(B_T) = 0$, or $\mathbb{P}(|B_T^2| \geq 1) = K_p^p \mathbb{E}|B_T^1|^p$. It remains to note that the martingales $X = (B_{T \wedge t}^1)_{t \geq 0}$ and $Y = (B_{T \wedge t}^2)_{t \geq 0}$ are orthogonal and Y is differentially subordinate to X: $[X, X]_t = T \wedge t = [Y, Y]_t$. □

On the search of the suitable majorant. In comparison with the continuous-time inequalities considered so far, we are in a more difficult situation: the weak type estimate (6.5) does not have its discrete-time counterpart, for which we have appropriate tools. However, let us try to proceed as in the discrete-time setting and define

$$U^0(x, y) = \sup\{\mathbb{P}(|Y_\infty| \geq 1) - \beta_p^p \mathbb{E}|X_\infty|^p\},$$

where the supremum is taken over all pairs (X, Y) of orthogonal real-valued processes starting from (x, y) such that the process $([X, X]_t - [Y, Y]_t)_{t \geq 0}$ is nondecreasing as a function of t. Arguing as in previous weak type estimates, we see that $U^0(x, y) = 1 - \beta_p^p |x|^p$ if $|y| \geq 1$. What about the values inside the strip $S = \mathbb{R} \times [-1, 1]$? In the discrete setting at this step we did apply the concavity-type property 2°: we did split the domain into a number of regions and for each of them, we have conjectured linearity of U along the line segments of slope -1 or of slope 1. Here the analogue of 2° is played by the condition (6.1), which in the real-valued case reads

$$U_{xx}(x, y)h^2 + U_{yy}(x, y)k^2 \leq -c(x, y)(h^2 - k^2).$$

The key observation here is that if U is harmonic and satisfies $U_{yy} \geq 0$, then the inequality above holds with $c = U_{yy}$. Thus, we *conjecture* that on S, U is the harmonic lift of $V(x, y) = 1_{\{|y| \geq 1\}} - \beta_p^p |x|^p$ and choose β so that $U(0, 0) = 0$. This leads to the function given by (6.7). □

Remark 6.2. Let us mention here that the above formula for U does not work for $p > 2$: the majorization property does not hold for any point of the x-axis, different from $(0, 0)$. To see this, we use (6.8) and note that

$$U(x, 0) = 1 - \frac{K_p^p}{\pi} \int_{\mathbb{R}} \frac{\left|\frac{2}{\pi} \log|s| + x\right|^p}{s^2 + 1} \, ds$$

$$= 1 - \frac{K_p^p}{2\pi} \left[\int_{\mathbb{R}} \frac{\left|\frac{2}{\pi} \log|s| + x\right|^p}{s^2 + 1} \, ds + \int_{\mathbb{R}} \frac{\left|-\frac{2}{\pi} \log|s| + x\right|^p}{s^2 + 1} \, ds \right]$$

$$< 1 - \frac{K_p^p}{2\pi} \left[\int_{\mathbb{R}} \frac{2 \left| \frac{2}{\pi} \log |s| \right|^p}{s^2 + 1} ds + \int_{\mathbb{R}} \frac{2 |x|^p}{s^2 + 1} ds \right]$$

$$= U(0,0) - K_p^p |x|^p = -K_p^p |x|^p.$$

Here in the second equation we have used the substitution $s := 1/s$ and in the above inequality we have exploited the bound

$$|a + b|^p + |a - b|^p < 2|a|^p + 2|b|^p, \qquad a, b \in \mathbb{R}, \ a \neq b,$$

which holds for $p > 2$.

6.3 L^p inequalities for local martingales

6.3.1 Formulation of the result

We turn to the moment estimates. Recall that $p^* = \max\{p, p/(p-1)\}$ for $1 < p < \infty$.

Theorem 6.3. *Let X, Y be two orthogonal local real-valued martingales such that Y is differentially subordinate to X. Then*

$$|||Y|||_p \leq \cot\left(\frac{\pi}{2p^*}\right) |||X|||_p, \qquad 1 < p < \infty, \tag{6.12}$$

and the inequality is sharp. Furthermore, if $1 < p \leq 2$, then X may be taken to be $\mathcal{H}$-valued. If $p \geq 2$, then Y may be taken to be $\mathcal{H}$-valued. Finally, if $p \geq 3$, then X also may be taken to be $\mathcal{H}$-valued.

What happens if $1 < p < 3$ and both processes are Hilbert-space valued? Quite surprisingly, the constants do change; such phenomenon did not take place in the nonorthogonal setting. Let z_p denote the largest positive root of the parabolic cylinder function of parameter p (see Appendix).

Theorem 6.4. *Let X, Y be two orthogonal local martingales taking values in $\mathcal{H}$ such that Y is differentially subordinate to X. Then*

$$|||Y|||_p \leq C_p |||X|||_p, \qquad 1 < p < 3, \tag{6.13}$$

where

$$C_p = \begin{cases} z_p^{-1} & \text{if } 1 < p \leq 2, \\ z_p & \text{if } 2 \leq p \leq 3. \end{cases}$$

The constant C_p is the best possible.

6.3.2 On the method of proof

There are many similarities in the proofs of the above moment inequalities corresponding to different values of p and ranges of the processes. For clarity, we have decided to include here an outline of the reasoning which is common for all of them. This will be useful also in the later estimates presented in this chapter.

Suppose that X, Y are local martingales taking values in an appropriate Hilbert space (that is, in $\mathcal{H}$ or $\mathbb{R}$). We may and do assume that $|||X|||_p < \infty$ and hence also $|||Y|||_p < \infty$, by Burkholder's L^p-inequality. Thus, X and Y are martingales and all we need is to show that

$$\mathbb{E}V(X_t, Y_t) \leq 0 \text{ for all } t \geq 0,$$

where $V(x, y) = |y|^p - C_p^p |x|^p$. We search for the majorant U_p of V_p in the class of functions of the form

$$U_p(x, y) = W_p(|x|, |y|), \tag{6.14}$$

where $W_p : [0, \infty) \times [0, \infty) \to \mathbb{R}$ satisfies

$$|W_p(x, y)| \leq c_p(x^p + y^p) \tag{6.15}$$

for all x, $y \geq 0$ and some c_p depending only on p. The latter condition immediately implies the integrability of $\sup_s |W_p(|X_s|, |Y_s|)|$, by means of Doob's inequality, and will enable us to use Lebesgue's dominated convergence theorem in the limit procedure. We will use the orthogonal version of Theorem 5.3. Let us rephrase the requirements (6.1) and (6.2) in terms of the function W_p. We start with the second condition. It can be rewritten as follows: for all $(x, y) \in \bigcup_i D_i$ and $h \in \mathcal{H}$,

$$\left[W_{pxx}(|x|, |y|) - \frac{W_{px}(|x|, |y|)}{|x|} \right] (x' \cdot h)^2 + \frac{W_{px}(|x|, |y|)}{|x|} |h|^2 \leq 0. \tag{6.16}$$

The formula (6.14) transforms (6.1) into the following: for $(x, y) \in S_i$ with $|x||y| \neq 0$ and any $h, k \in \mathcal{H}$,

$$\begin{aligned} &\left[W_{pxx}(|x|, |y|) - \frac{W_{px}(|x|, |y|)}{|x|} \right] (x' \cdot h)^2 \\ &+ \left[W_{pyy}(|x|, |y|) - \frac{W_{py}(|x|, |y|)}{|y|} \right] (y' \cdot k)^2 \\ &+ \frac{W_{px}(|x|, |y|)|h|^2}{|x|} + \frac{W_{py}(|x|, |y|)|k|^2}{|y|} \leq -c_i(x, y)(|h|^2 - |k|^2). \end{aligned} \tag{6.17}$$

Furthermore, W must satisfy the following conditions: the initial

$$W_p(x, y) \leq 0 \qquad \text{for all } x \geq y \geq 0 \tag{6.18}$$

and the majorization

$$W_p(x, y) \geq y^p - C_p^p x^p \qquad \text{for all } x, y \geq 0. \tag{6.19}$$

Having constructed such a W_p, we let U_p be defined by (6.14). Then Theorem 5.3, together with limiting arguments similar to those presented in the previous chapter, implies

$$\mathbb{E}V_p(X_t, Y_t) \leq \mathbb{E}U_p(X_t, Y_t) \leq \mathbb{E}U_p(X_0, Y_0) \leq 0,$$

as claimed.

Summarizing, in order to establish the inequality (6.13), we need to construct a sufficiently smooth function $W_p : [0, \infty) \times [0, \infty) \to \mathbb{R}$, which satisfies (6.15), (6.16), (6.17), (6.18), (6.19) and for which the corresponding functions c_i obey the bound (6.3).

6.3.3 Proof of Theorem 6.3

Proof of (6.12) *in the case* $1 < p \leq 2$. We may assume that $|||X|||_p < \infty$; then also $|||Y|||_p < \infty$, due to the moment estimate for differentially subordinate local martingales. Let $W_p : [0, \infty) \times [0, \infty) \to \mathbb{R}$ be given by

$$W_p(x, y) = \begin{cases} \alpha_p R^p \cos(p\theta) & \text{if } \theta \leq \frac{\pi}{2p}, \\ y^p - \tan^p \frac{\pi}{2p} x^p & \text{if } \frac{\pi}{2p} < \theta \leq \frac{\pi}{2}, \end{cases} \qquad (6.20)$$

where

$$\alpha_p = -\frac{\sin^{p-1}\left(\frac{\pi}{2p}\right)}{\cos\left(\frac{\pi}{2p}\right)},$$

$x = R\cos\theta$, $y = R\sin\theta$ and $\theta \in [0, \pi/2]$. Let $U_p : \mathcal{H} \times \mathbb{R} \to \mathbb{R}$ be given by (6.14) and let

$$D_1 = \{(x, y) \in \mathcal{H} \times \mathbb{R} : 0 < |y| < C_p|x|\},$$
$$D_2 = \{(x, y) \in \mathcal{H} \times \mathbb{R} : |y| > C_p|x| > 0\}.$$

We shall now verify the properties listed in the previous subsection.

1. *Regularity.* It is easy to see that W_p is continuous, of class C^1 on $(0, \infty) \times (0, \infty)$ and of class C^2 on $(0, \infty) \times (0, \infty) \setminus \{(x, y) : y = \tan(\pi/(2p))x\}$. Furthermore, $W_{py}(x, 0+) = 0$ for all $x \geq 0$, which implies that U_p is of class C^1 on $\{\mathcal{H} \times \mathbb{R} \setminus \{(x, y) : x = 0\}$.

2. *The growth condition* (6.15). This is obvious.

3. *The inequality* (6.16). Assume first that $\theta < \pi/(2p)$. We easily derive that

$$W_{px}(x, y) = p\alpha_p R^{p-1}\cos((p-1)\theta), \qquad W_{pxx}(x, y) = p(p-1)\alpha_p R^{p-2}\cos((p-2)\theta)$$

and the condition can be rewritten in the form

$$p\alpha_p R^{p-2}\left[(p-1)\cos((p-2)\theta)(x' \cdot h)^2 + \frac{\cos((p-1)\theta)}{\cos\theta}(|h|^2 - (x' \cdot h)^2)\right] \leq 0.$$

This is clear: $\alpha_p < 0$ and both summands in the square bracket are nonnegative. Now suppose that $\theta > \pi/(2p)$. Then (6.16) is equivalent to

$$-pC_p^p x^{p-2}\left[(p-1)(x' \cdot h)^2 + (|h|^2 - (x' \cdot h)^2)\right] \le 0,$$

which is obvious.

4. *The concavity property* (6.17). The process Y is real-valued, so $\langle y' \cdot k \rangle^2 = k^2$ for $y \ne 0$. If $\theta < \pi/(2p)$, then the left-hand side of (6.17) equals $I + II$, where

$$I = -c_1(x,y)(|h|^2 - k^2) = p(p-1)\alpha_p R^{p-2}\cos((p-2)\theta)(|h|^2 - k^2),$$

$$II = p\alpha_p R^{p-2}(|h|^2 - (x' \cdot h)^2)\left[\frac{\cos((p-1)\theta)}{\cos\theta} - (p-1)\cos((p-2)\theta)\right].$$

Therefore, it suffices to prove that II is nonpositive, or

$$\cos((p-1)\theta) \ge (p-1)\cos((p-2)\theta)\cos\theta.$$

This is equivalent to $(2-p)\cos((p-2)\theta)\cos\theta + \sin((2-p)\theta)\sin\theta \ge 0$, which is trivial. Finally, if $\theta > \pi/(2p)$, the left-hand side of (6.17) equals

$$-C_p^p\left(p(p-2)x^{p-2}(x' \cdot h)^2 + px^{p-2}|h|^2\right) + p(p-1)y^{p-2}k^2.$$

This does not exceed

$$-pC_p^2(C_p x)^{p-2}|h|^2 + p(p-1)y^{p-2}|k|^2 \le -p(p-1)y^{p-2}(|h|^2 - k^2),$$

where we have used the inequality $y > C_p x$ (that is, $\theta > \pi/(2p)$) and the trivial bound $C_p^2 = \tan^2(\pi/(2p)) \ge 1 \ge p-1$. Thus (6.17) holds, with $c_2(x,y) = p(p-1)y^{p-2}$.

5. *The initial condition.* The inequality $x \ge y \ge 0$ implies $\theta \le \frac{\pi}{4} \le \frac{\pi}{2p}$. Then $\cos(p\theta) \ge 0$ and $W_p(x,y) \le 0$.

6. *The majorization.* It suffices to focus on the case when $0 \le \theta \le \frac{\pi}{2p}$. The majorization is equivalent to

$$\frac{-\sin^{p-1}\left(\frac{\pi}{2p}\right)\left(\cos\left(\frac{\pi}{2p}\right)\right)^{-1}R^p\cos(p\theta) - \sin^p\theta}{\cos^p\theta} \ge -\tan^p\left(\frac{\pi}{2p}\right).$$

Denoting the left-hand side by F, we derive that $F(\pi/(2p)) = 0$, so the inequality $U \ge V$ will be established if we show that $F'(\theta) \le 0$ for $\theta \in (0, \pi/(2p))$. To do this, note that

$$F'(\theta) = \frac{p\sin((p-1)\theta)}{\cos^{p+1}\theta}\left[\frac{\sin^{p-1}\left(\frac{\pi}{2p}\right)}{\cos\left(\frac{\pi}{2p}\right)} - \frac{\sin^{p-1}\theta}{\sin((p-1)\theta)}\right]$$

and the expression in the square bracket is nonpositive. Indeed, it vanishes at $\theta = \pi/(2p)$ and

$$\left(\frac{\sin^{p-1}\theta}{\sin((p-1)\theta)}\right)' = \frac{(p-1)\sin^{p-2}\theta\sin((p-2)\theta)}{\sin^2((p-1)\theta)} \leq 0.$$

Thus, the majorization holds.

7. *The bound* (6.3). This is evident from the formulas for c_1 and c_2 obtained in the proof of item 4 above. $\qquad\square$

Proof of (6.12) *in the case* $2 < p < \infty$. Let $W_p : [0, \infty) \times [0, \infty) \to \mathbb{R}$ be defined as follows:

$$W_p(x, y) = \begin{cases} \alpha_p R^p \cos\left(p\left(\frac{\pi}{2} - \theta\right)\right) & \text{if } \theta \geq \pi/2 - \pi/(2p), \\ y^p - \cot^p\left(\frac{\pi}{2p}\right) x^p & \text{if } \theta < \pi/2 - \pi/(2p). \end{cases}$$

Here

$$\alpha_p = \frac{\cos^{p-1}\left(\frac{\pi}{2p}\right)}{\sin\frac{\pi}{2p}}$$

and, as previously, we have used polar coordinates: $x = R\cos\theta$, $y = R\sin\theta$, with $\theta \in [0, \pi/2]$. Let U_p be given by (6.14); we assume that this function is given on $\mathcal{H} \times \mathcal{H}$ when $p \geq 3$ and on $\mathbb{R} \times \mathcal{H}$ when $2 < p < 3$. Finally, let

$$D_1 = \{(x, y) : |y| > C_p|x| > 0\}, \qquad D_2 = \{(x, y) : 0 < |y| < C_p|x|\},$$

be subsets of $\mathcal{H} \times \mathcal{H}$ or $\mathbb{R} \times \mathcal{H}$, depending on the value of p.

Some properties of W_p are gathered in the lemma below.

Lemma 6.2.

(i) *For any* $x, y > 0$ *with* $y > \cot(\pi/(2p))x > 0$ *we have*

$$W_{pyy}(x, y)y \geq W_{py}(x, y) \tag{6.21}$$

and, if $p \geq 3$,

$$W_{pxx}(x, y)x \geq W_{px}(x, y). \tag{6.22}$$

(ii) *For any* $x, y > 0$ *with* $y < \cot(\pi/(2p))x$ *we have*

$$W_{pxx}(x, y)x \leq W_{px}(x, y), \tag{6.23}$$

$$W_{pyy}(x, y)y \geq W_{py}(x, y) \tag{6.24}$$

and

$$xW_{pyy}(x, y) + W_{px}(x, y) \leq 0. \tag{6.25}$$

Proof. (i) We have

$$W_{py}(x, y) = p\alpha_p R^{p-1} \cos\left((p-1)\left(\frac{\pi}{2} - \theta\right)\right)$$

and

$$W_{pyy}(x, y) = p(p-1)\alpha_p R^{p-2} \cos\left((p-2)\left(\frac{\pi}{2} - \theta\right)\right),$$

so (6.21) can be rewritten in the form

$$(p-2)\cos\left((p-2)\left(\frac{\pi}{2} - \theta\right)\right)\cos\left(\frac{\pi}{2} - \theta\right) + \sin\left((p-2)\left(\frac{\pi}{2} - \theta\right)\right)\sin\left(\frac{\pi}{2} - \theta\right) \geq 0,$$

for $\theta \in (\pi/2 - \pi/(2p), \pi/2)$. Denoting the left-hand side by $F(\theta)$, we easily check that

$$F'(\theta) = (p-1)(p-3)\sin\left((p-2)\left(\frac{\pi}{2} - \theta\right)\right)\sin\theta.$$

To show that F is nonnegative, suppose first that $2 < p < 3$. Then F is decreasing and it suffices to note that $F(\pi/2) = p - 2 > 0$. On the other hand, if $p \geq 3$, then F is nondecreasing and

$$F\left(\frac{\pi}{2} - \frac{\pi}{2p}\right) = (p-3)\cos\left((p-2)\frac{\pi}{2p}\right)\cos\frac{\pi}{2p} + \cos\left((p-3)\frac{\pi}{2p}\right) \geq 0,$$

so (6.21) follows. To show (6.22), we derive that

$$W_{px}(x, y) = -\alpha_p p R^{p-1} \sin\left((p-1)\left(\frac{\pi}{2} - \theta\right)\right)$$

and

$$W_{pxx}(x, y) = -\alpha_p p(p-1) R^{p-2} \cos\left((p-2)\left(\frac{\pi}{2} - \theta\right)\right), \qquad (6.26)$$

so (6.22) is equivalent to

$$-(p-1)\cos\left((p-2)\left(\frac{\pi}{2} - \theta\right)\right)\sin\left(\frac{\pi}{2} - \theta\right) + \sin\left((p-1)\left(\frac{\pi}{2} - \theta\right)\right) \geq 0,$$

or

$$(p-2)\cos\left((\mu-2)\left(\frac{\pi}{2} - \theta\right)\right)\sin\left(\frac{\pi}{2} - \theta\right) + \sin\left((p-2)\left(\frac{\pi}{2} - \theta\right)\right)\cos\left(\frac{\pi}{2} - \theta\right) \geq 0$$

for $\theta \in (\pi/2 - \pi/(2p), \pi/2)$. This holds true, since the two sides are equal for $\theta = \pi/2$ and the derivative of the left-hand side equals

$$(3-p)(p-1)\sin\left((p-2)\left(\frac{\pi}{2} - \theta\right)\right)\sin\left(\frac{\pi}{2} - \theta\right) \leq 0.$$

This yields the claim. Note that (6.22) *does not* hold for $2 < p < 3$. This is the reason why the constant $\cot(\pi/(2p))$ is no longer sufficient in (6.12) for these values of p in the case when both processes are $\mathcal{H}$-valued: this will become clear from the reasoning below.

(ii) If $y < \cot(\pi/(2p))x$, then

$$xW_{pxx}(x,y) - W_{px}(x,y) = -p(p-2)\cot^p\left(\frac{\pi}{2p}\right)x^{p-1} \le 0$$

and

$$yW_{pyy}(x,y) - W_{py}(x,y) = p(p-2)y^{p-1} \ge 0.$$

Furthermore,

$$xW_{pyy}(x,y) + W_{px}(x,y) = px\left[(p-1)y^{p-2} - \cot^p\left(\frac{\pi}{2p}\right)x^{p-2}\right]$$

$$\le pxy^{p-2}\left(p - 1 - \cot^2\frac{\pi}{2p}\right)$$

$$= pxy^{p-2}\left[p - \left(\sin\frac{\pi}{2p}\right)^{-2}\right]$$

and the expression in the square brackets is nonpositive. This follows immediately from the elementary inequality

$$\sin^2\left(\frac{\pi}{2}x\right) \le x, \qquad x \in [0, 1/2],$$

which can be verified readily. $\square$

1. *Regularity.* It is easily checked that W_p is continuous, of class C^1 on $(0, \infty) \times (0, \infty)$ and of class C^2 on $(0, \infty) \times (0, \infty) \setminus \{(x,y) : y = C_p x\}$. Furthermore, $W_{py}(x, 0) = 0$ for any $x \ge 0$ and $W_{px}(x, 0) = 0$ for any $x \ge 0$. This implies that U_p, given by (6.14), is of class C^1.

2. *The growth condition* (6.15). This is evident.

3. *The inequality* (6.16). If $(x, y) \in \mathcal{H} \times \mathcal{H}$ satisfies $|y| > \cot(\pi/(2p))|x| > 0$, then by (6.22) and (6.26) we have that

$$\left[W_{pxx}(|x|, |y|) - \frac{W_{px}(|x|, |y|)}{|x|}\right](x' \cdot h)^2 + \frac{W_{px}(|x|, |y|)}{|x|}|h|^2 \le W_{pxx}(|x|, |y|)|h|^2 \le 0.$$

On the other hand, if $0 < |y| < \cot(\pi/(2p))|x|$, then, by (6.23),

$$\left[W_{pxx}(|x|, |y|) - \frac{W_{px}(|x|, |y|)}{|x|}\right](x' \cdot h)^2 + \frac{W_{px}(|x|, |y|)}{|x|}|h|^2$$

$$\le \frac{W_{px}(|x|, |y|)}{|x|}|h|^2 = -pC_p^p|x|^{p-2}|h|^2 \le 0,$$

as needed.

4. *The condition* (6.17). Suppose first that $2 < p < 3$. Then X takes values in $\mathbb{R}$, so the inequality simplifies. On the set $\{(x, y) \in \mathbb{R}^2_+ : y > \cot(\pi/(2p)x\}$, the function W_p is harmonic, so for $(x, y) \in D_1$,

$$W_{pxx}(|x|, |y|)|h|^2 + \left[W_{pyy}(|x|, |y|) - \frac{W_{py}(|x|, |y|)}{|y|} \right] (y' \cdot k)^2 + \frac{W_{py}(|x|, |y|)|k|^2}{|y|}$$
$$\leq -W_{pyy}(|x|, |y|)|h|^2 + W_{pyy}(x, y)|k|^2 = -c(x, y)(|h|^2 - |k|^2), \qquad (6.27)$$

where in the latter passage we have used (6.21). This is the desired bound, since $c = W_{pyy} > 0$. If we have $0 < |y| < \cot(\pi/(2p))|x|$, then by (6.23), (6.24) and (6.25) the left-hand side is not larger than

$$\frac{W_{px}(|x|, |y|)}{|x|}|h|^2 + W_{pyy}(|x|, |y|)|k|^2 \leq -W_{pyy}(|x|, |y|)(|h|^2 - |k|^2). \qquad (6.28)$$

Now let $p \geq 3$. If $|y| > \cot(\pi/(2p))|x| > 0$, then by (6.22) and (6.21), the left-hand side of (6.17) does not exceed

$$W_{pxx}(|x|, |y|)|h|^2 + W_{pyy}(|x|, |y|)|k|^2 \leq W_{pxx}(|x|, |y|)(|h|^2 - |k|^2), \qquad (6.29)$$

since $W_{pxx}(|x|, |y|) \leq 0$ and W_p is harmonic on $\{(x, y) : y > \cot(\pi/(2p)x > 0\}$. If $(x, y) \in D_2$, then we repeat the reasoning from the case $2 < p < 3$.

5. *The condition* (6.18). This is obvious: for $x \geq y \geq 0$ we have

$$W_p(x, y) = y^p - \cot^p\left(\frac{\pi}{2p}\right) x^p \leq x^p \left(1 - \cot^p\left(\frac{\pi}{2p}\right)\right) \leq 0.$$

6. *The majorization* (6.19). Clearly, it suffices to show this inequality on the set $\{(x, y) : \theta > \pi/2 - \pi/(2p)\}$, where it can be rewritten in the form

$$\sin^p \theta - \cot^p\left(\frac{\pi}{2p}\right) \cos^p \theta \leq \frac{\cos^{p-1}\left(\frac{\pi}{2p}\right)}{\sin\frac{\pi}{2p}} \cos\left(p\left(\frac{\pi}{2} - \theta\right)\right)$$

or, after the substitution $\beta = \pi/2 - \theta \in [0, \pi/(2p))$,

$$\frac{\cos^p \beta - \cot^p(\pi/(2p)) \sin^p \beta}{\cos(p\beta)} \leq \frac{\cos^{p-1}\left(\frac{\pi}{2p}\right)}{\sin\frac{\pi}{2p}}.$$

Since the two sides become equal to one another when we let $\beta \to \pi/(2p)$, it suffices to show that the left-hand side, as a function of β, is nondecreasing on $(0, \pi/(2p))$. Differentiating, we see that this is equivalent to

$$\frac{\cot^{p-1} \beta}{\cot((p-1)\beta)} \geq \cot^p\left(\frac{\pi}{2p}\right).$$

The two sides above are equal to one another for $\beta = \pi/(2p)$, so we will be done if we show that the left-hand side is decreasing as a function of $\beta \in (0, \pi/(2p))$. If we calculate the derivative, we see that this is equivalent to the inequality $\sin(2\beta) < \sin(2(p-1)\beta)$. But this follows immediately from the bounds $0 \leq 2\beta \leq 2(p-1)\beta < \pi - 2\beta$.

  7. *The bound* (6.3). It suffices to look at (6.27), (6.28) and (6.29): it is clear that the estimate (6.3) is satisfied. $\square$

Sharpness. We will only deal with the case $1 < p \leq 2$, for remaining p's the argumentation is similar. Let κ be a fixed positive number smaller than $\tan(\pi/(2p))$. Consider a two-dimensional Brownian motion $B = (B_t^{(1)}, B_t^{(2)})_{t \geq 0}$ starting from $(1, 0)$ and let τ be the first exit time of B from the set $E = \{(x, y) : |y| \leq \kappa|x|\}$; it is well known that $\tau < \infty$ with probability 1. Put $U_p(x, y) = W_p(x, |y|)$ for $x \geq 0$ and $y \in \mathbb{R}$, where W_p is given by (6.20). We have that U_p is harmonic on the complement of E and thus, by Itô's formula,

$$U_p(1, 0) = \mathbb{E}U_p(B_{\tau \wedge t}^{(1)}, B_{\tau \wedge t}^{(2)}) = \mathbb{E}\left[(B_{\tau \wedge t}^{(1)})^p U_p(1, B_{\tau \wedge t}^{(2)}/B_{\tau \wedge t}^{(1)})\right] \leq U_p(1, \kappa)\mathbb{E}(B_{\tau \wedge t}^{(1)})^p.$$

Here we have used the fact that $U_p(1, \cdot)$ is increasing on $[0, \infty)$. We see that $U_p(1, \kappa) < 0$, so the above estimate can be rewritten in the form

$$\mathbb{E}(B_{\tau \wedge t}^{(1)})^p \leq U_p(1, 0)/U_p(1, \kappa).$$

Now let $X_t = B_{\tau \wedge t}^{(1)}$ and $Y_t = B_{\tau \wedge t}^{(2)}$. Then Y is differentially subordinate to X and both processes are orthogonal. By the last inequality, we have $||X||_p < \infty$ and, by (6.12), $||Y||_p < \infty$. The final observation is that, by the definition of τ,

$$||Y||_p = ||B_\tau^{(2)}||_p = \kappa||B_\tau^{(1)}||_p = \kappa||X||_p.$$

But κ can be taken arbitrarily close to $\tan(\pi/(2p))$. Thus the inequality (6.12) is sharp. $\square$

On the search of the suitable majorant. As usual, we restrict ourselves to $\mathcal{H} = \mathbb{R}$. Furthermore, we focus on the case $1 < p \leq 2$; for $p > 2$ the reasoning is essentially the same. Suppose that the optimal constant in the moment estimate equals β_p and set $V(x, y) = |y|^p - \beta^p|x|^p$. It is instructive to come back to the moment estimate for differentially subordinated martingales (without the orthogonality assumption). Recall the least special function corresponding to this inequality: when $1 < p \leq 2$, then

$$\overline{U}(x, y) = \begin{cases} c_p(|y| - (p-1)^{-1}|x|)(|x| + |y|)^{p-1} & \text{if } |y| \leq (p-1)^{-1}|x|, \\ |y|^p - (p-1)^{-p}|x|^p & \text{if } |y| > (p-1)^{-1}|x|, \end{cases}$$

for some appropriate constant c_p. It is natural to conjecture that the special function U_p in the orthogonal case should also be of this form: that is, we impose the condition

(A1) $U(x, y) = V(x, y)$ when $|y| \geq \beta_p|x|$.

The second assumption is the regularity condition

(A2) U is of class C^1 on $(0,\infty) \times (0,\infty)$.

What about the value $U(x,y)$ on the set $D = \{(x,y) : |y| < \gamma_p|x|\}$? Arguing as in the case of the weak type inequality, we guess that

(A3) U is harmonic inside D

(in particular, by (A2), U is of class C^1 on $(0,\infty) \times \mathbb{R}$). Finally, since V is homogeneous of order p, we assume that

(A4) $U(\lambda x, \lambda y) = |\lambda|^p U(x,y)$ for x, y, $\lambda \in \mathbb{R}$.

The three conditions above lead to the special function above. To see this, note that by (A4), U has the form

$$U(x,y) = R^p g(\theta),$$

where R and θ are the corresponding polar coordinates as above and g is an even function on $[-\pi/2, \pi/2]$. The assumption (A3) is equivalent to $g''(\theta) + p^2 g(\theta) = 0$ on the interval $[-\arctan\gamma_p, \arctan\gamma_p]$, so $g(\theta) = a\sin(p\theta) + b\cos(p\theta)$ there. Since g is even, we have $a = 0$. In addition, by (A1), $U(x, \beta_p x) = 0$ and hence $\cos(p\arctan\beta_p) = 0$. Since $p\arctan\beta_p < \pi$, we deduce that $\beta_p = \tan(\pi/(2p))$ and by (A2),

$$b = -\frac{\sin^{p-1}\left(\frac{\pi}{2p}\right)}{\cos\left(\frac{\pi}{2p}\right)}.$$

This gives the function U_p considered in the above proof. $\qquad\square$

6.3.4 Proof of Theorem 6.4

We shall need here some information on the parabolic cylinder functions: it can be found in the Appendix below.

Proof of (6.13) for $1 < p \leq 2$. Let ϕ_p be given by (A.4) in the Appendix. Introduce $W_p : [0,\infty) \times [0,\infty) \to \mathbb{R}$ by the formula

$$W_p(x,y) = \begin{cases} \alpha_p y^p \phi_p(x/y) & \text{if } y \leq z_p^{-1}x, \\ y^p - z_p^{-p}x^p & \text{if } y \geq z_p^{-1}x, \end{cases}$$

where

$$\alpha_p = -(z_p \phi_{p-1}(z_p))^{-1}. \tag{6.30}$$

Furthermore, let

$$D_1 = \{(x,y) \in \mathcal{H} \times \mathcal{H} : |y| > z_p^{-1}|x| > 0\}$$

and

$$D_2 = \{(x,y) \in \mathcal{H} \times \mathcal{H} : 0 < |y| < z_p^{-1}|x|\}.$$

Next we verify the necessary properties of W_p.

1. *Regularity.* It is easy to see that W_p is continuous, of class C^1 on the set $(0, \infty) \times (0, \infty)$ and of class C^2 on the set $(0, \infty) \times (0, \infty) \setminus \{(x, y) : y = z_p^{-1}x\}$. In consequence, the function U_p defined by (3.153) has the required smoothness. It can also be verified readily that the first-order derivative of U_p is bounded on bounded sets not containing $0 \in \mathcal{H} \times \mathcal{H}$.

2. *The growth condition* (6.15). This follows immediately from the asymptotics (A.7).

3. *The condition* (6.16). If $z_p|y| > |x| > 0$, then we have

$$\left[W_{pxx}(|x|, |y|) - \frac{W_{px}(|x|, |y|)}{|x|} \right] (x' \cdot h)^2 = p(2 - p)z_p^{-p}|x|^{p-2}(x' \cdot h)^2$$
$$\leq p(2 - p)z_p^{-p}|x|^{p-2}|h|^2 \tag{6.31}$$

and

$$\frac{W_{px}(|x|, |y|)}{|x|}|h|^2 = -pz_p^{-p}|x|^{p-2}|h|^2, \tag{6.32}$$

so (6.2) is valid. When $|x| > z_p|y| > 0$, we compute that

$$\left[W_{pxx}(|x|, |y|) - \frac{W_{px}(|x|, |y|)}{|x|} \right] (x' \cdot h)^2 = \alpha_p|y|^{p-2} \left[\phi_p''(r) - r^{-1}\phi_p'(r) \right] (x' \cdot h)^2$$
$$\leq \alpha_p|y|^{p-2} \left[\phi_p''(r) - r^{-1}\phi_p'(r) \right] |h|^2, \tag{6.33}$$

where we have used the notation $r = |x|/|y|$. Furthermore,

$$\frac{W_{px}(|x|, |y|)}{|x|}|h|^2 = \alpha_p|y|^{p-2}r^{-1}\phi_p'(r)|h|^2. \tag{6.34}$$

Adding this to (6.33) yields (6.2), since $\phi_p''(r) \geq 0$: see Lemma A.11 (v).

4. *The condition* (6.17). If $z_p|y| > |x| > 0$, then

$$\left[W_{pyy}(|x|, |y|) - \frac{W_{py}(|x|, |y|)}{|y|} \right] (y' \cdot k)^2 = p(p - 2)|y|^{p-2}(y' \cdot k)^2 \leq 0$$

and

$$\frac{W_{py}(|x|, |y|)}{|y|}|k|^2 = p|y|^{p-2}|k|^2.$$

Therefore, combining this with (6.31) and (6.32) we see that the left-hand side of (6.17) is not larger than

$$-p|y|^{p-2}(|h|^2 - |k|^2) - p[(p - 1)z_p^{-p}|x|^{p-2} - |y|^{p-2}]|h|^2 \leq -p|y|^{p-2}(|h|^2 - |k|^2), \tag{6.35}$$

as needed. Here in the last passage we have used the estimates $z_p^{-1}|x| < |y|$ and $z_p^2 \leq p-1$ (see Corollary 3 in the Appendix). On the other hand, if $0 < z_p|y| < |x|$, then consider the function F_p given by (A.9). A little calculation yields

$$\left[W_{pyy}(|x|,|y|) - \frac{W_{py}(|x|,|y|)}{|y|} \right] (y' \cdot k)^2 = \alpha_p |y|^{p-2} F_p(r)(y' \cdot k)^2,$$

which is nonpositive thanks to Lemma A.12 (here $r = |x|/|y|$, as before). Furthermore, by (A.5),

$$\frac{W_{py}(|x|,|y|)}{|y|} = \alpha_p |y|^{p-2} \left[p\phi_p(r) - r\phi_p'(r) \right] |k|^2 = -\alpha_p |y|^{p-2} \phi_p''(r)|k|^2,$$

which, combined with (6.33) and (6.34) implies that the left-hand side of (6.17) does not exceed

$$\alpha_p |y|^{p-2} \phi_p''(r)(|h|^2 - |k|^2).$$

5. *The inequality* (6.18). Note that ϕ_p is increasing (by Lemma A.11 (v)) and $z_p \leq 1$ (by Corollary 3). Thus, for $0 < y \leq x$,

$$W_p(x,y) = \alpha_p y^p \phi_p(x/y) \leq \alpha_p y^p \phi_p(1) \leq 0.$$

6. *The inequality* (6.19). This is obvious for $y \geq z_p^{-1}x$, so we focus on the case $y < z_p^{-1}x$. Then the majorization is equivalent to

$$\alpha_p \phi_p(s) \leq 1 - z_p^{-p} s^p \qquad \text{for } s > z_p.$$

Both sides are equal when $s = z_p$, so it suffices to establish an appropriate estimate for the derivatives: $\alpha_p \phi_p'(s) \leq -pz_p^{-p} s^{p-1}$ for $s > z_p$. We see that again the two sides are equal to one another when $s = z_p$; thus we will be done if we show that the function $s \mapsto \phi_p'(s)/s^{p-1}$ is nondecreasing on (z_p, ∞). After differentiation, this is equivalent to

$$\phi_p''(s)s - (p-1)\phi_p'(s) \geq 0 \qquad \text{on } (z_p, \infty),$$

or, by Lemma A.11 (i), $\phi_p'''(s) \leq 0$ for $s > z_p$. This is shown in the part (v) of that lemma.

7. *The bound* (6.3). In view of the above reasoning, we have to take $c_1(x,y) = p|y|^{p-2}$ and $c_2(x,y) = -\alpha_p |y|^{p-2} \phi_p''(r)$. It is evident that the estimate (6.3) is valid for this choice of c_i, $i = 1, 2$. $\qquad \square$

Proof of (6.13) *for* $2 < p < 3$. Consider ϕ_p given by (A.4) below. Let $W_p: [0,\infty) \times [0,\infty) \to \mathbb{R}$ be given by

$$W_p(x,y) = \begin{cases} \alpha_p x^p \phi_p(y/x) & \text{if } y \geq z_p x, \\ y^p - z_p^p x^p & \text{if } y \leq z_p x, \end{cases}$$

where

$$\alpha_p = (z_p \phi_{p-1}(z_p))^{-1}.$$

Let

$$D_1 = \{(x, y) \in \mathcal{H} \times \mathcal{H} : 0 < |y| < z_p|x|\}$$

and

$$D_2 = \{(x, y) \in \mathcal{H} \times \mathcal{H} : |y| > z_p|x| > 0\}.$$

As in the previous case, we will verify that W_p enjoys the requirements listed in Subsection 3.2 above.

1. *Regularity.* Clearly, we have that W_p is continuous, of class C^1 on $(0, \infty) \times (0, \infty)$ and of class C^2 on $(0, \infty) \times (0, \infty) \smallsetminus \{(x, y) : y = z_p x\}$. Hence the function U_p given by (3.153) has the necessary smoothness. In addition, it is easy to see that the first-order derivative of U_p is bounded on bounded sets.

2. *The growth condition* (6.15). This is guaranteed by the asymptotics (A.7).

3. *The condition* (6.16). Note that if $0 < |y| < z_p|x|$, then

$$\left[W_{pxx}(|x|, |y|) - \frac{W_{px}(|x|, |y|)}{|x|} \right] (x' \cdot h)^2 = -p(p-2) z_p^p |x|^{p-2} (x' \cdot h)^2 \le 0 \quad (6.36)$$

and

$$\frac{W_{px}(|x|, |y|)}{|x|} |h|^2 = -p z_p^p |x|^{p-2} |h|^2 \le 0, \tag{6.37}$$

so (6.2) follows. Suppose then that $|y| > z_p|x| > 0$ and recall the function F_p introduced in Lemma A.12. By means of this lemma, after some straightforward computations, one gets

$$\left[W_{pxx}(|x|, |y|) - \frac{W_{px}(|x|, |y|)}{|x|} \right] (x' \cdot h)^2 = \alpha_p |x|^{p-2} F_p(r) (x' \cdot h)^2 \le 0, \quad (6.38)$$

where we have set $r = |y|/|x|$. Moreover, by (A.5),

$$\frac{W_{px}(|x|, |y|)}{|x|} |h|^2 = \alpha_p \left[p\phi_p(r) - r\phi_p'(r) \right] |x|^{p-2} |h|^2$$

$$= -\alpha_p \phi_p''(r) |x|^{p-2} |h|^2 \tag{6.39}$$

is nonpositive; this completes the proof of (6.2).

4. *The condition* (6.17). If $0 < |y| < z_p|x|$, then

$$\left[W_{pyy}(|x|, |y|) - \frac{W_{py}(|x|, |y|)}{|y|} \right] (y' \cdot k)^2 = p(p-2)|y|^{p-2}(y' \cdot k)^2$$

$$\le p(p-2)|y|^{p-2}|k|^2$$

and

$$\frac{W_{py}(|x|,|y|)}{|y|}|k|^2 = p|y|^{p-2}|k|^2.$$

Combining this with (6.36) and (6.37) we see that the left-hand side of (6.17) can be bounded from above by

$$p(p-1)|y|^{p-2}|k|^2 - pz_p^p|x|^{p-2}|h|^2 \le p(p-1)z_p^{p-2}|x|^{p-2}|k|^2 - pz_p^p|x|^{p-2}|h|^2$$
$$\le -pz_p^p|x|^{p-2}(|h|^2 - |k|^2). \tag{6.40}$$

Here in the latter passage we have used Corollary 3. If $|y| > z_p|x| > 0$, then, again using the notation $r = |y|/|x|$,

$$\left[W_{pyy}(|x|,|y|) - \frac{V_{py}(|x|,|y|)}{|y|}\right](y' \cdot k)^2 = \alpha_p|x|^{p-2}\left[\phi_p''(r) - r^{-1}\phi_p'(r)\right](y' \cdot k)^2$$
$$\le \alpha_p|x|^{p-2}\left[\phi_p''(r) - r^{-1}\phi_p'(r)\right]|k|^2$$

and

$$\frac{W_{py}(|x|,|y|)}{|y|}|k|^2 = \alpha_p|x|^{p-2}r^{-1}\phi_p'(r)|k|^2.$$

Combining this with (6.38) and (6.39), we get that the left-hand side of (6.17) does not exceed $-\alpha_p|x|^{p-2}\phi_p''(r)(|h|^2 - |k|^2)$.

5. *The inequality* (6.18). This is obvious: $W_p(x,y) \le x^p(1 - z_p^p) \le 0$ if $x \ge y \ge 0$.

6. *The majorization* (6.19). This can be established exactly in the same manner as in the previous case. The details are left to the reader.

7. *The bound* (6.3). By the above considerations, we are forced to take $c_1(x,y) = pz_p^p|x|^{p-2}$ and $c_2(x,y) = \alpha_p|x|^{p-2}\phi_p''(r)$ and it is clear that the condition is satisfied. □

Sharpness. We shall show that (6.12) yields some moment inequalities for stopping times of Brownian motion. Suppose first that $1 < p \le 2$ and let $B = (B^{(1)}, B^{(2)})$ be a standard two-dimensional Brownian motion. Fix a positive integer n, a stopping time τ of B satisfying $\tau \in L^{p/2}$ and consider the martingales X, Y given by $X_t = (B^{(1)}_{\tau \wedge t}, 0, 0, \ldots)$ and

$$Y_t^{(n)} = (B^{(2)}_{\tau \wedge t \wedge 2^{-n}}, B^{(2)}_{\tau \wedge t \wedge (2 \cdot 2^{-n})} - B^{(2)}_{\tau \wedge t \wedge 2^{-n}}, B^{(2)}_{\tau \wedge t \wedge (3 \cdot 2^{-n})} - B^{(2)}_{\tau \wedge t \wedge (2 \cdot 2^{-n})}, \ldots) \tag{6.41}$$

for $t \ge 0$. That is, the kth coordinate of $Y_t^{(n)}$ is equal to the increment

$$B^{(2)}_{\tau \wedge t \wedge (k \cdot 2^{-n})} - B^{(2)}_{\tau \wedge t \wedge ((k-1) \cdot 2^{-n})},$$

for $k = 1, 2, \ldots$. Obviously, X and $Y^{(n)}$ are orthogonal, since $B^{(1)}$ and $B^{(2)}$ are independent. In addition, Y is differentially subordinate to X, because $[X, X]_t =$

$[Y, Y]_t = \tau \wedge t$ for all $t \geq 0$. Therefore, we infer from the inequality (6.13) that for any $t \geq 0$,

$$\|Y_t^{(n)}\|_p \leq C_p \|X_t\|_p. \tag{6.42}$$

On the other hand, it is well known (see, e.g., proof of Theorem 1.3 in [182]) that for any $t \geq 0$,

$$|Y_t^{(n)}|^2 = \sum_{k=1}^{\infty} |B_{\tau \wedge t \wedge (k \cdot 2^{-n})}^{(2)} - B_{\tau \wedge t \wedge ((k-1) \cdot 2^{-n})}^{(2)}|^2 \xrightarrow{n \to \infty} [B^{(2)}, B^{(2)}]_{\tau \wedge t} = \tau \wedge t$$

in L^2. Therefore, letting $n \to \infty$ in (6.42) yields

$$\|(\tau \wedge t)^{1/2}\|_p \leq C_p \|B_{\tau \wedge t}^{(1)}\|_p \leq C_p \|B_\tau^{(1)}\|_p$$

and, consequently,

$$\|\tau^{1/2}\|_p \leq C_p \|B_\tau^{(1)}\|_p. \tag{6.43}$$

This implies $C_p \geq z_p^{-1}$: see Davis [59]. The case $2 < p < 3$ is dealt with exactly in the same manner.   □

On the search of the suitable majorant. We will focus on the case $1 < p \leq 2$. The function U_p considered for $\mathbb{R}$-valued processes Y does not work if we allow vector-valued processes. This gives the following hint: to obtain the optimal constant in the moment inequality in the Hilbert-space-valued setting, we should take Y "as infinite-dimensional as possible". This immediately suggests the processes of the form (6.41) and letting $n \to \infty$ leads to the estimate (6.43). This can be rephrased in the language of optimal stopping theory. Namely, let β_p be a fixed positive number. Put $G(t, x) = t^{p/2} - \beta_p^p |x|^p$ and let

$$U(t, x) = \sup_\tau \mathbb{E} G(t + \tau, x + B_\tau),$$

be the corresponding value function. Here B is a standard Brownian motion in $\mathbb{R}$, starting from 0, and the supremum is taken over all stopping times τ of B, which satisfy the condition $\mathbb{E}\tau^{p/2} < \infty$. The question is: what is the smallest possible value of β_p such that $U(0, 0) \leq 0$? And, for this particular value of β_p, what is the explicit formula for U?

We begin by observing that the function U satisfies the homogeneity condition

$$U(\lambda^2 t, \lambda x) = |\lambda|^p U(t, x), \qquad \text{for } (t, x) \in [0, \infty) \times \mathbb{R}, \ \lambda \neq 0, \tag{6.44}$$

which follows from scaling properties of Brownian motion. Next, recalling the form of the special function in the real-valued setting (see the previous subsection), we conjecture the following.

(A1) U is of class C^1 on $(0, \infty) \times \mathbb{R}$,
(A2) $U(t, x) = G(t, x)$ when $t^{1/2} \geq \beta_p |x|$,
(A3) $U_t + \frac{1}{2} U_{xx} = 0$ when $t^{1/2} < \beta_p |x|$.

Some remarks, which relate (A1), (A2) and (A3) to the theory of optimal stopping, are in order. From the general theory of optimal stopping of Markov processes (see, e.g., [170]), the state space $[0, \infty) \times \mathbb{R}$ is split into two sets: the stopping region $U = G$ and the continuation set $U > G$. Furthermore, on the continuation set, the value function U lies in the null set of the infinitesimal operator of the underlying Markov process. These two facts are precisely the assumptions (A2) and (A3): in the mean time we conjecture, using the homogeneity property, that the stopping set is given by $\{(t, x) : t^{1/2} \geq \beta_p |x|\}$. Finally, (A1) is related to the so-called principle of smooth fit, which states that U matches at the common boundary of the stopping set and the continuation set in a smooth way: the partial derivatives are continuous on the state space (see [170] for more on this).

Now we are ready to determine the formula for U. Directly from (6.44), (A1) and the form of the stopping problem, we infer that

$$U(t, x) = t^{p/2} g(x t^{-1/2}),$$

where $g : \mathbb{R} \to \mathbb{R}$ is a C^1 even function. By (A3), it satisfies the differential equation

$$g''(s) - s g(s) + p g(s) = 0 \qquad \text{for } |s| \leq 1/\beta_p,$$

which is connected to the parabolic cylinder functions: see Appendix. Furthermore, comparing the left-hand and right-hand limits of g and g' at $1/\beta_p$, we obtain the equalities $g(1/\beta_p) = 0$ and $g'(1/\beta_p) = -p\beta_p$. Hence, we take $1/\beta_p = z_p$ to be the largest zero of the parabolic cylinder function of parameter p and, for $t^{1/2} < \beta_p |x|$,

$$U(t, x) = t^{p/2} \cdot \frac{-p\beta_p \phi_p(|x| t^{-1/2})}{\phi'_p(z_p)} = -t^{p/2} \frac{\phi_p(|x| t^{-1/2})}{z_p \phi_{p-1}(z_p)},$$

where ϕ_p is given by (A.4). This leads directly to the function W_p studied above.

$\square$

6.4 Inequalities for moments of different order

6.4.1 Formulation of the result

It is natural to ask what is the optimal constants $K_{p,q}$ in the estimates

$$|||Y|||_p \leq K_{p,q} |||X|||_q,$$

for real-valued orthogonal local martingales X, Y such that Y is differentially subordinate to X. We will show the following partial result in this direction. Let

$$K_{p,\infty} = \begin{cases} 1 & \text{if } 1 < p \leq 2, \\ \left[\frac{2^{p+2}}{\pi^{p+1}} \Gamma(p+1) \sum_{k=0}^{\infty} \frac{(-1)^k}{(2k+1)^{p+1}} \right]^{1/p} & \text{if } p > 2 \end{cases}$$

and, for $1 < p < \infty$,

$$K_{1,p} = K_{p/(p-1),\infty}.$$

Theorem 6.5. *Let X and Y be two real-valued orthogonal local martingales such that Y is differentially subordinate to X. Then for $1 < p < \infty$,*

$$|||Y|||_1 \leq K_{1,p}|||X|||_p \tag{6.45}$$

and

$$|||Y|||_p \leq K_{p,\infty}|||X|||_\infty. \tag{6.46}$$

Both inequalities are sharp.

The case when $1 < p < q < 2$ or $2 < p < q < \infty$ seems to be much more difficult and the corresponding best constants are not known.

6.4.2   Proof of Theorem 6.5

Proof of (6.46). First, note that we may restrict ourselves to the case $p > 2$. Indeed, if $1 \leq p \leq 2$, then

$$|||Y|||_p \leq |||Y|||_2 \leq |||X|||_2 \leq |||X|||_\infty,$$

and, obviously, the equality $|||Y|||_p = |||X|||_\infty$ holds for the constant pair $(X, Y) \equiv (1, 1)$ of orthogonal martingales. So, assume that $p > 2$. The construction of an appropriate special function U is very similar to that presented in the proof of the weak type estimate. Let $\mathcal{H} = \mathbb{R} \times (0, \infty)$ denote the upper half-plane and let $\mathcal{U} = \mathcal{U}_p : \mathcal{H} \to \mathbb{R}$ be given by the Poisson integral

$$\mathcal{U}(\alpha, \beta) = \frac{2^p}{\pi^{p+1}} \int_{-\infty}^{\infty} \frac{\beta \, |\log |t||^p}{(\alpha - t)^2 + \beta^2} dt.$$

The function $\mathcal{U}$ is harmonic on $\mathcal{H}$ and satisfies

$$\lim_{(\alpha,\beta) \to (z,0)} \mathcal{U}(\alpha, \beta) = \left| \frac{2}{\pi} \log |z| \right|^p, \qquad z \neq 0.$$

Let S denote the strip $(-1, 1) \times \mathbb{R}$ and consider the conformal mapping $\varphi(z) = ie^{-i\pi z/2}$, or

$$\varphi(x, y) = \left(e^{\pi y/2} \sin\left(\frac{\pi}{2}x\right), e^{\pi y/2} \cos\left(\frac{\pi}{2}x\right) \right), \qquad (x, y) \in \mathbb{R}^2.$$

One easily verifies that φ maps S onto $\mathcal{H}$. Define $U = U_p$ on S by

$$U(x, y) = \mathcal{U}(\varphi(x, y)).$$

The function U is harmonic on S and can be extended to a continuous function on the closure $\overline{S}$ of S by $U(\pm 1, y) = |y|^p$.

Further properties of U are investigated in the lemma below.

Lemma 6.3.

(i) *The function U satisfies $U(x, y) = U(-x, y)$ on $\overline{S}$.*

(ii) *We have*
$$U(x, y) \geq |y|^p \quad \text{for all } (x, y) \in \overline{S}.$$

(iii) *For any $(x, y) \in S$ we have $U_{xx}(x, y) \leq 0$ and $U_{yy}(x, y) \geq 0$.*

(iv) *If $(x, y) \in S$ and $y > 0$, then $U_{yyy}(x, y) \geq 0$.*

(v) *For any $(x, y) \in \overline{S}$ such that $|y| \leq |x|$, we have $U(x, y) \leq K_{p,\infty}^p$.*

(vi) *For any $(x, y) \in S$ we have $U(x, y) \leq 2^{p-1}|y|^p + 2^{p-1}K_{p,\infty}^p$.*

We omit the proof, since it is similar to that of Lemma 6.1 above. It is easy to see that the orthogonal version of Theorem 5.4 immediately yields (6.46): indeed, with no loss of generality we may assume that $|||X|||_\infty = 1$; then $|||Y|||_p < \infty$ (by Burkholder's inequality) and Y is a martingale. Thus, for any t,

$$\mathbb{E}|Y_t|^p \leq \mathbb{E}U(X_t, Y_t) \leq \mathbb{E}U(X_0, Y_0) \leq K_{p,\infty}^p$$

and we are done. □

Proof of (6.45). Here the things are more complicated. First, we will not work with (6.45) directly, but rather with the following modification:

$$|||Y|||_1 \leq |||X|||_p^p + L, \tag{6.47}$$

where L is a fixed positive number. We have already seen such a recipe in the proof of Theorem 3.6 in Chapter 3. In order to establish (6.47) we will use the value function of the following optimal stopping problem. Let $B = (B^{(1)}, B^{(2)})$ be a two-dimensional Brownian motion starting from $(0, 0)$ and introduce $U : \mathbb{R}^2 \to (-\infty, \infty]$ by

$$U(x, y) = \sup \mathbb{E}G(x + B_\tau^{(1)}, y + B_\tau^{(2)}), \tag{6.48}$$

where $G(x, y) = |y| - |x|^p$ and the supremum is taken over all stopping times of B satisfying $\mathbb{E}\tau^{p/2} < \infty$.

The key properties of U are listed in the lemma below.

Lemma 6.4.

(i) *The function U is finite on $\mathbb{R}^2$.*

(ii) *The function U is a superharmonic majorant of G.*

(iii) *For any fixed $x \in \mathbb{R}$, the function $U(x, \cdot)$ is convex.*

(iv) *If $|y| \leq |x|$, we have*

$$U(x, y) \leq \left(\frac{K_{1,p}}{p}\right)^{p/(p-1)} \cdot (p-1). \tag{6.49}$$

Proof. (i) Take a stopping time $\tau \in L^{p/2}$ and note that the process $(B^{(2)}_{\tau \wedge t})$ is differentially subordinate and orthogonal to $(x + B^{(1)}_{\tau \wedge t})$. Therefore, by virtue of (6.12) we have, for any t,

$$\mathbb{E}|y + B^{(2)}_{\tau \wedge t}| \le |y| + \mathbb{E}|B^{(2)}_{\tau \wedge t}| \le |y| + c + [\cot(\pi/2p^*)]^{-p}||B^{(2)}_{\tau \wedge t}||^p_p$$
$$\le |y| + c + ||x + B^{(1)}_{\tau \wedge t}||^p_p,$$

where $c = [\cot(\pi/2p^*)/p]^{p/(p-1)} \cdot (p-1)$. Since $\tau \in L^{p/2}$, the Burkholder-Davis-Gundy inequality implies that the martingales $(B^{(1)}_{\tau \wedge t})$, $(B^{(2)}_{\tau \wedge t})$ converge in L^p to $B^{(1)}_\tau$ and $B^{(2)}_\tau$, respectively. Thus, letting $t \to \infty$ yields $U(x,y) \le |y| + c$.

(ii) The inequality $U \ge G$ follows immediately by considering in (6.48) the stopping time $\tau \equiv 0$. The superharmonicity can be established using standard Markovian arguments (see, e.g., Chapter I in [170]).

(iii) Fix x, y_1, $y_2 \in \mathbb{R}$ and $\lambda \in (0,1)$. For any $\tau \in L^{p/2}$, by the triangle inequality,

$$\mathbb{E}G(x + B^{(1)}_\tau, \lambda y_1 + (1-\lambda)y_2 + B^{(2)}_\tau) \le \lambda \mathbb{E}G(x + B^{(1)}_\tau, y_1 + B^{(2)}_\tau)$$
$$+ (1-\lambda)\mathbb{E}G(x + B^{(1)}_\tau, y_2 + B^{(2)}_\tau)$$
$$\le \lambda U(x, y_1) + (1-\lambda)U(x, y_2).$$

It remains to take the supremum over τ to get the claim.

(iv) Fix a stopping time $\tau \in L^{p/2}$ and $t > 0$. We have

$$\mathbb{E}\left|y + B^{(2)}_{\tau \wedge t}\right| = \mathbb{E}\left(y + B^{(2)}_{\tau \wedge t}\right) \operatorname{sgn}\left(y + B^{(2)}_{\tau \wedge t}\right).$$

Consider the martingale $\zeta^t = (\zeta^t_r)_{r \ge 0}$ given by $\zeta^t_r = \mathbb{E}\left[\operatorname{sgn}\left(y + B^{(2)}_{\tau \wedge t}\right) \big| \mathcal{F}_{\tau \wedge r}\right]$. There exists an $\mathbb{R}^2$-valued predictable process $A = (A^{(1)}_r, A^{(2)}_r)_r$ such that for all r,

$$\zeta^t_r = \mathbb{E}\zeta^t_t + \int_0^{\tau \wedge r} A_s dB_s = \mathbb{E}\operatorname{sgn}\left(y + B^{(2)}_{\tau \wedge t}\right) + \int_0^{\tau \wedge r} A_s dB_s$$

(see, e.g., Chapter V in Revuz and Yor [182]). Therefore, using the properties of stochastic integrals, we may write

$$\mathbb{E}\left|y + B^{(2)}_{\tau \wedge t}\right| = y\mathbb{E}\operatorname{sgn}\left(y + B^{(2)}_{\tau \wedge t}\right) + \mathbb{E}B^{(2)}_{\tau \wedge t}\int_0^{\tau \wedge t} A_s dB_s \qquad (6.50)$$
$$= y\mathbb{E}\operatorname{sgn}\left(y + B^{(2)}_{\tau \wedge t}\right) + \mathbb{E}\int_0^{\tau \wedge t}(0,1)dB_s \int_0^{\tau \wedge t} A_s dB_s$$
$$= y\mathbb{E}\operatorname{sgn}\left(y + B^{(2)}_{\tau \wedge t}\right) + \mathbb{E}\int_0^{\tau \wedge t} A^{(2)}_s ds$$
$$= y\mathbb{E}\operatorname{sgn}\left(y + B^{(2)}_{\tau \wedge t}\right) + \mathbb{E}\int_0^{\tau \wedge t}(1,0)dB_s \int_0^{\tau \wedge t}(A^{(2)}_s, -A^{(1)}_s)dB_s$$

$$\leq |x| \left| \mathbb{E} \operatorname{sgn} \left(y + B^{(2)}_{\tau \wedge t} \right) \right| + \mathbb{E} B^{(1)}_{\tau \wedge t} \int_0^{\tau \wedge t} (A_s^{(2)}, -A_s^{(1)}) \mathrm{d}B_s$$

$$= \mathbb{E} \left(x + B^{(1)}_{\tau \wedge t} \right) \left[\operatorname{sgn} x \left| \mathbb{E} \operatorname{sgn} \left(y + B^{(2)}_{\tau \wedge t} \right) \right| + \int_0^{\tau \wedge t} (A_s^{(2)}, -A_s^{(1)}) \mathrm{d}B_s \right]$$

$$\leq \left\| x + B^{(1)}_{\tau \wedge t} \right\|_p \left\| \operatorname{sgn} x | \mathbb{E} \operatorname{sgn}(y + B^{(2)}_{\tau \wedge t})| + \int_0^{\tau \wedge t} (A_s^{(2)}, -A_s^{(1)}) \mathrm{d}B_s \right\|_{\frac{p}{p-1}} .$$

Observe that the martingale

$$(\eta_r^t)_{r \geq 0} = \left(\operatorname{sgn} x \left| \mathbb{E} \operatorname{sgn} \left(y + B^{(2)}_{\tau \wedge t} \right) \right| + \int_0^{\tau \wedge r} (A_s^{(2)}, -A_s^{(1)}) \mathrm{d}B_s \right)_{r \geq 0}$$

is differentially subordinate and orthogonal to ζ^t. Furthermore, we have $\|\zeta^t\|_\infty = \|\operatorname{sgn}(y + B^{(2)}_{\tau \wedge t})\|_\infty = 1$, so, by (6.46), we see that $\|\eta^t\|_{p/(p-1)} \leq K_{p/(p-1),\infty}$. Consequently,

$$\mathbb{E}|y + B^{(2)}_{\tau \wedge t}| \leq K_{p/(p-1),\infty} \|x + B^{(1)}_{\tau \wedge t}\|_p \leq \mathbb{E}|x + B^{(1)}_{\tau \wedge t}|^p + \left(\frac{K_{p/(p-1),\infty}}{p} \right)^{p/(p-1)} \cdot (p-1)$$

and it suffices to let $t \to \infty$ to obtain (6.49), using the argument based on the Burkholder-Davis-Gundy inequality. $\qquad \square$

Now we are ready to establish (6.45). Fix $\delta > 0$, $\varepsilon > \delta\sqrt{2}$, and convolve G and U with a nonnegative C^∞ function g^δ, supported on the ball with center $(0,0)$ and radius δ, satisfying $\|g^\delta\|_1 = 1$. In this way we obtain C^∞ functions G^δ and U^δ, such that $G^\delta \leq U^\delta$ and U^δ is superharmonic. Furthermore, by Lemma 6.4 (iii), we have $U^\delta_{yy} \geq 0$ and, by superharmonicity, $U^\delta_{xx} \leq 0$. Applying the orthogonal version of Theorem 5.3 yields

$$\mathbb{E}U^\delta(X_t, Y_t) = \mathbb{E}U^\delta(X_0, Y_0),$$

whence

$$\mathbb{E}G^\delta(X_{\tau \wedge t}, Y_{\tau \wedge t}) \leq \mathbb{E}U^\delta(X_0, Y_0).$$

Obviously, we have $|G^\delta(x, y)| \leq |x| + ||y| + \delta|^p \leq 2^{p-1}(|y|^p + \delta^p)$. Hence, by Lebesgue's dominated convergence theorem, if we let $\varepsilon \to 0$ and $\delta \to 0$, we get

$$\mathbb{E}|Y_{\tau \wedge t}| \leq \mathbb{E}|X_{\tau \wedge t}|^p + \left(\frac{K_{1,p}}{p} \right)^{p/(p-1)} \cdot (p-1).$$

By the Burkholder-Davis-Gundy inequalities, we may replace $\tau \wedge t$ by τ in the above estimate. Applying it to the pair $(X', Y') = (X/\lambda, Y/\lambda)$ with

$$\lambda = \frac{\|X\|_p p^{1/(p-1)}}{K_{1,p}^{1/(p-1)}}$$

(clearly, the differential subordination and the orthogonality remain valid) yields (6.45). $\qquad \square$

Sharpness. When $p > 2$ and $q = \infty$, then equality is attained for the pair $B = (B^{(1)}_{\tau \wedge t}, B^{(2)}_{\tau \wedge t})$, where $(B^{(1)}, B^{(2)})$ is a standard Brownian motion starting from $(0,0)$ and $\tau = \inf\{t \geq 0 : |B^{(1)}_t| = 1\}$. An analogous reasoning was already used in the case of the weak type estimates, so we omit the details. If $p = 1$ and $1 < q < 2$, we use a duality argument similar to the one above. Namely, suppose that (6.45) holds with some constant C_q. Let B be as above. Then

$$\mathbb{E}|B^{(2)}_{\tau \wedge t}|^{q/(q-1)} = \mathbb{E}B^{(2)}_{\tau \wedge t} \cdot |B^{(2)}_{\tau \wedge t}|^{q/(q-1)-2} B^{(2)}_{\tau \wedge t} = \mathbb{E}B^{(2)}_{\tau \wedge t}\xi = \mathbb{E}B^{(1)}_{\tau \wedge t}\zeta,$$

where (ξ, ζ) is a corresponding pair of orthogonal processes starting from $(0,0)$ and such that ζ is differentially subordinate to ξ. Consequently, since $B^{(1)}_{\tau \wedge t}$ is bounded by 1,

$$\mathbb{E}|B^{(2)}_{\tau \wedge t}|^{q/(q-1)} \leq \mathbb{E}|\zeta| \leq C_q \|\xi\|_q = C_q \left[\mathbb{E}|B^{(2)}_{\tau \wedge t}|^{q/(q-1)}\right]^{1/q}$$

or $\|B^{(2)}_\tau\|_{q/(q-1)} \leq C_q$. Thus $C_q \geq K_{q/(q-1),\infty}$, as desired. $\quad\square$

On the search of the suitable majorant, the case $q = \infty$. This is straightforward: write

$$U^0(x, y) = \sup\{\mathbb{E}|Y_t|^p\},$$

where the supremum is taken over all pairs (X, Y) of orthogonal martingales starting from (x, y) such that $([X, X]_t - [Y, Y]_t)_{t \geq 0}$ is nondecreasing and $\|X\|_\infty \leq 1$. This makes clear which pairs should give the supremum: we take any X and wait until it reaches ± 1; the corresponding Y satisfies $[Y, Y] - y^2 = [X, X] - x^2$. By a standard time-change argument, we may assume that X and hence also Y, are stopped one-dimensional Brownian motions. Hence, U^0 is necessarily the harmonic lift of the function $(x, y) \mapsto |y|^p$, $(x, y) \in S$. This is precisely the function used above. $\quad\square$

6.5 Logarithmic estimates

The special functions considered in the preceding section can be used to obtain the following sharp logarithmic estimate. Introduce the function $\Psi : [0, \infty) \to \mathbb{R}$ by the formula $\Psi(t) = (t + 1) \log(t + 1) - t$.

Theorem 6.6. *Suppose that X, Y are orthogonal real-valued martingales such that Y is differentially subordinate to X. Then for $K > 2/\pi$,*

$$\|Y\|_1 \leq \sup_t \mathbb{E}\Psi(K|X_t|) + 8 \int_1^\infty \frac{t^{2/(K\pi)} - \frac{2\log t}{K\pi} - 1}{t^2 + 1}dt. \tag{6.51}$$

The inequality is sharp for each K.

For a related estimate with the more natural function $\Psi(t) = t \log^+ t$, see [163] and the bibliographic notes at the end of this chapter.

6.5.1 Proof of Theorem 6.6

Proof of (6.52). First we shall establish the following dual estimate. Let $\Phi :$ $[0, \infty) \to \mathbb{R}$ be defined by $\Phi(t) = e^t - t - 1$. We will show that if $||X||_\infty \leq 1$, then, for $\gamma < \pi/2$,

$$\sup_t \mathbb{E}\Phi(\gamma|Y_t|) \leq 8 \int_1^\infty \frac{t^{2\gamma/\pi} - \frac{2\gamma}{\pi}\log t - 1}{t^2 + 1} dt \qquad (6.52)$$

and the constant on the right is the best possible. This is straightforward. For any $k = 2, 3, \ldots$ we have, by (6.46),

$$||Y||_k^k \leq K_{k,\infty}^k = \frac{2^{k+1}}{\pi^{k+1}} \int_0^\infty \frac{|\log|t||^k}{t^2 + 1} dt = \frac{4}{\pi} \int_1^\infty \frac{\left(\frac{2}{\pi}\log t\right)^k}{t^2 + 1} dt. \qquad (6.53)$$

Hence, for $\gamma < \pi/2$,

$$\mathbb{E}\Phi(\gamma|Y_t|) = \sum_{k=2}^\infty \frac{\gamma^k ||Y_t||_k^k}{k!} \leq \frac{4}{\pi} \int_1^\infty \frac{t^{2\gamma/\pi} - \frac{2\gamma}{\pi}\log t - 1}{t^2 + 1} dt$$

and it remains to take the supremum over t to get (6.52). To see that the bound on the right is the best possible, consider the two-dimensional Brownian motion started at $(0,0)$ and stopped at the boundary of the strip $[-1, 1] \times \mathbb{R}$. Then we have equality in (6.53) for all $k \geq 2$ (see the previous section) and hence (6.52) is sharp too. Now it suffices to repeat the arguments which enabled us above to deduce (6.45) from (6.46). To do this, we need to consider an appropriate optimal stopping problem and rewrite (6.50) accordingly. The crucial property of Φ and Ψ is that they are conjugate in the sense that Φ' is the inverse to Ψ' on $(0, \infty)$. We omit the further details, as the proof goes along the same lines. $\square$

Sharpness. This follows from duality; see sharpness of (6.45).

6.6 Moment inequalities for nonnegative martingales

6.6.1 Formulation of the result

We shall now study the moment estimates in the particular case when both processes are real valued and one of them is nonnegative. Let

$$C_p = \begin{cases} \tan(\pi/2p) & \text{if } 1 < p \leq 2, \\ (1 + \cos^p(\pi/p))^{1/p} (\sin(\pi/p))^{-1} & \text{if } 2 < p < \infty. \end{cases}$$

and

$$E_p = \begin{cases} 1 & \text{if } p = 1, \\ \left[\frac{\sin^{p-1}(\phi_p)\sin((p-1)\phi_p)}{1 - \cos^{p-1}(\phi_p)\cos((p-1)\phi_p)}\right]^{1/p} & \text{if } 1 < p < 2, \\ \cot(\pi/2p) & \text{if } 2 \leq p < \infty. \end{cases} \qquad (6.54)$$

Here ϕ_p is the unique number from the interval $(\pi/4, \pi/2)$ such that

$$\sin\left((p-1)\phi - \frac{p\pi}{4}\right) + \sin\frac{p\pi}{4}\cos^{p-1}\phi = 0. \tag{6.55}$$

Theorem 6.7. *Let X, Y be real-valued orthogonal local martingales such that Y is differentially subordinate to X.*

(i) *If X is nonnegative, then*

$$|||Y|||_p \le C_p|||X|||_p, \qquad 1 < p < \infty, \tag{6.56}$$

and the inequality is sharp.

(ii) *If Y is nonnegative, then*

$$|||Y|||_p \le E_p|||X|||_p, \qquad 1 \le p < \infty, \tag{6.57}$$

and the inequality is sharp.

Therefore, if we compare (6.56) and (6.57) to (6.12), we see that in the case when X is nonnegative, the best constant in the L^p inequality does not change for $1 < p \le 2$ and decreases for $p > 2$; if Y is assumed to be nonnegative, then the best constant remains the same for $p \ge 2$ and decreases for $1 \le p < 2$. Observe that we have the same behavior of the optimal constants in the nonorthogonal case.

6.6.2   Proof of Theorem 6.7

It suffices to establish (6.56) for $p > 2$ and (6.57) for $1 < p < 2$. The reasoning is the same as in the proof of Theorem 6.3, so we will only present the formulas for the special functions. To prove the first estimate for $p > 2$, let $W_p : [0, \infty) \times [0, \infty) \to \mathbb{R}$ be defined by

$$W_p(x, y) = \begin{cases} R^p\left[\cos(p\pi/2 - p\theta) \right. \\ \left. + \sin(p\pi/2 - p\theta) \cdot \cot(\pi/p)(\cos^{p-2}(\pi/p) + 1)\right] & \text{if } \theta > \pi/2 - \pi/p \\ y^p - C_p^p x^p & \text{if } \theta \le \pi/2 - \pi/p, \end{cases}$$

where, as previously, R and θ are the polar coordinates. Finally, the special function W_p corresponding to (6.57) for $1 < p < 2$ is given by

$$W_p(x, y) = \begin{cases} E_p^p R^p\left[\cot(\pi p/4)\sin(p\theta) - \cos(p\theta)\right] & \text{if } 0 \le \theta \le \phi_p, \\ y^p - E_p^p x^p & \text{if } \phi_p < \theta \le \pi/2. \end{cases}$$

6.7 Weak type inequalities for submartingales

6.7.1 Formulation of the result

We will prove a statement which can be regarded as the orthogonal version of Theorem 4.4. Let $c = 3.375\ldots$ be given by (6.61) below.

Theorem 6.8. *Assume that X is a submartingale such that $X_0 = 0$ and Y is a real-valued semimartingale which is orthogonal and strongly differentially subordinate to X. Then*

$$\mathbb{P}(|Y|^* \geq 1) \leq c\|X\|_1 \qquad (6.58)$$

and the constant c is the best possible.

6.7.2 Proof of Theorem 6.8

The estimate (6.58) follows from the existence of a function $U : \mathbb{R}^2 \to \mathbb{R}$ such that

(i) U is continuous and superharmonic,

(ii) for any $(x, y) \in \mathbb{R}^2$, the functions $t \mapsto U(x + t, y + t)$ and $t \mapsto U(x + t, y - t)$ are nonincreasing on $\mathbb{R}$,

(iii) U is concave along the horizontal lines,

(iv) $U(0, 0) = -c^{-1}$,

(v) $U(x, y) \geq U(0, 0)1_{\{|y|<1\}} - x^+$ for all $x, y \in \mathbb{R}$.

Indeed, convolving U with a smooth radial function g which has the usual properties, we see that the resulting function U^g satisfies $\Delta U^g \leq 0$, $U^g_{xx} \leq 0$ and $U^g_x + |U^g_y| \leq 0$. These conditions imply that for X, Y as in the statement of the theorem, the process $(U^g(X_t, Y_t))_{t \geq 0}$ is a local supermartingale; then the properties (iv) and (v) play the role of the initial condition and the majorization property. We refer the reader to [158] for the detailed explanation of this argument.

To prove the existence of a function U satisfying (i)–(v), we will construct some auxiliary objects first. For the sake of convenience, in the considerations below we identify $\mathbb{R}^2$ with the complex plane. Let $D = \{z \in \mathbb{C} : |z| < 1\}$ be the open unit disc of $\mathbb{C}$ and put

$$\begin{aligned} J &= \{z \in \mathbb{C} : |\operatorname{Im} z| \leq 1, |\operatorname{Im} z| \leq \operatorname{Re} z\}, \\ K &= \{z \in \mathbb{C} : |\operatorname{Re} z| \leq 1 \text{ or } |\operatorname{Im} z| \leq 1\}. \end{aligned} \qquad (6.59)$$

Define $h : \partial K \to \mathbb{R}$ by setting $h(x, \pm 1) = 1 - |x|$ if $|x| \geq 1$ and $h(\pm 1, y) = |y| - 1$ if $|y| \geq 1$. Consider the function F given by

$$F(z) = \alpha \int_0^z \frac{\sqrt{1 - w^4}}{1 + w^4}\, dw,$$

where

$$\alpha = \sqrt{2i} \left(\int_0^1 \frac{\sqrt{1-t^4}}{1+t^4} \, dt \right)^{-1}.$$

The function F is a conformal mapping of D onto the interior of K and it sends the arcs $\{e^{i\theta} : |\theta| < \pi/4\}$, $\{e^{i\theta} : \theta \in (\pi/4, 3\pi/4)\}$, $\{e^{i\theta} : \theta \in (3\pi/4, 5\pi/4)\}$, $\{e^{i\theta} : \theta \in (5\pi/4, 7\pi/4)\}$ onto the sets $\partial K \cap (0, \infty)^2$, $\partial K \cap (-\infty, 0) \times (0, \infty)$, $\partial K \cap (-\infty, 0)^2$, $\partial K \cap (0, \infty) \times (-\infty, 0)$, respectively. Finally, we have $F(e^{\pm\pi i/4}) = F(e^{\pm 3\pi i/4}) = \infty$. Let G be the inverse of F and define $u : D \to \mathbb{R}$ by the Poisson integral

$$u(re^{i\theta}) = \frac{1}{2\pi} \int_0^{2\pi} \frac{1-r^2}{1 - 2r\cos(t-\theta) + r^2} h(F(e^{it})) dt$$

for $r \in [0, 1)$ and $\theta \in [0, 2\pi)$ (this is well defined, see (6.63) below). To describe the optimal constant c in (6.58), let $R = 0.541\ldots$ be the unique solution to the equation

$$\int_0^R \frac{\sqrt{1+s^4}}{1-s^4} \, ds = \frac{\sqrt{2}}{2} \int_0^1 \frac{\sqrt{1-s^4}}{1+s^4} \, ds \tag{6.60}$$

and put

$$c = - \left[\frac{1}{2\pi} \int_0^{2\pi} \frac{1 - R^2}{1 - 2R\cos\left(t + \frac{\pi}{4}\right) + R^2} h(F(e^{it})) dt \right]^{-1}. \tag{6.61}$$

Computer simulations show that $c = 3.375\ldots$. Finally, let

$$\mathcal{U}(x, y) = u(G(x, y)) \qquad \text{for } (x, y) \in K.$$

Lemma 6.5. *The function $\mathcal{U}$ enjoys the following properties:*

(i) *It is continuous on K and harmonic in the interior of K.*

(ii) *The function $(x, y) \mapsto \mathcal{U}(x, y) + x$ is bounded on J.*

(iii) *$\mathcal{U}$ has the following symmetry: if $(x, y) \in K$, then*

$$\mathcal{U}(x, y) = \mathcal{U}(x, -y) = \mathcal{U}(-x, y) \quad \text{and} \quad \mathcal{U}(x, y) = -\mathcal{U}(y, x).$$

Proof. (i) Clearly, $\mathcal{U}$ is harmonic in the interior of K, since it is the real part of an analytic function there. To see that $\mathcal{U}$ is continuous on K, observe that the function $t \mapsto h(F(e^{it}))$ is continuous on $[-\pi, \pi] \setminus \{\pm\pi/4, \pm 3\pi/4\}$. This implies that u is continuous on $\overline{D} \setminus \{e^{\pm\pi i/4}, e^{\pm 3\pi i/4}\}$ and the latter set is precisely $G(K)$.

(ii) First we prove the identity

$$\frac{1}{2\pi} \int_0^{2\pi} \frac{1-r^2}{1 - 2r\cos(t-\theta) + r^2} F(e^{it}) \, dt = F(re^{i\theta}) \tag{6.62}$$

for $r \in [0,1)$, $\theta \in [0, 2\pi)$. To do this, apply Fubini's theorem to obtain

$$\frac{1}{2\pi} \int_0^{2\pi} \frac{1 - r^2}{1 - 2r\cos(t - \theta) + r^2} F(e^{it})\, dt$$

$$= \frac{1}{2\pi} \int_0^{2\pi} \frac{1 - r^2}{1 - 2r\cos(t - \theta) + r^2} \cdot \alpha \int_0^{e^{it}} \frac{\sqrt{1 - z^4}}{1 + z^4}\, dz\, dt$$

$$= \int_0^1 \frac{1}{2\pi} \int_0^{2\pi} \frac{1 - r^2}{1 - 2r\cos(t - \theta) + r^2} \cdot \alpha \frac{\sqrt{1 - s^4 e^{4it}}}{1 + s^4 e^{4it}} e^{it}\, dt\, ds.$$

For any fixed $s \in [0, 1)$, the expression under the outer integral is the Poisson formula for the function

$$f_s : z \mapsto \alpha z \frac{\sqrt{1 - s^4 z^4}}{1 + s^4 z^4},$$

which is continuous on $\overline{D}$ and analytic on D. Thus, this expression equals $f_s(re^{i\theta})$ and

$$\frac{1}{2\pi} \int_0^{2\pi} \frac{1 - r^2}{1 - 2r\cos(t - \theta) + r^2} F(e^{it})\, dt = \alpha \int_0^1 \frac{\sqrt{1 - s^4 r^4 e^{4it}}}{1 + s^4 r^4 e^{4it}} re^{it}\, ds$$

$$= F(re^{i\theta}).$$

To see that the above use of Fubini's theorem is permitted, note that for any fixed r and θ as above and some positive κ_1, κ_2,

$$\frac{\alpha}{2\pi} \int_0^{2\pi} \int_0^1 \left| \frac{1 - r^2}{1 - 2r\cos(t - \theta) + r^2} \frac{\sqrt{1 - s^4 e^{4it}}}{1 + s^4 e^{4it}} e^{it} \right| ds\, dt$$

$$\leq \kappa_1 \int_0^{2\pi} \int_0^1 \frac{1}{|1 + s^4 e^{4it}|}\, ds\, dt$$

$$\leq \kappa_2 \int_0^{2\pi} \int_0^1 \frac{1}{|1 + se^{4it}|}\, ds\, dt \qquad (6.63)$$

$$= 8\kappa_2 \int_0^{\pi/4} \int_0^1 \frac{1}{\sqrt{\left(\frac{s + \cos 4t}{\sin 4t}\right)^2 + 1}} \frac{ds\, dt}{\sin 4t}$$

$$= 2\kappa_2 \int_0^\pi \log\left(1 + \frac{1}{\cos(t/2)}\right) dt < \infty.$$

Therefore, if $(x, y) \in J$ and $G(x, y) = re^{i\theta}$, then we may write

$$\mathcal{U}(x, y) + x - 1$$

$$= u(G(x, y)) + \operatorname{Re} F(G(x, y)) - 1$$

$$= \frac{1}{2\pi} \int_0^{2\pi} \frac{1 - r^2}{1 - 2r\cos(t - \theta) + r^2} \big[h(F(e^{it})) + \operatorname{Re} F(e^{it}) - 1 \big] dt$$

$$= \frac{1}{2\pi} \int_0^{3\pi/2} \frac{1 - r^2}{1 - 2r\cos(t - \theta) + r^2} \big[h(F(e^{it})) + \operatorname{Re} F(e^{it}) - 1 \big] dt,$$

because $h(x,y) + x - 1 = 0$ for $(x,y) \in \partial K \cap \{z \in \mathbb{C} : \operatorname{Re} z > 1\}$. Now if we take $(x,y) \in J$ with x sufficiently large, say, $x \geq 2$, then θ lies in a proper closed subinterval of $(-\pi/2, 0)$ and thus the Poisson kernel is bounded uniformly in $r \in [0,1)$ and $t \in [0, 3\pi/2]$. It suffices to note that, by (6.63),

$$\int_0^{3\pi/2} \left| h(F(e^{it})) + \operatorname{Re} F(e^{it}) - 1 \right| dt < \infty.$$

(iii) We have $F(iz) = iF(z)$ for any $z \in D$, so $G(iz) = iG(z)$ for all $z \in K$; that is, if $(x,y) \in K$, then $G(-y,x) = iG(x,y)$. Consequently, if $G(x,y) = re^{i\theta}$, then

$$\mathcal{U}(-y,x) = u(iG(x,y)) = u(re^{i(\pi/2+\theta)})$$

$$= \frac{1}{2\pi} \int_0^{2\pi} \frac{1 - r^2}{1 - 2r\cos\left(t - \frac{\pi}{2} - \theta\right) + r^2} h(F(e^{it}))dt$$

$$= \frac{1}{2\pi} \int_{-\pi/2}^{3\pi/2} \frac{1 - r^2}{1 - 2r\cos(t - \theta) + r^2} h(F(ie^{it}))dt$$

$$= \frac{1}{2\pi} \int_{-\pi/2}^{3\pi/2} \frac{1 - r^2}{1 - 2r\cos(t - \theta) + r^2} h(iF(e^{it}))dt,$$

which is equal to $-\mathcal{U}(x,y)$, since $h(iz) = -h(z)$ for $z \in \partial K$. Therefore,

$$\mathcal{U}(x,y) = -\mathcal{U}(-y,x) = \mathcal{U}(x,y) \tag{6.64}$$

and it remains to show that $\mathcal{U}(x,y) = \mathcal{U}(x,-y)$ for $(x,y) \in J$. But this follows from the previous part: the function $(x,y) \mapsto \mathcal{U}(x,y) - \mathcal{U}(x,-y)$ is continuous and bounded on J, harmonic in the interior of this set and vanishes on its boundary. Thus it is identically 0. $\qquad\square$

In the next lemma we exhibit more properties of $\mathcal{U}$.

Lemma 6.6. *If $(x,y) \in J$, then*

$$|y| - x \leq \mathcal{U}(x,y) \leq \min\{1 - x, 0\}, \tag{6.65}$$

$$\mathcal{U}_x(x,y) \pm \mathcal{U}_y(x,y) \leq 0 \tag{6.66}$$

and

$$\mathcal{U}_{xx} \leq 0. \tag{6.67}$$

Proof. Note that the function $(x,y) \mapsto |y| - x$ is subharmonic and agrees with $\mathcal{U}$ on the boundary of J. Combining this with Lemma 6.5 (ii), we obtain the lower bound in (6.65). To show the upper bound, note that the functions $(x,y) \mapsto 1 - x$, $(x,y) \mapsto 0$ are harmonic and majorize $\mathcal{U}$ at ∂J; it remains to apply Lemma 6.5 (ii) to get (6.65).

We turn to (6.66). By the symmetry of $\mathcal{U}$ (see part (iii) of the previous lemma), it suffices to show that $\mathcal{U}_x(x, y) + \mathcal{U}_y(x, y) \leq 0$. Using the Schwarz reflection principle, $\mathcal{U}$ can be extended to a continuous function on $J \cup (J + 2i)$ and harmonic inside this set (here we have used the usual notation $J + w = \{z \in \mathbb{C} : z - w \in J\}$ for $w \in \mathbb{C}$). The extension is given by

$$\mathcal{U}(x, y) = 2 - 2x - \mathcal{U}(x, 2 - y)$$

for $(x, y) \in J + 2i$. Fix $k \in (0, 1)$ and define the function V on J by

$$V(x, y) = \mathcal{U}(x, y) - \mathcal{U}(x + k, y + k).$$

Using (6.65) several times, one can show that V is nonnegative on the boundary of J. Indeed: if $y = -x \in [0, 1]$, then $\mathcal{U}(x, y) = 0$ and $\mathcal{U}(x + k, y + k) \leq 0$; if $y = x \in [0, 1 - k]$, then $\mathcal{U}(x, y) = \mathcal{U}(x + k, y + k) = 0$; if $y = x \in (1 - k, 1]$, then $\mathcal{U}(x, y) = 0$ and

$$\mathcal{U}(x + k, y + k) = 2 - 2x - 2k - \mathcal{U}(x + k, 2 - y - k)$$
$$\leq 2 - 2x - 2k - (2 - y - k - x - k) = 0.$$

Next, $\mathcal{U}(x, -1) = 1 - x > 1 - x - k \geq \mathcal{U}(x + k, -1 + k)$; finally, $U(x, 1) = 1 - x$ and $\mathcal{U}(x + k, 1 + k) = 2 - 2k - 2x - \mathcal{U}(x + k, 1 - k) \leq 1 - x$. Furthermore, by Lemma 6.5, V is continuous, bounded on J and harmonic in the interior of J. Consequently, $V \geq 0$ and since $k \in (0, 1)$ was arbitrary, (6.66) follows.

To prove the concavity property (6.67), we proceed similarly. Fix $k \in (0, 1)$ and consider the function $V : J \to \mathbb{R}$ given by

$$V(x, y) = 2\mathcal{U}(x, y) - \mathcal{U}(x + k, y) - \mathcal{U}(x - k, y).$$

It suffices to prove that $V \geq 0$ on the boundary of J. Clearly, this is true on $[1 + k, \infty) \times \{-1, 1\}$ ($V \equiv 0$ there). When $x \in (1, 1 + k)$, then, using the symmetry of $\mathcal{U}$ and the lower bound from (6.65), we get

$$\mathcal{U}(x - k, \pm 1) = -\mathcal{U}(1, x - k) \leq 1 - (x - k),$$

so $V(x, \pm 1) \geq 0$. Similarly, by the symmetry of $\mathcal{U}$ we get, for $x \in [0, 1]$,

$$V(x, \pm x) = -\mathcal{U}(x + k, x) - \mathcal{U}(x - k, x) = -\mathcal{U}(x + k, x) + \mathcal{U}(x, x - k) \geq 0,$$

in virtue of (6.66). This completes the proof.   □

Now we are ready to prove the existence of U. Define

$$U(x, y) = \begin{cases} \mathcal{U}(x + 1, y) & \text{if } (x + 1, y) \in J, \\ 0 & \text{if } |y| \leq 1 \text{ and } (x + 1, y) \notin J, \\ -x^+ & \text{if } |y| > 1. \end{cases}$$

Let us verify the announced properties of this function.

(i) The continuity is straightforward. In addition, we have that U is harmonic in the interior of $J-1$ and U is majorized on $\mathbb{R}^2$ by the superharmonic function $(x,y) \mapsto -x^+$ (see (6.65)). Hence the mean-value inequality holds and U is superharmonic.

(ii) Clearly, we have the monotonicity outside the set $J-1$; thus the property follows from the continuity of U and (6.66).

(iii) The concavity is evident for $|y| \geq 1$. When $|y| < 1$, use (6.67) and the estimate $U(x,y) \leq 0$ on $J-1$ (see the upper bound in (6.65)).

(iv) This follows immediately from the equality $F(Re^{-i\pi/4}) = 1$, or $G(1,0) = Re^{-i\pi/4}$ (recall that R is given by (6.60))

(v) Both sides of the estimate are equal when $|y| \geq 1$, so let us assume that $y \in (-1,1)$. Then the majorization takes the form

$$U(x,y) \geq U(0,0) - x^+.$$

This is clear when $(x,y) \notin J-1$: the left-hand side equals 0 and the right-hand one is $U(0,0) = \mathcal{U}(1,0) \leq 0$ (see (6.65)). Next, suppose that $(x,y) \in J-1$. Since U is harmonic on this set, (6.67) implies that for any $x > -1$ the function $U(x,\cdot)$ is convex on $\{y : (x,y) \in J-1\}$ and hence $U(x,y) \geq U(x,0)$ by the symmetry of U. Thus, all we need is to verify that

$$U(x,0) \geq U(0,0) - x^+ \qquad \text{for } x > -1.$$

This follows at once from the fact that the two sides are equal to one another for $x = 0$ and that $U_x(x,0) \in [-1,0]$ for $x > -1$ (to see the latter, use the concavity of the function $U(\cdot,0)$ and the lower bound in (6.65)).

Sharpness. The example is similar to that constructed in the proof of the sharpness of (4.4), but, additionally, we have to take into account the orthogonality of the pair (X,Y). Let B be a two-dimensional Brownian motion starting from $(0,0)$, put $\tau = \inf\{t : B_t \in \partial J-1\}$ and consider the (random) interval $I = [\tau, \tau - B_\tau^{(1)}]$ if $B_\tau^{(1)} < 0$, and $I = \emptyset$ otherwise. Let X, Y be Itô processes defined by $X_0 = Y_0 = 0$ and, for $t > 0$,

$$dX_t = 1_{\{\tau \geq t\}}dB_t^{(1)} + 1_I(t)dt,$$
$$dY_t = 1_{\{\tau \geq t\}}dB_t^{(2)} + \operatorname{sgn} B_\tau^{(2)} 1_I(t)dt.$$

The pair (X,Y) behaves like B until it reaches the boundary of $J-1$ at the time τ. Then if $X_\tau \geq 0$, the pair stops; if $X_\tau < 0$ and $Y_\tau > 0$, then the pair moves along the line segment $\{(x-1,x) : x \in [0,1]\}$ until Y reaches 1; finally, if $X_\tau < 0$ and $Y_\tau < 0$, the pair moves along the line segment $\{(x-1,-x) : x \in [0,1]\}$ until Y reaches -1. Observe that Y is strongly differentially subordinate to X and (X,Y) is constant on the interval $[\tau+1,\infty)$. Now, the stopping time τ is exponentially integrable, so Itô's formula yields $\mathbb{E}U(X_\tau, Y_\tau) = U(0,0) = -c^{-1}$. However, we have $|Y_{\tau+1}| \geq 1$ and

$$U(X_\tau, Y_\tau) = U(X_{\tau+1}, Y_{\tau+1}) = -X_{\tau+1}^+ = -X_{\tau+1}$$

with probability 1, so

$$\mathbb{P}(|Y|^* \geq 1) = 1 \quad \text{and} \quad \mathbb{E}|X_{\tau+1}| = c^{-1}.$$

This does not finish the proof yet, since the quantity $\mathbb{E}|X_{\tau+1}|$ is strictly smaller than $||X||_1$ (X takes negative values). To overcome this problem, we make use of the "portioning argument", see the proof of the sharpness of (4.4) or consult [158]. □

On the search of the suitable majorant. Here the reasoning is the combination of the arguments appearing in the searches corresponding to (4.4) and (6.5). We write down the formula

$$U^0(x, y) = \sup\{\mathbb{P}(|Y_t| \geq 1) - c\mathbb{E}X_t^+\},$$

where the supremum is taken over all pairs (X, Y) of orthogonal real-valued processes such that $(X_0, Y_0) \equiv (x, y)$, X is a submartingale and Y is strongly differentially subordinate to X. Of course, we have $U^0(x, y) = 1 - x^+$ for all (x, y) such that $|y| \geq 1$. Furthermore, if $x < 0$, $|y| < 1$ and $|x| + |y| \leq 1$, then $U^0(x, y) = 1$, which can be seen by considering the (deterministic) processes

$$X_t = x + t \quad \text{and} \quad Y_t = y + t\,\mathrm{sgn}\, y \qquad \text{for } t \leq 1 - y$$

(we take $\mathrm{sgn}\, 0 = 1$). Thus it remains to determine U^0 in the interior of the set $J-1$ (J is given in (6.59)). A little experimentation suggests that U^0 should be harmonic on this set. Thus, the function $(x, y) \mapsto U^0(x - 1, y) - 1$ vanishes on $\{(x, y) : x = |y| \in [0, 1]\}$ and we expect it to be harmonic on J. Thus, applying the Schwarz reflection principle, it suffices to find the explicit formula for a continuous function on K which is harmonic in the interior of this set and equals h on the boundary of K. This was precisely our starting point in the above considerations. □

6.8 Moment inequalities for submartingales

Now we will establish the following moment inequality.

6.8.1 Formulation of the result

Let α be a fixed positive number and let $\varphi = \varphi(\alpha) \in [0, \pi/2)$ be given by $\alpha = \tan\varphi$. Set $p_0 = 2 - 2\varphi/\pi$ and

$$C_p = \begin{cases} \tan\frac{\pi}{2p} & \text{if } p \leq p_0, \\ \cot\frac{\pi - 2\varphi}{2p} & \text{if } p > p_0. \end{cases} \tag{6.68}$$

We will establish the following L^p inequality.

Theorem 6.9. *Suppose that $\alpha \geq 0$. Let X be a nonnegative submartingale and Y be an $\mathbb{R}$-valued process, which is α-strongly subordinate to X. If X and Y are orthogonal, then, for any $1 < p < \infty$,*

$$||Y||_p \leq C_p||X||_p \tag{6.69}$$

and the constant C_p is the best possible. Furthermore, if $p \geq 2$, then Y can be taken to be $\mathcal{H}$-valued.

6.8.2   Proof of Theorem 6.9

Proof of (6.69). The reasoning is similar to the one presented in Subsection 6.3.2 above: we search for a special function $U_p : [0, \infty) \times \mathbb{R} \to \mathbb{R}$ or $U_p : [0, \infty) \times \mathcal{H} \to \mathbb{R}$ of the form $U_p(x, y) = W_p(x, |y|)$. The only additional property which the corresponding function W_p must satisfy reads

$$W_{px} + \alpha|W_{py}| \leq 0.$$

This is needed to control the submartingale property; we have already seen in the previous chapter how the above condition works.

To avoid repetition of the same arguments, we will only present the formulas for the special functions and leave the details to the interested reader (or refer to [132]). We consider three cases: $1 < p < p_0$, $p_0 \leq p < 2$ and $p \geq 2$. In the first case, let W_p be given by (6.20): this is the function used in the proof of moment inequality for orthogonal martingales. When $p_0 \leq p < 2$, the special function is given as follows. Let $\psi_p = \frac{\pi}{2p}$ and $\phi_p = \frac{\pi}{2} - \frac{\pi - 2\varphi}{2p}$. Note that $\psi_p \leq \phi_p$: this is equivalent to $p \geq p_0$. We define $W_p(x, y)$ by

$$\begin{cases} R^p \sin^{p-1}(\phi_p) \, (\cos \phi_p)^{-1} \cos(p(\frac{\pi}{2} - \theta) + \varphi) & \text{if } \phi_p \leq \theta \leq \frac{\pi}{2}, \\ y^p - C_p^p x^p & \text{if } \psi_p < \theta \leq \phi_p, \\ R^p(\tan^p(\psi)_p - \tan^p \phi_p) \cos^p \theta - R^p \frac{\sin^{p-1}(\pi/2p)}{\cos(\pi/2p)} \cos(p\theta) & \text{if } 0 < \theta \leq \psi_p. \end{cases}$$

Finally, if $p \geq 2$, then $W_p : \mathbb{R}_+ \times \mathbb{R}_+ \to \mathbb{R}$ is defined by

$$W_p(x, y) = \begin{cases} \sin^{p-1}(\phi_p) \, (\cos \phi_p)^{-1} \cos(p\theta + \varphi) & \text{if } \phi_p \leq \phi \leq \frac{\pi}{2}, \\ h_p(\phi) & \text{if } 0 < \phi \leq \phi_p. \end{cases}$$

Here, as in the previous case, $\phi_p = \frac{\pi}{2} - \frac{\pi - 2\varphi}{2p}$. □

Sharpness. We only study the case $p \geq p_0$. Let $(B^{(1)}, B^{(2)})$ be two-dimensional Brownian motion starting from $(0, 1)$ and let A denote the local time of $B^{(1)}$ at 0. That is, we have, for all $t \geq 0$,

$$|B_t^{(1)}| = \int_{0+}^t \mathrm{sgn}(B_s^{(1)}) \mathrm{d}B_s^{(1)} + A_t.$$

Let $\tau = \tau_\psi$ denote the exit time of B from the cone $K_\psi = \{(x, y) : y > \tan \psi |x|\}$. Fix $\alpha \geq 0$ and, for $t \geq 0$, define

$$X_t = |B^{(1)}_{\tau \wedge t}|, \qquad Y_t = B^{(2)}_{\tau \wedge t} + \alpha A_{\tau \wedge t}.$$

Then $Y - 1$ is α-strongly subordinate to X and X, $Y - 1$ are orthogonal. Now, suppose that $\psi < \frac{\pi}{2} - \psi_p$ and fix a (small) positive number a. Consider the function f defined on the cone $\{-a \leq \theta \leq \psi\}$ by

$$f(x, y) = R^p \cdot \sin^{p-1} \phi_p \, (\cos \phi_p)^{-1} \cos(p\theta + \varphi).$$

Then f is harmonic in the interior of its domain and satisfies the condition

$$f_x(0, y) + \alpha f_y(0, y) = 0 \tag{6.70}$$

for all $y > 0$. Applying Itô's formula, we obtain, for any $t \geq 0$,

$$f(X_t, Y_t) = f(0, 1) + \int_{0+}^t f_x(X_s, Y_s) \mathrm{d}X_s + \int_{0+}^t f_y(X_s, Y_s) \mathrm{d}Y_s. \tag{6.71}$$

But the measure $\mathrm{d}A_s$ is concentrated on the set $\{s : X_s = 0\}$. Thus, by (6.70), we get

$$\int_{0+}^t f_x(X_s, Y_s) \mathrm{d}A_s + \int_{0+}^t f_y(X_s, Y_s) \mathrm{d}(\alpha A_s) = 0$$

and taking expectation in (6.71) yields

$$\mathbb{E} f(X_t, Y_t) = f(0, 1). \tag{6.72}$$

Since on K_ψ we have $f(x, y) = U_p(x, y) \geq |y|^p - C_p^p x^p \geq (\cot^p \psi - \tan^p \phi_p) x^p$, the equation above yields

$$f(0, 1) \geq (\cot^p \psi - \cot^p \phi_p) \mathbb{E} X_t^p = (\cot^p \psi - \tan^p \phi_p) \mathbb{E} X_t^p,$$

so $\mathbb{E} X_t^p \leq f(0, 1)/(\cot^p \psi - \tan^p \phi_p)$ and, consequently, $X \in L^p$. Therefore, if we let $t \to \infty$ in (6.72), we get, by Lebesgue's dominated convergence theorem,

$$\mathbb{E} X_\tau^p = f(0, 1)/f(1, \tan \psi).$$

Thus we see that if $\psi \to \frac{\pi}{2} - \phi_p$, then $||X||_p \to \infty$.

Now, by the definition of τ,

$$\cot \psi ||X||_p = \cot \psi ||X_\tau||_p = ||Y_\tau||_p \leq ||Y - 1||_p + 1$$

and all the terms are finite. It suffices to let $\psi \uparrow \frac{\pi}{2} - \phi_p$ to get the claim. □

On the search of the suitable majorant. This is similar to the search in the non-orthogonal setting. See also the reasoning leading to the special functions of Theorem 6.3.

6.9   Inequalities for orthogonal smooth functions

Here we return to estimates for smooth functions under differential subordination or α-subordination, initiated in Section 5.6 above. So, let D be a domain in $\mathbb{R}^n$ and fix $\xi \in D$. Let u, v be two C^2 functions on D, taking values in $\mathcal{H}$.

Definition 6.2. We say that u and v are orthogonal, if for any positive integers i, j, the gradients of u^i and v^j are orthogonal: $\nabla u^i \cdot \nabla v^j = 0$ on D.

For example, this is the case when D is a unit disc of $\mathbb{R}^2$ and u, v are two real-valued harmonic functions which satisfy the Cauchy-Riemann equations. Of course, any pair (u, v) of orthogonal functions gives rise to a pair of orthogonal semimartingales: we compose u and v with Brownian motion in $\mathbb{R}^n$ and apply Itô's formula. All the results obtained in this chapter can be easily translated to corresponding statements for orthogonal functions. We shall formulate here only two results: the sharp versions of inequalities of Riesz and Kolmogorov, comparing the sizes of harmonic functions and their conjugates.

Theorem 6.10. *Let* D, ξ *be as above. Suppose that* u, v *are orthogonal harmonic functions taking values in* $\mathcal{H}$ *such that* v *is differentially subordinate to* u *and* $|v(\xi)| \leq |u(\xi)|$. *Then for* $1 < p < \infty$, $||v||_p \leq \cot\left(\frac{\pi}{2p^*}\right) ||u||_p$ *and for* $1 \leq p \leq 2$,

$$||v||_{p,\infty} \leq \left[\frac{1}{\Gamma(p+1)} \cdot \frac{\pi^{p-1}}{2^{p-1}} \cdot \frac{1 + \frac{1}{3^2} + \frac{1}{5^2} + \frac{1}{7^2} + \cdots}{1 - \frac{1}{3^{p+1}} + \frac{1}{5^{p+1}} - \frac{1}{7^{p+1}} + \cdots} \right]^{1/p} ||u||_p.$$

Both inequalities are sharp even when $n = 2$, $\xi = (0, 0)$, D *is the unit disc of* $\mathbb{R}^2$ *and* u, v *satisfy the Cauchy-Riemann equations.*

Proof. It suffices to show the optimality of the constants. This can be easily extracted from the examples studied above. To show that $\cot(\pi/(2p^*))$ is the best possible, consider first the case $1 < p \leq 2$. For any $\varepsilon > 0$ there is a sector $\mathcal{A}$ of the form $\{(x, y) \in [0, \infty) \times \mathbb{R} : |y| \leq cx\}$ such that if $B = (B^1, B^2)$ is a Brownian motion started at $(1, 0)$ and stopped at the exit of $\mathcal{A}$, then

$$||B^2||_p > \left[\cot(\pi/(2p^*)) - \varepsilon \right] ||B^1||_p.$$

Let $F = (F^1, F^2)$ be a conformal map, which maps the unit disc D onto $\mathcal{A}$ such that $F(0, 0) = (1, 0)$. Then the functions $u = F^1$, $v = F^2$ satisfy the Cauchy-Riemann equations on D and

$$||v||_p = ||F^2||_p = ||B^2||_p > \left[\cot(\pi/(2p^*)) - \varepsilon \right] ||B^1||_p$$
$$= \left[\cot(\pi/(2p^*)) - \varepsilon \right] ||F^1||_p = \left[\cot(\pi/(2p^*)) - \varepsilon \right] ||u||_p.$$

The optimality of the constant appearing in the weak type inequality is dealt with in a similar manner: this time we use a conformal map which sends unit disc onto the strip $\mathbb{R} \times [-1, 1]$. We omit the further details. $\square$

6.10 L^p estimates for conformal martingales and Bessel processes

In the final section of this chapter we discuss a slightly different setting, in which the orthogonality does not affect the interplay between the processes X and Y, but concerns the inner structure of the processes X and Y. We say that an $\mathbb{R}^d$-valued, continuous-path martingale X is *conformal* (or *analytic*) if for any $1 \le i < j \le d$, the coordinates X^i, X^j are orthogonal and satisfy $[X^i, X^i] = [X^j, X^j]$. Two-dimensional conformal martingales arise naturally from the composition of analytic functions and Brownian motion in the complex plane. They also appear in the martingale study of the Beurling-Ahlfors operator; see the bibliographical notes at the end of the chapter.

Fix $p > 0$ and $d > 1$. We will be interested in the moment estimates

$$||Y||_p \le C_{p,d}||X||_p, \tag{6.73}$$

where X, Y are conformal martingales, taking values in $\mathbb{R}^d$, such that Y is differentially subordinate to X. Since a two-dimensional analytic martingale is just a time-changed planar Brownian motion, its norm is a time-changed Bessel process of dimension 2. Recall that a real-valued process is a Bessel process of dimension d, if it satisfies the stochastic differential equation

$$\mathrm{d}R_t = \mathrm{d}B_t + \frac{d-1}{2}\frac{\mathrm{d}t}{R_t},$$

where B denotes the standard one-dimensional Brownian motion. See, e.g., [182] for more on the subject. The above observation connecting conformal martingales to Bessel processes leads to the extension of (6.73) to the case when d is an arbitrary positive number, not necessarily an integer. In fact, we will consider an even more general setting. Let X, Y be two nonnegative, continuous-path submartingales and let

$$X = X_0 + M + A, \qquad Y = Y_0 + N + B \tag{6.74}$$

be their Doob-Meyer decompositions, uniquely determined by $M_0 = A_0 = N_0 = B_0 = 0$ and the further condition that A, B are predictable. Assume that X and Y satisfy the following condition: for a fixed $d > 1$ and all $t > 0$,

$$X_t\mathrm{d}A_t \ge \frac{d-1}{2}\mathrm{d}[X,X]_t, \qquad Y_t\mathrm{d}B_t \le \frac{d-1}{2}\mathrm{d}[Y,Y]_t. \tag{6.75}$$

For example, if $\overline{X}$, $\overline{Y}$ are conformal martingales in $\mathbb{R}^d$, then $|\overline{X}|$, $|\overline{Y}|$ are submartingales and by Itô's formula, their martingale and finite variation parts are

$$M_t = \sum_{j=1}^{d} \int_{0+}^{t} \frac{\overline{X}_s^j}{|\overline{X}_s|} d\overline{X}_s^j, \qquad A_t = \frac{d-1}{2} \int_{0+}^{t} \frac{1}{|\overline{X}_s|} d[\overline{X}^1, \overline{X}^1]_s,$$

$$N_t = \sum_{j=1}^{d} \int_{0+}^{t} \frac{\overline{Y}_s^j}{|\overline{Y}_s|} d\overline{Y}_s^j, \qquad B_t = \frac{d-1}{2} \int_{0+}^{t} \frac{1}{|\overline{Y}_s|} d[\overline{Y}^1, \overline{Y}^1]_s, \qquad t \geq 0.$$

Hence (6.75) is satisfied; in fact, both inequalities become equalities in this case. To give another example, consider adapted d-dimensional Bessel processes R, S and let τ be the stopping time; then $X = (R_{\tau \wedge t})_{t \geq 0}$, $Y = (S_{\tau \wedge t})_{t \geq 0}$ enjoy the property (6.75).

6.10.1   Formulation of the result

For a given $0 < p < \infty$ and $d > 1$ such that $p + d > 2$, let $z_0 = z_0(p, d)$ be the smallest root in $[-1, 1)$ of the solution to the equation (6.80) below and let

$$C_{p,d} = \begin{cases} \frac{1+z_0}{1-z_0}, & \text{if } (2-d)_+ < p \leq 2, \\ \frac{1-z_0}{1+z_0}, & \text{if } 2 < p < \infty. \end{cases} \tag{6.76}$$

Theorem 6.11. *Let X, Y be two nonnegative submartingales satisfying (6.75) and such that Y is differentially subordinate to X. Then for $(2 - d)_+ < p < \infty$, $d > 1$, we have*

$$||Y||_p \leq C_{p,d}||X||_p \tag{6.77}$$

and the constant $C_{p,d}$ is the best possible. If $0 < p \leq (2 - d)_+$, then the moment inequality does not hold with any finite $C_{p,d}$.

As an application, we have the following bound for conformal martingales and Bessel processes.

Corollary 1. *Assume that X, Y are conformal martingales in $\mathbb{R}^d$, $d \geq 2$, such that Y is differentially subordinate to X. Then for any $0 < p < \infty$,*

$$||Y||_p \leq C_{p,d}||X||_p \tag{6.78}$$

and the constant $C_{p,d}$ is the best possible.

Corollary 2. *Assume that R, S are d-dimensional Bessel processes, $d > 1$, driven by the same Brownian motion. Then for any $(2 - d)_+ < p < \infty$ and any stopping time $\tau \in L^{p/2}$, we have*

$$||S_\tau||_p \leq C_{p,d}||R_\tau||_p \tag{6.79}$$

and the constant $C_{p,d}$ is the best possible. If $0 < p \leq (2 - d)_+$, then the moment inequality does not hold with any finite $C_{p,d}$.

6.10.2 An auxiliary differential equation

We will present the proof of Theorem 6.11 only for $p \geq 1$; for the general case the interested reader is referred to [7]. So let $1 \leq p < \infty$, $d > 1$ be given and fixed. Consider the auxiliary differential equation

$$(1 - s^2)g''(s) - 2(d - 1)sg'(s) + p(d - 1)g(s) = 0. \tag{6.80}$$

As shown in [7], there is a continuous function $g = g_{p,d} : [-1,1) \to \mathbb{R}$ with $g(-1) = -1$, satisfying (6.80) for $s \in (-1,1)$; furthermore, this solution is shown to have at least one root in $(-1,1)$. Denoting by $z_0 = z_0(p,d)$ the smallest root of $g_{p,d}$, we have the following statement.

Lemma 6.7. *The function g enjoys the following properties:*

(i) *If $s \in (-1, z_0]$, then $g'(s) > 0$.*

(ii) *If $p \leq 2$, then g is convex on $[-1, z_0)$. If $p \geq 2$, then g is concave on $[-1, z_0)$. If $p \neq 2$, then the convexity/concavity is strict.*

(iii) *We have $z_0 > 0$ for $p < 2$, $z_0 = 0$ for $p = 2$, and $z_0 < 0$ for $p > 2$.*

The proof rests on a careful analysis of the differential equation (6.80): consult [7] for details. Next, introduce the function $v = v_p : [-1, 1] \to \mathbb{R}$ by

$$v(s) = \left(\frac{1+s}{2}\right)^p - \left(\frac{1+z_0}{1-z_0}\right)^p \left(\frac{1-s}{2}\right)^p.$$

An easy calculation shows that

$$v''(s) = \frac{p(p-1)}{2^p}\left[(1+s)^{p-2} - \left(\frac{1+z_0}{1-z_0}\right)^p(1-s)^{p-2}\right].$$

For $p \neq 2$, let $s_1 = s_1(p)$ denote the unique root of the expression in the square brackets above. It is easy to verify that $s_1 < 0$ and $s_1 < z_0$, using assertion (iii) of the above lemma. Let $c = c(p)$ be the unique positive constant for which $cg'(z_0) = v'(z_0)$. It is readily verified that

$$c = \frac{2p(1+z_0)^{p-1}}{2^p g'(z_0)(1-z_0)}.$$

Lemma 6.8.

(i) *Let $1 \leq p \leq 2$. Then for $s \in [-1, z_0]$ we have*

$$cg(s) \geq v(s). \tag{6.81}$$

(ii) *Let $p \geq 2$. Then for $s \in [-1, z_0]$ we have*

$$cg(s) \leq v(s). \tag{6.82}$$

Proof. For $p = 2$ we have $cg(s) = v(s)$, so both (6.81) and (6.82) are valid and hence we may assume that $p \neq 2$. We treat (i) and (ii) in a unified manner and show that

$$c(2 - p)g(s) \geq (2 - p)v(s)$$

for $s \in [-1, z_0]$. Observe that $(2-p)v''(s) \geq 0$ for $s \in (-1, s_1)$ and $(2-p)v''(s) \leq 0$ for $s \in (s_1, 1)$. Since $(2-p)g$ is a strictly convex function, we see that (6.81) holds on $[s_1, z_0]$ and is strict on $[s_1, z_0)$. Suppose that the set $\{s < z_0 : cg(s) = v(s)\}$ is nonempty and let s_0 denote its supremum. Then $s_0 < s_1$, $cg(s_0) = v(s_0)$ and $(2-p)cg(s) > (2-p)v(s)$ for $s \in (s_0, z_0)$, which implies $(2-p)cg'(s_0) \geq (2-p)v'(s_0)$. Hence, by (6.80),

$$
\begin{aligned}
0 &< (1 - s_0^2)(2 - p)cg''(s_0) \\
&= (d - 1)(2 - p)(2s_0 cg'(s_0) - pg(s_0)) \\
&\leq (d - 1)(2 - p)(2s_0 v'(s_0) - pv(s_0)) \\
&= -\frac{p(2 - p)(d - 1)(1 - s_0^2)}{2^p}\left[(1 + s_0)^{p-2} - \left(\frac{1 + z_0}{1 - z_0}\right)^p (1 - s_0)^{p-2}\right].
\end{aligned}
$$

This yields $s_0 \geq s_1$ (see the definition of s_1), a contradiction.   □

Finally, let us list some further properties of g that will be needed below. Again, the proof is based on a careful analysis of the differential equation (6.80). We omit the details and refer the reader to [7].

Lemma 6.9. *Assume that $s \in (-1, z_0]$.*

(i) *We have*

$$(2 - p)(1 - s^2)g''(s) - 2(p - 1)(p - 2)sg'(s) + p(p - 1)(p - 2)g(s) \geq 0. \quad (6.83)$$

(ii) *We have*

$$s(1 - s^2)g''(s) - [p + d - 2 + (d - p)s^2]g'(s) + p(d - 1)sg(s) \leq 0. \quad (6.84)$$

(iii) *If $p \leq 2$, then*

$$pg(s) + (1 - s)g'(s) \geq 0. \qquad (6.85)$$

(iv) *If $p \geq 2$, then*

$$pg(s) - (1 + s)g'(s) \leq 0. \qquad (6.86)$$

6.10.3   Proof of Theorem 6.11

Assume that $1 \leq p < \infty$, $d > 1$ are given and fixed. Recall the numbers $c = c(p)$, $z_0 = z_0(p, d)$ and the constant $C_{p,d}$ introduced in the previous two sections. We start by defining special functions $U = U_{p,d} : \mathbb{R}_+^2 \to \mathbb{R}$. For $1 \leq p \leq 2$, let

$$
U_{p,d}(x, y) = \begin{cases} c(x + y)^p g_{p,d}\left(\dfrac{y-x}{x+y}\right) & \text{if } y \leq \frac{1+z_0}{1-z_0}x, \\ y^p - C_{p,d}^p x^p & \text{if } y > \frac{1+z_0}{1-z_0}x, \end{cases}
$$

while for $p > 2$, we put

$$U_{p,d}(x,y) = \begin{cases} -cC_{p,d}^p(x+y)^p g_{p,d}\left(\frac{x-y}{x+y}\right) & \text{if } y \geq \frac{1-z_0}{1+z_0}x, \\ y^p - C_{p,d}^p x^p & \text{if } y < \frac{1-z_0}{1+z_0}x. \end{cases}$$

Moreover, let $V_{p,d}(x,y) = y^p - C_{p,d}^p x^p$ for any p. We will skip the lower indices and write U, V instead of $U_{p,d}$, $V_{p,d}$.

Using Lemma 6.9, one easily shows the following statement.

Lemma 6.10. *For any x, $y > 0$ and h, $k \in \mathbb{R}$ we have*

$$\left[U_{xx}(x,y) + \frac{(d-1)U_x(x,y)}{x}\right]h^2 + 2U_{xy}(x,y)hk$$

$$+ \left[U_{yy}(x,y) + \frac{(d-1)U_y(x,y)}{y}\right]k^2$$

$$\leq w(x,y)\cdot(h^2 - k^2), \tag{6.87}$$

where

$$w(x,y) = U_{xx}(x,y) + \frac{(d-1)U_x(x,y)}{x} - \left[U_{yy}(x,y) + \frac{(d-1)U_y(x,y)}{y}\right] \leq 0.$$

The remainder of the proof is standard. We convolve U, V with a smooth function g supported on a ball of radius δ and satisfying the usual assumptions. Denoting the results by U^δ and V^δ, we apply Itô's formula to U^δ and the pair $(X + 2\delta, Y + 2\delta)$ of submartingales satisfying the assumptions of the theorem. A careful analysis of the terms arising from the formula, combined with the inequalities (6.75), (6.87) and the differential subordination, implies the estimate

$$\mathbb{E}U^\delta(2\delta + X_{\tau_n \wedge t}, 2\delta + Y_{\tau_n \wedge t}) \leq \mathbb{E}U^\delta(X_0 + 2\delta, Y_0 + 2\delta) + \kappa(\delta), \qquad t \geq 0.$$

Here $\kappa(\delta) = o(1)$ as $\delta \to 0$ and $(\tau_n)_{n \geq 0}$ is a certain nondecreasing sequence of stopping times, depending only on X and Y, which converges almost surely to ∞. By (6.81) and (6.82), $U_{p,d} \geq V_{p,d}$ and hence $U^\delta \geq V^\delta$; thus, plugging this in the preceding inequality and letting $\delta \to 0$, $n \to \infty$ yields

$$\mathbb{E}|Y_t|^p \leq C_{p,d}^p \mathbb{E}|X_t|^p,$$

which immediately gives the claim. $\qquad\square$

Sharpness. We present the reasoning only in the case $p < 2$; then $z_0 > 0$ by virtue of Lemma 6.7 (iii). It suffices to show that the constant $C_{p,d}$ is the best in (6.79). Suppose that R, S are Bessel processes of dimension d, starting from 1, satisfying the stochastic differential equations

$$dR_t = dB_t + \frac{d-1}{2}\frac{dt}{R_t},$$

$$dS_t = -dB_t + \frac{d-1}{2}\frac{dt}{S_t} \tag{6.88}$$

(here B is a standard one-dimensional Brownian motion). Introduce the function $W : (0, \infty) \times [0, \infty) \to \mathbb{R}$ by the formula

$$W(x, y) = (x + y)^p g\left(\frac{y - x}{x + y}\right)$$

and, for any $a \in (0, z_0)$, consider the stopping time

$$\tau^a = \inf\left\{t \geq 0 : S_t \geq \frac{1 + a}{1 - a} R_t\right\}. \tag{6.89}$$

Applying Itô's formula, it is not difficult to check that $(W(R_{\tau^a \wedge t}, S_{\tau^a \wedge t}))_{t \geq 0}$ is a martingale. Consequently, for any $t \geq 0$,

$$W(1, 1) = \mathbb{E}W(R_{\tau^a \wedge t}, S_{\tau^a \wedge t}) \leq \sup_{[-1, a]} g \cdot \mathbb{E}(R_{\tau^a \wedge t} + S_{\tau^a \wedge t})^p.$$

But g has no roots in $[-1, a]$ and hence the number $\sup_{[-1,a]} g$ is negative. This yields

$$\mathbb{E}R^p_{\tau^a \wedge t} \leq W(1, 1) / \sup_{[-1, a]} g < \infty. \tag{6.90}$$

By the Burkholder-Gundy inequality for Bessel processes (see [60]) we obtain that $\tau^a \in L^{p/2}$ and R_{τ^a}, $S_{\tau^a} \in L^p$. Finally, since R, S start from 1, we have $\tau^a > 0$ almost surely and thus the expectations $\mathbb{E}R^p_{\tau^a}$ and $\mathbb{E}S^p_{\tau^a}$ are strictly positive. It suffices to use the definition of τ^a and the fact that $a < z_0$ is arbitrary to conclude that the constant $C_{p,d}$ is indeed the best possible. $\square$

On the search of the suitable majorant. We will describe the reasoning which leads to the sharp version of the estimate

$$\|S_\tau\|_p \leq C_{p,d} \|R_\tau\|_p \tag{6.91}$$

for stopped Bessel processes of dimension d. We will present the arguments only in the case $p < 2$; for $p \geq 2$ the ideas are similar. First, we assume that both R, S start from 1. A priori, these processes satisfy the stochastic differential equations

$$dR_t = dB_t + \frac{d - 1}{2}\frac{dt}{R_t},$$

$$dS_t = d\overline{B}_t + \frac{d - 1}{2}\frac{dt}{S_t},$$

where B, $\overline{B}$ are two adapted one-dimensional Brownian motions. Thus, during the search for the optimal $C_{p,d}$, we have to deal with two questions: what is the best choice for the pair $(B, \overline{B})$ and what is the optimal stopping time τ. The key is to look at the one-dimensional case, the solution to which follows immediately from the study of (3.59) (or rather (5.32)) above. Namely, if X, Y are adapted one-dimensional Brownian motions with $X_0 = Y_0 = 1$, then for $1 < p < 2$ we have $||Y_\tau||_p \leq (p-1)^{-1}||X_\tau||_p$; the sharpness is obtained by taking $\mathrm{d}Y_t = -\mathrm{sgn}\,(Y_t)\mathrm{d}X_t$ and the stopping time $\sigma^a = \inf\{t : |Y_t| \geq a|X_t|\}$, where $1 < a < (p-1)^{-1}$, and letting $a \to (p-1)^{-1}$. Thus, it is natural to conjecture that the extremal pairs (R, S) in (6.91) satisfy (6.88) and the asymptotically optimal stopping time is given by (6.89), where $1 < a < C_{p,d}$ is close to $C_{p,d}$. This gives rise to the function

$$U_{p,d}(x, y) = \begin{cases} W(x, y) & \text{if } y \leq C_{p,d}x, \\ y^p - C^p_{p,d}x^p & \text{if } y > C_{p,d}x, \end{cases}$$

for some unknown W. This function must be homogeneous of order p and satisfy the condition that $W((R_{\tau^a \wedge t}, S_{\tau^a \wedge t}))_{t \geq 0}$ is a martingale (for any a as above). Writing W in the form $W(x, y) = (x + y)^p g\left(\frac{y-x}{x+y}\right)$, the latter requirement, combined with (6.88) and Itô's formula, gives rise to the equation (6.80). □

6.10.4 Further results

Theorem 6.11 (or rather Corollary 1) deals with the case in which both martingales are conformal. One can consider similar results in which only one martingale has this property.

Theorem 6.12. *Suppose that X and Y are two $\mathbb{R}^2$-valued martingales on the filtration of 2-dimensional Brownian motion such that Y is differentially subordinate to X.*

(i) *If Y is conformal, then*

$$||Y||_p \leq \frac{a_p}{\sqrt{2}(1 - a_p)}||X||_p, \qquad 1 < p \leq 2,$$

where a_p is the least positive root in the interval $(0, 1)$ of the bounded Laguerre function L_p. This inequality is sharp.

(ii) *If X is conformal, then*

$$||Y||_p \leq \frac{\sqrt{2}(1 - a_p)}{a_p}||X||_p, \qquad 2 \leq p < \infty,$$

where a_p is the least positive root in the interval $(0, 1)$ of the bounded Laguerre function L_p. This inequality is sharp.

The proof is similar to that of Theorem 6.11. We omit the details and refer the interested reader to [14], [15].

6.11   Notes and comments

Section 6.1. The notion of orthogonality is classical and was already studied in the works of Itô. The modification of Burkholder's method leading to inequalities for differentially subordinated orthogonal martingales was introduced in the papers by Bañuelos and Wang [8], [9] and [10]. The appropriate version for submartingales was exploited by the author in [132].

Section 6.2. The special function corresponding to the weak type $(1,1)$ inequality was discovered by Davis [58] in his study of the corresponding result for conjugate harmonic functions on the unit disc of $\mathbb{C}$. Davis' approach was probabilistic in nature and used Brownian motion and Kakutani's theorem. Davis' result was rephrased in an analytic language by Baernstein [3]. This was taken up by Choi [50], who established a more general inequality for orthogonal harmonic functions given on Euclidean domains. The probabilistic counterpart of Davis' inequality, the weak type $(1,1)$ inequality for differentially subordinate orthogonal martingales, was established by Bañuelos and Wang [9]. The further extension to weak-type (p,p) estimates is due to Janakiraman [103] (see also [133]). The reader is also referred to the papers [190], [191] by Tomaszewski, which contain related weak type estimates for conjugate harmonic functions on the unit disc.

Section 6.3. The special functions corresponding to L^p estimates were constructed by Pichorides [171] during the study of the best constants in the M. Riesz's inequality for conjugate harmonic functions. See also Essén [72], [73] for related results. The extension to the martingale setting is due to Bañuelos and Wang [8], who also established some vector-valued extensions: namely, they proved that (6.12) holds for $\mathcal{H}$-valued X when $1 < p \leq 2$, and for $\mathcal{H}$-valued Y when $p \geq 2$. The above general statement in the vector-valued case is due to the author [155].

Bañuelos and Wang [8] used the L^p estimates for orthogonal martingales to derive the norms of Riesz transforms as operators acting on $L^p(\mathbb{R}^n)$. Their approach utilizes the so-called background radiation process introduced by Gundy and Varopoulos: see [86] and [192]. The norms turn out to be equal to the Pichorides-Cole constants $\cot(\pi/(2p^*))$, $1 < p < \infty$, as was first shown by Iwaniec and Martin [97] with the use of the method of rotations.

Section 6.4. The contents of that section is taken from [161].

Section 6.5. The logarithmic estimate for martingales comes from [161]. This $L \log L$ inequality is natural, especially in view of the corresponding result for harmonic functions, due to Zygmund [205]. There is a natural question about the best constant when Ψ is given by $\Psi(t) = t \log^+ t$. The following statement was established in [163]. Let

$$\Phi(t) = \begin{cases} t & \text{if } 0 \leq t \leq 1, \\ \exp(t-1) & \text{if } t > 1. \end{cases}$$

Theorem 6.13. *Let X, Y be real-valued orthogonal martingales such that Y is differentially subordinate to X and $Y_0 = 0$. Then for any $K > 2/\pi$ we have*

$$||Y||_1 \leq \sup_{t \geq 0} \mathbb{E}|X_t| \log^+ |X_t| + K^2 \int_0^\infty \frac{\Phi(s)}{\cosh(\pi K s/2)} \, ds$$

and the inequality is sharp.

Section 6.6. The results presented in this section are new, though in fact they base on the tools developed in Section 6.3. See also Pichorides [171] for related results for nonnegative harmonic functions and their conjugates.

Section 6.7 and Section 6.8. There is a natural question of how much of the work done in the martingale setting can be carried over to the submartingale case: compare the results of Chapter 3 and Chapter 4. This turns out to be a quite difficult problem: essentially, the only results in this direction are described in these sections. The weak type $(1,1)$ estimate was taken from [158] and the moment estimate of Section 6.8 comes from the author's paper [132].

Section 6.9. The literature on the subject is very rich, especially in the case when the domain is the unit disc of $\mathbb{C}$ and the harmonic functions are assumed to satisfy Cauchy-Riemann equations. The first results of this type are the classical theorems of Kolmogorov [109], Riesz [179], [180] and Zygmund [205]. For further results, see Bañuelos and Wang [8], Choi [50], Davis [58], Essén [72], [73], Essén, Shea and Stanton [74], [75], [76] and [77], Janakiraman [103], Pichorides [171] and Verbitsky [196]. See also Hollenbeck, Kalton and Verbitsky [93], Hollenbeck and Verbitsky [94], [95] for related results.

Section 6.10. Conformal martingales have been studied quite intensively in the literature, see, e.g., [83] for details. Their connection with the Beurling-Ahlfors operator has been investigated by a number of authors: see Bañuelos and Janakiraman [5], Bañuelos and Méndez-Hernandez [6], Bañuelos and Osękowski [7], Borichev, Janakiraman and Volberg [14], [15]. Theorem 6.11 presented above constitutes the main result of [7] (see also [15]), while Theorem 6.12 can be found in [14] and [15].

Chapter 7

Maximal Inequalities

We turn to another very important class of inequalities, involving the maximal functions and one-sided maximal functions of semimartingales.

7.1 Modification of Burkholder's method

7.1.1 A version for martingales in discrete time

We shall first show how to modify the method so that it yields maximal inequalities for ± 1-transforms of vector-valued martingales. Let $\mathcal{B}$ be a Banach space, put $D = \mathcal{B} \times \mathcal{B} \times [0, \infty) \times [0, \infty)$ and fix a function $V : D \to \mathbb{R}$, satisfying the condition

$$V(x, y, z, w) = V(x, y, |x| \vee z, |y| \vee w) \qquad \text{for all } (x, y, z, w) \in D. \tag{7.1}$$

Suppose that we want to establish the estimate

$$\mathbb{E}V(f_n, g_n, |f_n|^*, |g_n|^*) \leq 0, \qquad n = 0, 1, 2, \ldots, \tag{7.2}$$

for all simple martingales f, g taking values in $\mathcal{B}$ such that g is a ± 1-transform of f. As previously, one needs to consider a class of special functions: those which majorize V and yield a supermartingale when composed with the process $((f_n, g_n, |f_n|^*, |g_n|^*))_{n \geq 0}$. Consider a function $U : D \to \mathbb{R}$ which satisfies the following conditions.

$1°$ (Majorization property) If $(x, y, z, w) \in D$, then

$$V(x, y, z, w) \leq U(x, y, z, w). \tag{7.3}$$

$2°$ (Concavity along the lines of slope ± 1) Let $(x, y, z, w) \in D$ be such that $|x| \leq z$, $|y| \leq w$. If $\varepsilon \in \{-1, 1\}$ and $\alpha \in (0, 1)$, $t_1, t_2 \in \mathcal{B}$ satisfy $\alpha t_1 + (1 - \alpha)t_2 = 0$, then

$$\alpha U(x + t_1, y + \varepsilon t_1, z, w) + (1 - \alpha)U(x + t_2, y + \varepsilon t_2, z, w) \leq U(x, y, z, w). \tag{7.4}$$

3° (The initial condition) For any $x \in \mathcal{B}$ and $\varepsilon \in \{-1, 1\}$,

$$U(x, \varepsilon x, |x|, |x|) \leq 0. \tag{7.5}$$

4° (The "maximal process" condition) If $(x, y, z, w) \in D$, then

$$U(x, y, z, w) = U(x, y, |x| \vee z, |y| \vee w). \tag{7.6}$$

Clearly, the concavity condition is equivalent to saying that if d is a mean-zero random variable taking two values from $\mathcal{B}$, then

$$\mathbb{E}U(x+d, y+d, z, w) \leq U(x, y, z, w), \quad \mathbb{E}U(x+d, y-d, z, w) \leq U(x, y, z, w), \tag{7.7}$$

provided $|x| \leq z$ and $|y| \leq w$. By easy induction, if this holds, then (7.7) is also valid when d takes a finite number of values (but still $\mathbb{E}d = 0$).

The relation between functions satisfying 1°–4° and the inequality (7.2) is established in Theorems 7.1 and 7.2 below. We shall use the following notation: for any $(x, y) \in \mathcal{B} \times \mathcal{B}$, the class $M(x, y)$ consists of all pairs f, g of simple martingales satisfying $f_0 \equiv x$, $g_0 \equiv y$ and such that for any $n \geq 1$ we have $dg_n \equiv df_n$ or $dg_n \equiv -df_n$.

Theorem 7.1. *Suppose that U satisfies 1°, 2° and 4°. Then for any $(x, y, z, w) \in D$ and any $(f, g) \in M(x, y)$ we have*

$$\mathbb{E}V(f_n, g_n, |f_n|^* \vee z, |g_n|^* \vee w) \leq U(x, y, z, w), \quad n = 0, 1, 2, \ldots. \tag{7.8}$$

Consequently, if U satisfies 1°–4°, then (7.2) is valid.

Proof. It suffices to show that the process $(U(f_n, g_n, |f_n|^* \vee z, |g_n|^* \vee w))_{n \geq 0}$ is a supermartingale. By (7.6) we have, for $n \geq 1$,

$$U(f_n, g_n, |f_n|^* \vee z, |g_n|^* \vee w) = U(f_n, g_n, |f_n| \vee |f_{n-1}|^* \vee z, |g_n| \vee |g_{n-1}|^* \vee w)$$
$$= U(f_n, g_n, |f_{n-1}|^* \vee z, |g_{n-1}|^* \vee w).$$

Now, using the fact that $|f_{n-1}| \leq |f_{n-1}|^*$, $|g_{n-1}| \leq |g_{n-1}|^*$, we have, by the conditional form of (7.7),

$$\mathbb{E}\left[U(f_n, g_n, |f_{n-1}|^* \vee z, |g_{n-1}|^* \vee w) \big| \mathcal{F}_{n-1}\right]$$
$$= \mathbb{E}\left[U(f_{n-1} + df_n, g_{n-1} + dg_n, |f_{n-1}|^* \vee z, |g_{n-1}|^* \vee w) \big| \mathcal{F}_{n-1}\right]$$
$$\leq U(f_{n-1}, g_{n-1}, |f_{n-1}|^* \vee z, |g_{n-1}|^* \vee w).$$

This completes the proof.   □

We have also a result in the reverse direction.

Theorem 7.2. *If the inequality* (7.2) *is valid, then there is* $U : D \to \mathbb{R}$ *satisfying* $1°\text{–}4°$. *Furthermore, the least function with this property is given by*

$$U^0(x, y, z, w) = \sup\{\mathbb{E}V(f_\infty, g_\infty, |f_\infty|^* \vee z, |g_\infty|^* \vee w)\}, \tag{7.9}$$

where the supremum is taken over the class $M(x, y)$.

Proof. This can be proved using the splicing argument, similarly to the non-maximal case. We omit the easy proof. □

Remark 7.1. Before we proceed, let us make here two important observations.

(i) In fact, the reasoning above yields a slightly stronger statement. Namely, if U satisfies $1°\text{–}4°$, then (7.2) holds for all simple f, g such that g is a transform of f by a predictable sequence taking values in $\{-1, 1\}$. Similarly, suppose that in $2°$ we allow ε to take values from the interval $[-1, 1]$. Then we obtain inequalities in the case when g is a transform of f by a predictable sequence bounded in absolute value by 1.

(ii) The above technique can also be used in the situation when only one maximal function appears in the estimate under investigation. If this is the case, it suffices to omit the variable corresponding to the maximal function which is not involved. This results in simplifying of the problem, since then U depends only on three variables. We can go further and note that if the inequality (7.2) is non-maximal, that is, V is a function of x and y only, then the above approach is precisely Burkholder's method for ± 1-transforms described in Chapter 2.

Now we shall present the modification of the above approach for the case when f is a martingale and g is differentially subordinate to f. Then, for a given and fixed Borel $V : D \to \mathbb{R}$, one has to study functions U which satisfy the following conditions.

$1°$ If $(x, y, z, w) \in D$, then

$$V(x, y, z, w) \leq U(x, y, z, w). \tag{7.10}$$

$2°$ There are Borel functions A, $B : D \to \mathcal{B}^*$ such that if $(x, y, z, w) \in D$ satisfies $|x| \leq z$, $|y| \leq w$ and $h, k \in \mathcal{B}$ satisfy $|k| \leq |h|$, then

$$U(x + h, y + k, z, w) \leq U(x, y, z, w) + \langle A(x, y, z, w), h \rangle + \langle B(x, y, z, w), k \rangle. \tag{7.11}$$

$3°$ For any x, $y \in \mathcal{B}$ with $|y| \leq |x|$ we have

$$U(x, y, |x|, |y|) < 0. \tag{7.12}$$

$4°$ If $(x, y, z, w) \in D$, then

$$U(x, y, z, w) = U(x, y, |x| \vee z, |y| \vee w). \tag{7.13}$$

Theorem 7.3. *Suppose that $U, V : D \to \mathbb{R}$ are such that $1°$–$4°$ hold. Let f, g be $\mathcal{B}$-valued martingales such that g is differentially subordinate to f. If, in addition, f and g satisfy*

$$\mathbb{E}|V(f_n, g_n)| < \infty, \quad \mathbb{E}|U(f_n, g_n)| < \infty,$$
$$\mathbb{E}|A(f_n, g_n, |f_n|^*, |g_n|^*)||df_{n+1}| < \infty, \tag{7.14}$$
$$\mathbb{E}|B(f_n, g_n, |f_n|^*, |g_n|^*)||dg_{n+1}| < \infty$$

for any $n = 0, 1, 2, \ldots$, then (7.2) holds.

This can be established in the usual way: $(U(f_n, g_n, |f_n|^*, |g_n|^*))_{n \geq 0}$ is easily shown to be a supermartingale. The proof is straightforward and we shall not present it here.

Next, the method can be easily modified to yield inequalities involving one-sided maximal functions $f^* = \sup_{n \geq 0} f_n$, $g^* = \sup_{n \geq 0} g_n$. Clearly, these make sense only in the real-valued case. Let us focus on the extension of the technique for ± 1-transforms. Let $D = \mathbb{R}^4$ and let $V : D \to \mathbb{R}$ be a given function such that

$$V(x, y, z, w) = V(x, y, x \vee z, y \vee w) \qquad \text{for all} \quad (x, y, z, w) \in D.$$

Consider $U : D \to \mathbb{R}$ for which the following conditions are satisfied.

$1°$ If $(x, y, z, w) \in D$, then

$$V(x, y, z, w) \leq U(x, y, z, w).$$

$2°$ Let $(x, y, z, w) \in D$ be such that $x \leq z$, $y \leq w$. If $\varepsilon \in \{-1, 1\}$ and $\alpha \in (0, 1)$, $t_1, t_2 \in \mathbb{R}$ satisfy $\alpha t_1 + (1 - \alpha)t_2 = 0$, then

$$\alpha U(x + t_1, y + \varepsilon t_1, z, w) + (1 - \alpha)U(x + t_2, y + \varepsilon t_2, z, w) \leq U(x, y, z, w).$$

$3°$ For any $x \in \mathbb{R}$ and $\varepsilon \in \{-1, 1\}$,

$$U(x, \varepsilon x, x, \varepsilon x) \leq 0.$$

$4°$ If $(x, y, z, w) \in D$, then

$$U(x, y, z, w) = U(x, y, x \vee z, y \vee w).$$

The existence of such a function is equivalent to the validity of the estimate

$$\mathbb{E}V(f_n, g_n, f_n^*, g_n^*) \leq 0, \qquad n = 0, 1, 2, \ldots,$$

for any simple martingale f and its ± 1-transform g. The inequalities for differentially subordinate martingales can be studied in the similar manner. Of course, the approach above can be easily adjusted to "mixed" estimates, which involve both one-sided and two-sided maximal functions.

7.1.2 A version for sub- and supermartingales in discrete time

The above technique can be also implemented in the case when the dominating process f is a sub- or supermartingale. We shall focus on the submartingale case, replacing f by $-f$ if necessary. Let $D = \mathbb{R} \times \mathbb{R} \times [0, \infty) \times [0, \infty)$ and $V : D \to \mathbb{R}$ be a given function satisfying (7.1). Suppose we are interested in showing that

$$\mathbb{E}V(f_n, g_n, |f_n|^*, |g_n|^*) \leq 0, \qquad n = 0, 1, 2, \dots, \qquad (7.15)$$

for any simple submartingale f and any simple g which is a transform of f by a predictable sequence bounded in absolute value by 1. This problem is equivalent to finding a function U as in the martingale case, with 2° replaced by the following condition.

2° Let $(x, y, z, w) \in D$ be such that $|x| \leq z$, $|y| \leq w$. If $\varepsilon \in [-1, 1]$ and $\alpha \in (0, 1)$, $t_1, t_2 \in \mathbb{R}$ satisfy $\alpha t_1 + (1 - \alpha)t_2 \geq 0$, then

$$\alpha U(x+t_1, y+\varepsilon t_1, z, w)+(1-\alpha)U(x+t_2, y+\varepsilon t_2, z, w) \leq U(x, y, z, w). \quad (7.16)$$

The proof of this fact is the same as in the martingale setting. Similarly, suppose that V is given on $\mathbb{R} \times \mathcal{B} \times [0, \infty) \times [0, \infty)$ and one wants to study (7.15) for a submartingale f and its α-subordinate g taking values in $\mathcal{B}$ ($\alpha \geq 0$ is a fixed number). Consider the assumptions of Theorem 7.3, with 2° replaced by

2° There are Borel functions $A : D \to \mathbb{R}$, $B : D \to \mathcal{B}^*$ such that if $(x, y, z, w) \in D$ satisfies $|x| \leq z$, $|y| \leq w$ and $h \in \mathbb{R}$, $k \in \mathcal{B}$ satisfy $|k| \leq |h|$, then

$$U(x + h, y + k, z, w) \leq U(x, y, z, w) + \langle A(x, y, z, w), h \rangle + \langle B(x, y, z, w), k \rangle. \tag{7.17}$$

Furthermore, for any $(x, y, z, w) \in D$ such that $|x| \leq z$, $|y| \leq w$ and any $h \geq 0$, $k \in \mathcal{B}$ with $|k| \leq \alpha h$ we have

$$U(x + h, y + k, z, w) \leq U(x, y, z, w).$$

If there is U with the properties 1°–4°, then (7.15) is valid for any f, g satisfying (7.14).

There are similar statements for estimates involving one-sided maximal functions. The modification is straightforward and the details are left to the reader.

7.1.3 The passage to the continuous time

The reasoning is essentially the same as in the nonmaximal case. There are two ways of handling this problem. First, one can try to use discretization arguments to approximate continuous-time semimartingales by appropriate discrete-time sequences. The second approach is to apply the convolution argument and Itô's formula to the special function coming from the discrete-time case. We shall not state here general theorems as in Chapter 5 in order to avoid unnecessary and technical repetitions. It seems to be more convenient to illustrate the ideas on concrete examples. It should be stressed here that as in the non-maximal setting, the main difficulty lies in proving the estimate in the discrete-time case; the passage to the continuous-time case is a matter of standard technical arguments.

7.2   Doob-type inequalities

7.2.1   Formulation of the results

We shall show how the technique described above can be used to prove classical Doob's inequalities as well as certain extensions of them. We start with the weak type and moment inequalities.

Theorem 7.4. *Assume that $X = (X_t)_{t \geq 0}$ is a nonnegative submartingale. Then*

$$||X^*||_{p,\infty} \leq ||X||_p, \quad 1 \leq p \leq \infty \tag{7.18}$$

and the inequality is sharp.

Theorem 7.5. *Let $X = (X_t)_{t \geq 0}$ be a nonnegative submartingale. Then we have*

$$||X^*||_p \leq \frac{p}{p-1}||X||_p, \qquad p > 1, \tag{7.19}$$

and the inequality is sharp, even if X is assumed to be a nonnegative martingale.

The inequality (7.19) fails to hold when $p \leq 1$ and a natural question is what can be said for these values of p. We shall prove the following more general result. Let $\Phi : [0, \infty) \to [0, \infty)$ be an increasing function such that $\int_1^\infty \Phi(s)/s^2 ds < \infty$ and let

$$\Psi(x) = \begin{cases} x \int_x^\infty \frac{\Phi(s)}{s^2} ds & \text{if } x > 0, \\ \Phi(0) & \text{if } x = 0. \end{cases}$$

Theorem 7.6. *If X is a nonnegative supermartingale, then*

$$\mathbb{E}\Phi(X^*) \leq \mathbb{E}\Psi(X_0). \tag{7.20}$$

The inequality is sharp, even if X is a nonnegative martingale.

For example, if $\Phi(s) = s^p$, $0 < p < 1$, then $\Psi(x) = x^p/(1-p)$ is a concave function; hence the estimate (7.20) yields the following extension of Doob's inequality: if X is a nonnegative supermartingale, then

$$||X^*||_p \leq (1-p)^{-1/p}||X_0||_p = (1-p)^{-1/p}||X||_p \tag{7.21}$$

and the constant $(1-p)^{-1/p}$ is the best possible.

Next, we turn to logarithmic estimates.

Theorem 7.7. *Let* $X = (X_t)_{t \geq 0}$ *be a nonnegative submartingale. Then for* $K > 1$ *we have*

$$\|X^*\|_1 \leq K \sup_t \mathbb{E} X_t \log X_t + L(K), \tag{7.22}$$

where $L(K) = \frac{K^2}{(K-1)e}$. *The constant is the best possible, even if we restrict ourselves to nonnegative martingales. For* $K \leq 1$ *the logarithmic estimate does not hold with any finite* $L(K)$.

The final result of this section is the sharp comparison of moments of different order. We need the following auxiliary fact.

Lemma 7.1. *Let* $1 \leq p < q < \infty$. *Then there is a unique solution* $\gamma = \gamma_{p,q}$ *to the differential equation*

$$\gamma'(s) = \frac{ps^{p-1}}{q(q-1)\gamma(s)^{q-2}(s - \gamma(s))} \tag{7.23}$$

(extended to its maximal domain $[s_*, \infty)$), *which satisfies* $\gamma(s_*) = 0$, $\gamma(s) < s$ *and* $\lim_{s \to \infty} \gamma(s)/s = 1$.

See [169], [170] for details (the equation is studied there in a slightly different form). Unfortunately, there seems to be no closed-form formula for $\gamma_{p,q}$ except for some special cases. For example, it can be easily checked that $\gamma = \gamma_{1,q}$ can be written in the closed form

$$s \exp(-q\gamma(s)^{q-1}) = q(q-1) \int_{\gamma(s)}^{\infty} t^{q-1} \exp(-qt^{q-1}) dt,$$

from which we infer that

$$s_* = q(q-1) \int_0^{\infty} t^{q-1} \exp(-qt^{q-1}) dt = q^{1/(1-q)} \Gamma\left(\frac{q}{q-1}\right).$$

Theorem 7.8. *Let* $X = (X_t)_{t \geq 0}$ *be a nonnegative submartingale. Then for* $1 \leq p < q < \infty$ *we have*

$$\|X^*\|_p \leq C_{p,q} \|X\|_q, \tag{7.24}$$

where

$$C_{p,q} = s_*^{1-p/q} \left(\frac{q}{q-p}\right)^{1/p} \left(\frac{p}{q-p}\right)^{-1/q}.$$

The constant is the best possible, even if we restrict ourselves to discrete-time nonnegative martingales.

In particular, when $p = 1$, the inequality (7.24) can be rewritten in the more compact form

$$\|X^*\|_1 \leq \Gamma\left(\frac{2q-1}{q-1}\right)^{1-1/q} \|X\|_q. \tag{7.25}$$

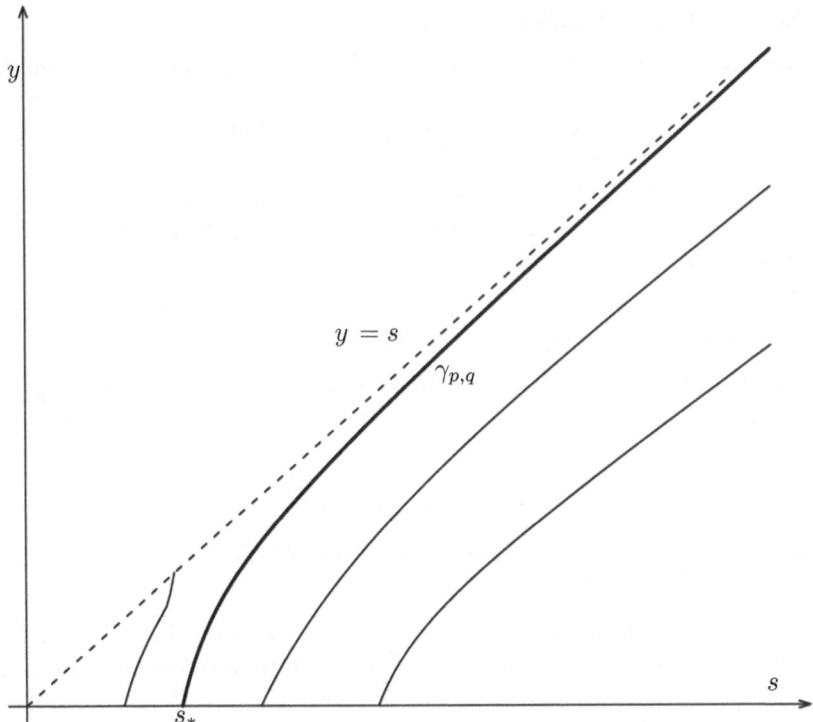

Figure 7.1: The function $\gamma_{p,q}$ (bold curve) is the unique solution γ to (7.23) which satisfies $\gamma(s)/s \to 1$ as $s \to \infty$.

7.2.2   Proof of Theorem 7.4

Proof of (7.18). Although the estimate follows immediately from Chebyshev's inequality and the stopping-time argument, it is instructive to see how Burkholder's method works here. Using standard approximation, it suffices to establish the bound for discrete-time simple submartingales. By homogeneity, it is enough to show that

$$\mathbb{P}(f_n^* \geq 1) \leq \mathbb{E}f_n^p, \qquad n = 0,\, 1,\, 2,\, \ldots.$$

Observe that the estimate depends only on one process: the sequence g does not appear at all. Consequently, the problem is two-dimensional, that is, the functions we shall study depend on x and z only. For any x, $z \geq 0$, let $V(x, z) = 1_{\{z \geq 1\}} - x^p$ and

$$U(x, z) = \begin{cases} 1 - x^p & \text{if } x \vee z \geq 1, \\ 0 & \text{if } x \vee z < 1. \end{cases}$$

We will verify the conditions $1°$–$4°$. The majorization is immediate. To check the concavity property, note that

$$U(x + d, (x + d) \vee z) \leq U(x, z) + U_x(x+, z)d,$$

valid for all $0 \leq x \leq z$ and $d \geq -x$. This estimate implies $2°$: here it is important that $U_x \leq 0$. Finally, the conditions $3°$ and $4°$ are evident; this proves (7.18). $\quad\square$

Sharpness. Obvious.

7.2.3   Proof of Theorem 7.5

Proof of (7.19). As previously, it suffices to focus on the discrete-time case. Let $V(x, z) = (x \vee z)^p - \left(\frac{p}{p-1}\right)^p x^p$ for $x, z \geq 0$. The special function $U : [0, \infty) \times [0, \infty) \to \mathbb{R}$ is given by the formula

$$U(x, z) = p(x \vee z)^{p-1}\left(x \vee z - \frac{p}{p-1}x\right).$$

Let us verify the conditions $1°$–$4°$. The majorization follows immediately from the mean value property of the function $t \mapsto t^p$. To show $2°$, note for any $0 \leq x \leq z$ and $d \geq -x$,

$$U(x + d, z) \leq U(x, z) - \frac{p^2}{p-1}z^{p-1}d. \tag{7.26}$$

Indeed, if $x + d \leq z$, then both sides are equal; if $x + d > z$, then we obtain, after some manipulations, the equivalent estimate

$$z^{p-1}(x + d) \leq \frac{(p-1)z^p}{p} + \frac{(x+d)^p}{p},$$

which follows from Young's inequality. Finally, the properties $3°$ and $4°$ are obvious and the inequality follows. $\quad\square$

Sharpness of (7.19). We shall use Theorem 7.2. Let $p > 1$ be fixed and suppose that the inequality (7.19) holds for all nonnegative martingales with some constant $\beta > 0$. For $x, z \geq 0$, let $V(x, z) = (x \vee z)^p - \beta^p x^p$ and

$$U^0(x, z) = \sup\{\mathbb{E}V(f_\infty, f^* \vee z)\},$$

where the supremum is taken over all simple nonnegative martingales f starting from x. Then U^0 satisfies appropriate versions of the conditions $1°$–$4°$. Furthermore, by the very definition, this function is homogeneous of order p. Now, fix $\delta > 0$ and use $2°$ with $x = z = 1$, $\varepsilon = 1$, $t_1 = \delta$, $t_2 = -1/p$ and $\alpha = (1 + p\delta)^{-1}$ to get

$$U^0(1, 1) \geq \frac{1}{1 + p\delta}U^0(1 + \delta, 1 + \delta) + \frac{p\delta}{1 + p\delta}U^0(1 - 1/p, 1)$$

$$\geq \frac{(1 + \delta)^p}{1 + p\delta}U^0(1, 1) + \frac{p\delta}{1 + p\delta}V(1 - 1/p, 1),$$

where in the last passage we have used the homogeneity and the majorization property. This estimate can be rewritten in the equivalent form

$$\frac{1 + p\delta - (1 + \delta)^p}{p\delta} U^0(1, 1) \geq 1 - \beta^p \left(\frac{p-1}{p}\right)^p.$$

Letting $\delta \to 0$ makes the left-hand side vanish; this shows that right-hand side is nonpositive. But this means that $\beta \geq p/(p-1)$ and that (7.19) is sharp, even for nonnegative martingales.   □

7.2.4   Proof of Theorem 7.6

Proof of (7.20). As previously, by approximation, it suffices to focus on the discrete-time case. Let $V : [0, \infty) \times [0, \infty) \to \mathbb{R}$ be defined by $V(x, z) = \Phi(x \vee z)$ and consider the function U given by $U(x, 0) = \Psi(x)$ and

$$U(x, z) = \left(1 - \frac{x}{x \vee z}\right) \Phi(x \vee z) + \frac{x}{x \vee z} \Psi(x \vee z)$$

for $x \geq 0$ and $z > 0$. We shall now verify the properties 1°–4°. To check the majorization, observe that $U(0, 0) = V(0, 0)$ and, for $x \vee z > 0$,

$$U(x, z) - \Phi(x \vee z) = x \int_{x \vee z}^{\infty} \frac{\Phi(s)}{s^2} ds - x \frac{\Phi(x \vee z)}{x \vee z}$$

$$\geq x \int_{x \vee z}^{\infty} \frac{\Phi(x \vee z)}{s^2} ds - x \frac{\Phi(x \vee z)}{x \vee z} = 0.$$

To show the concavity property, we may assume that $z > 0$ (otherwise the condition is trivial). For $x \in [0, z]$ and $d \geq -x$ we have

$$U(x + d, z) \leq U(x, z) + \left[-\frac{\Phi(z)}{z} + \int_{z}^{\infty} \frac{\Phi(s)}{s^2} ds\right] d. \qquad (7.27)$$

Indeed, if $x + d \leq z$, then the two sides are equal to one another; if $x + d > z$, then

$$U(x + d, z) = (x + d) \int_{x+d}^{\infty} \frac{\Phi(s)}{s^2} ds = (x + d) \int_{z}^{\infty} \frac{\Phi(s)}{s^2} ds - (x + d) \int_{z}^{x+d} \frac{\Phi(s)}{s^2} ds$$

$$\leq (x + d) \int_{z}^{\infty} \frac{\Phi(s)}{s^2} ds - (x + d) \int_{z}^{x+d} \frac{\Phi(z)}{s^2} ds$$

$$= U(x, z) + \left[-\frac{\Phi(z)}{z} + \int_{z}^{\infty} \frac{\Phi(s)}{s^2} ds\right] d.$$

The inequality (7.27) implies 2° since the expression in the square brackets is nonnegative. The initial condition does not hold and needs to be modified: we have $U(x,x) = \Psi(x)$ whenever $x \geq 0$. Finally, $4°$ is obvious and Burkholder's method yields

$$\mathbb{E}V(f_n, f_n^*) \leq \mathbb{E}U(f_n, f_n^*) \leq \mathbb{E}U(f_0, f_0) \leq \mathbb{E}\Psi(f_0),$$

which is exactly what we need. □

Sharpness of (7.20). We will construct an appropriate example. Fix $\delta > 0$ and let $X_1, X_2, \cdots$ be a sequence of independent random variables such that $X_k \in \{0, 1+\delta\}$ and $\mathbb{E}X_k = 1$, $k = 1, 2, \ldots$. Let f be given by $f_0 \equiv 1$ and $f_n = X_1X_2\ldots X_n$, $n = 1, 2, \ldots$. Then f is a nonnegative martingale such that f^* takes values in the set $\{1, 1+\delta, (1+\delta)^2, \ldots\}$. To be more precise, for any integer N we have $f_\infty^* = (1+\delta)^N$ if and only if $X_1 = X_2 = \cdots = X_N = 1+\delta$ and $X_{N+1} = 0$. Therefore,

$$\mathbb{E}\Phi(f^*) = \delta \sum_{N=0}^{\infty} \frac{\Phi((1+\delta)^N)}{(1+\delta)^{N+1}} \geq \frac{\delta}{1+\delta} \sum_{N=0}^{\infty} \frac{\Phi((1+\delta)^{N+1})}{(1+\delta)^{N+1}}$$

$$= \frac{1}{1+\delta} \sum_{N=0}^{\infty} \Phi((1+\delta)^{N+1}) \int_{(1+\delta)^N}^{(1+\delta)^{N+1}} s^{-2}ds$$

$$\geq \frac{1}{1+\delta} \sum_{N=0}^{\infty} \int_{(1+\delta)^N}^{(1+\delta)^{N+1}} \frac{\Phi(s)}{s^2}ds = \frac{1}{1+\delta} \int_1^{\infty} \frac{\Phi(s)}{s^2}ds = \frac{\mathbb{E}\Psi(f_0)}{1+\delta}.$$

Since $\delta > 0$ was arbitrary, the proof is complete. □

7.2.5 Proof of Theorem 7.7

Proof of (7.22). We can restrict ourselves to discrete-time submartingales f. Suppose that $K > 1$ and let $V : [0, \infty) \times [0, \infty) \to \mathbb{R}$ be given by $V(x, z) = (x \vee z) - Kx \log x - L(K)$. The corresponding special function $U : [0, \infty) \times [0, \infty) \to \mathbb{R}$ is given by

$$U(x, z) = \begin{cases} K(x \vee z) - Kx \log\left(\frac{K-1}{K}e(x \vee z)\right) - L(K) & \text{if } x \vee z \geq \frac{K}{(K-1)e}, \\ 0 & \text{if } x \vee z < \frac{K}{(K-1)e}. \end{cases}$$

Let us verify the conditions $1°$–$4°$. If $x \vee z \geq \frac{K}{(K-1)e}$, the majorization is equivalent to $s - x \geq x(\log s - \log x)$, where $s = (K-1)(x \vee z)/K$. This inequality is obvious for $x = 0$, while for $x > 0$ it can be rewritten in the form $s/x \geq \log(s/x) + 1$, which is also trivial. For $x \vee z < K/((K-1)e)$, the condition $1°$ is equivalent to an elementary bound $x \log x \geq e^{-1}$. To show $2°$, one easily verifies that for any $x, z \geq 0$ and $d \geq -x$ we have

$$U(x + d, z) \leq U(x, z) + A(x, z)d,$$

where $A : [0, \infty) \times [0, \infty) \to (-\infty, 0]$ is given by

$$A(x, z) = \begin{cases} -K \log \left(\frac{K-1}{K} e(x \vee z) \right) & \text{if } x \vee z \geq \frac{K}{(K-1)e}, \\ 0 & \text{if } x \vee z < \frac{K}{(K-1)e}. \end{cases}$$

The initial condition is straightforward: the function $x \mapsto U(x, x)$ is equal to 0 for $x \leq \frac{K}{(K-1)e}$ and is decreasing for $x > \frac{K}{(K-1)e}$. Finally, $4°$ is obvious. This completes the proof of (7.22). $\qquad\qquad\square$

Sharpness. Let $K > 1$ be fixed. To show the optimality of the constant $L(K)$ for nonnegative martingales, we use a version of Theorem 7.2. Let

$$U_0(x, z) = \sup\{\mathbb{E}(f_n^* \vee z) - K\mathbb{E}f_n \log f_n\}, \qquad x, z \geq 0,$$

where the supremum is taken over all n and all nonnegative martingales f starting from x. Then U_0 majorizes $V(x, z) = (x \vee z) - Kx \log x$, and satisfies $2°$ and $4°$. Furthermore, arguing as in the proof of Theorem 3.9 in Chapter 3, we can show that U_0 enjoys the homogeneity property

$$U_0(\lambda x, \lambda z) = \lambda U_0(x, z) - K\lambda x \log \lambda, \qquad \text{for } x, z \geq 0, \lambda > 0.$$

Let $\delta > 0$. Using this property together with $2°$ and $4°$ gives

$$\begin{aligned} U_0(1, 1) &\geq \frac{\delta}{K^{-1} + \delta} \left(1 + (K - 1) \log \left(\frac{K - 1}{K} \right) \right) + \frac{K^{-1}}{K^{-1} + \delta} U_0(1 + \delta, 1) \\ &= \frac{\delta}{K^{-1} + \delta} \left(1 + (K - 1) \log \left(\frac{K - 1}{K} \right) \right) + \frac{K^{-1}}{K^{-1} + \delta} U_0(1 + \delta, 1 + \delta) \\ &= \frac{\delta}{K^{-1} + \delta} \left(1 + (K - 1) \log \left(\frac{K - 1}{K} \right) \right) \\ &\quad + \frac{K^{-1}(1 + \delta)}{K^{-1} + \delta} \left[U_0(1, 1) - K \log(1 + \delta) \right]. \end{aligned}$$

Subtracting $U_0(1, 1)$ from both sides, dividing throughout by δ and letting $\delta \to 0$ yields $U_0(1, 1) \geq K \log \left(\frac{K}{K-1} \right)$. Consequently, again by homogeneity,

$$U_0 \left(\frac{K}{(K-1)e}, \frac{K}{(K-1)e} \right) \geq \frac{K^2}{(K-1)e} = L(K),$$

which is the claim, directly from the definition of U_0.

Finally, the fact that (7.22) does not hold for $K \leq 1$ with any $L(K) < \infty$ is an immediate consequence of the fact that $\lim_{K \downarrow 1} L(K) = \infty$. See the proof of Theorem 3.9 in Chapter 3 for details. $\qquad\qquad\square$

7.2.6 Proof of Theorem 7.8

Proof of (7.24). We can restrict ourselves to the discrete-time case. Clearly, we may assume that $||f||_q > 0$, otherwise there is nothing to prove. Let $1 \leq p < q < \infty$ be fixed, let s_* be the corresponding zero of $\gamma = \gamma_{p,q}$ and introduce $V : [0,\infty) \times [0,\infty) \to \mathbb{R}$ by $V(x,z) = (x \vee z)^p - x^q - s_*^p$. The special function U is given by

$$U(x,z) = \begin{cases} (x \vee z)^p + (q-1)[\gamma(x \vee z)]^q - qx[\gamma(x \vee z)]^{q-1} - s_*^p & \text{if } x \vee z \geq s_*, \\ 0 & \text{if } x \vee z < s_*. \end{cases}$$

Let us check the properties $1°-4°$. If $x \vee z \geq s_*$, then the inequality $U(x,z) \geq V(x,z)$ is equivalent to

$$[\gamma(x \vee z)]^q - x^q \leq q[\gamma(x \vee z)]^{q-1}(\gamma(x \vee z) - x)$$

and follows from the mean value property. On the other hand, if $x \vee z < s_*$, then

$$V(x,z) = (x \vee z)^p - x^q - s_*^p \leq s_*^p - s_*^p = 0 = U(x,z).$$

To show $2°$, it suffices to establish the pointwise bound

$$U(x+d,z) \leq U(x,z) + A(x,z)d,$$

where $A(x,z) = -q[\gamma(x \vee z)]^{q-1} 1_{\{x \vee z \geq s_*\}}$. This estimate is straightforward and we leave the details to the reader. To prove $3°$, observe that the function $x \mapsto U(x,x)$ equals 0 on $[0, s_*]$ and then decreases. Finally, $4°$ is obvious and Burkholder's method gives

$$||f^*||_p^p \leq ||f||_q^q + s_*^p.$$

It suffices to apply this bound to the submartingale λf, with

$$\lambda = \left(\frac{p}{q-p} s_*^p\right)^{1/q} ||f||_q^{-1}$$

and (7.24) follows. $\qquad\square$

Sharpness. Here we use a different argument. Fix $1 \leq p < q < \infty$ and let $\gamma = \gamma_{p,q}$. Let X be a one-dimensional Brownian motion, starting from s_*, stopped at $\tau = \inf\{t \geq 0 : X_t = \gamma(X_t^*)\}$. By standard approximation, if we show that

$$||X^*||_p = C_{p,q}||X||_q, \tag{7.28}$$

we will be done. First, since $\gamma(s)/s \to 1$ as $s \to \infty$, we have that X belongs to L^q (see, e.g., Wang [198]). The special function U used above is of class C^2 on $\{(x,z) : z > s_*\}$, so application of Itô's formula gives

$$\mathbb{E}U(X_t, X_t^*) = U(s_*, s_*) = 0$$

for any $t \geq 0$. However,

$$U(X_t, X_t^*) = U(X_t, X_t^*)1_{\{\tau \leq t\}} + U(X_t, X_t^*)1_{\{\tau > t\}}$$
$$= V(X_t, X_t^*)1_{\{\tau \leq t\}} + U(X_t, X_t^*)1_{\{\tau > t\}}$$

and, since $\gamma(s) \leq s$ for all $s \geq s_*$,

$$|U(X_t, X_t^*)1_{\{\tau > t\}}| \leq c(X_t^p + (X_t^*)^q)$$

for some absolute constant $c > 0$. Consequently, if we let $t \to \infty$ and use the boundedness of X in L^q, we obtain $\mathbb{E}V(X_\infty, X^*) = 0$, which is precisely (7.28). This completes the proof.   □

7.3   Maximal L^1 estimates for martingales

7.3.1   Formulation of the result.

We turn to inequalities which involve two processes. We start with the following fact.

Theorem 7.9. *For any real-valued martingales X, Y such that Y is differentially subordinate to X we have*

$$||Y||_1 \leq \beta|| \, |X|^*||_1, \tag{7.29}$$

where $\beta = 2.536\ldots$ is the unique positive number satisfying

$$\beta = 3 - \exp \frac{1 - \beta}{2}.$$

The constant β is the best possible.

If X is assumed to be nonnegative, then the constant in the above inequality decreases to $2 + (3e)^{-1} = 2.1226\ldots$. In addition, this can be further extended to cover the case of nonnegative supermartingales and their α-subordinates. Precisely, we have the following.

Theorem 7.10. *Let X be a nonnegative supermartingale and let Y be α-subordinate to X.*

(i) *If $\alpha \in [0, 1]$, then*

$$||Y||_1 \leq (2 + (3e)^{-1})||X^*||_1 = 2.1226\ldots||X^*||_1. \tag{7.30}$$

   The constant is the best possible. It is already the best possible if X is assumed to be a nonnegative martingale.

(ii) *If $\alpha > 1$, then*

$$||Y||_1 \leq (\alpha + 1 + ((2\alpha + 1)e)^{-1})||X^*||_1. \tag{7.31}$$

   The constant is the best possible.

A very natural and interesting question is what is the best constant if X is assumed to be a nonnegative submartingale and Y is α-subordinate to X. Unfortunately, we have been able to answer it only in the case when both processes have continuous paths; see below.

7.3.2 Proof of Theorem 7.9

For reader's convenience, we have decided to split the proof of (7.29) into two parts. First we shall establish this estimate in discrete time and when Y is a ± 1-transform of X. As already mentioned above, this is the heart of the matter. Then, in the second step we shall use the convolution argument and Itô's formula to obtain the result in full generality.

Proof of (7.29) for ± 1-transforms of simple martingales. Introduce the function $V : \mathbb{R} \times \mathbb{R} \times [0, \infty) \to \mathbb{R}$ by $V(x, y, z) = |y| - \beta |x| \vee z$. We shall show that for any integer n and any $\varepsilon > 0$ we have

$$\mathbb{E}V(f_n, g_n, \varepsilon \vee |f_n|^*) \leq 0, \tag{7.32}$$

where f, g are simple real-valued martingales such that g is a ± 1-transform of f. This clearly yields the claim by virtue of Lebesgue's dominated convergence theorem. Let S denote the strip $[-1, 1] \times \mathbb{R}$. Consider the following subsets of S:

$$D_0 = \{(x, y) : 0 \leq y \leq x \leq 1\},$$
$$D_1 = \{(x, y) : 0 \leq x \leq 1, \, x < y \leq x + \beta - 1\},$$
$$D_2 = \{(x, y) : 0 \leq x \leq 1, \, y > x + \beta - 1\}.$$

First let us introduce an auxiliary function u on S. It is given by the symmetry condition

$$u(x, y) = u(-x, y) = u(x, -y) \quad \text{for all } (x, y) \in S \tag{7.33}$$

and the formula

$$u(x, y) = \begin{cases} \frac{2}{\beta - 1}\left[-x - \frac{1}{3} - \frac{1}{3}(2 - 2x - y)(1 - x + y)^{1/2}\right] & \text{on } D_0, \\ \frac{2}{\beta - 1}\left[y - 3 + (2 - x)\exp\left(\frac{x - y}{2}\right)\right] & \text{on } D_1, \\ y - \beta + \frac{3 - \beta}{\beta - 1}(1 - x)\exp(x - y + \beta - 1) & \text{on } D_2. \end{cases}$$

We shall need the following technical fact.

Lemma 7.2. *The function u enjoys the following properties:*
(i) $u(1, y) \geq |y| - \beta$ *for all $y \in \mathbb{R}$,*
(ii) u *is diagonally concave,*
(iii) $u(1, \cdot)$ *is convex,*
(iv) *for any $y \in \mathbb{R}$,*

$$\lim_{\delta \downarrow 0} \frac{u(1, y) - u(1 - \delta, y \pm \delta)}{\delta} \geq u(1, 1).$$

Proof. It is easy to check that $u(1, \cdot)$ is of class C^1 on $\mathbb{R}$ and $u(1, y) = |y| - \beta$ provided $|y| \geq \beta$. Thus item (i) follows from (iii), which is evident. To show (ii), note that u is of class C^1 in the interior of the strip S and hence it suffices to check the diagonal concavity on $[0, 1] \times [0, \infty)$. We easily derive that $u_{xy}(x, y) \geq 0$, when (x, y) lies in the interior of D_0, D_1 or D_2, and hence

$$u_{xx}(x, y) + 2|u_{xy}(x, y)| + u_{yy}(x, y) = u_{xx}(x, y) + 2u_{xy}(x, y) + u_{yy}(x, y).$$

However, the right-hand side is equal to zero: the function u is linear along the line segments of slope 1 contained in $[0, 1] \times [0, \infty)$. This gives (ii). Finally, to check (iv), we compute

$$\lim_{\delta \downarrow 0} \frac{u(1, y) - u(1 - \delta, y + \delta)}{\delta} = \lim_{\delta \downarrow 0} \frac{u(1, -y) - u(1 - \delta, -y - \delta)}{\delta}$$

$$= \begin{cases} u(1, 1) + \frac{\beta+1}{\beta-1} - \frac{3-\beta}{\beta-1} \exp(-|y| + \beta) & \text{if } y < -\beta, \\ u(1, 1) + \frac{2}{\beta-1} \left(2 - \exp\left(\frac{1-|y|}{2} \right) \right) & \text{if } y \in [-\beta, -1], \\ u(1, 1) + \frac{2}{\beta-1} |y|^{1/2} & \text{if } y \in (-1, 0), \\ u(1, 1) & \text{if } y \in [0, \beta], \\ u(1, 1) - \frac{3-\beta}{\beta-1} \left[\exp(-y + \beta) - 1 \right] & \text{if } y > \beta \end{cases}$$

and note that all expressions are not smaller than $u(1, 1)$. The proof is complete.
□

The special function $U : \mathbb{R} \times \mathbb{R} \times (0, \infty) \to \mathbb{R}$ (note that $z = 0$ is not allowed) is defined by the formula

$$U(x, y, z) = (|x| \vee z)\, u \left(\frac{x}{|x| \vee z}, \frac{y}{|x| \vee z} \right).$$

Let us check that U has the properties 1°–4°. To show the majorization, note that we may assume $z = 1$, due to homogeneity. Then the inequality reads $u(x, y) \geq |y| - \beta$. By part (ii) of Lemma 7.2, it suffices to establish this for $|y| = 1$ and here part (i) comes into play. We turn to 2°. By (7.33), we will be done if we show this for $\varepsilon = -1$. Furthermore, by homogeneity, we may assume that $z = 1$. Let $x \in [-1, 1]$, $y \in \mathbb{R}$ and put $\Phi(t) = U(x + t, y - t, z)$. We shall prove that

$$\Phi(t) \leq \Phi(0) + A(x, y)t \tag{7.34}$$

for some A and all $t \in \mathbb{R}$; this immediately gives 2°. The inequality (7.34) is a consequence of the following three properties:

$$\Phi \text{ is concave on } [-1 - x, 1 - x], \tag{7.35}$$

$$\Phi \text{ is convex on each of the intervals } (-\infty, -1 - x], [1 - x, \infty), \tag{7.36}$$

$$\lim_{t \to -\infty} \frac{\Phi(t)}{t} \geq \Phi'(-1 - x+), \qquad \lim_{t \to \infty} \frac{\Phi(t)}{t} \leq \Phi'(1 - x-). \tag{7.37}$$

The condition (7.35) follows from part (ii) of Lemma 7.2. To show (7.36) fix $\alpha_1, \alpha_2 > 0$ satisfying $\alpha_1 + \alpha_2 = 1$, choose $t_1, t_2 \in [1 - x, \infty)$ and let $t = \sum \alpha_k t_k$. We have

$$\sum \alpha_k \Phi(t_k) = \sum \alpha_k U(x + t_k, y + t_k, 1)$$
$$= \sum \alpha_k \left[(x + t_k) u \left(-1, \frac{y + t_k}{x + t_k} \right) \right]$$
$$= (x + t) \sum \frac{\alpha_k (x + t_k)}{x + t} u \left(1, \frac{y + t_k}{x + t_k} \right).$$

However, by part (iii) of Lemma 7.2, this can be bounded from below by

$$(x + t) u \left(1, \sum \frac{y + t_k}{x + t_k} \cdot \frac{\alpha_k (x + t_k)}{x + t} \right) = (x + t) u \left(1, \frac{y + t}{x + t} \right) = \Phi(t).$$

The convexity on $(-\infty, 1]$ follows from the symmetry condition (7.33). Finally, to show (7.37), we note that for $t < -1 - x$,

$$\frac{\Phi(t)}{t} = \frac{x + t}{t} u \left(1, \frac{y - t}{x + t} \right) \xrightarrow{t \to \infty} u(1, 1) \leq \Phi'(1 - x-),$$

where in the last passage we have used part (iv) of Lemma 7.2. The second inequality in (7.37) follows from the symmetry condition (7.33). The condition $3°$, due to the appearance of ε in (7.32), needs to be slightly modified. The statement reads

$3°$ $U(x, \pm x, z) \leq 0$ for all $z > 0$ and $0 \leq x \leq z$,

and this follows by simply noting that $u(x, \pm x) = -\frac{2}{\beta - 1} < 0$ for $x \in [-1, 1]$. Finally, the property $4°$ follows directly from the definition. This completes the proof of (7.29) for ± 1-transforms. $\qquad \square$

Proof of (7.29) *in the general case.* We may and do assume that $|| \, |X|^* ||_1$ is finite, since otherwise the claim is obvious. For a fixed positive integer N, consider the stopping time $\tau = \inf\{t \geq 0 : |X_t| + |Y_t| \geq N\}$. We have, by the differential subordination,

$$|Y_{\tau \wedge t}| \leq |Y_{\tau \wedge t-}| + |\Delta Y_{\tau \wedge t}| \leq N + |\Delta X_{\tau \wedge t}|$$

(with the convention $Y_{0-} = 0$), so $|Y_\tau|^*$ belongs to L^1.

Observe that the function U satisfies the inequality

$$U_z(x, y, z) \leq 0 \tag{7.38}$$

for any $z > 0$ and any $x, y \in \mathbb{R}$ such that $|x| \leq z$. Furthermore, for such (x, y, z) and $h, k \in \mathbb{R}$ satisfying $|k| \leq |h|$, we have

$$U(x + h, y + k, z) \leq U(x, y, z) + U_x(x, y, z) h + U_y(x, y, z) k. \tag{7.39}$$

Indeed, we have shown this above in the case $h = \pm k$, but the reasoning works for $k \in [-|h|, |h|]$ as well.

Fix $\delta > 0$ and let $g : \mathbb{R}^3 \to [0, \infty)$ be a C^∞ function, supported on the unit ball of $\mathbb{R}^3$, satisfying $\int_{\mathbb{R}^3} g = 1$. Define a smooth $\overline{U} : \mathbb{R} \times \mathbb{R} \times (0, \infty) \to \mathbb{R}$ by the convolution

$$\overline{U}(x, y, z) = \int_{\mathbb{R}^3} U(x + \delta u_1, y + \delta u_2, z + 2\delta + \delta u_3) g(u_1, u_2, u_3) du_1 du_2 du_3.$$

Observe that when $|x| \leq z$, then also $|x + \delta u_1| \leq z + 2\delta + \delta u_3$ for (u_1, u_2, u_3) lying in the unit ball of $\mathbb{R}^3$. Consequently, $\overline{U}$ inherits the following properties from U:

(i) for any fixed $z > 0$, $\overline{U}(\cdot, \cdot, z)$ is diagonally concave on $[-z, z] \times \mathbb{R}$.

(ii) $\overline{U}(x, \cdot, z)$ is convex for any fixed $x \in [-z, z]$, $z > 0$.

(iii) $\overline{U}_z(x, y, z) \leq 0$ for any $z > 0$, $x \in [-z, z]$ and $y \in \mathbb{R}$.

(iv) $\overline{U}(x, y, z) \leq c(|y| + |x| \vee z)$ for all $(x, y, z) \in D$ and some absolute c (not depending on δ).

(v) $\overline{U}$ satisfies (7.39).

We may write, for $|x| < z$ and $y \in \mathbb{R}$,

$$\overline{U}_{xx}(x, y, z)h^2 + 2\overline{U}_{xy}(x, y, z)hk + \overline{U}_{yy}(x, y, z)k^2$$
$$\leq \overline{U}_{xx}(x, y, z)h^2 - \Delta\overline{U}(x, y, z)\frac{h^2 + k^2}{2} + \overline{U}_{yy}(x, y, z)k^2 \qquad (7.40)$$
$$= -c(x, y, z)(h^2 - k^2).$$

Here

$$c(x, y, z) = \frac{-\overline{U}_{xx}(x, y, z) + \overline{U}_{yy}(x, y, z)}{2} \geq \overline{U}_{yy}(x, y, z) \geq 0,$$

where in the first inequality we have used the superharmonicity of $\overline{U}$ (which follows from (i)), while in the second we have exploited (ii). Let $Z_t = (X_{\tau \wedge t}, Y_{\tau \wedge t}, |X_{\tau \wedge t}|^* \vee \varepsilon)$ for $t \geq 0$ and $\varepsilon > 0$. Application of Itô's formula gives

$$\overline{U}(Z_t) = \overline{U}(Z_0) + I_1 + I_2 + \frac{1}{2}I_3 + I_4,$$

where

$$I_1 = \int_{0+}^{\tau \wedge t} \overline{U}_x(Z_{s-})dX_s + \int_{0+}^{\tau \wedge t} \overline{U}_y(Z_{s-})dY_s,$$

$$I_2 = \int_{0+}^{\tau \wedge t} \overline{U}_z(Z_{s-})d(|X_s|^* \vee \varepsilon),$$

$$I_3 = \int_{0+}^{\tau \wedge t} \overline{U}_{xx}(Z_{s-})d[X,X]_s^c + 2\int_{0+}^{\tau \wedge t} \overline{U}_{xy}(Z_{s-})d[X,Y]_s^c + \int_{0+}^{\tau \wedge t} \overline{U}_{yy}(Z_{s-})d[Y,Y]_s^c,$$

$$I_4 = \sum_{0 < s \leq \tau \wedge t} \left[\overline{U}(Z_s) - \overline{U}(Z_{s-}) - \overline{U}_x(Z_{s-})|\Delta X_s| - \overline{U}_y(Z_{s-})|\Delta Y_s|\right].$$

Using the localizing argument if necessary, we may assume that $\mathbb{E}I_1 = 0$. By (iii), $I_2 \leq 0$. Reasoning as in the nonmaximal setting, we have that (7.40) implies $I_3 \leq 0$. Finally $I_4 \leq 0$ due to (v). Consequently, we obtain

$$\mathbb{E}\overline{U}(X_{\tau \wedge t}, Y_{\tau \wedge t}, |X_{\tau \wedge t}|^* \vee \varepsilon) \leq \mathbb{E}\overline{U}(X_0, Y_0, |X_0| \vee \varepsilon).$$

Now let δ go to 0. By (iv) and Lebesgue's dominated convergence theorem, we obtain

$$\mathbb{E}U(X_{\tau \wedge t}, Y_{\tau \wedge t}, |X_{\tau \wedge t}|^* \vee \varepsilon) \leq \mathbb{E}U(X_0, Y_0, |X_0| \vee \varepsilon) \leq 0,$$

so, using the majorization,

$$\mathbb{E}|Y_{\tau \wedge t}| \leq \beta \mathbb{E}(|X_{\tau \wedge t}|^* \vee \varepsilon).$$

It suffices to let $\varepsilon \to 0$, $N \to \infty$ and then $t \to \infty$ to get the claim. $\qquad\square$

Sharpness of (7.29). The proof will be based on Theorem 7.2. Suppose that $\beta_0 > 0$ is such that

$$\|g\|_1 \leq \beta_0 \| |f|^* \|_1$$

for all simple real-valued martingales f and their ± 1-transforms g. For x, $y \in \mathbb{R}$ and $z \geq 0$, let $V(x, y, z) = |y| - \beta_0(|x| \vee z)$ and

$$U^0(x, y, z) = \sup\{\mathbb{E}V(f_\infty, g_\infty, |f_\infty|^* \vee z)\},$$

where the supremum runs over all $(f, g) \in M(x, y)$. Then U^0 satisfies $1°$–$4°$ and, by the very definition, it is homogeneous: for any $\lambda \neq 0$,

$$U^0(\lambda x, \pm \lambda y, |\lambda| z) = |\lambda| U^0(x, y, z). \tag{7.41}$$

We shall use the notation

$$A(y) = U^0(0, y, 1) \qquad \text{and} \qquad B(y) = U^0(1, y, 1)$$

for $y \in \mathbb{R}$. It will be convenient to divide the proof into a few steps.

Step 1. The first observation is that the function B is convex. To see this, we repeat the argument from Lemma 5.7 above. Namely, let y_1, $y_2 \in \mathbb{R}$ and $\lambda \in (0, 1)$. Let $(f, g) \in M(1, \lambda y_1 + (1 - \lambda)y_2)$. If we set

$$g^{(1)} = g + (1 - \lambda)(y_1 - y_2) \qquad \text{and} \qquad g^{(2)} = g + \lambda(y_2 - y_1),$$

then, clearly, $(f, g^{(1)}) \in M(1, y_1)$ and $(f, g^{(2)}) \in M(1, y_2)$. Hence, by the triangle inequality,

$$\begin{aligned}
\mathbb{E}(|g_\infty| - \beta_0|f_\infty|) &= \mathbb{E}(|\lambda g_\infty^{(1)} + (1 - \lambda)g_\infty^{(2)}| - \beta_0|f_\infty|) \\
&\leq \lambda \mathbb{E}(|g_\infty^{(1)}| - \beta_0|f_\infty|) + (1 - \lambda)\mathbb{E}(|g_\infty^{(2)}| - \beta_0|f_\infty|) \\
&\leq \lambda B(y_1) + (1 - \lambda)B(y_2).
\end{aligned}$$

Now taking the supremum over all (f, g) as above yields the convexity of B. In particular, this shows that the function B is continuous.

Step 2. The next move is to establish the inequality

$$A(0) \geq B(1). \tag{7.42}$$

This is straightforward: we apply $2°$ with $x = y = 0$, $z = 1$, $\varepsilon = 1$, $t_1 = -t_2 = -1$ and $\alpha = 1/2$ to get

$$\frac{1}{2}U^0(-1, -1, 1) + \frac{1}{2}U^0(1, 1, 1) \leq U^0(0, 0, 1).$$

It remains to use the equality $U_0(-1, -1, 1) = U_0(1, 1, 1)$, which follows from (7.41).

Step 3. Now we will establish the much more involved estimate

$$A(y - 1) \geq B(y) + B(1)\left(1 - \exp\left(\frac{1 - y}{2}\right)\right), \tag{7.43}$$

valid for all $y \geq 1$. To do this, fix $\delta > 0$ and apply $2°$ with $x = z = 1$, y, $\varepsilon = -1$, $t_1 = -\delta$, $t_2 = T$ and $\alpha = T/(T + \delta)$. We obtain

$$U^0(1, y, 1) \geq \frac{T}{T + \delta}U^0(1 - \delta, y + \delta, 1) + \frac{\delta}{T + \delta}U^0(1 + T, y - T, 1 + T)$$

$$= \frac{T}{T + \delta}U^0(1 - \delta, y + \delta, 1) + \frac{\delta(1 + T)}{T + \delta}U^0\left(1, \frac{y - T}{1 + T}, 1\right),$$

where the latter passage is due to (7.41). Now let $T \to \infty$ and use the continuity of B established in $1°$ to obtain

$$B(y) \geq U^0(1 - \delta, y + \delta, 1) + \delta U^0(1, -1, 1) = U^0(1 - \delta, y + \delta, 1) + \delta B(1), \quad (7.44)$$

the last equality being a consequence of (7.41): $U^0(1, -1, 1) = U^0(1, 1, 1)$. Now we use the property $2°$ again, this time with $x = 1 - \delta$, $y + \delta$, $z = 1$, $\varepsilon = 1$, $t_1 = \delta - 1$, $t_2 = \delta$, $\alpha = \delta$. We arrive at

$$U^0(1 - \delta, y + \delta, 1) \geq \delta U^0(0, y + 2\delta - 1, 1) + (1 - \delta)U^0(1, y + 2\delta, 1)$$

$$= \delta A(y + 2\delta - 1) + (1 - \delta)B(y + 2\delta)$$

and plugging this into (7.44) yields

$$B(y) \geq (1 - \delta)B(y + 2\delta) + \delta A(y + 2\delta - 1) + \delta B(1). \tag{7.45}$$

Using a similar argumentation, one can establish the estimate

$$A(y + 2\delta - 1) \geq \frac{\delta}{1 + \delta}B(y) + \frac{\delta}{1 + \delta}B(y + 2\delta) + \frac{1 - \delta}{1 + \delta}A(y - 1). \tag{7.46}$$

Indeed, it suffices to combine the following inequalities: first,

$$A(y + 2\delta - 1) \geq \frac{1}{1 + \delta}U^0(\delta, y + \delta - 1, 1) + \frac{\delta}{1 + \delta}B(y + 2\delta),$$

coming from 2° with $x = 0$, $y+2\delta$, $z = 1$, $\varepsilon = -1$, $t_1 = \delta$, $t_2 = -1$ and $\alpha = 1/(1+\delta)$ (and the fact that $U^0(-1, y + 2\delta, 1) = U^0(1, y + 2\delta, 1) = B(y + 2\delta)$); second,

$$U^0(\delta, y + \delta - 1, 1) \geq (1 - \delta)A(y - 1) + \delta B(y),$$

a consequence of 2° with $x = \delta$, $y + \delta - 1$, $z = 1$, $\varepsilon = 1$, $t_1 = -\delta$, $t_2 = 1 - \delta$ and $\alpha = 1 - \delta$.

Multiply (7.45) throughout by $1/(1 + \delta)$ and add the result to (7.46). After some manipulations, we get

$$A(y + 2\delta - 1) - B(y + 2\delta) - B(1) \geq (1 - \delta)(A(y - 1) - B(y) - B(1)).$$

Therefore, by induction, we see that for any nonnegative integer N,

$$A(y + 2N\delta - 1) - B(y + 2N\delta) - B(1) \geq (1 - \delta)^N (A(y - 1) - B(y) - B(1)).$$

Now fix $s > 1$ and set $y = 1$, $\delta = (s - 1)/(2N)$. Letting $N \to \infty$ yields

$$A(s - 1) - B(s) - B(1) \geq (A(0) - 2B(1)) \exp\left(\frac{1 - s}{2}\right),$$

and using (7.42) we arrive at (7.43).

Step 4. This is the final part. We come back to (7.45) and insert the bound (7.43) there: as a result, we obtain

$$B(y) \geq B(y + 2\delta) + 2\delta B(1) - \delta B(1) \exp\left(\frac{1 - y - 2\delta}{2}\right).$$

By induction, we get, for any nonnegative integer N,

$$B(y) \geq B(y + 2N\delta) + 2N\delta B(1) - \delta B(1) \sum_{k=1}^{N} \exp\left(\frac{1 - y - 2k\delta}{2}\right)$$

$$= B(y + 2N\delta) + 2N\delta B(1) - \delta B(1) \frac{1 - e^{-N\delta}}{1 - e^{-\delta}} \exp\left(\frac{1 - y - 2\delta}{2}\right).$$

As previously, fix $s > 1$ and set $y - 1$, $\delta - (s - 1)/(2N)$. Letting $N \to \infty$ gives

$$B(1)\left(3 - s - \exp\left(\frac{1 - s}{2}\right)\right) \geq B(s)$$

and hence $B(\beta) \leq 0$. Since U^0 majorizes V, we have that

$$\beta - \beta_0 = V(1, \beta, 1) \leq U^0(1, \beta, 1) = B(\beta) \leq 0,$$

that is, $\beta_0 \geq \beta$. This shows that the constant β is indeed the best possible. $\qquad\square$

On the search of the suitable majorant. Let β_0 be a given positive number and set $V(x, y, z) = |y| - \beta_0 |x| \vee z$,

$$U^0(x, y, z) = \sup\{\mathbb{E}V(f_n, g_n, |f_n|^*)\},$$

where the supremum is taken over all n and all $(f, g) \in M(x, y)$. This formula gives us some crucial information. First, obviously, U^0 is homogeneous and symmetric, so in fact we need to find the function $U^0(\cdot, \cdot, 1)$ on the strip $[0, 1] \times [0, \infty)$. The second observation concerns the set $J = \{(x, y) : U^0(x, y, 1) = V(x, y, 1)\}$. This set is clearly symmetric with respect to the x and y axes; furthermore, we have $J \subseteq \{(\pm 1, y) : y \in \mathbb{R}\}$. Indeed, for any $x \in (-1, 1)$, $y \in \mathbb{R}$, one easily constructs a pair $(f, g) \in M(x, y)$ with $\|f\|_\infty \leq 1$ and $\|g\|_1 > y$. This pair yields the claimed inclusion, because $U^0(x, y, 1) \geq \mathbb{E}V(f_n, g_n, |f_n|^*) > |y| - \beta_0 = V(x, y, 1)$. A little experimentation leads to the following assumption on J:

(A1)　$J = \{(\pm 1, y) : |y| \geq \gamma\}$ for some $\gamma > 1$.

In particular, this implies

$$U(\pm 1, y, 1) = |y| - \beta_0 \qquad \text{for } |y| \geq \gamma. \tag{7.47}$$

Furthermore, we make the following conjectures concerning the special function (which from now on is denoted by U). Let, as above, $A(y) = U(0, y, 1)$, $B(y) = U(1, y, 1)$; furthermore, put $C(x) = U(x, 0, 1)$.

(A2)　$U(\cdot, \cdot, 1)$ is of class C^1 on the strip $(-1, 1) \times \mathbb{R}$.

(A3)　For any $x \in (0, 1)$ and $y \geq x$ we have

$$U(x, y, 1) = xB(-x + y + 1) + (1 - x)A(-x + y).$$

(A4)　For any $x \in (0, 1)$ and $0 \leq y \leq x$,

$$U(x, y, 1) = \frac{y}{-x + y + 1} B(-x + y + 1) + \frac{-x + 1}{-x + y + 1} C(x - y).$$

The next assumption is based on (7.44): we guess that

(A5)　$U_x(1-, y) - U_y(1, y) = B(1)$ for $y \in [0, \gamma]$.

The conditions (A1)–(A5) lead to the special function U studied in the proof above, as we shall show now. Since $U(x, y, 1) = U(-x, y, 1)$ for all $x \in [-1, 1]$ and $y \in \mathbb{R}$, we have, by (A2) and (A3),

$$A'(y) = -A(y) + B(y + 1). \tag{7.48}$$

This, together with (7.47), yields

$$A(y) = y - \beta_0 + c_1 e^{-y}, \qquad y \geq \gamma - 1, \tag{7.49}$$

where c_1 is a certain positive constant. Next, combining (A3) and (A5) gives that for $y \in [1, \gamma]$

$$2B'(y) = B(y) - A(y-1) - B(1)$$

and, by (7.48), implies $2B'(y) = A'(y-1) - B(1)$, or

$$2B(y) = A(y-1) - B(1)y + c_2 \tag{7.50}$$

for some constant c_2. Plugging this again into (7.48) and solving the resulting differential equation gives

$$A(y) = -B(1)y + c_2 + 2c_3 \exp\left(-\frac{y+1}{2}\right), \qquad y \in [0, \gamma-1],$$

for some $c_3 \in \mathbb{R}$; this implies

$$B(y) = B(1)(1-y) + c_2 + c_3 e^{-y/2}. \tag{7.51}$$

Plugging $y = 1$ and using the equations $A(\gamma-1-) = A(\gamma-1+)$ and $A'(\gamma-1-) = A'(\gamma-1+)$, we derive that $c_1 = (1 + B(1))e^{\gamma-1} - B(1)e^{(\gamma-1)/2}$, $c_2 = 2B(1)$, $c_3 = -B(1)e^{1/2}$ and

$$\beta_0 = \gamma - B(1)\left[3 - \gamma - \exp\left(\frac{1-\gamma}{2}\right)\right]. \tag{7.52}$$

Inserting these parameter values into (7.51), we obtain

$$B(y) = B(1)\left[3 - y - \exp\left(\frac{1-y}{2}\right)\right]$$

and, since $B'(\gamma) = 1$, we get $B(1) = 2\left(\exp(\frac{1-\gamma}{2}) - 2\right)^{-1}$. Plugging this into (7.52) gives

$$\beta_0 = \gamma + 2 + \frac{2(1-\gamma)}{2 - \exp\left(\frac{1-\gamma}{2}\right)}.$$

The right-hand side, as a function of $\gamma \in [1, \infty)$, attain its minimum β at $\gamma = \beta$. The final assumption is that this equality takes place. Then the formulas for A and B we have just obtained allow us to derive the values of $U(\cdot, \cdot, 1)$ on the set $\{(x, y) : |y| \geq |x|\}$. What we get coincides with the special function u above. To get $U(x, y, 1)$ for $|y| < |x|$, we use (A2), (A4) and the symmetry of $U(\cdot, \cdot, 1)$ with respect to the y-axis. Namely, we have $U_y(x, 0, 1) = 0$, which implies

$$C'(x) = \frac{B(1-x) - C(x)}{1 - x} \qquad \text{for } x \in [0, 1).$$

Now, by (A4) and (A5),

$$2B'(y) = \frac{B(y) - C(1-y)}{y} - B(1) \qquad \text{for } y \in [0, 1).$$

Consequently, $C(x) = -2B(1-x) + B(1)x + c_4$ for some constant c_4; plugging this into one of the above differential equations and solving gives

$$C(x) = B(1)x - \frac{2}{3}B(1) + \frac{c_4}{3} + c_5(1-x)^{3/2} \qquad \text{for } x \in [0,1),$$

where c_5 is another constant. Using the equalities $C'(0) = 0$ and $C(0) = 0$ gives explicit formulas for B and C on $[0,1]$. These, in turn, by virtue of (A4), yield the function U on $\{(x,y) : |y| < |x| \le 1\}$: we get precisely the function u defined above. $\square$

7.3.3   Proof of Theorem 7.10

Quite unexpectedly, this result is closely related to the moment inequality for bounded submartingales studied in Theorem 4.16. Let u_α be the special function invented there.

Proof of (7.30) *and* (7.31). We shall only sketch the proof of the first estimate. We focus on the discrete-time version for ± 1-transforms of simple supermartingales; the passage to the continuous-time can be carried out in the same manner as above. Let $D = [0, \infty) \times \mathbb{R} \times (0, \infty)$. Define $V, U : D \to \mathbb{R}$ by $V(x, y, z) = |y| - (2 + (3e)^{-1})(|x| \vee z)$ and

$$U(x, y, z) = (x \vee z)\left[u_\alpha\left(1 - \frac{x}{x \vee z}, \frac{y}{x \vee z}\right) - u_\alpha(0, 1)\right].$$

Repeating the above reasoning, we conclude that the function

$$u(x, y) = u_\alpha(1 - x, y) - u_\alpha(0, 1)$$

satisfies the properties listed in Lemma 7.2 (with β replaced by $2 + (3e)^{-1}$). In addition, we have that for any $y \in \mathbb{R}$ and $|\gamma| \le 1$, the function $t \mapsto u(t, y + \gamma t)$ is increasing on $[0, 1]$. This implies that U satisfies the conditions 1°–4° and the estimate follows. $\square$

Sharpness of (7.31). Let $\alpha \ge 1$. Suppose that the inequality

$$||g||_1 \le \beta_0 ||f^*||_1$$

holds for all nonnegative supermartingales f and their α-subordinates g. Let

$$U^0(x, y, z) = \sup\{\mathbb{E}\big[|g_\infty| - \beta_0|f_\infty^*|^* \vee z\big]\},$$

where the supremum is taken over all simple f, g satisfying appropriate properties. By a supermartingale version of Theorem 7.2, the function

$$u = U^0(\cdot, \cdot, 1) : [0, 1] \times \mathbb{R} \to \mathbb{R}$$

has the following properties:

$$u(x,y) \geq |y| - \beta_0, \qquad (7.53)$$
$$u \text{ is diagonally concave,} \qquad (7.54)$$
$$u(1,y) \geq u(1-t, y \pm \alpha t) \text{ for any } t \in [0,1], \qquad (7.55)$$
$$u(x, \pm x) \leq 0 \qquad \text{for all } x \in [0,1]. \qquad (7.56)$$

Let $B(y) = u(1,y)$ for $y \in \mathbb{R}$. Furthermore, for $y \geq \alpha/(2\alpha+1)$, denote

$$C(y) = u\left(\frac{\alpha+1}{2\alpha+1}, y - \frac{\alpha}{2\alpha+1}\right).$$

It is convenient to split the proof into a few intermediate steps. Throughout, δ is a sufficiently small positive number.

Step 1. We will show that for any $y \geq \alpha/(2\alpha+1)$,

$$B(y) \geq \frac{\delta(2\alpha+1)}{\alpha} C(y + (\alpha+1)\delta) + \frac{\alpha - \delta(2\alpha+1)}{\alpha} B(y + (\alpha+1)\delta). \qquad (7.57)$$

To prove this, note that by (7.55) we have

$$B(y) = u(1,y) \geq u(1 - \delta, y + \alpha\delta)$$

and, by (7.54),

$$u(1-\delta, y+\alpha\delta) \geq \frac{\delta(2\alpha+1)}{\alpha} u\left(\frac{\alpha+1}{2\alpha+1}, y - \frac{\alpha}{2\alpha+1} + (\alpha+1)\delta\right)$$
$$+ \frac{\alpha - \delta(2\alpha+1)}{\alpha} u(1, y + (\alpha+1)\delta)$$
$$= \frac{\delta(2\alpha+1)}{\alpha} C(y + (\alpha+1)\delta) + \frac{\alpha - \delta(2\alpha+1)}{\alpha} B(y + (\alpha+1)\delta).$$

Combining these two facts yields (7.57).

Step 2. Next we show that for $y \geq \alpha/(2\alpha+1)$,

$$C(y + (\alpha+1)\delta) \geq \frac{\delta(2\alpha+1)}{2 + \delta(2\alpha+1)}\left(y + \frac{1}{2\alpha+1} - \beta_0\right) \qquad (7.58)$$
$$+ \frac{(\alpha+1)(2\alpha+1)\delta}{\alpha(2 + \delta(2\alpha+1))} B(y) + \frac{2\alpha - (\alpha+1)(2\alpha+1)\delta}{\alpha(2 + \delta(2\alpha+1))} C(y).$$

The proof is similar to that of (7.57). By (7.54), we have

$$C(y + (\alpha+1)\delta) \geq \frac{\delta(2\alpha+1)}{2 + \delta(2\alpha+1)} u\left(0, y + \frac{1}{2\alpha+1} + (\alpha+1)\delta\right) + \frac{2A}{2 + \delta(2\alpha+1)},$$

where
$$A = u\left(\frac{\alpha+1}{2\alpha+1} + \frac{(\alpha+1)\delta}{2}, y - \frac{\alpha}{2\alpha+1} + \alpha\delta - \frac{(\alpha+1)\delta}{2}\right).$$

Furthermore, again by (7.54),

$$A \geq \frac{(\alpha+1)(2\alpha+1)\delta}{2\alpha} B(y) + \frac{2\alpha - (\alpha+1)(2\alpha+1)\delta}{2\alpha} C(y).$$

In addition, by (7.53),

$$u\left(0, y + \frac{1}{2\alpha+1} + (\alpha+1)\delta\right) \geq y + \frac{1}{2\alpha+1} + (\alpha+1)\delta - \beta_0.$$

Combining the three inequalities above gives (7.58).

Step 3. Multiply both sides of (7.58) by

$$\lambda_+ = \frac{2\alpha + 3 + \sqrt{(2\alpha+1)^2 - 4\delta(\alpha+1)(2\alpha+1)}}{2(\alpha+1)}$$

and add it to (7.57). After some lengthy but straightforward computations we get

$$C(y + (\alpha+1)\delta) - p_{\alpha,\delta}B(y + (\alpha+1)\delta) \geq q_{\alpha,\delta}\big(C(y) - p_{\alpha,\delta}B(y)\big)$$
$$+ r_{\alpha,\delta}\left(y + \frac{1}{2\alpha+1} - \beta_0\right), \qquad (7.59)$$

where

$$p_{\alpha,\delta} = \frac{\alpha - \delta(2\alpha+1)}{\lambda_+\alpha - \delta(2\alpha+1)},$$
$$q_{\alpha,\delta} = \frac{\lambda_+(2\alpha - (\alpha+1)(2\alpha+1)\delta)}{(2 + \delta(2\alpha+1))(\lambda_+\alpha - \delta(2\alpha+1))},$$
$$r_{\alpha,\delta} = \frac{\lambda_+\alpha(2\alpha+1)\delta}{(2 + \delta(2\alpha+1))(\lambda_+\alpha - \delta(2\alpha+1))}.$$

By induction, (7.59) gives that for any positive integer N,

$$C(y + N(\alpha+1)\delta) - p_{\alpha,\delta}B(y + N(\alpha+1)\delta) \geq I_1 + I_2, \qquad (7.60)$$

where
$$I_1 = q_{\alpha,\delta}^N\big(C(y) - p_{\alpha,\delta}B(y)\big)$$

and

$$I_2 = r_{\alpha,\delta} \sum_{k=0}^{N-1} q_{\alpha,\delta}^{N-1-k}\left(y + k(\alpha+1)\delta + \frac{1}{2\alpha+1} - \beta_0\right).$$

Now take $y_1 > y_2 \geq \alpha/(2\alpha + 1)$, put $y = y_2$, $\delta = (y_1 - y_2)/(N(\alpha + 1))$ in (7.60), and let $N \to \infty$. One easily checks that then $p_{\alpha,\delta} \to 1/2$ and

$$q_{\alpha,\delta} = 1 - \delta(2\alpha + 1) + o(\delta) = 1 - \frac{2\alpha + 1}{\alpha + 1}\frac{y_1 - y_2}{N} + o(N^{-1}),$$

so

$$I_1 \to \exp\left(-\frac{2\alpha + 1}{\alpha + 1}(y_1 - y_2)\right)\left(C(y) - \frac{B(y)}{2}\right).$$

Furthermore, $r_{\alpha,\delta} = (2\alpha + 1)\delta/2 + o(\delta)$ and

$$I_2 = \frac{(2\alpha + 1)\delta}{2}\left[\left(y_2 + \frac{1}{2\alpha + 1} - \beta_0\right)\frac{q_{\alpha,\delta}^N - 1}{q_{\alpha,\delta} - 1}\right.$$

$$\left. + (\alpha + 1)\delta\frac{(N - 1)q_{\alpha,\delta}^{-2} - Nq_{\alpha,\delta}^{-1} + q_{\alpha,\delta}^{N-2}}{(q_{\alpha,\delta} - 1)^2}\right] + o(1)$$

$$\to \frac{1}{2}\left[\exp\left(-\frac{2\alpha + 1}{\alpha + 1}(y_1 - y_2)\right) - 1\right]\left(-y_2 + \beta_0 + \frac{\alpha}{2\alpha + 1}\right) + \frac{y_1 - y_2}{2}.$$

Using all these facts in (7.60) and working a little bit yields

$$C(y_1) - \frac{B(y_1)}{2} - \frac{y_1}{2} + \frac{\beta_0}{2} + \frac{\alpha}{2(2\alpha + 1)} \tag{7.61}$$

$$\geq \exp\left[-\frac{2\alpha + 1}{\alpha + 1}(y_1 - y_2)\right]\left(C(y_2) - \frac{B(y_2)}{2} - \frac{y_2}{2} + \frac{\beta_0}{2} + \frac{\alpha}{2(2\alpha + 1)}\right).$$

Step 4. Similarly, multiply both sides of (7.58) by

$$\lambda_- = \frac{2\alpha + 3 - \sqrt{(2\alpha + 1)^2 - 4\delta(\alpha + 1)(2\alpha + 1)}}{2(\alpha + 1)},$$

add the result to (7.57) and proceed as in the previous step. What we obtain is that for all $y_1 > y_2 \geq \alpha/(2\alpha + 1)$,

$$C(y_1) - (\alpha + 1)B(y_1) + \alpha y_1 + \frac{2\alpha^3}{2\alpha + 1} + \alpha - \beta_0\alpha$$

$$\geq \exp\left[\frac{2\alpha + 1}{2\alpha(\alpha + 1)}(y_1 - y_2)\right]\left(C(y_2) - (\alpha + 1)B(y_2) + \alpha y_2 + \frac{2\alpha^3}{2\alpha + 1} + \alpha - \beta_0\alpha\right).$$

Note that this implies

$$C(y) - (\alpha + 1)B(y) + \alpha y + \frac{2\alpha^3}{2\alpha + 1} + \alpha - \beta_0\alpha \leq 0 \tag{7.62}$$

for all $y \geq \alpha/(2\alpha + 1)$. Indeed, if this estimate is not valid for some y, then use the preceding inequality with $y_1 > y_2 = y$ and let $y_1 \to \infty$. As the result, we get that $C - (\alpha + 1)B$ has exponential growth at infinity. However, this is impossible: arguing as in the proof of Lemma 5.7, we get that both B and C are Lipschitz.

Step 5. This is the final part. By (7.54) and then (7.56) we have

$$
\begin{aligned}
C\left(\frac{\alpha}{2\alpha+1}\right) &\geq \frac{\alpha+1}{2\alpha+1}B\left(\frac{\alpha}{2\alpha+1}\right) + \frac{\alpha}{2\alpha+1}u\left(0, -\frac{\alpha+1}{2\alpha+1}\right) \\
&\geq \frac{\alpha+1}{2\alpha+1}B\left(\frac{\alpha}{2\alpha+1}\right) + \frac{\alpha}{2\alpha+1}\left(\frac{\alpha+1}{2\alpha+1} - \beta_0\right).
\end{aligned}
\tag{7.63}
$$

Combining this with (7.62), applied to $y = \alpha/(2\alpha+1)$, yields, after some manipulations,

$$
B\left(\frac{\alpha}{2\alpha+1}\right) + \beta_0 \geq \frac{1}{2(2\alpha+1)} + \frac{2\alpha+1}{\alpha}.
\tag{7.64}
$$

Now, use (7.61) with $y_1 = 1$ and $y_2 = \alpha/(2\alpha+1)$ and plug the two estimates above to get

$$
C(1) - \frac{B(1)}{2} - \frac{1}{2} + \frac{\beta_0}{2} + \frac{\alpha}{2\alpha+1} \geq e^{-1}\left(\frac{B\left(\frac{\alpha}{2\alpha+1}\right) + \beta_0}{2(2\alpha+1)} + \frac{\alpha(\alpha+1)}{(2\alpha+1)^2}\right)
$$
$$
\geq (2e)^{-1},
$$

or

$$
C(1) - \frac{B(1)}{2} \geq \frac{1}{2} - \frac{\beta_0}{2} - \frac{\alpha}{2(2\alpha+1)} + (2e)^{-1}.
$$

On the other hand, by (7.62) applied to $y = 1$, and by (7.56),

$$
C(1) - \frac{B(1)}{2} \leq \left(\alpha + \frac{1}{2}\right)B(1) - 2\alpha - \frac{2\alpha^3}{2\alpha+1} + \beta_0\alpha \leq -2\alpha - \frac{2\alpha^3}{2\alpha+1} + \beta_0\alpha.
$$

Combining this with the previous inequality gives $\beta_0 \geq \alpha + 1 + ((2\alpha+1)e)^{-1}$, as desired. $\qquad\square$

Sharpness of (7.30) *in the martingale case.* Suppose that for any nonnegative martingale f and its ± 1-transform g we have

$$
\|g_n\|_1 \leq \beta_0\|f_n^*\|_1, \qquad n = 0, 1, 2, \ldots
$$

and let

$$
U^0(x, y, z) = \sup\{\mathbb{E}|g_\infty| - \beta_0\mathbb{E}f_\infty^* \vee z\},
$$

where the supremum is taken over appropriate sequences f and g. Now, repeating word by word the arguments from the proof of sharpness of (7.29), we conclude that (7.44) holds. That is, for any $y \in \mathbb{R}$ and $\delta \in [0, 1]$,

$$
U^0(1, y, 1) \geq U^0(1 - \delta, y + \delta, 1) + \delta U^0(1, 1, 1).
$$

Consider the function $u : [0,1] \times \mathbb{R} \to \mathbb{R}$ given by $u(x,y) = U^0(x,y,1) - U^0(1,1,1)x$. This function majorizes $v(x,y) = |y| - \beta_0$ (since $U^0(\cdot, \cdot, 1)$ does and $U^0(1,1,1) \leq 0$) and is diagonally concave. The above inequality can be rewritten in the form $u(1,y) \geq u(1-\delta, y+\delta)$. Finally, we have $u(1,1) = 0$. However, we have shown above that the existence of such a function implies that $\beta_0 \geq 2 + (3e)^{-1}$: repeat the previous proof with $\alpha = 1$. $\square$

On the search of the suitable majorant. The reasoning is similar to the one presented previously. We omit the details. $\square$

7.4 Maximal L^p inequalities

7.4.1 Formulation of the result

Now we shall study related L^p estimates. These can be stated as follows.

Theorem 7.11. *Let X, Y be Hilbert-space-valued martingales with Y being differentially subordinate to X. Then for any $p \geq 2$,*

$$|| |Y|^* ||_p \leq p ||X||_p \qquad (7.65)$$

and the constant p is the best possible. It is already the best possible in the following discrete-time inequality: if f is a martingale and g is its ± 1 transform, then

$$|| |g|^* ||_p \leq p || |f|^* ||_p. \qquad (7.66)$$

Note that the validity of the above estimates follows immediately from the moment inequality of Burkholder and Doob's maximal estimate. The nontrivial (and quite surprising) fact is that both (7.65) and (7.66) are sharp.

We shall also prove the following submartingale inequality.

Theorem 7.12. *Let α be a fixed number in the interval $[0,1]$. Let X be a nonnegative submartingale and let Y be real-valued and α-strongly subordinate to X. Then for any $p \geq 2$,*

$$|| |Y|^* ||_p \leq (\alpha + 1)p ||X||_p \qquad (7.67)$$

and the constant $(\alpha + 1)p$ is the best possible. It is already the best possible in the discrete-time estimate

$$||g^*||_p \leq (\alpha + 1)p ||f^*||_p \qquad (7.68)$$

(note that on the left we have the one-sided maximal function of g).

We shall establish this result only in the discrete-time case. A natural question is what is the best constant in the inequalities above in the case $1 < p < 2$. Unfortunately, we have been unable to answer it; our reasoning works only for the case $p \geq 2$.

7.4.2   Proof of Theorem 7.12 and sharpness of (7.66)

A DIFFERENTIAL EQUATION. To establish (7.67), first we need to study an auxiliary object. For a fixed $\alpha \in (0, 1]$ and $p \geq 2$, let $C = C_{p,\alpha} = [(\alpha + 1)p]^p(p - 1)$. Consider the following differential equation

$$\gamma'(x) = \frac{-1 + C(1 - \gamma(x))\gamma(x)x^{p-2}}{1 + C(1 - \gamma(x))x^{p-1}}. \tag{7.69}$$

Lemma 7.3. *There is a solution* $\gamma : [((\alpha + 1)p)^{-1}, \infty) \to \mathbb{R}$ *of* (7.69) *with the initial condition*

$$\gamma\left(\frac{1}{(\alpha + 1)p}\right) = 1 - \frac{1}{(\alpha + 1)p}. \tag{7.70}$$

This solution is nondecreasing, concave and bounded above by 1. Furthermore,

$$(p - 2)(1 - \gamma(x)) - x\gamma'(x) \leq 0. \tag{7.71}$$

For the proof, we refer the reader to the original paper [156]. We extend γ to the whole half-line $[0, \infty)$ by setting

$$\gamma(x) = [(p - 1)(\alpha + 1) - 1]x + \frac{1}{p}, \qquad x \in \left[0, \frac{1}{(\alpha + 1)p}\right).$$

It can be readily verified that γ is of class C^1 on $(0, \infty)$. See Figure 7.2 on top of the next page to gain some intuition.

Let $H : [((\alpha + 1)p)^{-1}, \infty) \to [1, \infty)$ be given by $H(x) = x + \gamma(x)$ and let h be the inverse to H. Clearly, we have

$$x - 1 \leq h(x) \leq x, \qquad x \geq 1. \tag{7.72}$$

Finally, let us provide a formula for h' to be used later. Since

$$h'(x) = \frac{1}{H'(h(x))} = \frac{1}{1 + \gamma'(h(x))}, \qquad x > 1, \tag{7.73}$$

one finds, using (7.69), that

$$h'(x) = \frac{1 + ((\alpha + 1)p)^p(p - 1)(h(x) - x + 1)h(x)^{p-1}}{((\alpha + 1)p)^p(p - 1)(h(x) - x + 1)h(x)^{p-2}x}. \tag{7.74}$$

Proof of (7.67). Let $D = [0, \infty) \times \mathbb{R} \times (0, \infty) \to \mathbb{R}$ and introduce $V : D \to \mathbb{R}$ by $V(x, y, w) = (|y| \vee w)^p - (p(\alpha + 1))^p x^p$. To define the corresponding special function, consider the following subsets of the strip $S = [0, \infty) \times [-1, 1]$:

$$D_0 = \{(x, y) \in S : |y| \leq \gamma(x)\},$$
$$D_1 = \{(x, y) \in S : |y| > \gamma(x), x + |y| \leq 1\},$$
$$D_2 = \{(x, y) \in S : |y| > \gamma(x), x + |y| > 1\}.$$

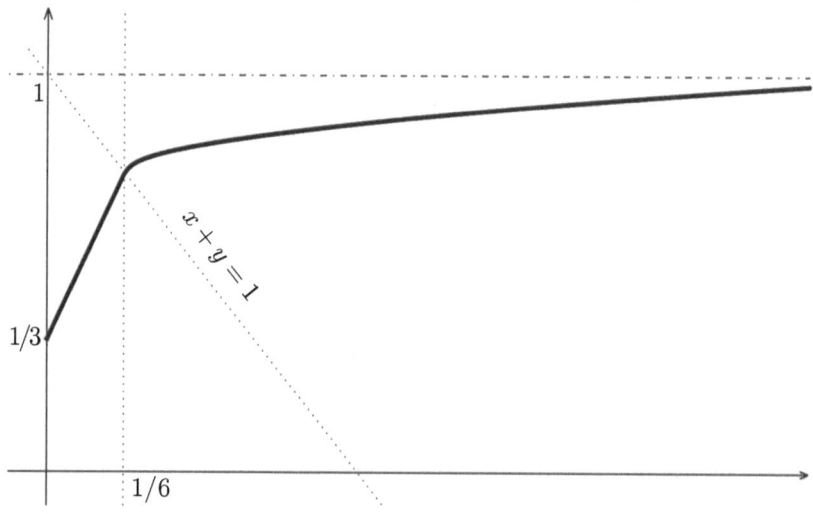

Figure 7.2: The graph of γ (bold curve) in the case $p = 3$, $\alpha = 1$. Note that γ is linear on $[0, 1/6]$ and solves (7.69) on $(1/6, \infty)$.

Introduce the function $u : S \to \mathbb{R}$ by

$$
u(x, y) = \begin{cases}
1 - [(\alpha + 1)p]^p x^p & \text{on } D_0, \\
1 - \left(\frac{px + p|y| - 1}{p - 1}\right)^{p-1}[p(p(\alpha + 1) - 1)x - p|y| + 1] & \text{on } D_1, \\
1 - [(\alpha + 1)p]^p h(x + |y|)^{p-1}[px - (p - 1)h(x + |y|)] & \text{on } D_2.
\end{cases}
$$

Finally, let $U : [0, \infty) \times \mathbb{R} \times (0, \infty) \to \mathbb{R}$ be given by

$$
U(x, y, w) = (|y| \vee w)^p u\left(\frac{x}{|y| \vee w}, \frac{y}{|y| \vee w}\right).
$$

One easily verifies that U is of class C^1. We shall prove that it satisfies the conditions $1°$–$4°$; due to the complexity of the calculations, we have decided to split the verification into a few intermediate lemmas.

Lemma 7.4. *For any $(x, y, z) \in [0, \infty) \times \mathbb{R} \times (0, \infty)$, we have*

$$
U(x, y, z) \geq V(x, y, z). \tag{7.75}
$$

Proof. The inequality is equivalent to $u(x, y) \geq 1 - [(\alpha + 1)p]^p x^p$ and we need to establish it only on D_1 and D_2. On D_1, the substitutions $X = px$ and $Y = p|y| - 1$ (note that $Y \geq 0$) transform it into

$$
(\alpha + 1)^p X^p \geq \left(\frac{X + Y}{p - 1}\right)^{p-1}[(p(\alpha + 1) - 1)X - Y].
$$

However, this is valid for all nonnegative X, Y. Indeed, observe that by homogeneity we may assume $X + Y = 1$, and then the estimate reads

$$F(X) := (\alpha + 1)^p X^p - (p - 1)^{-p+1}[p(\alpha + 1)X - 1] \geq 0, \qquad X \in [0, 1].$$

Now it suffices to note that F is convex on $[0, 1]$ and satisfies

$$F\left(\frac{1}{(p-1)(\alpha+1)}\right) = F'\left(\frac{1}{(p-1)(\alpha+1)}\right) = 0.$$

It remains to show the majorization on D_2. It is dealt with in a similar manner: setting $s = x + |y| > 1$, we see that (7.75) is equivalent to

$$G(x) := x^p - h(s)^{p-1}[px - (p - 1)h(s)] \geq 0, \qquad s - 1 < x < h(s).$$

It is easily verified that G is convex and satisfies $G(h(s)) = G'(h(s)) = 0$. This completes the proof of (7.75).   □

We turn to the concavity and initial condition.

Lemma 7.5. *For fixed y, z satisfying $z > 0$, $|y| \leq z$, and any $a \in [-1, 1]$, the function $\Phi = \Phi_{y,z,a} : [0, \infty) \to \mathbb{R}$ given by*

$$\Phi(t) = U(t, y + at, z)$$

is concave.

Remark 7.2. It is clear that Lemma 7.5 yields that if x, y, z, k_x, k_y satisfy x, $x + k_x \geq 0$, $z > 0$, $|y| \leq z$ and $|k_y| \leq |k_x|$, then

$$U(x + k_x, y + k_y, z) \leq U(x, y, z) + U_x(x, y, z)k_x + U_y(x, y, z)k_y \qquad (7.76)$$

(for $x = 0$ we replace $U_x(0, y, z)$ by the right-hand derivative $U_x(0+, y, z)$). In addition, Lemma 7.5 implies 3°: for any $x \geq 1$ we have

$$U(x, 1, 1) \leq 0. \qquad (7.77)$$

To see this, note that $\Phi_{0,1,x^{-1}}(0) = U(0, 0, 1) = 1$ and

$$\Phi_{0,1,x^{-1}}(((\alpha+1)p)^{-1}) = U(((\alpha+1)p)^{-1}, x^{-1}((\alpha+1)p)^{-1}, 1) = 0,$$

since $(((\alpha+1)p)^{-1}, x^{-1}((\alpha+1)p)^{-1}, 1) \in D_0$. But $x \geq 1 > ((\alpha+1)p)^{-1}$, so Lemma 7.5 gives $U(x, 1, 1) = \Phi_{0,1,x^{-1}}(x) \leq 0$.

Proof of Lemma 7.5. By homogeneity, we may assume $z = 1$. As Φ is of class C^1, it suffices to verify that $\Phi''(t) \leq 0$ for those t, for which $(t, y + at)$ lies in the interior of D_0, D_1, D_2 or outside the strip S. Since $U(x, y, z) = U(x, -y, z)$, we may restrict ourselves to the case $y + at \geq 0$. If $(t, y+at)$ belongs to D_0°, the interior of D_0, then $\Phi''(t) = -[(\alpha+1)p]^p \cdot p(p-1)t^{p-2} < 0$, while for $(t, y+at) \in D_1^\circ$ we have

$$\Phi''(t) = -\frac{p^3(pt + p(y+at) - 1)^{p-3}(1+a)}{(p-1)^{p-2}}(I_1 + I_2),$$

where

$$I_1 = pt[(p-2)(1+a)(p(\alpha+1) - 1) + 2(p(\alpha+1) - 1 - a)] \geq 0,$$
$$I_2 = (p(y+at) - 1)(2\alpha + 1 - a) \geq 0.$$

The remaining two cases are a bit more complicated. If $(t, y+at) \in D_2^\circ$, then

$$\frac{\Phi''(t)}{Cp(1+a)^2} = J_1 + J_2 + J_3,$$

where

$$J_1 = h(t+y+at)^{p-2}h''(t+y+at)[h(t+y+at) - t],$$
$$J_2 = h(t+y+at)^{p-3}[h'(t+y+at)]^2[(p-1)h(t+y+at) - (p-2)t],$$
$$J_3 = -\frac{2}{a+1}h(t+y+at)^{p-2}h'(t+y+at).$$

Now if we change y and t keeping $s = t+y+at$ fixed, then $J_1 + J_2 + J_3$ is a linear function of $t \in [s-1, h(s)]$. Therefore, to prove that it is nonpositive, it suffices to verify this for $t = h(s)$ and $t = s - 1$. For $t = h(s)$, we have

$$J_1 + J_2 + J_3 = h(s)^{p-2}h'(s)\left[h'(s) - \frac{2}{a+1}\right] \leq 0,$$

since $0 \leq h'(s) \leq 1$ (see (7.73)). If $t = s - 1$, rewrite (7.74) in the form

$$Cs(h(s) + 1 - s)h(s)^{p-2}h'(s) = 1 + C(h(s) + 1 - s)h(s)^{p-1}$$

and differentiate both sides; as a result, we obtain

$$Cs\left[J_1 + J_2 + J_3 + h(s)^{p-2}h'(s)\left(\frac{2}{a+1} - 1\right)\right]$$
$$= Ch(s)^{p-2}[(h'(s) - 1)h(s) + (p-2)(h(s) + 1 - s)h'(s)].$$

As $h' \geq 0$ and $2/(a+1) \geq 1$, we will be done if we show that the right-hand side is nonpositive. This is equivalent to

$$h'(s)[h(s) + (p-2)(h(s) + 1 - s)] \leq h(s).$$

Now use (7.73) and substitute $h(s) = r$, noting that $h(s) + 1 - s = 1 - \gamma(r)$, to obtain

$$r + (p-2)(1 - \gamma(r)) \leq r(1 + \gamma'(r)),$$

or $r\gamma'(r) \geq (p-2)(1 - \gamma(r))$, which is (7.71).

Finally, suppose that $y + at > 1$. For such t we have

$$\Phi(t) = (y + at)^p u(t/(y + at), 1)$$

and hence, setting $X = t/(y + t)$, $Y = y + at$, we easily check that $\Phi''(t)$ equals

$$Y^{p-2}\left[p(p-1)a^2u(X,1) + 2a(p-1)(1-aX)u_x(X,1) + (1-aX)^2u_{xx}(X,1)\right].$$

First let us derive the expressions for the partial derivatives. Using (7.74), we have

$$u_x(X,1) = \frac{p}{X+1}[1 + C(h(X+1) - X)h(X+1)^{p-1}] - \frac{Cph(X+1)^{p-1}}{p-1},$$

$$u_{xx}(X,1) = \frac{p(p-1)}{(X+1)^2}[1 + C(h(X+1) - X)h(X+1)^{p-1}]$$

$$- \frac{Cph(X+1)^{p-1}}{X+1} - \frac{Cph(X+1)^{p-2}h'(X+1)}{X+1}.$$

Now it can be checked that

$$\Phi''(t)Y^{2-p}/p = K_1 + K_2 + K_3,$$

where

$$K_1 = (p-1)\left(\frac{a+1}{X+1}\right)^2[1 + C(h(X+1) - X)h(X+1)^{p-1}],$$

$$K_2 = -\frac{Ch(X+1)^{p-1}}{X+1}(1 + 2a - a^2X),$$

$$K_3 = -\left(\frac{1-aX}{X+1}\right)^2 \cdot \frac{1 + C(h(X+1) - X)h(X+1)^{p-1}}{h(X+1) - X}$$

$$\leq -\left(\frac{1-aX}{X+1}\right)^2 \cdot Ch(X+1)^{p-1}.$$

Further,

$$K_2 + K_3 \leq -\frac{Ch(X+1)^{p-1}}{(X+1)^2}[(1 + 2a - a^2X)(X+1) + (1-aX)^2]$$

$$= -\frac{Ch(X+1)^{p-1}(a+1)}{(X+1)^2}[2 + X(1-a)]$$

$$\leq -\left(\frac{a+1}{X+1}\right)^2 Ch(X+1)^{p-1},$$

where in the last passage we have used $a \leq 1$. On the other hand, as h is nondecreasing,

$$1 = \frac{Ch(1)^p}{p-1} \leq \frac{Ch(X+1)^{p-1}h(1)}{p-1}.$$

Moreover, since $x \mapsto h(x+1) - x$ is nonincreasing (see (7.73)), we have $h(X+1) - X \leq h(1)$. Combining these two facts, we obtain

$$
\begin{aligned}
K_1 &\leq (p-1)\left(\frac{a+1}{X+1}\right)^2 [1 + Ch(1)h(X+1)^{p-1}] \\
&\leq \left(\frac{a+1}{X+1}\right)^2 Ch(X+1)^{p-1}[h(1) + (p-1)h(1)] \\
&\leq \left(\frac{a+1}{X+1}\right)^2 Ch(X+1)^{p-1},
\end{aligned}
$$

as $ph(1) = (\alpha+1)^{-1} \leq 1$. This implies $K_1 + K_2 + K_3 \leq 0$ and completes the proof. $\qquad\square$

We have established "half" of property 2°; here is the remaining part.

Lemma 7.6. *For any x, y, z such that $x \geq 0$, $z > 0$ and $|y| \leq z$ we have*

$$
U_x(x, y, z) \leq -\alpha |U_y(x, y, z)| \tag{7.78}
$$

(if $x = 0$, then U_x is replaced by the right-sided derivative).

Proof. It suffices to show that for fixed y, z, $|y| \leq z$, and $a \in [-\alpha, \alpha]$, the function $\Phi = \Phi_{y,z,a} : [0, \infty) \to \mathbb{R}$ given by $\Phi(t) = U(t, y + at, z)$ is nonincreasing. Since $\alpha \leq 1$, we know from the previous lemma that Φ is concave. Hence, all we need is to show that $\Phi'(0+) \leq 0$. By symmetry, we may assume that $y \geq 0$. If $y \leq 1/p$, then the derivative equals 0; in the remaining case, we have

$$
\Phi'(0+) = -\frac{p^2(py-1)^{p-1}}{(p-1)^{p-1}}(\alpha - a) \leq 0,
$$

so the claim is established. $\qquad\square$

Thus, we have checked 1°, 2° and 3°; property 4° is obvious. We are ready to complete the proof of (7.67). Clearly, we may and do assume that f is bounded in L^p; then, by Burkholder's moment inequality and Doob's maximal estimate, we get that $|g|^*$ is in L^p and all we need is the bound

$$
\mathbb{E}V(f_n, g_n, |g_n|^*) \leq 0.
$$

To apply the submartingale version of Theorem 7.3, we need to verify the required integrability (7.14). It suffices to show the bounds

$$
U(x, y, z) \leq K(x + |y| + z)^p \tag{7.79}
$$

and

$$
U_x(x, y, z) \leq K(x + |y| + z)^{p-1}, \qquad U_y(x, y, z) \leq K(x + |y| + z)^{p-1}, \tag{7.80}
$$

for all x, $y \in \mathbb{R}$ and $z > 0$, where K is a certain absolute constant. The inequality (7.79) is evident for those points (x, y, z), for which $(\frac{x}{|y| \vee z}, \frac{y}{|y| \vee z}) \in D_0 \cup D_1$; for the remaining (x, y, z) it suffices to use (7.72). Finally, the inequality (7.80) is clear if $(\frac{x}{|y| \vee z}, \frac{y}{|y| \vee z}) \in D_0 \cup D_1$. For the remaining points one applies (7.72) and (7.73), the latter inequality implying $h' < 1$. This completes the proof.   □

Sharpness of (7.66). We will show that the inequality is sharp in the discrete-time case, even if g is assumed to be a ± 1 transform of f. Suppose the best constant in this estimate equals $\beta > 0$. Let

$$U^0(x, y, z, w) = \sup\{\mathbb{E}(|g_n|^* \vee w)^p - \beta^p \mathbb{E}(|f_n|^* \vee z)^p\},$$

where the supremum is taken over all integers n and all martingales f, g satisfying $\mathbb{P}((f_0, g_0) = (x, y)) = 1$ and $df_k = \pm dg_k$, $k = 1, 2, \ldots$. Then U^0 satisfies 1°–4° and is homogeneous of order p: $W(\lambda x, \pm \lambda y, |\lambda| z, |\lambda| w) = |\lambda|^p W(x, y, z, w)$ for all $x, y \in \mathbb{R}$, $z, w \geq 0$ and $\lambda \neq 0$.

Let δ be a small number belonging to $(0, 1/p)$. By 2°, applied to $x = 0$, $y = w = 1$, $z = \delta/(1 + 2\delta)$, $\varepsilon = 1$ and $t_1 = \delta$, $t_2 = -1/p$, we obtain

$$U^0\left(0, 1, \frac{\delta}{1 + 2\delta}, 1\right) \geq \frac{p\delta}{1 + p\delta} U^0\left(-\frac{1}{p}, 1 - \frac{1}{p}, \frac{\delta}{1 + 2\delta}, 1\right) \tag{7.81}$$
$$+ \frac{1}{1 + p\delta} U^0\left(\delta, 1 + \delta, \frac{\delta}{1 + 2\delta}, 1 + \delta\right).$$

Now, by 1° and 4°,

$$U^0\left(-\frac{1}{p}, 1 - \frac{1}{p}, \frac{\delta}{1 + 2\delta}, 1\right) = U^0\left(-\frac{1}{p}, 1 - \frac{1}{p}, \frac{1}{p}, 1\right) \geq 1 - \left(\frac{\beta}{p}\right)^p. \tag{7.82}$$

Furthermore, by 4°,

$$U^0\left(\delta, 1 + \delta, \frac{\delta}{1 + 2\delta}, 1 + \delta\right) = U^0(\delta, 1 + \delta, \delta, 1 + \delta),$$

which, by 2° (with $x = z = \delta$, $y = w = 1 + \delta$, $\varepsilon = -1$ and $t_1 = -\delta$, $t_2 = \frac{1}{p} + \delta(\frac{1}{p} - 1)$) can be bounded from below by

$$\frac{p\delta}{1 + \delta} U^0\left(\frac{1 + \delta}{p}, 1 - \frac{1}{p} + \delta\left(2 - \frac{1}{p}\right), \delta, 1 + \delta\right) + \frac{1 + \delta - p\delta}{1 + \delta} U^0(0, 1 + 2\delta, \delta, 1 + \delta).$$

Using the majorization, we get

$$U^0\left(\frac{1 + \delta}{p}, 1 - \frac{1}{p} + \delta\left(2 - \frac{1}{p}\right), \delta, 1 + \delta\right) \geq (1 + \delta)^p \left[1 - \left(\frac{\beta}{p}\right)^p\right],$$

furthermore, by 4° and the homogeneity of W,

$$U^0(0, 1 + 2\delta, \delta, 1 + \delta) = W(0, 1 + 2\delta, \delta, 1 + 2\delta) = (1 + 2\delta)^p U^0\left(0, 1, \frac{\delta}{1 + 2\delta}, 1\right).$$

Now plug all the above estimates into (7.81) to get

$$U^0\left(0, 1, \frac{\delta}{1+2\delta}, 1\right)\left[1 - \frac{(1+\delta-p\delta)(1+2\delta)^p}{(1+\delta)(1+p\delta)}\right] \geq$$
$$\frac{p\delta}{1+p\delta}\left[1 - \left(\frac{\beta}{p}\right)^p\right](1 + (1+\delta)^{p-1}). \tag{7.83}$$

It follows from the definition of U^0 that

$$U^0\left(0, 1, \frac{\delta}{1+2\delta}, 1\right) \leq W(0, 1, 0, 1).$$

Furthermore, one easily checks that the function

$$F(s) = 1 - \frac{(1+s-ps)(1+2s)^p}{(1+s)(1+ps)}, \qquad s > -\frac{1}{p},$$

satisfies $F(0) = F'(0) = 0$. Hence

$$1 - \left(\frac{\beta}{p}\right)^p \leq \frac{U^0(0, 1, 0, 1) \cdot F(\delta) \cdot (1+p\delta)}{p\delta(1 + (1+\delta)^{p-1})}$$

and letting $\delta \to 0$ yields $1 - \left(\frac{\beta}{p}\right)^p \leq 0$, or $\beta \geq p$. $\qquad\square$

Sharpness of (7.68). Suppose the best constant in the estimate equals $\beta > 0$. Introduce the function $U^0 : [0, \infty) \times \mathbb{R} \times [0, \infty) \times [0, \infty) \to \mathbb{R}$ by

$$U^0(x, y, z, w) = \sup\{\mathbb{E}(g_n^* \vee w)^p - \beta^p\mathbb{E}(|f_n|^* \vee z)^p\},$$

where the supremum is taken over all integers n, all nonnegative submartingales f and all integrable sequences g satisfying $\mathbb{P}((f_0, g_0) = (x, y)) = 1$ and, for $k = 1, 2, \ldots,$

$$|df_k| \geq |dg_k|, \qquad \alpha\mathbb{E}(df_k|\mathcal{F}_{k-1}) \geq |\mathbb{E}(dg_k|\mathcal{F}_{k-1})|$$

with probability 1. We see that U^0 is homogeneous and satisfies 1°–4°. Note that this time 2° consists of two conditions: the concavity and the property

$$U^0(x+d, y+\alpha d, z, w) \leq U^0(x, y, z, w), \quad \text{if } 0 \leq x \leq z, \ y \leq w \text{ and } d \geq 0. \tag{7.84}$$

Fix $\delta \in (0, 1/p)$ and apply (7.84) with $x = 0$, $y = w = 1$, $z = \delta/(1 + (\alpha+1)p)$, $d = \delta$ and then use 4° to obtain

$$U^0\left(0, 1, \frac{\delta}{1+(\alpha+1)\delta}, 1\right) \geq U^0\left(\delta, 1+\alpha\delta, \frac{\delta}{1+(\alpha+1)\delta}, 1\right)$$
$$= U^0(\delta, 1+\alpha\delta, \delta, 1+\alpha\delta). \tag{7.85}$$

Using 1°, the concavity and the majorization as above, we get

$$U^0(\delta, 1 + \alpha\delta, \delta, 1 + \alpha\delta) \geq \frac{\delta(\alpha + 1)p}{1 + \alpha\delta}(1 + \alpha\delta)^p \left[1 - \left(\frac{\beta}{(\alpha + 1)p}\right)^p\right]$$
$$+ \frac{1 + \alpha\delta - \delta(\alpha + 1)p}{1 + \alpha\delta}(1 + (\alpha + 1)\delta)^p U^0\left(0, 1, \frac{\delta}{1 + (\alpha + 1)\delta}, 1\right),$$

which, combined with (7.85), gives

$$U^0\left(0, 1, \frac{\delta}{1 + (\alpha + 1)\delta}, 1\right)\left[1 - \frac{1 + \alpha\delta - \delta(\alpha + 1)p}{1 + \alpha\delta}(1 + (\alpha + 1)\delta)^p\right]$$
$$\geq \delta(\alpha + 1)p(1 + \alpha\delta)^{p-1}\left[1 - \left(\frac{\beta}{(\alpha + 1)p}\right)^p\right].$$

Now it suffices to use

$$U^0\left(0, 1, \frac{\delta}{1 + (\alpha + 1)\delta}, 1\right) \leq U^0(0, 1, 0, 1)$$

and the fact that the function

$$G(s) = 1 - \frac{1 + \alpha s - s(\alpha + 1)p}{1 + \alpha s}(1 + (\alpha + 1)s)^p, \quad s > -1/\alpha,$$

satisfies $G(0) = G'(0) = 0$, to obtain

$$1 - \left(\frac{\beta}{(\alpha + 1)p}\right)^p \leq \frac{U^0(0, 1, 0, 1)G(\delta)}{\delta(\alpha + 1)p(1 + \alpha\delta)^{p-1}}.$$

Letting $\delta \to 0$ gives $1 - \left(\frac{\beta}{(\alpha+1)p}\right)^p \leq 0$, or $\beta \geq (\alpha + 1)p$. This completes the proof.   □

On the search of the suitable majorant. We shall focus on the submartingale inequality (7.67) for $\alpha > 0$. As usual, we assume that the best constant is some $\beta > 0$, let $V(x, y, w) = (|y| \vee w)^p - \beta^p|x|^p$ and write the corresponding function U^0:

$$U^0(x, y, w) = \sup\{\mathbb{E}V(f_n, g_n, |g_n|^* \vee w)\}.$$

By homogeneity, it suffices to determine this function for $w = 1$. We start with the analysis of the set $\{(x, y) : U^0(x, y, 1) = V(x, y, 1)\}$. Observe that it contains no point of the form $(x, \pm 1)$. First we shall prove this for $x = 0$: fix $\delta > 0$ and consider a pair (f, g) such that $f_0 \equiv 0$, $f_1 = f_2 = \cdots \equiv \delta$ and $g_0 \equiv 1$, $g_1 = g_2 = \cdots \equiv 1 + \alpha\delta$. Then

$$U^0(0, 1, 1) \geq \mathbb{E}V(f_1, g_1, |g_1|^*) = (1 + \alpha\delta)^p - \beta^p\delta^p > 1 = V(0, 1, 1),$$

where the strict inequality holds if δ is sufficiently small. To show that $U^0(x, 1, 1) > V(x, 1, 1)$ for $x > 0$, consider a martingale pair (f, g) such that $f_0 \equiv x$, $g_0 \equiv 1$, $df_1 \in \{-\delta, \delta\}$, $df_2 = df_3 = \cdots \equiv 0$ and $dg_n = -df_n$ for $n \geq 1$. Then

$$U^0(x, 1, 1) \geq \mathbb{E}V(f_1, g_1, |g_1|^*) = \frac{1}{2}\left((1+\delta)^p + 1 - \beta^p\left[(x+\delta)^p + (x-\delta)^p\right]\right)$$
$$> 1 - \beta^p x^p = V(x, 1, 1),$$

provided δ is sufficiently small.

The next observation concerning the set $\{U^0 = V\}$ is the following. Fix $y \in [0, 1]$, let x be a large positive number and look at the definition of $U^0(x, y, 1)$. Intuitively, we search for such f and g, for which the pth moment of $|g|^* \vee 1$ is large in comparison with the size of the pth moment of f. However, if we want $|g|^* \vee 1$ to increase, we need to make g reach the endpoints of the interval $[-1, 1]$. Since $p > 2$, this may result in a significant increase of $\|f\|_p$, especially if x is large, and this will draw $\mathbb{E}V(f_n, g_n, |g_n|^* \vee 1)$ down. This leads to the hypothesis that for fixed $y \in (-1, 1)$ and large x, the best pair (f, g) is the constant one. In other words:

(A1)   There is a function $\gamma : \mathbb{R} \to [0, 1]$ such that if $|y| \leq \gamma(x)$, then $U^0(x, y, 1) = V(x, y, 1)$.

Furthermore, we assume that

(A2)   The function γ is strictly increasing and of class C^1.
(A3)   The function $U(\cdot, \cdot, 1)$ is of class C^1 on $[0, \infty) \times (-1, 1)$,
(A4)   On the set $\{(x, y) : \gamma(x) \leq y \leq 1\}$, the function $U(\cdot, \cdot, 1)$ is linear along the lines of slope -1.
(A5) $U_x(0+, y) + \alpha U_y(0, y) = 0$ for $y \geq \gamma(0)$.

The next assumption is based on the behavior of the function $U(\cdot, \cdot, 1)$ on the line $y = 1$. If we consider the pair f, g as above, we get

$$2U(x, 1, 1) \geq U(x - \delta, 1 + \delta, 1 + \delta) + U(x + \delta, 1 - \delta, 1)$$
$$= (1+\delta)^p U\left(\frac{x-\delta}{1+\delta}, 1, 1\right) + U(x + \delta, 1 - \delta, 1).$$

Subtracting $2U(x, 1, 1)$ from both sides, dividing throughout by δ and letting $\delta \to 0$ gives $xU_x(x, 1, 1) + U_y(x, 1, 1) - pU(x, 1, 1) \leq 0$. We conjecture that we have equality here.

(A6)   For $x > 0$,

$$xU_x(x, 1, 1) + U_y(x, 1, 1) = pU(x, 1, 1). \tag{7.86}$$

Let us show how the assumptions (A1)–(A4) lead to the differential equation (7.69). Observe that if $x \geq 0$ and $t \in [0, \min\{x, 1 - \gamma(x)\}]$, then

$$U(x - t, \gamma(x) + t, 1) = U(x, \gamma(x), 1) + [-U_x(x, \gamma(x), 1) + U_y(x, \gamma(x), 1)]t$$
$$= V(x, \gamma(x), 1) + [-V_x(x, \gamma(x), 1) + V_y(x, \gamma(x), 1)]t$$
$$= 1 - \beta^p x^p + p\beta^p x^{p-1}t.$$

Differentiating this equation with respect to x and t leads to the following formulas for U_x and U_y:

$$U_x(x - t, \gamma(x) + t, 1) = -p\beta^p x^{p-1} + \frac{p(p - 1)\beta^p x^{p-2} t}{1 + \gamma'(x)},$$

$$U_y(x - t, \gamma(x) + t, 1) = \frac{p(p - 1)\beta^p x^{p-2} t}{1 + \gamma'(x)}. \tag{7.87}$$

Now if $x + \gamma(x) \geq 1$, then plugging these two expressions into (7.86) yields (7.69), with C replaced by $\beta^p(p - 1)$. Next, use (7.87), this time with $x > 0$ satisfying $x + \gamma(x) < 1$ and take $t \to x$: by (A5), we obtain $\gamma'(x) = (p - 1)(1 + \alpha) - 1$. Now, let x_0 be the unique positive number satisfying $x_0 + \gamma(x_0) = 1$. Comparing $\gamma'(x_0-)$ with $\gamma'(x_0+)$ leads to the equation

$$(p - 1)(1 + \alpha) - 1 = \frac{-1 + (p - 1)\beta^p x_0^{p-1}(1 - x_0)}{1 + \beta^p(p - 1)x_0^p},$$

or, equivalently,

$$\beta^p = \frac{1 + \alpha}{x_0^{p-1} - x_0^p(p - 1)(1 + \alpha)}.$$

The right-hand side, as a function of $x_0 \in (0, ((p - 1)(1 + \alpha))^{-1})$, attains its minimum $(\alpha + 1)^p p^p$ at the point $(p(\alpha + 1))^{-1}$. This suggests the final assumption.

(A7) $\beta = (\alpha + 1)p$ and the condition (7.70) is satisfied.

Now it is straightforward to obtain the formula for $U(\cdot, \cdot, 1)$: we know the function γ and it remains to use (7.87).   □

7.5   Double maximal inequality

We have come to one of the most difficult inequalities of this monograph. We shall determine the optimal constant β in the maximal inequality

$$|| \, |Y|^* ||_1 \leq \beta || \, |X|^* ||_1,$$

under the assumption that X is a martingale and Y is differentially subordinate to X. First, we need an auxiliary section, where we introduce the parameters which are needed to describe the constant and define the corresponding special function.

7.5.1   Related differential equations

Let $w > 1$ be a fixed number. Standard argumentation yields the existence and uniqueness of $Y = Y^w : [1, w] \to \mathbb{R}$ satisfying

$$Y'(t) = \frac{1}{2}\left(1 + \frac{1}{w}\right)(1 + t)^{-2}\left[t^2 + 2(1 - t)\left(\exp\left(\frac{t - Y(t)}{2}\right) - 1\right)\right], \tag{7.88}$$

$t \in (1, w)$, with the terminal condition $Y(w) = w$.

Lemma 7.7. *Let $w > 1$. Then Y^w is nondecreasing and*

$$Y^w(t) \geq t \qquad \text{for all } t, \text{ with equality only for } t = w. \tag{7.89}$$

Proof. Note that $2(1-t)\exp(\frac{t-Y^w(t)}{2}) \leq 0$ for $t \in (1, w)$, which implies

$$(Y^w)'(t) \leq \frac{1+w^{-1}}{2(1+t)^2}(t^2 + 2t - 2)$$

$$= \frac{1+w^{-1}}{2}\left(t + \frac{3}{1+t}\right)'.$$

Consequently, since $t \leq w$ and $w > 1$,

$$Y^w(t) \geq Y^w(w) - \frac{1+w^{-1}}{2}\left[w + \frac{3}{1+w} - t - \frac{3}{1+t}\right]$$

$$\geq w - \frac{1+w^{-1}}{2}(w-t) \tag{7.90}$$

$$= t + \frac{1-w^{-1}}{2}(w-t),$$

which gives (7.89). This also implies $2(1-t)(\exp(\frac{t-Y^w(t)}{2}) - 1) \geq 0$ for $t \in (1, w)$ and in turn,

$$(Y^w)'(t) \geq \frac{1+w^{-1}}{2}\left(\frac{t}{1+t}\right)^2 \geq 0. \tag{7.91}$$

The proof is complete. $\qquad\qquad\qquad\qquad\qquad\qquad\qquad\qquad\qquad\qquad\square$

Let β be the positive number given by

$$\beta = \min_w \left\{ Y^w(1) + \frac{5}{4}\left(1 + \frac{1}{w}\right)\right\}. \tag{7.92}$$

It will be shown that this is the value of the best constant. Let us provide some approximation of β.

Lemma 7.8. *We have*

$$3.4142\ldots < \beta < 3.4358\ldots. \tag{7.93}$$

Proof. The number on the left is $2 + \sqrt{2}$. To prove this two-sided bound, take $w > 1$ and use the first line of (7.90) with $t = 1$ to obtain

$$Y^w(1) + \frac{5}{4}(1 + w^{-1}) \geq w - \frac{1+w^{-1}}{2}\left(w + \frac{3}{1+w} - \frac{5}{2}\right) + \frac{5}{4}(1 + w^{-1})$$

$$= 2 + w^{-1} + \frac{w}{2}. \tag{7.94}$$

The expression on the right, as a function of $w \in (1, \infty)$, attains its minimum $2 + \sqrt{2}$ for $w = \sqrt{2}$. This gives the left inequality in (7.93). To prove the right one, we proceed as previously, using a lower bound for $(Y^w)'$ coming from (7.91). After integration, we get

$$Y^w(1) + \frac{5}{4}\left(1 + w^{-1}\right) \le \left(1 + \frac{1}{w}\right)[2 - \log 2 + \log(1 + w)] + \frac{w}{2} - 1,$$

and the upper bound in (7.93) is the minimum of the expression on the right above.   □

It is clear that the function $w \mapsto Y^w(1) + 5/4(1 + 1/w)$ is continuous. In addition, it tends to $7/2 > \beta$ as $w \downarrow 1$ and, by (7.94), tends to infinity as $w \to \infty$. Hence the minimum defining β is attained for some w_0. To avoid the question about the uniqueness of w_0, let us take the smallest number with this property. Combining (7.94) with the right inequality in (7.93), we conclude that $1/w_0 + w_0/2 < 1.436$, so $1.18 < w_0 < 1.69$. To complete the discussion about the explicit values of β and w_0, let us record here that numerical approximation gives $\beta = 3.4351\ldots$ and $w_0 = 1.302\ldots$.

A few words about some auxiliary notation. Set $w_1 = Y^{w_0}(1)$, write $\gamma = -1 - 1/w_0$ and use the function $y_0 : [w_1, w_0] \to [1, w_0]$, the inverse of Y^{w_0}; we shall often skip the argument and write y_0 instead of $y_0(w)$. It can be easily checked that the function y_0 satisfies the differential equation

$$y_0' = -\frac{1}{\gamma} \cdot \frac{2(1 + y_0)^2}{y_0^2 + 2y_0 - 2 + 2(1 - y_0)\exp(\frac{y_0 - w}{2})} \tag{7.95}$$

for $w \in (w_1, w_0)$. Moreover, in view of (7.89), we have

$$y_0(w) \le w \text{ for } w \in [w_1, w_0], \text{ with equality only for } w = w_0. \tag{7.96}$$

We conclude this section with a technical fact to be needed later.

Lemma 7.9. *We have*

$$y_0' \ge 1 \tag{7.97}$$

and

$$y_0'(y_0 - 1) \le y_0 + 1. \tag{7.98}$$

Proof. The estimate (7.9) follows immediately from the fact that

$$y_0' \ge -\frac{2}{\gamma} \cdot \frac{(1 + y_0)^2}{y_0^2 + 2y_0 - 2}$$

and that both the factors are bigger than 1.

To prove (7.98), observe that, by (7.96), $2(1 - y_0) \exp\left(\frac{y_0 - w}{2}\right) \geq 2(1 - y_0)$. Plugging this into (7.95) gives

$$y_0'(y_0 - 1) \leq -\frac{1}{\gamma} \cdot \frac{2(1 + y_0)^2}{y_0^2} \cdot (y_0 - 1),$$

so we will be done if we show that $2(y_0^2 - 1) \leq -\gamma y_0^2$. But

$$2(y_0^2 - 1) + \gamma y_0^2 = y_0^2(2 + \gamma) - 2 \leq w_0^2(2 + \gamma) - 2 = w_0(w_0 - 1) - 2 \leq 0,$$

where the latter estimate comes from the bound $w_0 < 2$. □

7.5.2 Formulation of the result

We shall establish the following fact.

Theorem 7.13. *Assume that X, Y are real-valued martingales such that Y is differentially subordinate to X. Then*

$$|||Y|^*||_1 \leq \beta|||X|^*||_1, \tag{7.99}$$

where β is given by (7.92). The constant β is the best possible. It is already the best possible in the discrete-time setting, when Y is assumed to be a ± 1-transform of X.

As previously, we will show the estimate only in the discrete-time setting, for ± 1 transforms. The passage to the continuous time requires standard approximation arguments and we shall not present them.

7.5.3 Proof of Theorem 7.13

Proof of (7.99). We start with reduction of the dimension. Initially, we need to find a function $U(\cdot, \cdot, \cdot, \cdot)$ of four variables. However, due to the homogeneity of the inequality (7.99), it suffices to construct $u = U(\cdot, \cdot, 1, \cdot)$. To do this, consider the following subsets of $[0, 1] \times [0, \infty) \times (0, \infty)$:

$$D_1 = \{(x, y, w) : w \leq w_1, y \leq x\},$$
$$D_2 = \{(x, y, w) : w \leq w_1, x < y \leq x + w_1 - 1\},$$
$$D_3 = \{(x, y, w) : w \leq w_1, x + w_1 - 1 < y\},$$
$$D_4 = \{(x, y, w) : w_1 < w \leq w_0, y \leq x + y_0(w) - 1\},$$
$$D_5 = \{(x, y, w) : w_1 < w \leq w_0, x + y_0(w) - 1 < y \leq x + w - 1\},$$
$$D_6 = \{(x, y, w) : w_1 < w \leq w_0, x + w - 1 < y\},$$
$$D_7 = \{(x, y, w) : w > w_0, x + y \leq 1\},$$
$$D_8 = \{(x, y, w) : w > w_0, 1 < x + y \leq 1 + w - w_0\},$$
$$D_9 = \{(x, y, w) : w > w_0, 1 + w - w_0 < x + y\}.$$

Now we introduce the function $u : [-1, 1] \times \mathbb{R} \times (0, \infty) \to \mathbb{R}$. First we define it on the sets D_1–D_9.

$$u(x, y, w) = \begin{cases} -\frac{\gamma}{4}(y^2 - x^2 + 1) + \frac{5}{4}\gamma & \text{on } D_1, \\ 3\gamma - \gamma y + (x - 2)\gamma \exp\left(\frac{x-y}{2}\right) & \text{on } D_2, \\ \gamma(3 - y) + \gamma \exp\left(\frac{1-w_1}{2}\right)\left(-1 + y - w_1 - \frac{(y-w_1+1)^2 - x^2}{4}\right) & \text{on } D_3, \\ -\frac{\gamma}{2(1+y_0)}(y^2 - x^2 + 1) + w - \beta & \text{on } D_4, \\ \frac{2\gamma}{1+y_0} \exp\left(\frac{y_0 - y - 1 + x}{2}\right)(x - 2) + \alpha(y, w) & \text{on } D_5, \\ \frac{2\gamma}{1+y_0} \exp\left(\frac{y_0 - w}{2}\right)\left(-1 + y - w - \frac{(y-w+1)^2 - x^2}{4}\right) + \alpha(y, w) & \text{on } D_6, \\ \frac{\exp(w_0 - w)}{2w_0}(y^2 - x^2 + 1) + w - \beta & \text{on } D_7, \\ \frac{(1-x)}{w_0} \exp(x + y + w_0 - w - 1) + w - \beta & \text{on } D_8, \\ \frac{(y-w+w_0)^2 - x^2 + 1}{2w_0} + w - \beta & \text{on } D_9, \end{cases}$$

where

$$\alpha(y, w) = \gamma(1 - y) + \frac{\gamma y_0^2 + 2\gamma}{2(1 + y_0)} + w - \beta.$$

We extend u to its whole domain $[-1, 1] \times \mathbb{R} \times (0, \infty)$, setting

$$u(x, y, w) = u(-x, y, w) = u(x, -y, w) = u(-x, -y, w) \tag{7.100}$$

for all $x \in [0, 1]$, $y \geq 0$ and $w > 0$.

In the lemma below we list the main properties of the function u, which will be exploited while showing that U satisfies $1°$–$4°$. For the proof of these technical facts we refer the reader to the original paper [162].

Lemma 7.10.

(i) *The function u is continuous. In addition, it is of class C^1 on each of the sets $\{(x, y, w) : w < w_1\}$, $\{(x, y, w) : w \in (w_1, w_0)\}$, $\{(x, y, w) : w > w_0\}$.*

(ii) *For all $w > 0$ and $|y| \leq w$,*

$$\lim_{\delta \downarrow 0} \frac{u(1, y, w) - u(1 - \delta, y \pm \delta, w)}{\delta} \geq \gamma. \tag{7.101}$$

Furthermore, for all $w > 0$ and $x \in (-1, 1]$,

$$\lim_{\delta \downarrow 0} \frac{u(x, w, w) - u(x - \delta, w - \delta, w)}{\delta} \geq \gamma. \tag{7.102}$$

(iii) *For $x \in [-1, 1]$ and $w \in (0, \infty) \setminus \{w_0, w_1\}$, we have $u_w(x, w, w) \leq 0$.*

(iv) *For any x, the function $H_x : (-1 - x, 1 - x) \to \mathbb{R}$, given by $H_x(t) = u_x(x + t, t, t) + u_y(x + t, t, t)$, is nonincreasing.*

(v) *The function $J : (0, \infty) \to \mathbb{R}$ given by $J(y) = u(1, y, y)$ is convex.*

(vi) *For any fixed $w > 0$, the function $u(\cdot, \cdot, w)$, restricted to the rectangle $[-1, 1] \times [-w, w]$, is diagonally concave, i.e., concave along any line of slope ± 1.*

(vii) *For any fixed $w > 0$, the function $y \mapsto u(1, y, w)$ is nondecreasing on $[0, w]$.*

(viii) *For any $w > 0$ and $|y| \leq w$,*

$$u(1, y, w) - (y - 1)u_y(1, y, w) - wu_w(1, y, w) \leq \gamma. \tag{7.103}$$

(ix) *For any $w > 0$ and $|x| \leq 1$, $|y| \leq w$ we have*

$$u(x, y, w) \geq w - \beta.$$

(x) *If $w \in (0, 1]$ and $|x| = |y| \leq w$, then $u(x, y, w) \leq 0$.*

Now we introduce the special function $U : \mathbb{R} \times \mathbb{R} \times (0, \infty) \times (0, \infty) \to \mathbb{R}$ by

$$U(x, y, z, w) = (|x| \vee z)u\left(\frac{x}{|x| \vee z}, \frac{y}{|x| \vee z}, \frac{|y| \vee w}{|x| \vee z}\right) \tag{7.104}$$

and let $V(x, y, z, w) = |y| \vee w - \beta|x| \vee z$ for x, y, z, w as above. Let us verify the conditions $1°$–$4°$. The majorization follows immediately from (ix). To show $2°$, we fix x, y, z, w as in its statement. Since U is homogeneous, we may and do assume that $z = 1$. By the continuity of U, we are allowed to take $|x| < z$ and $|y| < w$. Moreover, since U satisfies $U(x, y, z, w) = U(x, -y, z, w)$, we may assume that $\varepsilon = 1$. Now let $\Phi(t) = U(x+t, y+t, z, w)$ for $t \in \mathbb{R}$ and observe that it suffices to prove that

$$\Phi(t) \leq \Phi(0) + \Phi'(0)t \tag{7.105}$$

for *positive* t. Indeed, applying this to the function $\overline{\Phi}(t) = U(-x+t, -y+t, z, w)$ we get, for $t < 0$,

$$\Phi(t) = \overline{\Phi}(-t) \leq \overline{\Phi}(0) + \overline{\Phi}'(0)(-t) = \Phi(0) + \Phi'(0)t.$$

Thus there is a linear function Ψ such that $\Phi \leq \Psi$ on $\mathbb{R}$ and $\Phi(0) = \Psi(0)$; this implies the desired concavity.

To show (7.105), we consider two cases: $y \geq x + w - 1$ and $y < x + w - 1$. We shall only present the details in the first case: the second one can be handled in a similar manner. Namely, we shall show that

$$\Phi \text{ is concave on the set } [0, w - y], \tag{7.106}$$
$$\Phi'(t+) \leq \Phi'((w - y)-) \text{ for } t \in (w - y, 1 \quad x), \tag{7.107}$$
$$\Phi \text{ is convex on } [1 - x, \infty), \tag{7.108}$$
$$\lim_{t \to \infty} \Phi'(t) \leq \Phi'((w - y)-). \tag{7.109}$$

These properties clearly yield (7.105). The first condition is a consequence of Lemma 7.10 (vi). Property (7.107) follows from items (iii) and (iv) of that lemma; indeed,

$$\Phi'(t+) = \lim_{s\downarrow t}(u_x + u_y + u_w)(x+s, y+s, y+s)$$

$$\leq \limsup_{s\downarrow t}(u_x + u_y)(x+s, y+s, y+s)$$

$$\leq \limsup_{s\uparrow(w-y)}(u_x + u_y)(x+s, y+s, y+s) = \Phi'((w-y)-).$$

The condition (7.108) is shown exactly in the same manner as in the proof of inequality (7.29); here we use item (v) of Lemma 7.10. Finally, we turn to (7.109). We have, for sufficiently large t,

$$\Phi(t) = U(x+t, y+t, x+t, y+t)$$

$$= (x+t)u\left(1, \frac{y+t}{x+t}, \frac{y+t}{x+t}\right)$$

$$= \begin{cases} \gamma(x+t) + \frac{1}{2}\gamma(x-y) - \frac{\gamma(y-x)^2}{4(x+t)} & \text{if } y < x, \\ 3\gamma(x+t) - \gamma(y+t) - \gamma(x+t)\exp\left(\frac{x-y}{2(x+t)}\right) & \text{if } y \geq x, \end{cases}$$

from which we infer that $\lim_{t\to\infty}\Phi'(t) = \gamma$. It suffices to use (7.102) and the condition 2° is proved. To show 3°, observe that by the concavity we have just proved,

$$U(x, y, z, |y|) \leq U(x, y, |x|, |y|) = \frac{U(x, y, |x|, |y|) + U(-x, -y, |x|, |y|)}{2}$$

$$\leq U(0, 0, |x|, |y|) \leq U(0, 0, |y|, |y|) \leq 0.$$

Finally, 4° is obvious. Consequently, Burkholder's method gives

$$\mathbb{E}V(f_n, g_n, |f_n|^* \vee \varepsilon, |g_n|^*) \leq \mathbb{E}U(f_n, g_n, |f_n|^* \vee \varepsilon, |g_n|^*) \leq 0$$

and letting $\varepsilon \to 0$ yields the claim. $\square$

7.5.4 Sharpness

This is really involved. Suppose that the inequality (7.99) holds with some constant γ: the set of such γ's forms an interval $[\beta', \infty)$. As in the previous estimates, we make use of Theorem 7.2 and exploit the properties of the function

$$U^\gamma(x, y, z, w) = \sup \mathbb{E}\big[|g_n|^* \vee w - \gamma|f_n|^* \vee z\big],$$

the supremum being taken over all n and all $(f, g) \in M(x, y)$.

Lemma 7.11. *The function $F : [\beta', \infty) \to \mathbb{R}$ given by $F(\gamma) = U^{\gamma}(1,1,1,1)$ is convex.*

Proof. This is straightforward. Fix $\lambda \in (0,1)$, $\gamma_1, \gamma_2 \geq \beta'$ and $(f,g) \in M(1,1)$. Then for any n,

$$\mathbb{E}\big[|g_n|^* \vee w - (\lambda\gamma_1 + (1-\lambda)\gamma_2)|f_n|^* \vee z\big] = \lambda\mathbb{E}\big[|g_n|^* \vee w - \gamma_1|f_n|^* \vee z\big]$$
$$+ (1-\lambda)\mathbb{E}\big[|g_n|^* \vee w - \gamma_2|f_n|^* \vee z\big]$$
$$\leq \lambda F(\gamma_1) + (1-\lambda)F(\gamma_2).$$

It suffices to take the supremum over f, g and n to complete the proof. □

Next suppose that the inequality (7.99) holds with some constant $\beta_0 < \beta$. By the previous lemma, enlarging β_0 if necessary, we may assume that $U^{\beta_0}(1,1,1,1) \leq U^{\beta}(1,1,1,1)+1/100$. However, we have $U^{\beta} \leq U$, because U^{β} is the *least* majorant of $(x,y,z,w) \mapsto |y| \vee w - \beta|x| \vee z$ satisfying $2°$–$4°$. Thus, in particular, we have $U^{\beta}(1,1,1,1) \leq U(1,1,1,1) = \gamma \leq -1 - (1.7)^{-1}$. The latter estimate follows from the bound $w_0 < 1.69$, see above. Consequently,

$$U^{\beta_0}(1,1,1,1) < -3/2. \tag{7.110}$$

From now on, we will work with the function U^{β_0}. It satisfies $1°$–$4°$, with $V = V_{\beta_0}$. Furthermore, it is homogeneous of order 1 and such that

$$U^{\beta_0}(x,y,z,w) = U^{\beta_0}(-x,y,z,w) = U^{\beta_0}(x,-y,z,w)$$

for all $x, y \in \mathbb{R}$, $z, w > 0$. We shall use the following notation: $u_0(x,y,w) = U^{\beta_0}(x,y,1,w)$, $A^w(y) = u_0(0,y,w)$, $B^w(y) = u_0(1,y,w)$ and $\gamma_0 = u_0(1,1,1)$.

Lemma 7.12. *For any $x \in [-1,1]$, $y_1, y_2 \in \mathbb{R}$ and $w_1, w_2 > 0$ we have*

$$|u_0(x,y_1,w_1) - u_0(x,y_2,w_2)| \leq \max\{|y_1 - y_2|, |w_1 - w_2|\}.$$

Proof. By the triangle inequality, for any numbers $a_1, a_2, \ldots, a_n, b_1, b_2, \ldots, b_n$,

$$|y_1 + a_1| \vee |y_1 + a_2| \vee \cdots \vee |y_1 + a_n| \vee w_1 - |y_1 + a_1| \vee |y_2 + a_2| \vee \cdots \vee |y_2 + a_n| \vee w_2$$
$$\leq \max\{|y_1 - y_2|, |w_1 - w_2|\}.$$

Therefore, if f, g are simple martingales such that f starts from x, g starts from 0 and $dg_n = \pm df_n$ for $n \geq 1$, then, by the definition of u_0,

$$\mathbb{E}((y_1 + g)_n^* \vee w_1 - \beta_0 f_n^* \vee 1) - u_0(x,y_2,w_2)$$
$$\leq \mathbb{E}\big[((y_1 + g)_n^* \vee w_1 - \beta_0 f_n^* \vee 1) - ((y_2 + g)_n^* \vee w_1 - \beta_0 f_n^* \vee 1)\big]$$
$$\leq \max\{|y_1 - y_2|, |w_1 - w_2|\}.$$

It suffices to take the supremum over f, y and n to obtain

$$u_0(x,y_1,w_1) - u_0(x,y_2,w_2) \leq \max\{|y_1 - y_2|, |w_1 - w_2|\},$$

and the claim follows by symmetry. □

Lemma 7.13. *For any* $w > 0$, $|y| \leq w$ *and* $\delta \in (0, 1)$,

$$B^w(y) \geq u_0(1 - \delta, y + \delta, w) + \delta\gamma_0. \tag{7.111}$$

Proof. This is shown exactly in the same manner as (7.44).   □

Lemma 7.14. *For any* $w > y \geq 1$ *and* $\delta \in (0, 1)$ *satisfying* $\delta \leq (w - y)/2$,

$$B^w(y) \geq \delta A^w(y + 2\delta - 1) + (1 - \delta)B^w(y + 2\delta) + \delta\gamma_0, \tag{7.112}$$

$$A^w(y + 2\delta - 1) \geq \frac{\delta}{1 + \delta}B^w(y + 2\delta) + \frac{1}{1 + \delta}u_0(-\delta, y + \delta - 1, w)$$
$$\geq \frac{\delta}{1 + \delta}B^w(y + 2\delta) + \frac{\delta}{1 + \delta}B^w(y) + \frac{1 - \delta}{1 + \delta}A^w(y - 1) \tag{7.113}$$

and

$$(1 - \delta)(B^w(y) - A^w(y - 1) + \gamma_0) \geq B^w(y + 2\delta) - A^w(y - 1 + 2\delta) + \gamma_0. \tag{7.114}$$

Proof. Repeat the argumentation leading to (7.45).   □

Lemma 7.15. *For any* $w > 1$,

$$B^w(w) \leq \gamma_0 \left[3 - w - \exp\left(\frac{1 - w}{2}\right) \right]. \tag{7.115}$$

Proof. Essentially, we have shown this in the proof of sharpness of (7.29). However, we shall present a quick proof since we shall need some estimates which will appear in between. First, we have $A^w(0) \geq B^w(1)$, by 2° applied to $(x, y, z, w) := (0, 0, 1, w)$, $\varepsilon = 1$ and $t_1 = -1$, $t_2 = 1$. Thus, using (7.114) and induction,

$$(1 - \delta)^N \gamma_0 \geq (1 - \delta)^N(B^w(1) - A^w(0) + \gamma_0) \geq B^w(1 + 2N\delta) - A^w(2N\delta) + \gamma_0.$$

Hence, if we put $\delta = (w - 1)/(2N)$ and let $N \to \infty$, we arrive at

$$A^w(w - 1) \geq B^w(w) + \gamma_0(1 - \exp((1 - w)/2)). \tag{7.116}$$

Arguments leading to (7.112), with y replaced by w, give

$$B^w(w) \geq (1 - \delta)B^{w+2\delta}(w + 2\delta) + \delta A^{w+\delta}(w + 2\delta - 1) + \delta\gamma_0,$$

so, by Lemma 7.12,

$$B^w(w) \geq (1 - \delta)B^{w+2\delta}(w + 2\delta) + \delta A^{w+2\delta}(w + 2\delta - 1) + \delta\gamma_0 - \delta^2. \tag{7.117}$$

Applying (7.116) yields

$$B^w(w) \geq B^{w+2\delta}(w + 2\delta) + \delta\gamma_0(2 - \exp((1 - w - 2\delta)/2) - \delta^2.$$

Using induction as in the proof of (7.116), this leads to

$$\gamma_0 = B^1(1) \geq B^w(w) + \gamma_0[w - 2 + \exp((1 - w)/2)],$$

which is the claim.   □

Lemma 7.16. *Suppose that $w \in (1,2)$ and let $1 \le y \le w$. Then*

$$A^w(y-1)y^2 \ge B^w(y)(y^2 - 2y + 2) + 2(y-1)(w - \beta_0). \tag{7.118}$$

Proof. We apply $2°$ three times:

$$A^w(y-1) \ge \frac{y}{y+2} B^w(y) + \frac{2}{y+2} u_0(-y/2, y/2 - 1, w),$$

$$u_0(-y/2, y/2 - 1, w) \ge \frac{2-y}{y} u_0(-1 + y/2, -y/2, w) + \frac{2y-2}{y} u_0(-1, 0, w),$$

$$u_0(-1 + y/2, -y/2, w) = u_0(1 - y/2, y/2, w) \ge \frac{y}{2} A^w(y-1) + \frac{2-y}{2} B^w(y).$$

Combining these estimates with $u_0(-1, 0, w) \ge w - \beta_0$, a consequence of $1°$, we get (7.118). $\square$

Lemma 7.17. *Suppose that $w \in (1,2)$ and let $1 \le y \le w$. Then*

$$(A^w(w-1) - w + \beta_0) \left[2 \exp\left(\frac{w-y}{2}\right) - 1 \right] \tag{7.119}$$
$$\ge (B^w(y) - w + \beta_0)(1 - 2y^{-1} + 2y^{-2}) + (y - w)\gamma_0$$
$$+ 2(B^w(w) - w + \beta_0 + \gamma_0) \left[\exp\left(\frac{w-y}{2}\right) - 1 \right]$$

and

$$(B^w(y) - w + \beta_0) \left[1 + \frac{2(y-1)}{y^2} \left(1 - \exp\left(\frac{y-w}{2}\right) \right) \right] \tag{7.120}$$
$$\ge B^w(w) - w + \beta_0 + \gamma_0 \left[w - y - 1 + \exp\left(\frac{y-w}{2}\right) \right].$$

In addition,

$$B^{w-2\delta}(w - 2\delta) \ge B^w(w) + 2\delta\gamma_0 - \frac{\delta \exp(\frac{y-w}{2})}{1 + 2y^{-2}(y-1)(1 - \exp(\frac{y-w}{2}))} \tag{7.121}$$
$$\times \left\{ 2y^{-2}(y-1)(B^w(w) - w + \beta_0 + \gamma(w-y)) + \gamma_0 \right\} - \delta^2.$$

Proof. Using (7.114) inductively yields the following estimate: for $1 \le y'' \le y' \le w$,

$$\exp\left(\frac{y''-y'}{2}\right)(B^w(y'') - A^w(y'' - 1) + \gamma_0) \ge B^w(y') - A^w(y' - 1) + \gamma_0. \tag{7.122}$$

Take $y' = w$ and note that $B^w(y'' + 2\delta) \ge B^w(y'') - 2\delta$, a consequence of Lemma 7.12. Plug these two estimates into (7.113) to get

$$A^w(y'' + 2\delta - 1)$$
$$\ge A^w(y'' - 1) + \frac{2\delta}{1+\delta} \left[\exp\left(\frac{w-y''}{2}\right)(B^w(w) - A^w(w-1) + \gamma_0) - \gamma_0 - \delta \right].$$

Now set $\delta = (w - y)/(2N)$, write the above estimates for $y'' = y$, $y'' = y + 2\delta$, ...,
$y'' = y + (2N - 2)\delta$ and sum them up. We obtain

$$A^w(w - 1) = A^w(y + 2N\delta - 1) \geq A^w(y - 1) - \frac{2\delta}{1 + \delta} N(\gamma_0 + \delta)$$
$$+ \frac{2\delta}{1 + \delta}(B^w(w) - A^w(w - 1) + \gamma_0) \exp\left(\frac{w - y}{2}\right) \frac{1 - \exp(-N\delta)}{1 - \exp(-\delta)}.$$

Letting $N \to \infty$ gives

$$A^w(w-1) \geq A^w(y-1) + 2(B^w(w) - A^w(w-1) + \gamma_0)\left[\exp\left(\frac{w - y}{2}\right) - 1\right] + \gamma_0(y - w)$$

and combining this with (7.118) yields the first estimate. We skip the proof of
(7.120), since it can be established using similar argumentation. To get (7.121),
plug (7.120) into (7.119) to obtain

$$A^w(w - 1) \geq B^w(w) + \gamma_0 - \frac{\exp(\frac{y-w}{2})}{1 + 2y^{-2}(y - 1)(1 - \exp(\frac{y-w}{2}))}$$
$$\times \left\{2y^{-2}(y - 1)(B^w(w) - w + \beta_0 + \gamma(w - y)) + \gamma_0\right\}.$$

It suffices to make use of (7.117) (with w replaced by $w - 2\delta$) to complete the
proof.   □

The final estimate we will need is the following. It can be established essentially in the same manner as above; we omit the tedious and lengthy calculations.

Lemma 7.18. *For any $w \geq 1$,*

$$B^w(w) \geq \frac{w^2(1 + \gamma_0)}{2} + 2w - \beta_0. \tag{7.123}$$

Now we are ready to complete the proof.

Sharpness of (7.99). The first observation is that $\gamma_0 \in (-2, -3/2)$. The inequality
$\gamma_0 < -3/2$ is precisely (7.110). To get the lower bound, apply (7.123) to $w = 1$ to
obtain $\gamma_0 \geq 5 - 2\beta_0 > -2$ (we have $\beta_0 < \beta < 3.5$). Now let $v_0 = -(1 + \gamma_0)^{-1} \in$
$(1, 2)$, define Y^{v_0} as in Section 7.3 and let $\bar{y}_0$ be the inverse to Y^{v_0}. Finally, let
$v_1 = Y^{v_0}(1)$ and

$$C(w) = -\frac{2\gamma_0}{1 + \bar{y}_0} \exp\left(\frac{\bar{y}_0 - w}{2}\right) + \gamma_0(1 - w) + \frac{\gamma_0(\bar{y}_0^2 + 2)}{2(1 + \bar{y}_0)} + w - \beta_0$$

for $w \in [v_1, v_0]$. Observe that since $\bar{y}_0(v_0) = v_0$, we have, after some manipulations,

$$C(v_0) = \frac{3}{2}v_0 - \beta_0 = \frac{v_0^2(1 + \gamma_0)}{2} + 2v_0 - \beta_0 \leq B^{v_0}(v_0), \tag{7.124}$$

by virtue of Lemma 7.18. Furthermore, it can be verified that C satisfies the differential equation

$$C'(w) = -\gamma_0 + \frac{\exp\left(\frac{\overline{y}_0 - w}{2}\right)}{2\left[1 + 2\overline{y}_0^{-2}(\overline{y}_0 - 1)(1 - \exp(\frac{\overline{y}_0 - w}{2}))\right]}$$
$$\times \left\{2\overline{y}_0^{-2}(\overline{y}_0 - 1)(C(w) - w + \beta_0 + \gamma(w - \overline{y}_0)) + \gamma_0\right\},$$

for $w \in (v_1, v_0)$. Note that C'' is bounded on (v_1, v_0). To see this, observe that the solution Y^{v_0} to (7.88) can be extended to an increasing C^∞ function on a certain open interval I containing $[1, v_0]$. Consequently, y_0, y_0', y_0'' are bounded on (v_1, v_0) and hence C'' also has this property. Therefore for some absolute constant r,

$$C(w - 2\delta) \le C(w) + 2\delta\gamma_0 - \frac{\delta \exp\left(\frac{\overline{y}_0 - w}{2}\right)}{1 + 2\overline{y}_0^{-2}(\overline{y}_0 - 1)(1 - \exp(\frac{\overline{y}_0 - w}{2}))}$$
$$\times \left\{2\overline{y}_0^{-2}(\overline{y}_0 - 1)(C(w) - w + \beta_0 + \gamma(w - \overline{y}_0)) + \gamma_0\right\} + r\delta^2. \tag{7.125}$$

Combining this with (7.121), applied to $y = \overline{y}_0$ (which is allowed, since $\overline{y}_0 \in (1, 2)$), yields

$$B^{w - 2\delta}(w - 2\delta) - C(w - 2\delta) \ge (B^w(w) - C^w(w)) \cdot R(\delta, w) - (r + 1)\delta^2,$$

where $R(\delta, w)$ is a certain constant lying in $[0, 1]$. By induction and (7.124), we obtain $B(v_1) - C(v_1) \ge 0$, which implies, by (7.115), that

$$\gamma_0 \left[3 - v_1 - \exp\left(\frac{1 - v_1}{2}\right)\right] \ge C(v_1).$$

This is equivalent to

$$\beta_0 \ge v_1 + \frac{5}{4}\left(1 + \frac{1}{v_0}\right) = Y^{v_0}(1) + \frac{5}{4}\left(1 + \frac{1}{v_0}\right)$$

and gives $\beta_0 \ge \beta$, by virtue of (7.92). This contradicts the assumption $\beta_0 < \beta$ and completes the proof. $\qquad\square$

7.6   Double maximal inequality for nonnegative martingales

7.6.1   Formulation of the result

If the dominated martingale is nonnegative, then the constant in (7.99) decreases. We have the following fact.

Theorem 7.14. *Suppose that X, Y are real-valued martingales such that X is nonnegative and Y is differentially subordinate to X. Then*

$$|||Y|^*||_1 \le 3||X^*||_1 \qquad (7.126)$$

and the constant is the best possible.

As in the proof of the preceding estimate, we focus on the version for ± 1-transforms in the discrete-time case.

7.6.2   Proof of Theorem 7.14

A special function. Arguing as above, it suffices to construct $U(\cdot, \cdot, 1, \cdot)$. Let $S = \{(x, y, w) : x \in [0,1], |y| \le w\}$ and consider the following subsets of S:

$$D_1 = \{(x, y, w) \in S : |y| \le x\},$$
$$D_2 = \{(x, y, w) \in S : x \le |y| \le x + w - 1\},$$
$$D_3 = \{(x, y, w) \in S : x + w - 1 < |y| \le w\}.$$

Let $u : S \to \mathbb{R}$ be given as follows. First, if $w \ge 1$, then $u(x, y, w)$ equals

$$\begin{cases} \frac{2}{3} \exp[\frac{1}{2}(1 - w)]\{2 + (2x + |y| - 2)(-x + |y| + 1)^{1/2}\} + w - 3 & \text{on } D_1, \\ 2x \exp[\frac{1}{2}(-x + |y| - w + 1)] + w - 3 & \text{on } D_2, \\ 2x - x \log(x - |y| + w) + w - 3 & \text{on } D_3 \end{cases}$$

(with the convention $0 \log 0 = 0$). If $w < 1$, then we set $u(x, y, w) = u(x, y, 1)$.

The special function $U : [0, \infty) \times \mathbb{R} \times (0, \infty) \times (0, \infty) \to \mathbb{R}$ is given by

$$U(x, y, z, w) = (x \vee z)u\left(\frac{x}{x \vee z}, \frac{y}{x \vee z}, \frac{|y| \vee w}{x \vee z}\right).$$

This function satisfies 1°–4°, which can be seen by repeating the argumentation presented in the previous section. We omit the details and refer the interested reader to [152]. □

Sharpness. We shall be brief. Suppose that $\beta > 0$ is the best constant in (7.126) in the discrete-time case, when Y is assumed to be a ± 1-transform of X. Let U_0

be the function guaranteed by Theorem 7.2. The function U_0 satisfies $1°$–$4°$, is homogeneous and for any $x \geq 0$, $z > 0$ and $y_1, y_2 \in \mathbb{R}$, $w_1, w_2 > 0$ we have

$$|U_0(x, y_1, z, w_1) - U_0(x, y_2, z, w_2)| \leq \max\{|y_1 - y_2|, |w_1 - w_2|\}. \tag{7.127}$$

Arguing as in the proof of (7.44) and (7.45), we get, for $w > 0$ and $\delta \in (0,1)$,

$$U_0(1, w, 1, w) \geq U_0(1 - \delta, w + \delta, 1, w + \delta) + \delta U_0(1, 1, 1, 1) \tag{7.128}$$

and

$$U_0(1 - \delta, w + \delta, 1, w + \delta) \geq (1 - \delta)U_0(1, w + 2\delta, 1, w + 2\delta) + \delta(w + \delta - \beta). \tag{7.129}$$

Combining (7.128) and (7.129), we get

$$U_0(1, w, 1, w) \geq (1 - \delta)U_0(1, w + 2\delta, 1, w + 2\delta) + \delta U_0(1, 1, 1, 1) + \delta(w + \delta - \beta).$$

Substituting $F(w) = U_0(1, w, 1, w) - U_0(1, 1, 1, 1) - (w - \beta + 2)$, we rewrite the above inequality in the form $F(w) \geq (1 - \delta)F(w + 2\delta) - \delta^2$. This, by induction, yields

$$F(w) \geq (1 - \delta)^n F(w + 2n\delta) - n\delta^2.$$

Now fix $z > 1$ and take $w = 1$, $\delta = (z - 1)/(2n)$ (here n must be sufficiently large so that $\delta < 1$). Letting $n \to \infty$ gives

$$\beta - 3 = F(1) \geq F(z) \exp\left(\frac{1-z}{2}\right).$$

However, by (7.127), F has at most linear growth; thus, letting $z \to \infty$, we obtain $\beta - 3 \geq 0$. This completes the proof. $\qquad\square$

7.7 Bounds for one-sided maximal functions

7.7.1 Formulation of the result

We turn to related L^1 estimates which involve one-sided maximal function of the dominated process. Let $\beta_0 = 2.0856\ldots$ be the positive solution to the equation

$$2 \log\left(\frac{8}{3} - \beta_0\right) = 1 - \beta_0$$

and $\beta_0^+ = \frac{14}{9} = 1.555\ldots$.

Theorem 7.15. *Let X, Y be real-valued martingales such that Y is differentially subordinate to X.*

(i) *We have*

$$\|Y^*\|_1 \leq \beta_0\||X|^*\|_1, \tag{7.130}$$

and the constant β_0 is the best possible. It is already the best possible in the discrete-time setting, even when Y is assumed to be a ± 1-transform of X.

(ii) *If X is nonnegative, then*

$$||Y^*||_1 \leq \beta_0^+ ||X^*||_1, \tag{7.131}$$

and the constant β_0^+ is the best possible. It is already the best possible in the discrete-time setting, even when Y is assumed to be a ± 1-transform of X.

In the case when the dominating sequence is a nonnegative supermartingale, we have the following result.

Theorem 7.16. *Let X be a nonnegative supermartingale and let Y be 1-subordinate to f. Then*

$$||Y^*||_1 \leq 3|| |X|^* ||_1 \tag{7.132}$$

and the constant 3 is the best possible. It is already the best possible in the discrete-time case, when Y is assumed to be a ± 1 transform of X.

We shall only prove discrete-time versions of the above results and restrict ourselves to ± 1-transforms.

7.7.2 Proof of Theorem 7.15

Proof of (7.130). A very interesting feature of the L^1 estimates for one-sided maximal functions is that there is an additional reduction of the dimension of the problem. First, it is not difficult to see that we may assume that f_0 is constant. Second, we may assume that $g_0 \geq 0$ almost surely: if this is not the case, we replace v_0 by $\operatorname{sgn} f_0$ and the new sequence g we obtain has a larger one-sided maximal function. Now we search for a function on $D = \mathbb{R} \times \mathbb{R} \times [0, \infty) \times \mathbb{R}$, given by the formula

$$U^0(x, y, z, w) = \sup\{\mathbb{E}g_n^* \vee w - \beta\mathbb{E}|f_n|^* \vee z\},$$

with the supremum taken over the usual parameters. We have a reduction of the dimension of this problem, coming from the fact that U^0 is homogeneous. A new argument is that for any $d > 0$,

$$
\begin{aligned}
U^0(x, y + d, z, w + d) &= \sup_{M(x, y+d)} \{\mathbb{E}g_n^* \vee (w + d) - \beta\mathbb{E}|f_n|^* \vee z\} \\
&= \sup_{M(x, y)} \{\mathbb{E}(g_n + d)^* \vee (w + d) - \beta\mathbb{E}|f_n|^* \vee z\} \\
&= d + \sup_{M(x, y)} \{\mathbb{E}g_n^* \vee w - \beta\mathbb{E}|f_n|^* \vee z\} \\
&= d + U^0(x, y, z, w).
\end{aligned}
$$

Consequently, it suffices to find the appropriate $U(\cdot, \cdot, 1, 0)$ and then extend it to the whole D by putting

$$U(x, y, z, w) = y \vee w + (|x| \vee z)U\left(\frac{x}{|x| \vee z}, \frac{y - (y \vee w)}{|x| \vee z}, 1, 0\right).$$

Let us introduce an auxiliary parameter. The equation

$$2 \log \left(2 - \frac{2}{3a} \right) = \frac{a-2}{3a}, \qquad a > \frac{1}{3}, \tag{7.133}$$

has a unique solution $a = 0.46986\ldots$, related to β_0 by the identity

$$\beta_0 = \frac{2a+2}{3a}.$$

Let S denote the strip $[-1,1] \times (-\infty, 0]$ and consider the following subsets of S.

$$D_1 = \{(x,y) : |x| + y > 0\},$$
$$D_2 = \{(x,y) : 0 \geq |x| + y > 1 - \beta_0\},$$
$$D_3 = \{(x,y) : |x| + y \leq 1 - \beta_0\}.$$

Introduce $u : S \to \mathbb{R}$ by

$$u(x,y) = \begin{cases} a(2|x| - y - 2)(1 - |x| - y)^{1/2} - 3a|x| + y & \text{if } (x,y) \in D_1, \\ 3a(2 - |x|) \exp(\frac{1}{2}(|x| + y)) + (1 - 3a)y - 8a & \text{if } (x,y) \in D_2, \\ \frac{9a^2}{4(3a-1)}(1 - |x|) \exp(|x| + y) - \beta_0 & \text{if } (x,y) \in D_3. \end{cases}$$

We have the following fact, an analogue of Lemmas 7.2 and 7.9. The proof is straightforward, so we shall not present the details, referring the reader to [138].

Lemma 7.19. *The function u has the following properties:*

$$u(1, \cdot) \text{ is convex}, \tag{7.134}$$
$$u(1, y) \geq -\beta_0, \tag{7.135}$$
$$u(x, 0) \geq -\beta_0, \tag{7.136}$$
$$u \text{ is diagonally concave}. \tag{7.137}$$

Define $U : \mathbb{R} \times \mathbb{R} \times (0, \infty) \times (0, \infty) \to \mathbb{R}$ by

$$U(x, y, z, w) = y \vee w + (|x| \vee z)u \left(\frac{x}{|x| \vee z}, \frac{y - (y \vee w)}{|x| \vee z} \right). \tag{7.138}$$

Using Lemma 7.19, we check that U satisfies the properties $1°$–$4°$, just repeating the argumentation from the proofs of (7.29) and (7.99). □

Proof of (7.131). We shall only present the special function leading to this estimate. Let S^+ denote the strip $[0, 1] \times (-\infty, 0]$ and let

$$D_1 = \left\{ (x,y) \in S^+ : x - y > \tfrac{2}{3}, x \leq \tfrac{2}{3} \right\},$$
$$D_2 = \left\{ (x,y) \in S^+ : x + y < \tfrac{2}{3}, x > \tfrac{2}{3} \right\},$$
$$D_3 = \left\{ (x,y) \in S^+ : x + y \geq \tfrac{2}{3} \right\},$$
$$D_4 = \left\{ (x,y) \in S^+ : x - y \leq \tfrac{2}{3} \right\}.$$

Introduce the function $u^+ : S^+ \to \mathbb{R}$ by

$$u^+(x,y) = \begin{cases} x\exp[\frac{3}{2}(-x+y)+1] - \beta_0^+, & \text{if } (x,y) \in D_1, \\ (\frac{4}{3} - x)\exp[\frac{3}{2}(x+y)-1] - \beta_0^+, & \text{if } (x,y) \in D_2, \\ -x+y - \frac{1}{\sqrt{3}}(1-x-y)^{1/2}(2-2x+y), & \text{if } (x,y) \in D_3, \\ x - x\log(\frac{3}{2}(x-y)) - \beta_0^+, & \text{if } (x,y) \in D_4. \end{cases}$$

The special function U is then given by (7.138), with u replaced by u^+.   □

Sharpness of (7.130). Suppose that the best constant in (7.130), restricted to ± 1-transforms, equals β and let U_0 be the function guaranteed by Theorem 7.2. By definition, U_0 satisfies

$$U_0(tx, ty, tz, tw) = tU_0(x, y, z, w) \quad \text{for } t > 0, \tag{7.139}$$

and

$$U_0(x, y, z, w) = U_0(x, y+t, z, w+t) - t \quad \text{for } t > -w. \tag{7.140}$$

Introduce the functions $A, B : (-\infty, 0] \to \mathbb{R}$, $C : [0, 1] \to \mathbb{R}$ by

$$A(y) = U_0(0, y, 1, 0), \; B(y) = U_0(1, y, 1, 0) = U_0(-1, y, 1, 0), \; C(x) = U_0(x, 0, 1, 0).$$

Step 1. We start with the observation that for $x \in (0, 1]$ and $\delta \in (0, x]$, the property (7.64) gives

$$C(x) \geq \frac{2\delta}{1-x+2\delta}B(x-1) + \frac{1-x}{1-x+2\delta}(C(x-2\delta)+\delta)$$

$$\geq \frac{2\delta}{1-x+2\delta}B(x-1-2\delta) + \frac{1-x}{1-x+2\delta}(C(x-2\delta)+\delta),$$

where the latter inequality follows from the fact that B is nondecreasing (by its definition). Furthermore,

$$B(x-1) \geq \delta + \delta B(0) + \frac{\delta}{1-x+2\delta}C(x-2\delta) + \frac{1-x+\delta}{1-x+2\delta}B(x-1-2\delta).$$

Equivalently,

$$C(x) - C(x-2\delta) \geq 2\delta\left[\frac{B(x-1-2\delta)}{1-x+2\delta} - \frac{C(x-2\delta)}{1-x+2\delta}\right] + \frac{2\delta(1-x)}{1-x+2\delta},$$

$$2B(x-1) - 2B(x-1-2\delta) \geq 2\delta\left[\frac{C(x-2\delta)}{1-x+2\delta} - \frac{B(x-1-2\delta)}{1-x+2\delta}\right] + 2\delta(1+B(0)).$$

Adding these two estimates gives

$$C(x)+2B(x-1)-C(x-2\delta)-2B(x-1-2\delta) \geq 2\delta(2+B(0)) - \frac{4\delta^2}{1-x+2\delta}. \tag{7.141}$$

Now fix an integer n, substitute $\delta = 1/(2n)$, $x = k/n$, $k = 1, 2, \ldots, n$ in (7.141) and sum the resulting inequalities; we get

$$C(1) + 2B(0) - C(0) - 2B(-1) \geq 2 + B(0) - \frac{1}{n^2} \sum_{k=1}^{n} \frac{1}{1 - \frac{k-1}{n}}.$$

Passing to the limit $n \to \infty$ and using the equalities $C(1) = B(0)$, $C(0) = A(0)$ we arrive at

$$2B(0) - A(0) - 2B(-1) \geq 2. \tag{7.142}$$

Step 2. Now let us show that

$$A(0) \geq B(-1) + 1. \tag{7.143}$$

To do this, use the property 2° twice to obtain

$$A(0) \geq \frac{\delta}{1+\delta} B(-1) + \frac{1}{1+\delta}(C(\delta) + \delta)$$

$$\geq \frac{\delta}{1+\delta} B(-1) + \frac{1}{1+\delta}(\delta B(-1) + (1-\delta)(\delta + A(0)) + \delta),$$

or, equivalently, $A(0) \geq B(-1) + 1 - \frac{\delta}{2}$. As δ is arbitrary, (7.143) follows.

Step 3. The property 2°, used twice, yields

$$A(y - 2\delta) \geq \frac{\delta}{1+\delta} B(y - 2\delta - 1) + \frac{1}{1+\delta} U_0(-\delta, y - \delta, 1, 0)$$
$$\geq \frac{\delta}{1+\delta} B(y - 2\delta - 1) + \frac{\delta}{1+\delta} B(y-1) + \frac{1-\delta}{1+\delta} A(y) \tag{7.144}$$

if $\delta < 1$ and $y \leq 0$. Moreover, if $y < 0$, $\delta \in (0,1)$ and $t > -y + 1$, then

$$B(y - 1) \geq \frac{t}{t+\delta} U_0(1 - \delta, y - 1 - \delta, 1, 0) + \frac{\delta}{t+\delta} U_0(1 + t, y - 1 + t, 1, 0)$$
$$= \frac{t}{t+\delta} U_0(1 - \delta, y + 1 - \delta, 1, 0) + \frac{\delta(1+t)}{t+\delta}\left(\frac{y-1+t}{1+t} + U_0(1, 0, 1, 0)\right),$$

which upon taking $t \to \infty$ gives

$$B(y - 1) \geq U_0(1 - \delta, y - 1 - \delta, 1, 0) + \delta(1 + B(0)). \tag{7.145}$$

Combining this estimate with the following consequence of 2°:

$$U_0(1 - \delta, y - 1 - \delta, 1, 0) \geq \delta A(y - 2\delta) + (1 - \delta)B(y - 1 - 2\delta)$$

gives

$$B(y - 1) \geq \delta A(y - 2\delta) + (1 - \delta)B(y - 1 - 2\delta) + \delta(1 + B(0)). \tag{7.146}$$

Now multiply (7.144) throughout by $1 + \delta$ and add it to (7.146) to obtain

$$A(y - 2\delta) - B(y - 1 - 2\delta) \geq (1 - \delta)(A(y) - B(y - 1)) + \delta(1 + B(0)),$$

which, by induction, leads to the estimate

$$A(-2n\delta) - B(-2n\delta - 1) - 1 - B(0) \geq (1 - \delta)^n(A(0) - B(-1) - 1 - B(0)),$$

valid for any nonnegative integer n. Fix $y < 0$, $\delta = -y/(2n)$ and let $n \to \infty$ to obtain

$$A(y) - B(y - 1) - 1 - B(0) \geq e^{y/2}(A(0) - B(-1) - 1 - B(0)) \geq -B(0)e^{y/2}, \quad (7.147)$$

where the last estimate follows from (7.143).

Now we come back to (7.146) and write it in the equivalent form

$$B(y - 1) - B(y - 1 - 2\delta) \geq \delta(A(y - 2\delta) - B(y - 1 - 2\delta)) + \delta(1 + B(0)).$$

By (7.147), we get

$$B(y - 1) - B(y - 1 - 2\delta) \geq \delta(-e^{y/2}B(0) + 2 + 2B(0)).$$

This gives, by induction,

$$B(-1) - B(-2n\delta - 1) = \sum_{k=0}^{n} [B(-2k\delta - 1) - B(-2k\delta - 1 - 2\delta)]$$

$$\geq n\delta(2 + 2B(0)) - \delta B(0)\frac{1 - e^{-n\delta}}{1 - e^{-\delta}}.$$

Now fix $y < 0$, take $\delta = -y/(2n)$ and let $n \to \infty$ to obtain

$$B(-1) - B(y - 1) \geq -y(1 + B(0)) - B(0)(1 - e^{y/2}). \quad (7.148)$$

Now, by (7.142) and (7.143),

$$B(-1) = \frac{1}{3}B(-1) + \frac{2}{3}B(-1) \leq \frac{1}{3}A(0) + \frac{2}{3}B(-1) + \frac{1}{3} \leq \frac{2}{3}B(0) - 1.$$

Furthermore, by the definition of B we have $B(y - 1) \geq -\beta$. Plugging these estimates into (7.148) yields

$$\beta \geq -y(1 + B(0)) - B(0)(1 - e^{y/2}) + 1 - \frac{2}{3}B(0).$$

This implies that $1 + B(0) \leq 0$, otherwise we would obtain $\beta = \infty$ (note that $y < 0$ is arbitrary). Take $y \in (-\infty, 0]$ satisfying

$$e^{y/2} = \frac{2}{B(0)} + 2.$$

We get

$$\beta \geq -2(1 + B(0)) \log \left(2 + \frac{2}{B(0)} \right) + 3 + \frac{1}{3} B(0)$$

and the right-hand side, as a function of $B(0) \in (-\infty, -1]$, attains its minimum β_0 at $B(0) = -3a$ (where a is given by (7.133)). Hence $\beta \geq \beta_0$ and the proof is complete. □

Sharpness of (7.131) *for nonnegative martingales.* Suppose that for any nonnegative martingale f and its ± 1 transform g we have

$$\|g^*\|_1 \leq \beta \|f^*\|_1$$

and let U_0^+ be the corresponding special function coming from Theorem 7.2. From its definition it follows that

$$U_0^+(tx, ty, tz, tw) = t U_0^+(x, y, z, w) \quad \text{for } t > 0, \tag{7.149}$$

and

$$U_0^+(x, y, z, w) = U_0^+(x, y + t, z, w + t) - t \quad \text{for } t > -w.$$

Furthermore,

$$\text{the function } U_0^+(1, \cdot, 1, 0) \text{ is nondecreasing.} \tag{7.150}$$

It will be convenient to work with the functions

$$A(y) = U_0^+ \left(\frac{2}{3}, y, 1, 0 \right), \; B(y) = U_0^+(1, y, 1, 0), \; C(x) = U_0^+(x, 0, 1, 0).$$

As previously, we divide the proof into a few intermediate steps.

Step 1. First let us note that the arguments leading to the estimate (7.141) are valid for these functions and hence so is (7.141) itself. For a fixed positive integer n, let us write (7.141) for $\delta = 1/(6n)$, $x = \frac{2}{3} + 2k\delta$, $k = 1, 2, \ldots, n$ and sum all these inequalities to obtain

$$C(1) + 2B(0) - C \left(\frac{2}{3} \right) - 2B \left(-\frac{1}{3} \right) \geq \frac{1}{3}(1 + B(0)) - \frac{1}{9n^2} \sum_{k=1}^{n} \frac{1}{\frac{1}{3} - \frac{k-1}{3n}}.$$

Now let $n \to \infty$ and use $C(1) = B(0)$ to get

$$3B(0) \geq C \left(\frac{2}{3} \right) + 2B \left(-\frac{1}{3} \right) + \frac{1}{3}(1 + B(0)). \tag{7.151}$$

Step 2. We will show that

$$C \left(\frac{2}{3} \right) \geq \frac{2}{3} B \left(-\frac{1}{3} \right) + \frac{4}{9} - \frac{\beta}{3}. \tag{7.152}$$

To this end, note that, using (7.64) twice, for $\delta < 1/3$,

$$
\begin{aligned}
C\left(\frac{2}{3}\right) &\geq \frac{3\delta}{1+3\delta}B\left(-\frac{1}{3}\right) + \frac{1}{1+3\delta}\left[\delta + C\left(\frac{2}{3} - \delta\right)\right] \\
&\geq \frac{3\delta}{1+3\delta}B\left(-\frac{1}{3}\right) + \frac{1}{1+3\delta}\left\{\delta + \frac{3\delta}{2}(-\beta) + \frac{2-3\delta}{2}\left[\delta + C\left(\frac{2}{3}\right)\right]\right\}.
\end{aligned}
$$

This is equivalent to

$$
C\left(\frac{2}{3}\right) \geq \frac{2}{3}B\left(-\frac{1}{3}\right) + \frac{2}{9}\left(2 - \frac{3}{2}\delta\right) - \frac{\beta}{3}
$$

and it suffices to let $\delta \to 0$.

Step 3. Using the property (7.64), we get, for $y < -1/3$ (see (7.145) and the arguments leading to it),

$$
B(y) \geq U_0^+(1 - \delta, y - \delta, 1, 0) + \delta(1 + B(0)).
$$

Furthermore, again by (7.64),

$$
U_0^+(1 - \delta, y - \delta, 1, 0) \geq (1 - 3\delta)B(y - 2\delta) + 3\delta A\left(y + \frac{1}{3} - 2\delta\right)
$$

and hence

$$
B(y) \geq (1 - 3\delta)B(y - 2\delta) + 3\delta A\left(y + \frac{1}{3} - 2\delta\right) + \delta(1 + B(0)). \tag{7.153}
$$

Moreover,

$$
\begin{aligned}
A\left(y + \frac{1}{3} - 2\delta\right) &\geq \frac{3\delta}{2 + 3\delta}U_0^+\left(0, y - \frac{1}{3} - 2\delta, 1, 0\right) \tag{7.154} \\
&\quad + \frac{2}{2 + 3\delta}U_0^+\left(\frac{2}{3} + \delta, y + \frac{1}{3} - \delta, 1, 0\right) \\
&\geq \frac{3\delta}{2 + 3\delta}(-\beta) + \frac{2}{2 + 3\delta}\left[3\delta B(y) + (1 - 3\delta)A\left(y + \frac{1}{3}\right)\right].
\end{aligned}
$$

Step 4. Now we will combine (7.153) and (7.154) and use them several times. Multiply (7.154) by $\gamma > 0$ (to be specified later) and add it to (7.153). We obtain

$$
\begin{aligned}
B(y) &\cdot \left(1 - \frac{6\gamma\delta}{2 + 3\delta}\right) - A\left(y + \frac{1}{3}\right) \cdot \frac{(2 - 6\delta)\gamma}{2 + 3\delta} \\
&\geq B(y - 2\delta) \cdot (1 - 3\delta) - A\left(y + \frac{1}{3} - 2\delta\right) \cdot (\gamma - 3\delta) + \delta\left(1 + B(0) - \frac{3\beta\gamma}{2 + 3\delta}\right) \\
&\geq B(y - 2\delta) \cdot (1 - 3\delta) - A\left(y + \frac{1}{3} - 2\delta\right) \cdot (\gamma - 3\delta) + \delta\left(1 + B(0) - \frac{3\beta\gamma}{2}\right).
\end{aligned}
$$

Now the choice $\gamma = (5 - \sqrt{9 - 24\delta})/4$ allows to write the inequality above in the form

$$F(y) \geq Q_\delta F(y - 2\delta) + \delta \left(1 + B(0) - \frac{3\beta\gamma}{2}\right), \qquad (7.155)$$

where

$$F(y) = B(y) \cdot \left(1 - \frac{6\gamma\delta}{2 + 3\delta}\right) - A\left(y + \frac{1}{3}\right) \cdot \frac{(2 - 6\delta)\gamma}{2 + 3\delta}$$

and

$$Q_\delta = (1 - 3\delta) \left(1 - \frac{6\gamma\delta}{2 + 3\delta}\right)^{-1}.$$

The inequality (7.155), by induction, leads to

$$F(-1/3) \geq Q_\delta^n F(-1/3 + 2n\delta) + \delta \left(1 + B(0) - \frac{3\beta\gamma}{2}\right) \cdot \frac{Q_\delta^n - 1}{Q_\delta - 1}.$$

Now fix $Y < -1/3$, take $\delta = (Y + 1/3)/(2n)$ and let $n \to \infty$. Then

$$\gamma \to \frac{1}{2}, \quad Q_\delta^n \to \exp\left(\frac{3}{4}\left(Y + \frac{1}{3}\right)\right)$$

and we arrive at

$$B\left(-\frac{1}{3}\right) - \frac{1}{2}A(0) \geq \exp\left(\frac{3}{4}\left(Y + \frac{1}{3}\right)\right)\left(B(Y) - \frac{1}{2}A\left(Y + \frac{1}{3}\right)\right)$$
$$+ \frac{2}{3}\left(1 + B(0) - \frac{3\beta}{4}\right)\left[\exp\left(\frac{3}{4}(Y + \frac{1}{3})\right) - 1\right].$$

Now we have $B(Y) \geq -\beta$ and $A(Y + \frac{1}{3}) \leq A(0)$. Hence, letting $Y \to -\infty$ yields

$$F(-1/3) \geq -\frac{2}{3}\left(1 + B(0) - \frac{3\beta}{4}\right). \qquad (7.156)$$

Now combine (7.151), (7.152) and (7.156) to obtain $\beta \geq 14/9$. The proof is complete. $\qquad \square$

On the search of the suitable majorant. We shall focus on the inequality (7.130). The reader is encouraged to compare the argumentation below with the corresponding search for the inequality (7.29); there are many similarities.

Suppose that the best constant in this estimate equals β and let

$$V(x, y, z, w) = (y \vee w) - \beta(|x| \vee z).$$

First, write the formula for the corresponding function U^0:

$$U^0(x, y, z, w) = \sup\{\mathbb{E}(g_n \vee w) - \beta\mathbb{E}(|f_n|^* \vee z)\},$$

where the supremum is taken over all n and all $(f, g) \in M(x, y)$. This function satisfies (7.139) and (7.140), which implies that it is enough to find $U(\cdot, \cdot, 1, 0)$: $[-1, 1] \times (-\infty, 0] \to \mathbb{R}$. Furthermore, by symmetry, it suffices to construct this function on $[0, 1] \times (-\infty, 0]$. First, note that the set $J = \{(x, y) \in [0, 1] \times (-\infty, 0] : U^0(x, y, 1, 0) = V(x, y, 1, 0)\}$ is contained in $\{1\} \times (-\infty, 0]$. Indeed, if $x \in (0, 1)$ and $y \leq 0$, then there is a pair $(f, g) \in M(x, y)$ such that $\|f\|_\infty \leq 1$ and $g^* > 0$ with non-zero probability. Consequently,

$$U^0(x, y, 1, 0) \geq \mathbb{E} V(f_\infty, g_\infty, 1, 0) > -\beta = V(x, y, 1, 0).$$

After some experimentation, we are led to the assumption

(A1) $J = \{1\} \times (-\infty, \gamma]$ for some $\gamma < -1$.

In particular, this gives $U(1, y, 1, 0) = -\beta$ for $y \leq \gamma$ (as usual, we have passed from U^0 to U). Next, we impose the regularity assumption

(A2) $U(\cdot, \cdot, 1, 0)$ is of class C^1 on $(-1, 1) \times (-\infty, 0)$; the functions $U(1, \cdot, 1, 0)$ and $U(\cdot, 0, 1, 0)$ are of class C^1 on $(-\infty, 0)$ and $(-1, 1)$, respectively.

To handle $U(1, y, 1, 0)$ for $y \in (\gamma, 0)$, we make the following conjecture, based on (7.145).

(A3) $U_x(1-, y, 1, 0) + U_y(1, y, 1, 0) = 1 + U(1, 0, 1, 0)$ for $y < 0$.

The next condition concerns the behavior of $U(\cdot, \cdot, 1, 0)$ in the interior of $(-1, 1) \times (-\infty, 0)$.

(A4)   When restricted to $[0, 1] \times (-\infty, 0]$, the function $U(\cdot, \cdot, 1, 0)$ is linear along the line segments of slope -1.

Now we proceed in the same manner as in the search corresponding to (7.29). Let $A = U(0, \cdot, 1, 0)$, $B = U(1, \cdot, 1, 0)$ and $C = U(\cdot, 0, 1, 0)$. The assumption (A2) and the symmetry of U with respect to the first coordinate imply $U_x(0, y, 1, 0) = 0$ for $y < 0$, which, by (A4), yields the differential equation

$$A'(y) = A(y) - B(y - 1) \qquad \text{for } y < 0. \tag{7.157}$$

If $y < \gamma + 1$, then (A1) gives $B(y - 1) = -\beta$ and we get that

$$A(y) = c_1 e^y - \beta,$$

for some constant c_1. Next, if $y \in (\gamma, -1)$, then (A3) and (A4) imply

$$2B'(y) = A(y + 1) - B(y) + 1 + B(0),$$

which, combined with (7.157), gives

$$2B(y) = A(y + 1) + (1 + B(0))y + c_2$$

for some c_2. Plugging this into (7.157) and solving the obtained differential equation yields that

$$A(y) = c_2 + (1 + B(0))(y + 1) + c_3 e^{y/2} \qquad \text{for } y \in (\gamma + 1, 0),$$

where c_3 is another constant. Next we deal with B on $(-1, 0)$. Using (A3) and (A4), we get

$$2B'(y) = \frac{B(y) - C(y+1)}{y} + 1 + B(0). \tag{7.158}$$

Next, for any $x \in (0, 1)$ and sufficiently small $\delta > 0$ we have, by (7.140) and 2°,

$$U(x, 0, 1, 0) \geq \frac{1}{2} [U(x - \delta, \delta, 1, 0) + U(x + \delta, -\delta, 1, 0)]$$
$$= \frac{1}{2} [U(x - \delta, \delta, 1, \delta) + U(x + \delta, -\delta, 1, 0)].$$

If we subtract $U(x, 0, 1, 0)$ from both sides, divide throughout by δ and let $\delta \to 0$, we are led to the following assumption

(A5) $U_y(x, 0, 1, 0) = 1$ for $x \in (0, 1)$.

This equality, together with (A4), implies

$$C'(x) = \frac{B(x-1) - C(x)}{1-x} + 1 \qquad \text{for } x \in [0, 1).$$

Combining this with (7.158) yields $2B'(y) = -C'(y+1) + 2 + B(0)$, or $2B(y) = -C(y+1) + (2 + B(0))y + c_4$; plugging this into (7.158) again and solving the obtained differential equation, we get

$$B(y) = \frac{c_4}{3} + y + c_5(-y)^{3/2}.$$

The final observation is that the symmetry condition, together with (A2), gives $C'(0) = 0$. Now, setting $y = 0$ in the equality above, we get $c_4 = 3B(0)$; in addition, the equations $B(-1-) = B(-1+)$, $B'(1-) = B'(1+)$, $B(\gamma-) = B(\gamma+)$, $B'(\gamma-) = B(\gamma+)$ give further connections between c_2, c_3, c_5, γ, β and $B(0)$. Solving the obtained complicated system of equations, one gets

$$\beta = (1 + B(0)) \left(\frac{5}{3} + 2\log \frac{B(0)}{2(1 + B(0))} - \frac{4}{3} \frac{B(0)}{1 + B(0)} \right).$$

The right-hand side, as a function of $B(0) \in (-\infty, -1)$, attains its maximum β_0 for $B(0) = -3a$, with a defined by (7.133). This gives rise to the final assumption

(A6) $B(0) = -3a$,

from which all the remaining parameters (c_1, c_2 and so on) can be easily derived. The obtained function $U(\cdot, \cdot, 1, 0)$ coincides with the function u above.   □

7.7.3   Proof of Theorem 7.16

Proof of (7.132). As before, we shall only present the formula for the special function. We introduce $V : [0, \infty) \times \mathbb{R} \times [0, \infty) \times \mathbb{R} \to \mathbb{R}$ by setting $V(x, y, z, w) =$

$(y \vee w) - 3(x \vee z)$ and define $u : [0,1] \times (-\infty, 0] \to \mathbb{R}$ by

$$u(x,y) = \begin{cases} 2x - x\log(x-y) - 3 & \text{if } x - y \le 1, \\ 2x \exp\left[\tfrac{1}{2}(-x+y+1)\right] - 3 & \text{if } x - y > 1, \end{cases}$$

with the convention $0 \log 0 = 0$. The function $U : [0,\infty) \times \mathbb{R} \times (0,\infty) \times \mathbb{R} \to \mathbb{R}$ is given by

$$U(x,y,z,w) = (y \vee w) + (x \vee z)u\left(\frac{x}{x \vee z}, \frac{y - (y \vee w)}{x \vee z}\right).$$

It suffices to verify the conditions 1°–4°, which is done in a similar manner as above. The interested reader is referred to the original paper [154] for details. □

Sharpness. See the proof of the optimality of the constant 3 in (7.126). In fact the argumentation presented there can be repeated, we only need to replace (7.127) by the equality

$$U_0(1, w_1, 1, w_1) = U_0(1, w_2, 1, w_2) + w_1 - w_2.$$

This guarantees that the corresponding function F has linear growth and completes the proof. □

7.8 L^p bounds for one-sided maximal function

7.8.1 Formulation of the result

For any $p \in (1, \infty]$, let

$$C_p = \begin{cases} \Gamma\left(\dfrac{2p-1}{p-1}\right)^{1-1/p} & \text{if } 1 < p \le 2, \\[3ex] \left(2^{p/(p-1)} - \dfrac{p}{p-1}\displaystyle\int_1^2 s^{1/(p-1)} e^{s-2} ds\right)^{1-1/p} & \text{if } 2 < p < \infty, \\[3ex] 1 + e^{-1} & \text{if } p = \infty. \end{cases}$$

We shall establish the following fact.

Theorem 7.17. *Suppose that X, Y are real-valued martingales such that Y is differentially subordinate to X. If $1 < p \le \infty$, then*

$$\|Y^*\|_1 \le C_p \|X\|_p \tag{7.159}$$

and the constant C_p is the best possible. It is already the best possible in the discrete-time setting and when Y is assumed to be a ± 1-transform of X.

The second result of this section can be stated as follows.

Theorem 7.18. *Suppose that X is a nonnegative submartingale and let Y be a real-valued process which is 1-subordinate to X. Then*

$$||Y^*||_p \leq 2\Gamma(p+1)^{1/p}||X||_\infty, \qquad 1 \leq p < \infty, \tag{7.160}$$

and the inequality is sharp. It is already the best possible in the discrete-time setting, even when Y is a ± 1-transform of X.

We shall only present the proof of these results in the discrete-time case for ± 1-transforms.

7.8.2 Proof of Theorem 7.17 for $1 < p \leq 2$

Proof of (7.159). First we define the function $\gamma_p : [0, \infty) \to (-\infty, 0]$ by

$$\gamma_p(t) = - \exp(pt^{p-1}) \int_t^\infty \exp(-ps^{p-1})ds. \tag{7.161}$$

Since

$$\gamma_p(t) = - \int_0^\infty \exp\left\{-p(p-1) \int_0^s (t+u)^{p-2}du\right\} ds,$$

the function γ_p is nonincreasing on $[0, \infty)$. Let $G_p : (-\infty, \gamma_p(0)] \to [0, \infty)$ denote the inverse to the function $t \mapsto \gamma_p(t) - t$, $t \geq 0$. We will need the following estimate.

Lemma 7.20. $G_p G_p'' + (p-2)(G_p')^2 \leq 0$.

Proof. The inequality to be proved is equivalent to $(G_p/G_p')' \geq p-1$. Since $\gamma_p'(t) = p(p-1)t^{p-2}\gamma_p(t) + 1$, we obtain

$$G_p'(x) = (\gamma_p(G_p(x)) - 1)^{-1} = [p(p-1)G_p^{p-2}(x)(x + G_p(x))]^{-1}$$

and

$$1 + G_p'(x) = \frac{\gamma_p(G_p(x))}{p(p-1)G_p^{p-2}(x)\gamma_p(G_p(x))}.$$

Therefore

$$\left(\frac{G_p(x)}{G_p'(x)}\right)' = [p(p-1)G_p^{p-1}(x)(x + G_p(x))]' = p - 1 + \frac{G_p(x)\gamma_p'(G_p(x))}{\gamma_p(G(x))} \geq p - 1,$$

because $G_p(x) \geq 0$ and $\gamma_p(G_p(x)) < 0$, $\gamma_p'(G_p(x)) \leq 0$. $\square$

Let $D = \{(x, y, w) \in \mathbb{R}^3 : y \leq w\}$ and let $V : D \to \mathbb{R}$ be given by $V(x, y, w) = (y \vee w) - |x|^p + \gamma_p(0)$. We are ready to introduce the special function. Set

$$D_1 = \{(x, y, w) \in D : y - w - |x| \geq \gamma_p(0)\},$$
$$D_2 = \{(x, y, w) \in D : y - w - |x| < \gamma_p(0) \text{ and } |x| \geq G_p(y - w - |x|)\},$$
$$D_0 = D \setminus (D_1 \cup D_2).$$

Let $U_p : D \to \mathbb{R}$ be given by

$$U_p(x, y, w) = \begin{cases} -\frac{(y-w)^2 - x^2}{2\gamma_p(0)} + \frac{\gamma_p(0)}{2} + y & \text{on } D_1, \\ w + \gamma_p(0) \\ \quad + (p-1)G_p(y - w - |x|)^p - p|x|G_p(y - w - |x|)^{p-1} & \text{on } D_2, \\ w - |x|^p + \gamma_p(0) & \text{on } D_0. \end{cases}$$

We will now verify that U_p satisfies $1°$–$4°$ and thus establishes (7.159). To do this, it suffices to show the following facts.

Lemma 7.21.

(i) *The function U_p is of class C^1 in the interior of D.*

(ii) *For any $\varepsilon \in \{-1, 1\}$ and $(x, y, w) \in D$, the function $F = F_{\varepsilon, x, y, w} : (-\infty, w - y] \to \mathbb{R}$ given by $F(t) = U_p(x + \varepsilon t, y + t, w)$ is concave.*

(iii) *For any $\varepsilon \in \{-1, 1\}$ and x, y, $h \in \mathbb{R}$,*

$$U_p(x + \varepsilon t, y + t, (y + t) \vee y) \le U_p(x, y, y) + \varepsilon U_{px}(x, y, y)t + t. \qquad (7.162)$$

(iv) *We have*

$$U_p(x, y, w) \ge V_p(x, y, w) \quad \text{for } (x, y, w) \in D. \qquad (7.163)$$

(v) *We have*

$$U_p(x, y, w) \le 0 \qquad \text{when } y \le w \le |x|. \qquad (7.164)$$

Proof. (i) This is straightforward: U_p is of class C^1 in the interior of D_0, D_1 and D_2, so the claim reduces to a tedious verification that the partial derivatives U_{px}, U_{py} and U_{pw} match at the common boundaries of D_0, D_1 and D_2.

(ii) In view of (i), it suffices to show that $F''(t) \le 0$ for those t, for which the second derivative exists. By virtue of the translation property $F_{\varepsilon, x, y, w}(u) = F_{\varepsilon, x+\varepsilon s, y+s, w}(u - s)$, valid for all u and s, it suffices to check that $F''(t) \le 0$ for $t = 0$. Furthermore, since $U_{px}(0, y, w) = 0$ and $U_p(x, y, w) = U_p(-x, y, w)$, we may restrict ourselves to $x > 0$.

If $\varepsilon = 1$, then we easily verify that $F''(0) = 0$ if (x, y, w) lies in the interior $(D_1 \cup D_2)°$ of $D_1 \cup D_2$ and $F''(0) = -p(p-1)x^{p-2} \le 0$ if $(x, y, w) \in D_0°$. Thus it remains to check the case $\varepsilon = -1$. We start from the observation that $F''(0) = 0$ if (x, y, w) belongs to $D_1°$. If $(x, y, w) \in D_2°$, then

$$F''(0) = 4p(p-1)G_p^{p-3}[G_p G_p'(G_p' + 1) + (G_p - x)((p-2)(G_p')^2 + G_p G_p'')],$$

where all the functions on the right are evaluated at $x_0 = y - w - x$. Since $y \le w$, we have $x \le -x_0$ and, in view of Lemma 7.20,

$$F''(0) \le 4p(p-1)G_p^{p-3}(x_0)[G_p(x_0)G_p'(x_0)(G_p'(x_0) + 1)$$
$$+ (G_p(x_0) + x_0)((p-2)(G_p'(x_0))^2 + G_p(x_0)G_p''(x_0))] \qquad (7.165)$$
$$= 0.$$

Here in the latter passage we have used the equality

$$G_p(x)G_p''(x) + (p-2)(G_p'(x))^2 = -\frac{G_p(x)G_p'(x)(G_p'(x)+1)}{G_p(x)+x},$$

which can be easily extracted from the proof of Lemma 7.20. Thus we are done with D_2°. Finally, if (x, y, w) belongs to D_0°, then $F''(0) = -p(p-1)x^{p-2} \le 0$.

(iii) We may assume that $x \ge 0$, due to the symmetry of the function U_p. Note that $U_{py}(x, y-, y) = 1$; therefore, if $t \le 0$, then the estimate follows from the concavity of U_p along the lines of slope ± 1, established in the previous part. If $t > 0$, then

$$U_p(x + \varepsilon t, y + t, (y+t) \vee y) = U_p(x, y+t, y+t) = y + t + U_p(x + \varepsilon t, 0, 0),$$

and hence we will be done if we show that the function $s \mapsto U_p(s, 0, 0)$ is concave on $[0, \infty)$. However, its second derivative equals $1/\gamma_p(0) < 0$ for $s < \gamma_p(0)$ and

$$p(p-1)G_p^{p-3}(-s)[(G_p(-s) - s)((p-2)(G_p'(-s))^2 + G_p(-s)^{p-2}G_p''(-s))$$
$$+ G_p(-s)G_p'(-s)(G_p'(-s)+2)]$$
$$= p(p-1)G_p(-s)^{p-2}G_p'(-s) \le 0$$

for $s > \gamma_p(0)$. Here we have used the equality from (7.165), with $x_0 = -s$.

(iv) Again, it suffices to deal only with nonnegative x. On the set D_0 both sides of (7.163) are equal. To prove the majorization on D_2, let $\Phi(s) = \gamma_p(0) - s^p$ for $s \ge 0$. Observe that

$$U_p(x, y, w) = w + \Phi(G_p(y - w - x)) + \Phi'(G_p(y - w - x))(x - G_p(y - w - x)),$$

which, by the concavity of Φ, is not smaller than $w + \Phi(x)$. Finally, the estimate for $(x, y, w) \in D_1$ is a consequence of the fact that

$$U_{py}(x, y-, w) = \frac{\gamma_p(0) - (y - w)}{\gamma_p(0)} \ge 0,$$

so

$$U_p(x, y, w) - V_p(x, y, w) \ge U_p(x, y_0, w) - V_p(x, y_0, w) \ge 0.$$

Here $(x, y_0, w) \in \partial D_2$ and the latter bound follows from the majorization on D_2, which we have just established.

(v) It suffices to establish the bound for $y = w$. We have

$$U_p(x, y, y) = U_p(|x|, 0, 0) + y \le U_p(|x|, 0, 0) + |x|.$$

As shown in the proof of (iii), $s \mapsto U_p(s, 0, 0)$, $s \ge 0$, is concave, hence so is the function $s \mapsto U_p(s, 0, 0) + s$, $s \ge 0$. It suffices to note that its derivative vanishes at $-\gamma_p(0)$, so the value at this point (which is equal to 0), is the supremum we are searching for. $\quad\square$

Sharpness. As we have already seen (cf. (7.25)), the constant C_p is already optimal in the inequality $||f^*||_1 \le C_p||f||_p$. In other words, the estimate (7.159) is sharp, even if $X = Y$. $\quad\square$

7.8.3   The proof of Theorem 7.17 for $p > 2$

Proof of (7.159). Here the argumentation is similar, so we will be brief. Suppose that p is finite and let $\gamma_p : [0, \infty) \to (-\infty, 0)$ be given by

$$\gamma_p(t) = \exp(-pt^{p-1}) \left[-\int_{p^{-1/(p-1)}}^t \exp(ps^{p-1}) ds - p^{-1/(p-1)} e \right]$$

$$= -t + p(p-1) \exp(-pt^{p-1}) \int_{p^{-1/(p-1)}}^t s^{p-1} \exp(ps^{p-1}) ds$$

if $t > p^{-1/(p-1)}$, and

$$\gamma_p(t) = (p-2)(t - p^{-1/(p-1)}) - p^{-1/(p-1)}$$

if $t \in [0, p^{-1/(p-1)}]$. It is easy to check that γ_p is of class C^1 and nondecreasing. This implies that the function $t \mapsto \gamma_p(t) + t$, $t \geq p^{-1/(p-1)}$, is invertible: let G_p denote its inverse. We have the following version of Lemma 7.20.

Lemma 7.22. $G_p G_p''' + (p-2)(G_p')^2 \geq 0.$

Define

$$M_p = \frac{p-1}{p^{p/(p-1)}} \left[2^{p/(p-1)} - \frac{p}{p-1} \int_1^2 s^{1/(p-1)} e^{s-2} ds \right] \tag{7.166}$$

and let $V : D \to \mathbb{R}$ be given by

$$V(x, y, w) = (y \vee w) - |x|^p + M_p.$$

Introduce $H_p : \mathbb{R}^2 \to \mathbb{R}$ by

$$H_p(x, y) = (p-1)^{1-p}(-(p-1)|x| + |y|)(|x| + |y|)^{p-1}$$

and let

$$D_1 = \{(x, y, w) \in D : y - w \geq \gamma_p(x), \ x + y - w \leq 0\},$$
$$D_2 = \{(x, y, w) \in D : y - w \geq \gamma_p(x), \ x + y - w > 0\},$$
$$D_0 = D \setminus (D_1 \cup D_2).$$

The special function $U_p : D \to \mathbb{R}$ is defined by

$$U_p(x, y, w) = \begin{cases} w + H_p(x, y - w + (p-1)p^{-1/(p-1)}) - M_p & \text{on } D_1, \\ w - M_p \\ \quad + (p-1)G_p(|x| + y - w)^p - p|x|G_p(|x| + y - w)^{p-1} & \text{on } D_2, \\ w - |x|^p - M_p & \text{on } D_0. \end{cases}$$

Here is the analogue of Lemma 7.21: it implies that U_p satisfies the conditions 1°–4°. We omit the proof.

Now consider a martingale $f = (f_n)_{n=1}^N$, starting from t_0 which satisfies the following condition: if $0 \leq n \leq N-1$, then on the set $\{f_n = t - n\delta\}$, the difference df_{n+1} takes the values $-\delta$ and $-\gamma_p(G_p(f_n(\omega)))$; on the complement of this set, $df_{n+1} \equiv 0$. Let g be a ± 1 transform of f, given by $g_0 = f_0$ and $dg_n = -df_n$, $n = 1, 2, \ldots, N$. The key fact about the pair (f,g) is that

$$\mathbb{E}U_p(f_n, g_n, g_n^*) \leq \mathbb{E}U_p(f_{n+1}, g_{n+1}, g_{n+1}^*) + R\delta^2, \quad n = 0, 1, 2, \ldots, N-1. \quad (7.176)$$

This is an immediate consequence of Lemma 7.24 (applied conditionally with respect to $\mathcal{F}_n$) and the fact that $U_p(f_n, g_n, g_n^*) \neq U_p(f_{n+1}, g_{n+1}, g_{n+1}^*)$ if and only if $f_n = t - n\delta$, or $g_n = t + n\delta = g_n^*$.

Another property of the pair (f,g) is that if $f_N \neq 0$, then $U_p(f_N, g_N, g_N^*) = V_p(f_N, g_N, g_N^*)$. Indeed, $f_N \neq 0$ implies $df_n > 0$ for some $n \geq 1$ and then, by construction,

$$g_N^* - g_N = g_n^* - g_n = -dg_n = df_n = \gamma_p(f_n) = \gamma_p(f_N).$$

Thus we may write

$$\begin{aligned} M_p &= U_p(t_0, t_0, t_0) \\ &\leq \mathbb{E}U_p(f_N, g_N, g_N^*) + RN\delta^2 \quad\quad\quad\quad\quad\quad\quad (7.177) \\ &= \mathbb{E}V_p(f_N, g_N, g_N^*)1_{\{f_N \neq 0\}} + U_p(0, 2t_0, 2t_0)\mathbb{P}(f_N = 0) + RN\delta^2, \end{aligned}$$

since $g_N = g_N^* = 2t_0$ on the set $\{f_N = 0\}$.

Step 3. Now let us extend the pair (f,g) as follows. Fix $\kappa > 0$ and put $f_N = f_{N+1} = f_{N+2} = \cdots$ and $g_N = g_{N+1} = g_{N+2} = \cdots$ on $\{f_N \neq 0\}$, while on $\{f_N = 0\}$, let the conditional distribution of $(f_n, g_n)_{n \geq N}$ with respect to $\{f_N = 0\}$ be that of the pair $(f^{\kappa, 2t_0}, g^{\kappa, 2t_0})$, obtained at the end of Step 1. The resulting process (f,g) consists of simple martingales and, by (7.174) and (7.177),

$$M_p \leq \mathbb{E}V_p(f_\infty, g_\infty, g_\infty^*) + RN\delta^2 + \kappa\mathbb{P}(f_N = 0).$$

Now it suffices to note that choosing N sufficiently large and κ sufficiently small, we can make the expression $RN\delta^2 + \kappa\mathbb{P}(f_N = 0)$ arbitrarily small. This shows that M_p is indeed the smallest C which is allowed in (7.170). $\quad\square$

Sharpness, $p = \infty$. We may assume that $\|f\|_\infty = 1$. The proof will be based on a version of Theorem 7.2. Namely, let let $U_0 : \{(x,y,z) : |x| \leq 1, y \leq z\} \to \mathbb{R}$ be given by

$$U_0(x,y,z) = \mathbb{E}(g_\infty^* \vee z),$$

where the supremum is taken over the class of all pairs $(f,g) \in M(x,y)$ such that $\|f\|_\infty \leq 1$. Then U_0 satisfies the appropriate modifications of conditions $1°$, $2°$ and $4°$. Furthermore, note that the function U_0 satisfies (7.171) (with obvious restriction to x lying in $[-1, 1]$).

Now we shall exploit the properties of this function. First we will show that

$$U_0(0,0,0) \geq 1. \tag{7.178}$$

To prove this, take $\delta \in (0,1)$ and use 2° to obtain

$$U_0(0,0,0) \geq \frac{1}{1+\delta}U_0(\delta,\delta,\delta) + \frac{\delta}{1+\delta}U_0(-1,-1,0).$$

We have $U_0(-1,-1,0) \geq 0$ by 1°, and $U_0(\delta,\delta,\delta) = \delta + U(\delta,0,0)$ by (7.171). Thus,

$$U_0(0,0,0) \geq \frac{\delta + U_0(\delta,0,0)}{1+\delta}. \tag{7.179}$$

Similarly, using 2° and then 1°,

$$U(\delta,0,0) \geq (1-\delta)U_0(0,\delta,\delta) + \delta U_0(1,\delta-1,0) \geq (1-\delta)\big[\delta + U_0(0,0,0)\big].$$

Plug this into (7.179), subtract $U_0(0,0,0)$ from both sides, divide throughout by δ and let $\delta \to 0$ to get (7.178).

Next, fix a positive integer N and set $\delta = (1-e^{-1})/N$. For any $k = 1, 2, \ldots, N$, we have, by 1°, 2° and (7.171),

$$U_0(k\delta,0,0) \geq \frac{\delta}{1-k\delta+\delta}U_0(1,k\delta-1,0) + \frac{1-k\delta}{1-k\delta+\delta}U_0((k-1)\delta,\delta,\delta)$$

$$\geq \frac{1-k\delta}{1-k\delta+\delta}\big[\delta + U_0((k-1)\delta,0,0)\big],$$

or, equivalently,

$$\frac{U_0(k\delta,0,0)}{1-k\delta} \geq \frac{U_0((k-1)\delta,0,0)}{1-(k-1)\delta} + \frac{\delta}{1-(k-1)\delta}.$$

By induction, it follows that

$$eU_0(1-e^{-1},0,0) = \frac{U_0(N\delta,0,0)}{1-N\delta} \geq U_0(0,0,0) + \sum_{k=1}^{N}\frac{\delta}{1-(k-1)\delta}.$$

Letting $N \to \infty$ and using (7.178), we arrive at

$$eU_0(1-e^{-1},0,0) \geq 1 + \int_0^{1-e^{-1}}\frac{\mathrm{d}x}{1-x} = 2,$$

and hence, by (7.171),

$$U_0(1-e^{-1},1-e^{-1},1-e^{-1}) = 1 - e^{-1} + U_0(1-e^{-1},0,0) \geq 1 + e^{-1}.$$

This yields the optimality of C_∞, by the very definition of U_0. $\square$

On the search of the suitable majorant. The search is based on exploiting the properties of the function

$$U^0(x,y,z) = \sup\{\mathbb{E}\big((g_n^* \vee z) - |f_n|^p\big)\}.$$

The reasoning is similar to the ones presented previously; we omit the details. $\square$

7.8.4 Proof of Theorem 7.18

We may and do assume that $||f||_\infty \leq 1$.

Proof of (7.160) *for* $p = 1$. Let U be the special function introduced in the proof of Theorem 7.16. Let $W_1 : [0,1] \times \mathbb{R} \times \mathbb{R} \to \mathbb{R}$ be given by the formula

$$W_1(x, y, w) = U(1 - x, y, 1, w) + 3.$$

It can be verified that W_1 has the following properties:

$$\text{if } \varepsilon \in \{-1, 1\} \text{ and } y, w \in \mathbb{R}, \text{ then the function} \atop t \mapsto W_1(t, y + \varepsilon t, w), \, t \in [0, 1], \text{ is concave and nonincreasing;} \qquad (7.180)$$

$$W_1(x, y, w) \geq y \vee w. \qquad (7.181)$$

Property (7.180) implies that the process $(W_1(f_n, g_n, g_n^*))_{n \geq 0}$ is a supermartingale. Combining this with (7.181) and (7.180) again gives

$$\mathbb{E}g_n^* \leq \mathbb{E}W_1(f_n, g_n, g_n^*) \leq \mathbb{E}W_1(f_0, g_0, g_0^*) \leq \mathbb{E}W_1(0, 0, 0) = 2,$$

which is the desired bound. □

Proof of (7.160) *for* $p > 1$. This time the special function $W_p : [0, 1] \times \mathbb{R} \times \mathbb{R} \to \mathbb{R}$ is given by

$$W_p(x, y, w) = p(p - 1) \int_0^\infty a^{p-2} W_1(x, y - a, (w - a) \vee 0) da.$$

By (7.180), for any $\varepsilon \in \{-1, 1\}$ and $y, w \in \mathbb{R}$ and any $a > 0$, the function

$$t \mapsto W_1(t, y - a + \varepsilon t, (w - a) \vee 0)$$

is concave and nonincreasing. This implies that $t \mapsto W_p(t, y + t, w)$ has the analogous property, and consequently the process $(W_p(f_n, g_n, g_n^*))_{n \geq 0}$ is a supermartingale. On the other hand, by (7.181), we have

$$W_p(x, y, w) \geq p(p - 1) \int_0^\infty a^{p-2} [(y - a) \vee (w - a) \vee 0] \, da$$

$$= p(p - 1) \int_0^{y \vee w} a^{p-2} ((y \vee w) - a) da = (y \vee w)^p.$$

Therefore,

$$\mathbb{E}(g_n^*)^p \leq \mathbb{E}W_p(f_n, g_n, g_n^*) \leq \mathbb{E}W_p(f_0, g_0, g_0^*) \leq W_p(0, 0, 0)$$

$$= p(p - 1) \int_0^\infty a^{p-2} W_1(0, -a, 0) da = 2^p \Gamma(p + 1),$$

which is the claim. □

Sharpness. Fix $\delta \in (0,1)$ and consider the following example. Let ξ_1, ξ_2, ... be a sequence of independent mean zero random variables which take values in the set $\{-\delta, 1-\delta\}$. Let $S = (S_n)_{n\geq 0}$ be given as follows. Set $S_0 \equiv 0$ and, for $n \geq 1$,

$$S_{2n-1} = \xi_1 + \xi_2 + \cdots + \xi_{n-1} + n\delta,$$
$$S_{2n} = \xi_1 + \xi_2 + \cdots + \xi_n + n\delta.$$

It is very easy to check that S is a nonnegative submartingale. Introduce the stopping time $\tau = \inf\{n : S_n = 1\}$ and let $f_n = S_{\tau \wedge n}$, $n = 0, 1, 2, \ldots$. By Doob's optional sampling theorem, f is also a nonnegative submartingale. Furthermore, it can be easily verified that f is bounded by 1. Let g_n be given by $dg_n = (-1)^{n+1} df_n$, $n = 0, 1, 2, \ldots$. Directly from the definition, we see that g has the following behavior: it starts from 0 and then increases in δ at each step, until it finally drops, at time τ, from $(\tau - 1)\delta$ to $\tau\delta - 1$, and stays there forever. Consequently,

$$\mathbb{P}(g^* = (n-1)\delta) = \mathbb{P}(\tau = n) = (1-\delta)^{n/2-1}\delta, \qquad n = 2, 4, 6, \ldots$$

and hence

$$\mathbb{E}(g^*)^p = \sum_{k=1}^{\infty} [(2k-1)\delta]^p (1-\delta)^{k-1}\delta.$$

It is not difficult to show that the right-hand side converges to $2^p\Gamma(p+1)$ as $\delta \to 0$. This shows that the inequality (7.160) is sharp. $\square$

7.9   Burkholder's method for continuous-path semimartingales

Another very interesting problem is to study the above estimates for continuous-path semimartingales. It turns out that, in contrast with the non-maximal setting, the passage from the general to the continuous-path case results in a decrease of the values of optimal constants. This raises the question of how to refine Burkholder's method so that one can exploit the additional regularity of trajectories, and we will address it now.

Let $\mathcal{H}$ be a separable Hilbert space (we may and do assume that it is equal to ℓ_2), set $D = \mathcal{H} \times \mathcal{H} \times [0,\infty) \times [0,\infty)$ and fix a Borel function $V : D \to \mathbb{R}$ satisfying (7.1), which is bounded on bounded sets. Suppose we are interested in the inequality

$$\mathbb{E}V(X_t, Y_t, |X_t|^*, |Y_t|^*) \leq 0, \tag{7.182}$$

to be valid for all $t \geq 0$ and the class of all pairs (X, Y) of continuous and bounded $\mathcal{H}$-valued martingales such that Y is differentially subordinate to X. Suppose that $U : D \to \mathbb{R}$ is of class C^2 and satisfies the following conditions.

$1°$ $U \geq V$ on D.

$2°$ There is a function $M : D \to [0, \infty)$, bounded on any set of the form $\{(x, y, z, w) : |x|, |y| \leq L, 1/L \leq z, w \leq L\}$ for some $L > 0$, such that for any $(x, y, z, w) \in D$ and $h, k \in \mathcal{H}$,

$$(U_{xx}(x, y, z, w)h, h) + 2(U_{xy}(x, y, z, w)h, k) + (U_{yy}(x, y, z, w)k, k)$$
$$\leq M(x, y, z, w)(|k|^2 - |h|^2).$$

$3°$ For any $(x, y, z, w) \in D$ such that $w = |y| \leq |x| \leq z$

$$U(x, y, z, w) \leq 0.$$

$4°$ For any $x \neq 0$ and $|y| \leq w$ we have $U_z(x, y, |x|, w) \leq 0$. For any $y \neq 0$ and $|x| \leq z$, $U_w(x, y, z, |y|) \leq 0$.

Theorem 7.19. *If U satisfies $1°$–$4°$, then (7.182) holds for all bounded $\mathcal{H}$-valued continuous-path martingales X, Y such that Y is differentially subordinate to X.*

Proof. It suffices to show that

$$\mathbb{E}U(X_t, Y_t, |X_t|^*, |Y_t|^*) \leq 0 \tag{7.183}$$

for any t. Furthermore, since X, Y are bounded and U is continuous, we may assume that

$$\mathbb{P}(|X_t|, |Y_t| \text{ belong to } (\delta, \delta^{-1}) \text{ for all } t \geq 0) = 1,$$

where $\delta > 0$ is a fixed small number: simply replace X, Y by (δ, X), (δ, Y) if necessary. Let

$$\overline{M} = \sup\{M(x, y, z, w) : |x|, |y| \in (\delta, \delta^{-1})\} < \infty$$

and, as usual, denote the ith and jth coordinate of X and Y by X^i and Y^j, respectively.

For a fixed $t \geq 0$ and $\varepsilon > 0$, there is $D = D(\varepsilon) \geq 1$ such that if $d \geq D$, then

$$\mathbb{E} \sum_{k>d} [X^k, X^k]_t = \mathbb{E} \sum_{k>d} |X_t^k|^2 < \varepsilon. \tag{7.184}$$

For $0 \leq s \leq t$, let

$$X_s^{(d)} = (X_s^1, X_s^2, \dots, X_s^d, 0, 0, \dots),$$
$$Y_s^{(d)} = (Y_s^1, Y_s^2, \dots, Y_s^d, 0, 0, \dots)$$

and

$$Z_s^{(d)} = (X_s^{(d)}, Y_s^{(d)}, |X_s^{(d)}|^*, |Y_s^{(d)}|^*).$$

Since $X^{(d)}$, $Y^{(d)}$ take values in a finite-dimensional subspace of $\mathcal{H}$, Itô's formula is applicable. Thus, using 3°, we obtain

$$U(Z_t^{(d)}) = U(Z_0^{(d)}) + I_1 + I_2 + I_3 \leq I_1 + I_2 + I_3, \qquad (7.185)$$

with

$$I_1 = \int_0^t U_x(Z_s^{(d)}) dX_s^{(d)} + \int_0^t U_y(Z_s^{(d)}) dY_s^{(d)},$$

$$I_2 = \int_0^t U_z(Z_s^{(d)}) d|X_s^{(d)}|^* + \int_0^t U_w(Z_s^{(d)}) d|Y_s^{(d)}|^*,$$

$$I_3 = \int_0^t U_{xx}(Z_s^{(d)}) d[X^{(d)}, X^{(d)}]_s + 2 \int_0^t U_{xy}(Z_s^{(d)}) d[X^{(d)}, Y^{(d)}]_s$$

$$+ \int_0^t U_{yy}(Z_s^{(d)}) d[Y^{(d)}, Y^{(d)}]_s.$$

The random variable I_1 has null expectation. The term I_2 is nonpositive: by 4°, we have $U_z(Z_s^{(d)*}) \leq 0$ on $\{s : X_s^{(d)} = X_s^{(d)*}\}$, and this set is precisely the support of $dX_s^{(d)*}$. The second integral is dealt with likewise and it remains to handle I_3. Let $0 \leq s_0 < s_1 \leq t$. For any $j \geq 0$, let $(\eta_i^j)_{1 \leq i \leq i_j}$ be a sequence of nondecreasing finite stopping times with $\eta_0^j = s_0$, $\eta_{i_j}^j = s_1$, such that $\lim_{j \to \infty} \max_{1 \leq i \leq i_j - 1} |\eta_{i+1}^j - \eta_i^j| = 0$. Keeping j fixed, we apply, for each $i = 0, 1, 2, \ldots, i_j$, the property 2° to

$$x = X_{s_0}^{(d)}, \quad y = Y_{s_0}^{(d)}, \quad z = |X_{s_0}^{(d)}|^*, \quad w = |Y_{s_0}^{(d)}|^*$$

and

$$h = h_i^j = X_{\eta_{i+1}^j}^{(d)} - X_{\eta_i^j}^{(d)}, \quad k = k_i^j = Y_{\eta_{i+1}^j}^{(d)} - Y_{\eta_i^j}^{(d)}.$$

Summing the obtained $i_j + 1$ inequalities and letting $j \to \infty$ yields

$$\sum_{m=1}^d \sum_{n=1}^d \left[U_{x_m x_n}(Z_{s_0}^{(d)})[(X^{(d)})^m, (X^{(d)})^n]_{s_0}^{s_1} + 2U_{x_m y_n}(Z_{s_0}^{(d)})[(X^{(d)})^m, (Y^{(d)})^n]_{s_0}^{s_1} \right.$$

$$\left. + U_{y_m y_n}(Z_{s_0}^{(d)})[(Y^{(d)})^m, (Y^{(d)})^n]_{s_0}^{s_1} \right]$$

$$\leq \overline{M} \sum_{k=1}^d \left([(Y^{(d)})^k, (Y^{(d)})^k]_{s_0}^{s_1} - [(X^{(d)})^k, (X^{(d)})^k]_{s_0}^{s_1} \right).$$

If we approximate I_3 by discrete sums, we see that the inequality above leads to

$$I_3 \leq \overline{M} \sum_{k=1}^d \left([(Y^{(d)})^k, (Y^{(d)})^k]_0^t - [(X^{(d)})^k, (X^{(d)})^k]_0^t \right)$$

$$\leq -\overline{M} \sum_{k>d} \left([Y^k, Y^k]_0^t - [X^k, X^k]_0^t \right),$$

where the last passage is due to the differential subordination.

Now take the expectation of both sides of (7.185) and use (7.184) to obtain $\mathbb{E}U(Z_t^{(d)}) \leq \overline{M}\varepsilon$. Then let $d \to \infty$ to get $\mathbb{E}U(X_t, Y_t, |X_t|^*, |Y_t|^*) \leq \overline{M}\varepsilon$, by Lebesgue's dominated convergence theorem. Since ε was chosen arbitrarily, (7.183) follows.   □

When the dominating process is a sub- or supermartingale, the statement is similar. We shall formulate the conditions $1°$–$4°$ when X is assumed to be a submartingale, the passage to $-X$ yields corresponding versions for supermartingales. So, suppose we want to prove (7.182) for all submartingales X and all $\mathcal{H}$-valued processes Y that are α-subordinate to X. Let $D = \mathbb{R} \times \mathcal{H} \times [0, \infty) \times [0, \infty)$. Then we search for functions $U : D \to \mathbb{R}$ that enjoy the following properties:

$1°$ $U \geq V$ on D.

$2°$ We have $U_x(x, y, z, w) + \alpha|U_y(x, y, z, w)| \leq 0$ provided $|x| \leq z$ and $|y| \leq w$. Furthermore, there is a function $M : D \to [0, \infty)$, bounded on any set of the form $\{(x, y, z, w) : |x|, |y|, w, z \leq L\}$ for some $L > 0$, such that for any $(x, y, z, w) \in D$ and $h \in \mathbb{R}$, $k \in \mathcal{H}$,

$$U_{xx}(x, y, z, w)h^2 + 2(U_{xy}(x, y, z, w)h, k) + (U_{yy}(x, y, z, w)k, k)$$
$$\leq M(x, y, z, w)(|k|^2 - h^2).$$

$3°$ For any $(x, y, z, w) \in D$ such that $w = |y| \leq |x| \leq z$

$$U(x, y, z, w) \leq 0.$$

$4°$ For any $x \neq 0$ and $|y| \leq w$ we have $U_z(x, y, |x|, w) \leq 0$. For any $y \neq 0$ and $|x| \leq z$, $U_w(x, y, z, |y|) \leq 0$.

Theorem 7.20. *If there is U satisfying the conditions $1°$–$4°$ above, then (7.182) holds for all bounded continuous-path submartingales X and all bounded continuous-path processes Y which are α-subordinate to X.*

We omit the analogous proof.

The problem of showing the reverse implication of Theorems 7.19 and 7.20 is in general much more delicate. It is natural to proceed as previously; let us first focus on the martingale setting. For any x, $y \in \mathbb{R}$, let $M(x, y)$ denote the class of all bounded continuous-path martingales (X, Y) such that X starts from x, Y starts from y and we have $Y_t = y + \int_{0+}^t H_s dX_s$ for some predictable H taking values in $\{-1, 1\}$. Obviously, if $(X, Y) \in M(x, y)$, then the pointwise limit (X_∞, Y_∞) exists almost surely. For a given V as above, define

$$U_0(x, y, z, w) - \sup\{\mathbb{E}V(X_\infty, Y_\infty, |X|^* \vee z, |Y|^* \vee w)\},$$

where the supremum is taken over $M(x, y)$. The main difficulty is that there is no reason for the function U_0 to be of class C^2. However, as we shall see below, one may still recover some properties of U_0, which are useful in proving the sharpness. While dealing with submartingales, one fixes $\alpha \geq 0$, $x \geq 0$ and $y \in \mathbb{R}$ and denotes by $M_\alpha(x, y)$ the class of all pairs (X, Y) of bounded real-valued Itô processes of the form

$$X_t = x + \int_{0+}^t \phi_s \mathrm{d}B_s + \int_{0+}^t \psi_s \mathrm{d}s, \qquad Y_t = y + \int_{0+}^t \zeta_s \mathrm{d}B_s + \int_{0+}^t \xi_s \mathrm{d}s, \qquad (7.186)$$
$$\text{with } |\zeta_s| = |\phi_s| \text{ and } |\xi_s| = \alpha|\psi_s| \text{ for all } s \geq 0.$$

Then the function U_0 is defined analogously: again, it is not difficult to see that the pointwise limits X_∞, Y_∞ exist with probability 1 (for example, using escape estimates). The same problems appear: in general, we are not able to deduce any regularity properties of this function (however, for some specific V, we will overcome this).

7.10 Maximal bounds for continuous-path martingales

7.10.1 Formulation of the result

We start with the following martingale inequality.

Theorem 7.21. *Suppose X, Y are $\mathcal{H}$-valued continuous martingales such that Y is differentially subordinate to X. Then the inequalities*

$$||Y||_p \leq \sqrt{\frac{2}{p}} \, |||X|^*||_p, \qquad 1 \leq p < 2, \qquad (7.187)$$

and

$$||Y||_p \leq (p - 1)|||X|^*||_p, \quad 2 \leq p < \infty \qquad (7.188)$$

hold and the constants are the best possible. They are the best possible even when X is assumed to take values in $\mathbb{R}$ and Y is the Itô integral, with respect to X, of some predictable H taking values in $\{-1, 1\}$.

7.10.2 Proof of Theorem 7.21

Proof of (7.187). For $p \in [1, 2)$, we will write $\beta = \beta_p = \sqrt{2/p}$. Let $U, V : \mathcal{H} \times \mathcal{H} \times (0, \infty) \to \mathbb{R}$ (note that $z = 0$ is not allowed) be given by

$$U(x, y, z) = \frac{p}{2}\beta^{p-2}(|y|^2 - |x|^2 - (\beta^2 - 1)z^2)z^{p-2} \qquad (7.189)$$

and $V(x, y, z) = |y|^p - \beta^p z^p$.

Lemma 7.25. *The function U satisfies the conditions $1°$–$4°$.*

Proof. To prove the majorization, observe that we may assume that $z = 1$, due to homogeneity. Now, by the mean value property of the concave function $t \mapsto t^{p/2}$,

$$U(x, y, 1) = \frac{p}{2}\beta^{p-2}(|y|^2 - |x|^2 - (\beta^2 - 1)) \geq \frac{p}{2}\beta^{p-2}(|y|^2 - \beta^2)$$
$$= \frac{p}{2}(\beta^2)^{p/2-1}(|y|^2 - \beta^2) \geq |y|^p - \beta^p = V(x, y, 1).$$

To check $2°$, note that

$$(U_{xx}(x, y, z)h, h) + 2(U_{xy}(x, y, z)h, k) + (U_{yy}(x, y, z)k, k) = p\beta^{p-2}z^{p-2}(|k|^2 - |h|^2)$$

and the function $M(x, y, z) = p\beta^{p-2}z^{p-2}$ has the required boundedness property. Condition $3°$ is evident. Finally, we have

$$U_z(x, y, |x|) = \frac{p(p-2)}{2}\beta^{p-2}|y|^2|x|^{p-3} \leq 0,$$

which gives $4°$. $\qquad\square$

Now the proof is immediate: using localization, reduce the inequality to bounded martingales. That is, for a fixed n, we consider the stopping time $T_n = \inf\{t : |X_t| \geq n \text{ or } |Y_t| \geq n\}$ and define the martingales $(X_t^{(n)})$ and $(Y_t^{(n)})$ by

$$X_t^{(n)} = \begin{cases} X_{T_n \wedge t} & \text{if } T_n > 0, \\ 0 & \text{if } T_n = 0, \end{cases}$$

with a similar definition for $Y_t^{(n)}$. Since U and $X^{(n)}$, $Y^{(n)}$ satisfy the conditions of Theorem 7.19, we have, for any $t \geq 0$,

$$\mathbb{E}|Y_t^{(n)}|^p \leq \beta^p \mathbb{E}(|X_t^{(n)}|^*)^p \leq \beta^p \mathbb{E}(|X|^*)^p.$$

Now let $n \to \infty$ to obtain $||Y_t||_p \leq \beta|||X|^*||_p$, by means of Fatou's lemma. It now suffices to take the supremum over t to get the claim. $\qquad\square$

Remark. The function U leading to the inequality (7.187) is not unique. For example, one can try to work with the following alternative. First introduce the auxiliary $\Phi : [\beta - 1, \infty) \to \mathbb{R}$ by

$$\Phi(t) = e^{-t}\left[-p\int_{\beta-1}^t e^s(s+1)^{p-1}ds + p\beta^{p-2}(1-\beta)e^{\beta-1}\right]. \qquad (7.190)$$

and let $u : [0, 1] \times [0, \infty) \to \mathbb{R}$ be given by

$$u(x, y) = \begin{cases} (y - x + 1)^p + (1 - x)\Phi(y - x) - \beta^p & \text{if } y - x \geq \beta - 1, \\ \frac{1}{2}p\beta^{p-2}(y^2 - x^2 - \beta^2 + 1) & \text{if } y - x < \beta - 1. \end{cases} \qquad (7.191)$$

The special function $U : \mathbb{R}^d \times \mathbb{R}^d \times (0, \infty)$ is defined by

$$U(x, y, z) = (|x| \vee z)^p u \left(\frac{|x|}{|x| \vee z}, \frac{|y|}{|x| \vee z} \right). \tag{7.192}$$

This function has almost all the required properties; however, it is not of class C^2 on the set $\{(x, y, z) : |y| - |x| = (\beta - 1)(|x| \vee z)\}$ and one needs to use smoothing arguments to overcome this difficulty. We omit the details (but see the next section).

7.10.3 Sharpness

Let $1 \le p < \infty$ be fixed and suppose that κ is the best constant in the inequality

$$||Y||_p \le \kappa |||X|^*||_p, \tag{7.193}$$

to be valid for all $(X, Y) \in \mathcal{M}(x, y)$ such that $y = \pm x$. Recall that for any $x, y \in \mathbb{R}$, the class $M(x, y)$ consists of all bounded continuous path martingales (X, Y) such that X starts from x, Y starts from y and we have $Y_t = y + \int_{0+}^t H_s dX_s$ for some predictable H taking values in $\{-1, 1\}$. Let $\kappa' > \kappa$ and consider the function $U_0 : \mathbb{R} \times \mathbb{R} \times [0, \infty) \to \mathbb{R}$, given by

$$U_0(x, y, z) = \sup_{\mathcal{M}(x,y)} \{\mathbb{E}|Y_\infty|^p - (\kappa')^p \mathbb{E}(|X_\infty|^* \vee z)^p\}.$$

Lemma 7.26. *The function U_0 has the following properties.*

(i) $U_0(x, y, z) < \infty$ *for any* $(x, y, z) \in \mathbb{R} \times \mathbb{R} \times [0, \infty)$,

(ii) *For any* $(x, y, z) \in \mathbb{R} \times \mathbb{R} \times [0, \infty)$,

$$U_0(\lambda x, \pm \lambda y, |\lambda| z) = |\lambda|^p U_0(x, y, z) \text{ for } \lambda \ne 0. \tag{7.194}$$

(iii) *For any* $(x, y, z) \in \mathbb{R} \times \mathbb{R} \times [0, \infty)$,

$$U_0(x, y, z) \ge |y|^p - (\kappa')^p(|x| \vee z)^p. \tag{7.195}$$

(iv) *For any* $(x, y, z) \in \mathbb{R} \times \mathbb{R} \times [0, \infty)$, $|x| \le z$, $\varepsilon \in \{-1, 1\}$, *if* $\alpha_1, \alpha_2 \in (0, 1)$ *and* $t_1, t_2 \in (-x - z, -x + z)$ *satisfy* $\alpha_1 + \alpha_2 = 1$ *and* $\alpha_1 t_1 + \alpha_2 t_2 = 0$, *then*

$$U_0(x, y, z) \ge \alpha_1 U_0(x + t_1, y + \varepsilon t_1, z) + \alpha_2 U_0(x + t_2, y + \varepsilon t_2, z).$$

(v) *If* $y \in \mathbb{R}$, $\delta > 0$ *and* $a \in (0, 2)$, *then*

$$U_0(1, y, 1) \ge \frac{\delta}{a + \delta} U_0(1 - a, y + a, 1 + \delta) + \frac{a}{a + \delta} U_0(1 + \delta, y - \delta, 1 + \delta)$$

$$\ge \frac{\delta}{a + \delta} [U_0(1 - a, y + a, 1) + (\kappa')^p(1 - (1 + \delta)^p)]$$

$$+ \frac{a}{a + \delta}(1 + \delta)^p U_0\left(1, \frac{y - \delta}{1 + \delta}, 1\right). \tag{7.196}$$

Proof. (i) Let $(x, y, z) \in \mathbb{R} \times \mathbb{R} \times [0, \infty)$ and $(X, Y) \in M(x, y)$. Since $(Y - y + x, X) \in M(x, x)$, we have, by the triangle inequality and (7.193),

$$||Y_\infty||_p \leq ||(Y - y + x)_\infty||_p + |y - x| \leq \kappa |||X|^*||_p + |y - x|$$
$$\leq ((\kappa')^p |||X|^* \vee z||_p^p + C \cdot |y - x|^p)^{1/p},$$

where C depends only on κ, κ' and p. Thus $U_0(x, y, z) \leq C|y - x|^p$.

(ii) This is quick: $(X, Y) \in M(x, y)$ if and only if $(\lambda X, \pm \lambda Y) \in M(\lambda x, \pm \lambda y)$.

(iii) The pair (x, y) of constant martingales belongs to $M(x, y)$.

(iv) We will use the continuous analogue the "splicing argument". Suppose that $(X^i, Y^i) \in \mathcal{M}(x + t_i, y + \varepsilon t_i)$ and let H^i be corresponding predictable processes, $i = 1, 2$. Intuitively speaking, we will "glue" the two pairs using Brownian motion and obtain a pair belonging to $M(x, y)$. To this end, we may and do assume that these processes are given on the same probability space equipped with the same filtration. Suppose B is a Brownian motion starting from x, independent of these two pairs. Let $\tau = \inf\{t : B_t \in \{x + t_1, x + t_2\}\}$ and set

$$X_t = \begin{cases} B_t & \text{if } t \leq \tau, \\ X^i_{t-\tau} & \text{if } t > \tau \text{ and } B_\tau = x + t_i, \end{cases}$$

$$H_t = \begin{cases} \varepsilon & \text{if } t \leq \tau, \\ H^i_{t-\tau} & \text{if } t > \tau \text{ and } B_\tau = x + t_i \end{cases}$$

and $Y_t = y + \int_{(0,t]} H_s dX_s$. It is easy to see that X is a martingale with respect to the natural filtration and $(X, Y) \in M(x, y)$. With probability 1,

$$Y_\infty = Y^1_\infty 1_{\{B_\tau = x + t_1\}} + Y^2_\infty 1_{\{B_\tau = x + t_2\}} \tag{7.197}$$

and, since $|x + t_1|, |x + t_2| \leq z$,

$$|X|^* \vee z = (|X^1|^* \vee z)1_{\{B_\tau = x + t_1\}} + (|X^2|^* \vee z)1_{\{B_\tau = x + t_2\}}. \tag{7.198}$$

Therefore

$$W(x, y, z) \geq \mathbb{E}|Y_\infty|^p - (\kappa')^p \mathbb{E}(|X|^* \vee z)^p$$

$$= \sum_{i=1}^2 (\mathbb{E}|Y^i_\infty|^p - (\kappa')^p \mathbb{E}(|X^i|^* \vee z)^p) \mathbb{P}(B_\tau = x + t_i) \tag{7.199}$$

$$= \sum_{i=1}^2 \alpha_i (\mathbb{E}|Y^i_\infty|^p - (\kappa')^p \mathbb{E}(|X^i|^* \vee z)^p).$$

Now take supremum on the right-hand side over the classes $M(x + t_1, y + \varepsilon t_1)$ and $M(x + t_2, y + \varepsilon t_2)$ to obtain the desired estimate.

(v) We repeat the argumentation from the previous part, with $x = z = 1$, $\varepsilon = -1$, $t_1 = -a$ and $t_2 = \delta$. Equation (7.197) remains valid; however, (7.198) is no longer true: we still have $(|X|^* \vee 1)1_{\{B_\tau=1+\delta\}} = (|X^2|^* \vee 1)1_{\{B_\tau=1+\delta\}}$, but the equality $(|X|^* \vee 1)1_{\{B_\tau=1-a\}} = (|X^1|^* \vee 1)1_{\{B_\tau=1-a\}}$ does not hold in general. Nonetheless, we have the inequality

$$|X|^* \vee 1 \leq (|X^1|^* \vee (1+\delta))1_{\{B_\tau=1-a\}} + (|X^2|^* \vee 1)1_{\{B_\tau=1+\delta\}}. \qquad (7.200)$$

But, by the very definition of U_0,

$$U_0(x,y,z_1) \leq U_0(x,y,z_2) \quad \text{for } z_1 \geq z_2, \qquad (7.201)$$

so arguing as in (7.199), we get the first inequality in (7.196). To deal with the second one, we need to compare $U_0(1-a, y+a, 1+\delta)$ and $U_0(1-a, y+a, 1)$. Note that for x, $\delta \geq 0$ we have $(x \vee (1+\delta))^p - (x \vee 1)^p \leq (1+\delta)^p - 1$. Thus, for any $(X,Y) \in M(1-a, y+a)$,

$$U_0(1-a, y+a, 1+\delta) \geq \mathbb{E}|Y_\infty|^p - (\kappa')^p\mathbb{E}\big(|X|^* \vee (1+\delta)\big)^p$$
$$\geq \mathbb{E}|Y_\infty|^p - (\kappa')^p\mathbb{E}\big(|X|^* \vee 1\big)^p + (\kappa')^p(1 - (1+\delta)^p).$$

Taking supremum over all such (X,Y) yields

$$U_0(1-a, y+a, 1+\delta) \geq U_0(1-a, y+a, 1) + (\kappa')^p(1 - (1+\delta)^p).$$

Thus the second inequality in (7.196) follows, since, by homogeneity of U_0,

$$U_0(1+\delta, y-\delta, 1+\delta) = (1+\delta)^p U_0\Big(1, \frac{y-\delta}{1+\delta}, 1\Big).$$

This completes the proof. $\square$

Now we are ready to prove that the inequalities (7.187) and (7.188) are sharp.

Sharpness of (7.187). We keep the notation $\beta = \beta_p = \sqrt{2/p}$. Apply (iv) with $x = \beta/2$, $y = 1 - \beta/2$, $z = 1$, $\varepsilon = -1$, $t_1 = 1 - \beta$ and $t_2 = 1 - \beta/2$ (α_1 and α_2 are uniquely determined by t_1 and t_2) to get

$$U_0\Big(\frac{\beta}{2}, 1 - \frac{\beta}{2}, 1\Big) \geq \frac{2-\beta}{\beta}U_0\Big(1 - \frac{\beta}{2}, \frac{\beta}{2}, 1\Big) + \frac{2\beta-2}{\beta}U_0(1,0,1). \qquad (7.202)$$

The condition (iv) with $x = 1 - \beta/2$, $y = \beta/2$, $z = 1$, $\varepsilon = 1$, $t_1 = \beta/2$ and $t_2 = -1$ implies

$$U_0\Big(1 - \frac{\beta}{2}, \frac{\beta}{2}, 1\Big) \geq \frac{2}{\beta+2}U_0(1, \beta, 1) + \frac{\beta}{\beta+2}U_0\Big(-\frac{\beta}{2}, -1 + \frac{\beta}{2}, 1\Big). \qquad (7.203)$$

By homogeneity, $U_0\Big(-\frac{\beta}{2}, -1 + \frac{\beta}{2}, 1\Big) = U_0\Big(\frac{\beta}{2}, 1 - \frac{\beta}{2}, 1\Big)$. Furthermore, (iii) implies $U_0(1, \beta, 1) \geq \beta^p - (\kappa')^p$; combining this with (7.202) and (7.203) gives

$$\frac{\beta^2}{2}U_0\Big(\frac{\beta}{2}, 1 - \frac{\beta}{2}, 1\Big) \geq \Big(1 - \frac{\beta}{2}\Big)(\beta^p - (\kappa')^p) + \Big(\frac{\beta^2}{2} + \frac{\beta}{2} - 1\Big)U_0(1,0,1). \quad (7.204)$$

Now exploit (v), with $y = 0$, $a = 1 - \beta/2$ and $\delta > 0$, to obtain

$$U_0(1,0,1) \geq \frac{2\delta}{2 - \beta + 2\delta} \left[U_0 \left(\frac{\beta}{2}, 1 - \frac{\beta}{2}, 1 \right) + (\kappa')^p \left(1 - (1 + \delta)^p \right) \right]$$
$$+ \frac{2 - \beta}{2 - \beta + 2\delta} (1 + \delta)^p U_0 \left(1, -\delta/(1 + \delta), 1 \right). \tag{7.205}$$

By (ii), $U_0 \left(1, -\delta/(1 + \delta), 1 \right) = U_0(1, \delta/(1 + \delta), 1)$; moreover, if we use the first inequality in (7.196), with y, a, δ replaced by the numbers $\delta/(1+\delta)$, $(1 - \frac{\beta}{2}) \frac{1+2\delta}{1+\delta} - \frac{\delta}{1+\delta}$ and $\delta/(1 + \delta)$, respectively, we obtain

$$U_0 \left(1, \frac{\delta}{1 + \delta}, 1 \right) \geq \frac{2\delta}{(2 - \beta)(1 + 2\delta)} \left(\frac{1 + 2\delta}{1 + \delta} \right)^p U_0 \left(\frac{\beta}{2}, \left(1 - \frac{\beta}{2} \right), 1 \right)$$
$$+ \frac{(2 - \beta)(1 + 2\delta) - 2\delta}{(2 - \beta)(1 + 2\delta)} \left(\frac{1 + 2\delta}{1 + \delta} \right)^p U_0 (1, 0, 1).$$

Plug this into (7.205), subtract $U_0(1, 0, 1)$ from both sides, divide throughout by 2δ and let $\delta \to 0$ to obtain

$$0 \geq \frac{2U_0(\beta/2, 1 - \beta/2, 1)}{2 - \beta} + \left(p - \frac{2}{2 - \beta} \right) U_0(1, 0, 1).$$

Combining this with (7.204) and using $\beta = \sqrt{2/p}$ we get

$$\frac{1}{p} U_0 \left(\frac{\beta}{2}, 1 - \frac{\beta}{2}, 1 \right) \geq \left(1 - \frac{1}{\sqrt{2p}} \right) \left(\left(\frac{2}{p} \right)^{p/2} - (\kappa')^p \right) + \frac{1}{p} U_0 \left(\frac{\beta}{2}, 1 - \frac{\beta}{2}, 1 \right)$$

or $\kappa' \geq \sqrt{2/p}$. Since $\kappa' > \kappa$ was arbitrary, we conclude that the constant $\sqrt{2/p}$ can not be replaced in (7.187) by a smaller one. □

Sharpness of (7.188). Apply (iv) with $x = 0$, $y = p$, $z = 1$, $\varepsilon = -1$, $t_1 = 1$ and $t_2 = -\delta$ to get

$$U_0(0, p, 1) \geq \frac{\delta}{1 + \delta} U_0(1, p - 1, 1) + \frac{1}{1 + \delta} U_0(-\delta, p + \delta, 1)$$
$$\geq \frac{\delta}{1 + \delta} ((p - 1)^p - (\kappa')^p) + \frac{1}{1 + \delta} U_0(\delta, p + \delta, 1), \tag{7.206}$$

where we have used the majorization (iii) and the homogeneity. Using again (iv), this time with $x = \delta$, $y = p + \delta$, $z - 1$, $\varepsilon - -1$, $t_1 = 1 - \delta$ and $t_2 = -\delta$, we get

$$U_0(\delta, p + \delta, 1) \geq \delta U_0(1, p + 2\delta - 1, 1) + (1 - \delta)U_0(0, p + 2\delta, 1)$$
$$\geq \delta U_0(1, p + 2\delta - 1, 1) + (1 - \delta)U_0(0, p + 2\delta, 1 + 2\delta/p), \tag{7.207}$$

where in the last passage we have exploited (7.201). Now by (v), with y, a, δ replaced by $p + 2\delta - 1$, 1 and $2\delta/p$, respectively, together with the majorization and homogeneity, we have

$$
\begin{aligned}
U_0(1, p + 2\delta - 1, 1) &\geq \frac{2\delta}{p + 2\delta} U_0(0, p + 2\delta, 1 + 2\delta/p) \\
&\quad + \frac{p}{p + 2\delta} U_0(1 + 2\delta/p, (1 + 2\delta/p)(p - 1), 1 + 2\delta/p) \\
&\geq \frac{2\delta}{p + 2\delta}(1 + 2\delta/p)^p U_0(0, p, 1) \\
&\quad + \frac{p}{p + 2\delta}(1 + 2\delta/p)^p((p - 1)^p - (\kappa')^p).
\end{aligned}
\tag{7.208}
$$

Now combine (7.207) with (7.208), and insert the obtained lower bound for $U_0(\delta, p + \delta, 1)$ into (7.206). We obtain an estimate, which, after subtracting $U_0(0, p, 1)$ from both sides, dividing throughout by δ and letting $\delta \to 0$, becomes

$$
0 = U_0(0, p, 1) \cdot \lim_{\delta \to 0} \left[1 - \frac{1 - \delta}{1 + \delta} \left(1 + \frac{2\delta}{p} \right)^p \right] \geq 2((p - 1)^p - (\kappa')^p).
$$

This implies $\kappa' \geq p - 1$ and, consequently, $\kappa \geq p - 1$. $\qquad\square$

On the search of the suitable majorant. Let us sketch an argument which leads to the right choice of the optimal constant β_p, $1 < p < 2$, and the right guess of the special function U used in the proof of (7.187).

First we will indicate how to construct the function described in the remark at the end of Section 7.3. Let $p \in [1, 2)$ be fixed and write down the desired inequality

$$
\|Y\|_p \leq \beta \| |X|^* \|_p,
$$

with the optimal $\beta \geq 1$ to be determined; some experimentation shows that β should be smaller than 2. Let us restrict ourselves to the real-valued martingales: $\mathcal{H} = \mathbb{R}$. Since the function $V(x, y, z) = |y|^p - \beta^p(|x| \wedge z)^p$ is homogeneous of order p, we assume that U also has this property. Let $u(x, y) = U(x, y, 1)$ for $x \in [-1, 1]$ and $y \in \mathbb{R}$. In addition, we assume that U is of class C^1 on $\mathbb{R} \times \mathbb{R} \times (0, \infty)$.

It is natural to expect that there should be some similarities between U and the special function used by Burkholder in [25]: both functions concern essentially the same maximal inequality (strictly speaking, this is the case if $p = 1$; but for $p > 1$, the conditions imposed below also lead to the right function). Burkholder's majorant suggests that we should search for u in the class of functions that satisfy the following assumptions (A1)–(A3):

(A1)   For all $x \in [-1, 1]$ and $y \in \mathbb{R}$,

$$
u(x, y) = u(-x, y) = u(x, -y).
\tag{7.209}
$$

(A2)   If $0 < x < 1$ and $x \leq y$, then

$$u(x,y) = (1-x)A(-x+y) + xB(1-x+y) \qquad (7.210)$$

and if $0 < y < x < 1$, then

$$u(x,y) = \frac{1-x}{1-x+y}C(x-y) + \frac{y}{1-x+y}B(1-x+y), \qquad (7.211)$$

where $A = u(0,\cdot)$, $B = u(1,\cdot)$ and $C = u(\cdot,0)$.

(A3)   For all $y \geq \beta$ we have $u(1,y) = U(1,y,1) = V(1,y,1)$.

Lemma 7.27. *If u satisfies conditions* (A1)–(A3), *then for $x \in [0,1]$ and $y \geq x + \beta - 1$ we have*

$$u(x,y) = (y-x+1)^p - \beta^p + (1-x)e^{x-y}\left[A(\beta-1)e^{\beta-1} - p\int_{\beta-1}^{-x+y} e^s(s+1)^{p-1}\mathrm{d}s\right].$$

Proof. By (7.210), for any $y \geq \beta - 1$ and $\delta \in (0,1)$ we have

$$u(\delta, y+\delta) = (1-\delta)A(y) + \delta B(1+y).$$

Subtracting $A(y)$ from both sides, dividing by δ and letting $\delta \to 0$ yields $u_x(0,y) + u_y(0,y) = -A(y) + B(1+y)$. But $u_y(0,y) = A'(y)$ and, by (7.209), $u_x(0,y) = 0$, so we obtain

$$A'(y) = -A(y) + (y+1)^p - \beta^p.$$

Solving this differential equation gives

$$A(y) = -pe^{-y}\int_{\beta-1}^{y} e^s(s+1)^{p-1}\mathrm{d}s + (y+1)^p - \beta^p + A(\beta-1)e^{-y+\beta-1},$$

and plugging this into (7.210) yields the claim. $\square$

Now we will find the function u on the remaining part of the domain. It is easy to see that the property (iv) from the definition of $\mathcal{U}(V)$ implies that the function $w : s \mapsto u(s, 1-s)$, $s \in [0,1]$, is concave. From Lemma 7.27, we know the explicit form of w on the interval $[0, 1 - \beta/2]$. Some experimentation suggests the following assumption, which is *not* satisfied by Burkholder's majorant, but in our case leads to the right function. This is the key condition.

(A4) w is linear on $[1 - \beta/2, 1]$.

Lemma 7.28. *Under the assumptions* (A1)–(A4), *for all $x, y \in [0,1]$ such that $x + y \leq 1$ and $-x + y \leq \beta - 1$, we have,*

$$u(x,y) = \frac{p\beta^{p-2}}{2}(y^2 - x^2 - \beta^2 + 1). \qquad (7.212)$$

Proof. Let

$$a_\beta = (\beta - 1)A(\beta - 1) + p\beta^{p-1}(\beta - 2). \tag{7.213}$$

Since u is of class C^1, we obtain that for $s \in [1 - \beta/2, 1]$,

$$w(s) = u(1 - \beta/2, \beta/2) + (u_x(1 - \beta/2, \beta/2) - u_y(1 - \beta/2, \beta/2))(s - 1 + \beta/2)$$

$$= \frac{\beta}{2}A(\beta - 1) + a_\beta(s - 1 + \beta/2).$$

Suppose that $y \in [0, \beta - 1]$. By (7.210), the function $s \mapsto u(s, y + s)$ is linear, so, for $0 < \delta < (1 - y)/2$,

$$u(\delta, y + \delta) = \frac{2\delta}{1 - y}w\left(\frac{1 - y}{2}\right) + \frac{1 - y - 2\delta}{1 - y}A(y).$$

Subtract $A(y)$ from both sides, divide throughout by δ and let $\delta \to 0$ to obtain

$$A'(y) = u_x(0, y) + u_y(0, y) = -\frac{2A(y)}{1 - y} + \frac{2}{1 - y}w\left(\frac{1 - y}{2}\right).$$

Solving the differential equation, we get

$$A(y) = \frac{\beta}{2}A(\beta - 1) + a_\beta\left(\frac{\beta}{2} - y\right) + \frac{A(\beta - 1) - a_\beta}{2(2 - \beta)}(1 - y)^2.$$

By (7.209), $A'(0) = 0$; this gives $A(\beta - 1) = a_\beta(\beta - 1)$, so, by (7.213),

$$a_\beta = -p\beta^{p-2} \qquad \text{and} \qquad A(y) = \frac{p\beta^{p-2}}{2}(y^2 + 1 - \beta^2).$$

This enables us to obtain (7.212) for $y \geq x$: it suffices to observe that

$$u(x, y) = \frac{2x}{1 + x - y}w\left(\frac{1 - y}{2}\right) + \frac{1 - x - y}{1 + x - y}A(y - x),$$

which follows directly from (7.210).

If $y < x$, we proceed similarly: by (7.211), we have, for $x \in [0, 1)$ and $0 < \delta < (1 - x)/2$,

$$u(x + \delta, \delta) = \frac{1 - x - 2\delta}{1 - x}C(x) + \frac{2\delta}{1 - x}w\left(\frac{1 + x}{2}\right).$$

Subtract $C(x)$ from both sides, divide by δ and let $\delta \to 0$ to obtain a differential equation for C. Solve it and use $C'(0) = 0$ to get $C(x) = p\beta^{p-2}(-x^2 - \beta^2 + 1)/2$. To obtain (7.212), apply the following consequence of (7.211):

$$u(x, y) = \frac{2y}{1 - x + y}w\left(\frac{1 + x - y}{2}\right) + \frac{1 - x - y}{1 - x + y}C(x - y).$$

The claim follows. $\square$

Lemma 7.29. *If u satisfies (A1)–(A4), then $\beta \geq \sqrt{2/p}$.*

Proof. Since U is homogeneous of order p, we have $x^p u(1,0) = U(x,0,x)$ for $x > 0$. Differentiating at 1 and using the property (iii) from the definition of the class $\mathcal{U}(V)$, we get

$$pC(1) = pu(1,0) = U_x(1,0,1) + U_z(1,0,1) \leq U_x(1,0,1) = C'(1),$$

which is the claim. $\qquad\square$

We impose the following condition.

(A5) We have $\beta = \sqrt{2/p}$.

To complete the description of u, it remains to guess its values on the set $E = \{(x,y) : x \in [0,1], 1 - x < y < x + \beta - 1\}$. Here is our final assumption.

(A6) For $(x,y) \in E$, the formula (7.212) is valid.

As one easily checks, in this way we have obtained the function u given by (7.191). The description of U is completed by (7.192). As already mentioned in the remark at the end of Section 7.3, this function is not sufficiently smooth, so Itô's formula is not directly applicable. However, as we have seen in the preceding chapters, in general the majorant corresponding to a given martingale inequality is not uniquely determined. When U is described by different expressions on pairwise disjoint subsets, it is useful to look at $\overline{U}$, given by one of these expressions on the whole domain. Fortunately, a similar phenomenon occurs also in our case. The function U defined by (7.189), much simpler and more regular than the one just obtained above, is sufficient for our purposes. $\qquad\square$

7.11 A maximal bound for continuous-path supermartingales

7.11.1 Formulation of the result

Theorem 7.22. *Let α be a fixed nonnegative number. Assume that X is a nonnegative continuous-path supermartingale and Y is α-subordinate to X. Then*

$$\|Y\|_1 \leq \beta \|X^*\|_1, \qquad (7.214)$$

where $\beta = \beta(possup) = \alpha + 1 + ((2\alpha + 1)e)^{-1}$. The constant is the best possible, even if we restrict ourselves to Itô processes of the form (7.186).

7.11.2 Proof of Theorem 7.22

Proof of (7.214). This inequality holds for general, not necessarily continuous-path processes: see Theorem 7.10 above.   □

Sharpness. Let $\alpha \geq 0$, $\beta > 0$ be fixed and suppose that we have

$$||Y||_1 \leq \beta ||X^*||_1 \qquad (7.215)$$

for any Itô processes X, Y of the form (7.186), with the additional condition that X is a nonnegative supermartingale. Let $D = [0, \infty) \times \mathbb{R} \times [0, \infty)$, define $V_\beta(x, y, z) = |y| - \beta(|x| \vee z)$ for $(x, y, z) \in D$ and let $U : D \to \mathbb{R}$ be the function given by

$$U_0(x, y, z) = \sup_{M_\alpha(x,y)} \{\mathbb{E}V_\beta(X_\infty, Y_\infty, |X|^* \vee z)\}.$$

Lemma 7.30. *The function U_0 has the following properties:*

$$U_0 \geq V_\beta \quad \text{on } D, \qquad (7.216)$$

$$U_0(x, y, z_1) \leq U_0(x, y, z_2) \quad \text{if } (x, y, z_i) \in D \text{ and } z_1 \geq z_2 \geq x, \qquad (7.217)$$

$$U_0(x, \cdot, z) : y \mapsto U_0(x, y, z) \quad \text{is convex for any fixed } 0 \leq x \leq z, \ z > 0. \qquad (7.218)$$

For all $\varepsilon \in \{-1, 1\}$, $\lambda_1, \lambda_2 \in (0, 1)$, $x \in [0, z]$, $y \in \mathbb{R}$, $z > 0$ and $t_1, t_2 \in [-x, z - x]$ such that $\lambda_1 + \lambda_2 = 1$ and $\lambda_1 t_1 + \lambda_2 t_2 = 0$,

$$U_0(x, y, z) \geq \lambda_1 U_0(x + t_1, y + \varepsilon t_1, z) + \lambda_2 U_0(x + t_2, y + \varepsilon t_2, z). \qquad (7.219)$$

Furthermore,

$$U_0(x, y, z) \geq U_0(x - d, y \pm \alpha d, z) \qquad \text{if } x \leq z \text{ and } 0 < d \leq x, \qquad (7.220)$$

Finally,

$$U_0(x, y, z) \leq 0 \qquad \text{if } |y| \leq x \leq z. \qquad (7.221)$$

Proof. The first thing which needs to be checked is the finiteness of U_0. This is straightforward: if $(x, y, z) \in D$ and $(X, Y) \in M_\alpha(x, y)$, then $Y - y$ is α-subordinate to X; hence, by the triangle inequality and (7.215),

$$\mathbb{E}V_\beta(X_t, Y_t, X^* \vee z) \leq |y| + \mathbb{E}V_\beta(X_t, Y_t - y, X^* \vee z) \leq |y|.$$

Since $(X, Y) \in M_\alpha(x, y)$ and $t \geq 0$ were arbitrary, we obtain $U_0(x, y, z) \leq |y| < \infty$. To show (7.216), it suffices to note that the constant pair (x, y) belongs to the class $M_\alpha(x, y)$. The condition (7.217) follows directly from the definition of U_0 and the fact that V_β also has this property. To prove (7.218), fix (x, y_1, z), $(x, y_2, z) \in D$, $\lambda \in (0, 1)$ and let $y = \lambda y_1 + (1 - \lambda)y_2$. Take (X, Y) from $M_\alpha(x, y)$.

Since $(X, y_1 - y + Y)$ and $(X, y_2 - y + Y)$ belong to $M_\alpha(x, y_1)$ and $M_\alpha(x, y_2)$, respectively, we have, by the triangle inequality,

$$\mathbb{E} V_\beta(X_\infty, Y_\infty, X^* \vee z) \le \lambda \mathbb{E} V_\beta(X_\infty, y_1 - y + Y_\infty, X^* \vee z)$$
$$+ (1 - \lambda) \mathbb{E} V_\beta(X_\infty, y_2 - y + Y_\infty, X^* \vee z)$$
$$\le \lambda U(x, y_1, z) + (1 - \lambda) U(x, y_2, z).$$

Now it suffices to take the supremum over X and Y and (7.218) follows.

We turn to (7.219). In fact, we will prove the following stronger version, which allows one of t_i's to take values larger than $z - x$. Namely, for all $\varepsilon \in \{-1, 1\}$, $\lambda_1, \lambda_2 \in (0, 1)$, $x \in [0, z]$, $y \in \mathbb{R}$, $z > 0$ and $t_1 < 0 < t_2$ such that $\lambda_1 + \lambda_2 = 1$, $\lambda_1 t_1 + \lambda_2 t_2 = 0$ and $x + t_1 \ge 0$,

$$U_0(x, y, z) \ge \lambda_1 \left[U_0(x + t_1, y + \varepsilon t_1, z) - \beta(x + t_2 - z)_+ \right]$$
$$+ \lambda_2 U_0(x + t_2, y + \varepsilon t_2, z). \tag{7.222}$$

This more general statement is needed in the proof of the optimality of the constant β(possub). We will prove (7.222) only for $\varepsilon = 1$; the argumentation for $\varepsilon = -1$ is similar. Let $x_i = x + t_i$, $y_i = y + t_i$ and take two pairs (X^1, Y^1) and (X^2, Y^2) from $M_\alpha(x_1, y_1)$ and $M_\alpha(x_2, y_2)$, respectively. Let ψ^i, ϕ^i, ζ^i and ξ^i denote the corresponding predictable processes in the decompositions of X^i and Y^i. We may and do assume that these processes are given on the same probability space equipped with the same filtration and are driven by the same Brownian motion W. Enlarging the probability space if necessary, we may assume that there is a Brownian motion B starting from x, which is independent of W. We use B to "glue" (X^1, Y^1) and (X^2, Y^2) into one Itô process (X, Y). Precisely, introduce the stopping time $\tau = \inf\{t : B_t \in \{x_1, x_2\}\}$ and set

$$X_t = \begin{cases} B_t & \text{if } t \le \tau, \\ X^i_{t-\tau} & \text{if } t > \tau \text{ and } B_\tau = x_i, \end{cases}$$

and

$$Y_t = \begin{cases} y - x + B_t & \text{if } t \le \tau \\ Y^i_{t-\tau} & \text{if } t > \tau \text{ and } B_\tau = x_i. \end{cases}$$

Then $(X, Y) \in M_\alpha(x, y)$, with

$$\phi_t = \begin{cases} 1 & \text{if } t \le \tau, \\ \phi^i_{t-\tau} & \text{if } t > \tau \text{ and } B_\tau = x_i, \end{cases} \qquad \psi_t = \begin{cases} 0 & \text{if } t \le \tau, \\ \psi^i_{t-\tau} & \text{if } t > \tau \text{ and } B_\tau = x_i \end{cases}$$

and

$$\zeta_t = \begin{cases} 1 & \text{if } t \le \tau, \\ \zeta^i_{t-\tau} & \text{if } t > \tau \text{ and } B_\tau = x_i, \end{cases} \qquad \xi_t = \begin{cases} 0 & \text{if } t \le \tau, \\ \xi^i_{t-\tau} & \text{if } t > \tau \text{ and } B_\tau = x_i. \end{cases}$$

We have, with probability 1,

$$Y_\infty = Y_\infty^1 1_{\{B_\tau = x_1\}} + Y_\infty^2 1_{\{B_\tau = x_2\}}$$

and, since $x_1 < x < x_2$,

$$(X^* \vee z)1_{\{B_\tau = x_1\}} \leq (X^{1*} \vee x_2 \vee z)1_{\{B_\tau = x_1\}}$$
$$\leq \left[(X^{1*} \vee z) + (x_2 - z)_+ \right]1_{\{B_\tau = x_1\}},$$
$$(X^* \vee z)1_{\{B_\tau = x_2\}} = (X^{2*} \vee z)1_{\{B_\tau = x_2\}}.$$

Therefore, we get

$$U_0(x, y, z) \geq \mathbb{E}|Y_\infty| - \beta\mathbb{E}(X^* \vee z)$$

$$\geq -\beta(x_2 - z)_+ \mathbb{P}(B_\tau = x_1) + \sum_{i=1}^{2} \left(\mathbb{E}|Y_\infty^i| - \beta\mathbb{E}(X^{i*} \vee z) \right)\mathbb{P}(B_\tau = x_i)$$

$$= -\lambda_1\beta(x_2 - z)_+ + \sum_{i=1}^{2} \lambda_i(\mathbb{E}|Y_\infty^i| - \beta\mathbb{E}(X^{i*} \vee z)).$$

Now take the supremum on the right-hand side over the classes $M_\alpha(x_1, y_1)$ and $M_\alpha(x_2, y_2)$ to obtain (7.222).

Next we establish (7.220). Take $(X, Y) \in M_\alpha(x - d, y - \alpha d)$ and consider the process (X', Y') defined by the formula

$$(X', Y')_t = \begin{cases} (x - t, y - \alpha t) & \text{if } t \leq d, \\ (X_{t-d}, Y_{t-d}) & \text{if } t > d, \end{cases}$$

with respect to the filtration $(\mathcal{F}'_t) = (\mathcal{F}_{(t-d)\vee 0})$. It is easy to see that $(X', Y') \in M_\alpha(x, y)$, $(X_\infty, Y_\infty) = (X'_\infty, Y'_\infty)$ and $X'^* \vee z = X^* \vee z$, so

$$\mathbb{E}V_\beta(X_\infty, Y_\infty, X^* \vee z) = \mathbb{E}V_\beta(X'_\infty, Y'_\infty, X'^*) \leq U(x, y, z).$$

Take the supremum over (X, Y) to get $U(x - d, y - \alpha d, z) \leq U(x, y, z)$. Finally, to show (7.221), suppose that the point $(x, y, z) \in D$ satisfies $|y| \leq x \leq z$. The first inequality implies that for any $(X, Y) \in M_\alpha(x, y)$, the process Y is α-subordinate to X, so

$$\mathbb{E}V_\beta(X_t, Y_t, X_t^* \vee z) \leq 0.$$

Therefore, the claim follows and the proof is complete.   □

Let $u(x, y) = U(x, y, 1)$ for $x \in [0, 1]$ and $y \in \mathbb{R}$. We will also use the notation $B(y) = u(1, y)$ for $y \in \mathbb{R}$. By Lemma 8.6, U belongs to $\mathcal{U}^{\text{sup}}(V_\beta)$ and satisfies $b(U) \leq 0$. In consequence, we have

$$u(x, y) \geq |y| - \beta \quad \text{for all } x \in [0, 1], y \in \mathbb{R}, \tag{7.223}$$
$$u \quad \text{is diagonally concave.} \tag{7.224}$$

and

$$u(x, y) \geq u(x - d, y - \alpha d) \quad \text{for } 0 \leq x - d \leq x \leq 1 \text{ and } y \in \mathbb{R} \qquad (7.225)$$
$$u(x, \pm x) \leq 0 \qquad \text{for } x \in [0, 1], \qquad (7.226)$$

Furthermore, it is easy to see that u satisfies

$$u(x, y) = u(x, -y) \qquad \text{for all } x \in [0, 1] \text{ and } y \in \mathbb{R}. \qquad (7.227)$$

We will show that the existence of $u : [0, 1] \times \mathbb{R} \to \mathbb{R}$ satisfying these properties implies $\beta \geq \beta(\text{possup})$. We consider two cases.

The case $\boldsymbol{\alpha = 0}$. Here the calculations are relatively simple. Take small $\delta > 0$ (in fact, $\delta \in (0, 1)$ is enough) and use (7.225) with $x = 1$, $y \in \mathbb{R}$ and $d = \delta$ to obtain

$$B(y) = u(1, y) \geq u(1 - \delta, y).$$

Next apply (7.224) to get

$$u(1 - \delta, y) \geq \delta u(0, y + 1 - \delta) + (1 - \delta)u(1, y - \delta) = \delta u(0, y + 1 - \delta) + (1 - \delta)B(y - \delta).$$

Combine the two estimates above with the following consequence of (7.223):

$$u(0, y + 1 - \delta) \geq (y + 1 - \delta) - \beta.$$

We obtain

$$B(y) \geq \delta(y + 1 - \delta - \beta) + (1 - \delta)B(y - \delta),$$

which can be rewritten

$$B(y) - (y - \beta) \geq (1 - \delta)\big[B(y - \delta) - (y - \delta - \beta)\big]. \qquad (7.228)$$

Write this estimate twice, with $y = \delta$ and $y = 0$:

$$B(\delta) - (\delta - \beta) \geq (1 - \delta)(B(0) + \beta),$$
$$B(0) + \beta \geq (1 - \delta)(B(-\delta) - (-\delta - \beta)).$$

But, by (7.227), B is an even function, so $B(-\delta) = B(\delta)$. Thus, combining the above two estimates yields

$$(B(0) + \beta)(2\delta - \delta^2) \geq 2\delta(1 - \delta).$$

Dividing throughout by δ and letting $\delta \to 0$ gives

$$B(0) + \beta \geq 1. \qquad (7.229)$$

Now we come back to (7.228). By induction, we get, for any integer N,

$$B(y) - (y - \beta) \geq (1 - \delta)^N \big[B(y - N\delta) - (y - N\delta - \beta)\big].$$

Let $y = 1$ and $\delta = 1/N$. If we pass with N to infinity and use (7.229), we get

$$B(1) - (1 - \beta) \geq e^{-1}(B(0) + \beta) \geq e^{-1}.$$

It suffices to apply (7.226) to get $\beta \geq 1 + e^{-1}$, as claimed.

The case $\alpha > 0$. See the sharpness of (7.31) and repeat the argumentation, word by word. In particular, compare (7.53)–(7.56) to (7.223)–(7.226). □

On the search of the suitable majorant. This is similar to the search in the martingale case. We omit the details. □

7.12 A maximal bound for continuous-path submartingales

7.12.1 Formulation of the result

Theorem 7.23. *Let α be a fixed nonnegative number. Assume that X is a nonnegative continuous-path submartingale and Y is α-subordinate to X. Then*

$$\|Y\|_1 \leq \beta \|X^*\|_1, \tag{7.230}$$

where $\beta = \beta(\mathrm{possub}) = \alpha + \left(\frac{2\alpha+2}{2\alpha+1}\right)^{1/2}$. The constant is the best possible, even if we restrict ourselves to Itô processes of the form (7.186).

7.12.2 Proof of Theorem 7.23

Proof of (7.230). First we introduce auxiliary parameters

$$\gamma = \left(\frac{2(\alpha+1)}{2\alpha+1}\right)^{1/2}, \qquad \overline{\gamma} = \gamma - \frac{1}{2\alpha+1}, \qquad \lambda = \frac{\gamma}{2}\exp\left(-1 + \frac{2}{\gamma}\right)$$

and let

$$\beta = \beta(\mathrm{possub}) = \alpha + \gamma.$$

Consider the subsets D_1, D_2, D_3 of $[0,1] \times \mathbb{R}$, defined by

$$D_1 = \left\{(x,y) : x \leq \frac{\alpha}{2\alpha+1}, \; x + |y| \geq \overline{\gamma}\right\},$$

$$D_2 = \left\{(x,y) : x > \frac{\alpha}{2\alpha+1}, \; -x + |y| \geq \gamma - 1\right\},$$

$$D_3 = ([0,1] \times \mathbb{R}) \setminus (D_1 \cup D_2).$$

As previously, first we introduce an auxiliary function $u : [0,1] \times \mathbb{R} \to \mathbb{R}$. This time it is defined as follows. On the set D_1, put

$$u(x,y) = -\alpha x + |y| + \alpha + \lambda \exp\left[-\frac{2\alpha+1}{\alpha+1}\left(x + |y| - \frac{\alpha}{2\alpha+1}\right)\right]\left(x + \frac{1}{2\alpha+1}\right) - \beta.$$

If $(x,y) \in D_2$, then set

$$u(x,y) = -\alpha x + |y| + \alpha + \lambda \exp\left[-\frac{2\alpha+1}{\alpha+1}\left(-x + |y| + \frac{\alpha}{2\alpha+1}\right)\right](1 - x) - \beta.$$

Finally, on D_3, let

$$u(x,y) = \frac{|y|^2 - x^2 - 1}{2\gamma} - \left(\alpha - \frac{\alpha}{\gamma(2\alpha+1)}\right)x.$$

One easily verifies that u is of class C^1 on $(0,1)\times\mathbb{R}$. The special function $U : D \to \mathbb{R}$ corresponding to (7.230) is given by the formula

$$U(x,y,z) = (x \vee z)u\left(\frac{x}{x \vee z}, \frac{y}{x \vee z}\right).$$

Furthermore, let $V(x,y,z) = |y| - \beta(|x| \vee z)$.

Lemma 7.31. *The function U satisfies* (7.216), (7.217), (7.218), (7.219),

$$U(x,y,z) \geq U(x+d, y \pm \alpha d, z) \qquad \text{if } 0 \leq x < x+d \leq z, \qquad (7.231)$$

and (7.221).

Proof. This is straightforward. We omit the tedious calculations. $\square$

The only problem is that U is not smooth enough and hence we need an additional convolution argument.

Lemma 7.32. *For any $\kappa > 0$ there is $\overline{U} = \overline{U}^\kappa$, which is of class C^∞ and satisfies* (7.216), (7.217), (7.218), (7.219), (7.231) *and*

$$\overline{U}(x,y,z) \leq \kappa \qquad \text{if } |y| \leq x \leq z. \qquad (7.232)$$

Proof. Let $g : \mathbb{R}^3 \to [0,\infty)$ be a C^∞ function, supported on the ball of center $(0,0,0)$ and radius 1, satisfying $\int_{\mathbb{R}^3} g = 1$ and such that

$$g(r,s,t) = g(r,-s,t) = g(r,s,-t) \qquad \text{for all } r, s \text{ and } t. \qquad (7.233)$$

Let $\delta = \kappa/(5\beta) > 0$ and let $\overline{U} = \overline{U}^\kappa : D \to \mathbb{R}$ be given by

$$\overline{U}(x,y,z) = \int_{[-1,1]^3} U(x + 2\delta - r\delta, y - s\delta, z + 5\delta - t\delta)g(r,s,t)dr\,ds\,dt + \kappa.$$

We will show that this function has the desired properties. Clearly, $\overline{U}$ if of class C^∞. To prove that $\overline{U}$ satisfies (7.216), we use the fact that U enjoys this majorization and hence, by the symmetry condition (7.233),

$$\overline{U}(x,y,z) \geq y - \beta(z+5\delta) + \delta \int_{[-1,1]^3} (-s+\beta t)g(r,s,t)dr\,ds\,dt + 5\beta\delta = y - \beta z.$$

The inequality $\overline{U}(x,y,z) \geq -y - \beta z$ is established in the same manner. The remaining properties (7.217), (7.218), (7.219) and (7.231) follow immediately from the definition of $\overline{U}$. Finally, note that if $|y| \leq x \leq z$, then

$$|y - s\delta| \leq x + 2\delta - r\delta \leq z + 5\delta - t\delta,$$

for any $r, s, t \in [-1, 1]$. Thus, for such x, y, z,

$$\overline{U}(x,y,z) \leq \kappa,$$

and the proof is complete. $\qquad\square$

Now let us translate the properties listed in the above lemma into their differential counterparts.

Lemma 7.33. *Let $\overline{U} = \overline{U}^\kappa$ be as in the previous lemma. We have the following.*

(i) *For any $x > 0$ and $y \in \mathbb{R}$,*

$$\overline{U}_z(x,y,x) \leq 0. \tag{7.234}$$

(ii) *For any $(x,y,z) \in D$, $0 < x \leq z$, we have*

$$\overline{U}_x(x,y,z) + \alpha|\overline{U}_y(x,y,z)| \leq 0. \tag{7.235}$$

(iii) *For all $(x,y,z) \in D$ with $0 < x \leq z$ there is $c = c(x,y,z) \geq 0$ such that if h, $k \in \mathbb{R}$, then*

$$\overline{U}_{xx}(x,y,z)h^2 + 2\overline{U}_{xy}(x,y,z)hk + \overline{U}_{yy}(x,y,z)k^2 \leq c(k^2 - h^2). \tag{7.236}$$

Proof. The property (i) follows from (7.217), while (ii) is an immediate consequence of (7.231). To show (iii), note that by (7.218),

$$\overline{U}_{yy}(x,y,z) \geq 0 \qquad \text{if } 0 < x \leq z, y \in \mathbb{R}. \tag{7.237}$$

By (7.219), for any fixed z the function $\overline{U}(\cdot, \cdot, z)$ is concave along any line segment of slope ± 1, contained in $[0, z] \times \mathbb{R}$. Therefore,

$$\overline{U}_{xx}(x,y,z) \pm 2\overline{U}_{xy}(x,y,z) + \overline{U}_{yy}(x,y,z) \leq 0 \qquad \text{for } 0 < x \leq z, y \in \mathbb{R}.$$

In particular this implies that $\overline{U}_{xx}(x,y,z) + \overline{U}_{yy}(x,y,z) \leq 0$ and hence, by (7.237),

$$\overline{U}_{xx}(x,y,z) \leq 0 \qquad \text{if } 0 < x \leq z, y \in \mathbb{R}. \tag{7.238}$$

Consequently, if h, $k \in \mathbb{R}$, then

$$\overline{U}_{xx}(x, y, z)h^2 + 2\overline{U}_{xy}(x, y, z)hk + \overline{U}_{yy}(x, y)k^2$$

$$\leq \overline{U}_{xx}(x, y, z)h^2 - (\overline{U}_{xx}(x, y, z) + \overline{U}_{yy}(x, y, z))\frac{h^2 + k^2}{2} + \overline{U}_{yy}(x, y)k^2$$

$$= \frac{\overline{U}_{yy}(x, y, z) - \overline{U}_{xx}(x, y, z)}{2}(k^2 - h^2).$$

Hence, by (7.237) and (7.238), (iii) holds and we are done. $\qquad\square$

In other words, the function $\overline{U}^\kappa$ satisfies the conditions $1°$, $2°$ and $4°$, while the condition $3°$ is replaced by (7.232). Consequently, Burkholder's method implies that for any bounded continuous-path X, Y such that X is a nonnegative submartingale and Y is α-subordinate to X we have

$$\mathbb{E}V_\beta(X_t, Y_t, X_t^* \vee \varepsilon) \leq \mathbb{E}\overline{U}(X_t, Y_t, X_t^* \vee \varepsilon) \leq \mathbb{E}\overline{U}(X_0, Y_0, X_0^* \vee \varepsilon) \leq \kappa.$$

It suffices to let $\kappa \to 0$ and use a localizing argument to get the claim. $\qquad\square$

Sharpness. As in the supermartingale case, one has to consider two possibilities: $\alpha = 0$ and $\alpha > 0$. We will focus on the second case, and leave the details of the first to the reader. So, fix positive α, β_0 and assume that

$$||Y||_1 \leq \beta_0||X^*||_1$$

for any nonnegative continuous submartingale X and any semimartingale Y which is α-subordinate to X. Let U_0 be given by

$$U_0(x, y, z) = \sup_{M_\alpha(x,y)} \{\mathbb{E}V_\beta(X_\infty, Y_\infty, X^*)\}$$

and set $u(x, y) = U(x, y, 1)$. Furthermore, let

$$B(y) = u(0, y) \quad \text{and} \quad C(y) = u\left(\frac{\alpha}{2\alpha + 1}, y - \frac{\alpha}{2\alpha + 1}\right) \quad \text{for } y \geq \alpha/(2\alpha + 1).$$

The function u satisfies (7.53), (7.54) and the following analogue of (7.55):

$$u(x, y) \geq u(x + d, y + \alpha d) \qquad \text{for } 0 \leq x \leq x + d \leq 1 \text{ and } y \in \mathbb{R}.$$

We split the proof into a few parts.

Step 1. Observe that the "reflected" function $\overline{u} : [0, 1] \times \mathbb{R} \to \mathbb{R}$ given by $\overline{u}(x, y) = u(1 - x, y)$ satisfies the conditions (7.53)–(7.55). Therefore, some of the calculations from Subsection 7.3.3 can be repeated. In particular, (7.62) yields

$$\overline{C}(y) - (\alpha + 1)\overline{B}(y) + \alpha y + \frac{2\alpha^3}{2\alpha + 1} + \alpha - \beta_0\alpha \leq 0,$$

where $\overline{B}$ and $\overline{C}$ are the corresponding restrictions of $\overline{u}$. Coming back to u, B, C just defined above, the latter inequality becomes

$$C(y) - (\alpha + 1)B(y) + \alpha y + \frac{2\alpha^3}{2\alpha + 1} + \alpha - \beta_0 \alpha \leq 0. \qquad (7.239)$$

Step 2. Here we will use a new argument. Applying (7.222) and the homogeneity of U_0, we get

$$u(1, 0) = U_0(1, 0, 1)$$

$$\geq \frac{2\delta}{\gamma + 2\delta} u\left(1 - \frac{\gamma}{2}, \frac{\gamma}{2}\right) - \frac{2\beta_0 \delta^2}{\gamma + 2\delta} + \frac{\gamma}{\gamma + 2\delta} U_0(1 + \delta, -\delta, 1 + \delta) \qquad (7.240)$$

$$= \frac{2\delta}{\gamma + 2\delta} u\left(1 - \frac{\gamma}{2}, \frac{\gamma}{2}\right) + \frac{\gamma(1 + \delta)}{\gamma + 2\delta} u\left(1, \frac{\delta}{1 + \delta}\right) - \frac{2\beta_0 \delta^2}{\gamma + 2\delta}.$$

Moreover, by the diagonal concavity,

$$u\left(1, \frac{\delta}{1 + \delta}\right) \geq \frac{2\delta}{\gamma(1 + \delta) + \delta} u\left(1 - \frac{\gamma}{2} + \frac{\delta}{2(1 + \delta)}, \frac{\gamma}{2} + \frac{\delta}{2(1 + \delta)}\right)$$

$$+ \frac{\gamma(1 + \delta) - \delta}{\gamma(1 + \delta) + \delta} \frac{1 + 2\delta}{1 + \delta} u(1, 0),$$

and

$$u\left(1 - \frac{\gamma}{2}, \frac{\gamma}{2}\right) \geq \frac{\gamma(2\alpha + 1)}{2(\alpha + 1)} C(\overline{\gamma}) + \frac{2(\alpha + 1) - \gamma(2\alpha + 1)}{2(\alpha + 1)} u(1, \gamma)$$

$$\geq \frac{\gamma(2\alpha + 1)}{2(\alpha + 1)} C(\overline{\gamma}) + \frac{2(\alpha + 1) - \gamma(2\alpha + 1)}{2(\alpha + 1)} (\gamma - \beta_0), \qquad (7.241)$$

where the first passage above was allowed due to $\gamma < 2(\alpha + 1)/(2\alpha + 1)$ and the second follows from the majorization. Plug these two estimates into (7.240) and combine the result with the following consequence of the diagonal concavity:

$$u(1 - \frac{\gamma}{2} + \frac{\delta}{2(1 + \delta)}, \frac{\gamma}{2} + \frac{\delta}{2(1 + \delta)}) \geq \frac{(\gamma(1 + \delta) + \delta)(2\alpha + 1)}{2(\alpha + 1)(1 + \delta)} C(\overline{\gamma})$$

$$+ \left(1 - \frac{(\gamma(1 + \delta) + \delta)(2\alpha + 1)}{2(\alpha + 1)(1 + \delta)}\right) (\gamma - \beta_0).$$

What we get is a rather complicated estimate of the form

$$u(1, 0) \geq a_1 C(\overline{\gamma}) + a_2(\gamma - \beta_0) + a_3 u(1, 0) - \frac{2\beta_0 \delta^2}{\gamma + 2\delta}, \qquad (7.242)$$

where the coefficients a_1, a_2 and a_3 depend on α and δ. We will not derive the explicit formulas for these; we will only need their asymptotic behavior as $\delta \to 0$:

$$a_1 = \frac{2\alpha + 1}{\alpha + 1} \delta + o(\delta), \quad a_2 = \frac{2(\alpha + 1) - \gamma(2\alpha + 1)}{\gamma(\alpha + 1)} \delta + o(\delta)$$

and

$$a_3 = 1 + \frac{\gamma - 2}{\gamma}\delta + o(\delta).$$

These equations can be easily derived from the above estimates. Now, subtracting $u(1,0)$ from both sides of (7.242), dividing throughout by δ and letting $\delta \to 0$ gives

$$0 \geq \frac{2\alpha + 1}{\alpha + 1}C(\overline{\gamma}) + \frac{2(\alpha + 1) - \gamma(2\alpha + 1)}{\gamma(\alpha + 1)}(\gamma - \beta_0) + \frac{\gamma - 2}{\gamma}u(1,0). \qquad (7.243)$$

Step 3. We use the diagonal concavity of u to obtain

$$C(\overline{\gamma}) \geq \frac{\gamma(2\alpha + 1)}{\gamma(2\alpha + 1) + 2\alpha}B(\overline{\gamma}) + \frac{2\alpha}{\gamma(2\alpha + 1) + 2\alpha}u\left(\frac{\gamma}{2} + \frac{\alpha}{2\alpha + 1}, \frac{\gamma}{2} - \frac{\alpha + 1}{2\alpha + 1}\right)$$

and

$$u\left(\frac{\gamma}{2} + \frac{\alpha}{2\alpha + 1}, \frac{\gamma}{2} - \frac{\alpha + 1}{2\alpha + 1}\right) \geq \frac{2(\alpha + 1) - \gamma(2\alpha + 1)}{\gamma(2\alpha + 1)}u\left(1 - \frac{\gamma}{2}, \frac{\gamma}{2}\right)$$
$$+ \frac{2\gamma(2\alpha + 1) - 2(\alpha + 1)}{\gamma(2\alpha + 1)}u(1,0).$$

Combining these two estimates and applying the lower bounds for $u(1 - \frac{\gamma}{2}, \frac{\gamma}{2})$ and $u(1,0)$ coming from (7.241) and (7.243), we obtain, after tedious but straightforward computations,

$$\frac{(\alpha + 1)(2\alpha + 1)(2 - \gamma)}{\alpha(2(\alpha + 1) - \gamma(2\alpha + 1))}(C(\overline{\gamma}) - (\alpha + 1)B(\overline{\gamma})) \geq (\gamma\alpha + \alpha + 1)(\gamma - \beta_0).$$

However, $\gamma < 2(\alpha + 1)/(2\alpha + 1) < 2$ and, by (7.239),

$$C(\overline{\gamma}) - (\alpha + 1)B(\overline{\gamma}) \leq -\alpha\overline{\gamma} - \frac{2\alpha^3}{2\alpha + 1} - \alpha + \beta_0\alpha.$$

Therefore, the preceding inequality yields

$$\beta_0 \geq \gamma + \frac{2\alpha(2 - \gamma)(\alpha + 1)^2}{(\alpha + 1)(2 - \gamma)(2\alpha + 1) + (\gamma\alpha + \alpha + 1)(2(\alpha + 1) - \gamma(2\alpha + 1))},$$

or, after some calculation, $\beta_0 \geq \gamma + \alpha$. This completes the proof. □

On the search of the suitable majorant. This is similar to the search in the martingale case. We omit the details. □

7.13   Notes and comments

Section 7.1. The method appeared first in Burkholder's paper [35]: it was shown there how to establish maximal inequalities for martingales and their transforms (however, [32] also contains some results in this direction). The version of the technique for supermartingales was described in the author's paper [131] and the modification for one-sided maximal functions can be found in [138]. See also [156], where the method was applied under strong differential subordination.

    Section 7.2. Doob's weak type and moment inequalities for submartingale maximal functions (for $1 \leq p \leq \infty$ and $1 < p \leq \infty$, respectively) can be found in [70]. The above approach is taken from [32]. The moment inequality (7.21) for nonnegative supermartingales was proved by Shao [184]; the extension presented above is a new result. The version of Doob's inequality for different orders of moments was studied for $p = 1 < q$ by Jacka [98], using probabilistic methods, and Gilat [84], using analytic methods. The general statement is due to Peskir [169], who proved the estimate using the theory of optimal stopping.

    Section 7.3. Theorem 7.9 was established by Burkholder [35] in the case when Y is a stochastic integral with respect to X of some predictable process taking values in $[-1, 1]$. The above statement for differentially subordinate martingales is new. So is the proof of the optimality of the constant $\beta = 2.536\ldots$: Burkholder showed this by constructing an appropriate example. Theorem 7.10 was partially studied by the author in [131], in the stochastic integral setting. The version for supermartingales and their α-subordinates presented above is new.

    Section 7.4. The material of that section is taken from [156].

    Section 7.5. The validity of the double maximal inequality with some absolute constant is trivial and follows, for example, from Davis' maximal inequality for the square function [57]. The optimal value of the constant was determined by the author in [162].

    Section 7.6. The results presented in that section come from [152].

    Section 7.7. The L^1-inequalities for one-sided maximal function (with some absolute constants) follow immediately from the results of the previous two sections. The optimal values were found in [138]: the approach presented above is taken from that paper.

    Section 7.8. The results from that section are taken from [147].

    Sections 7.9–7.12. The modification of Burkholder's method for continuous-path martingales was first described by the author in [153]. That paper contains also the proof of Theorem 7.21. The proofs of Theorem 7.22 and Theorem 7.23 are taken from [164].

Chapter 8

Square Function Inequalities

As we have mentioned in Chapter 2, the inequalities for differentially subordinated martingales imply related estimates for the square function. However, in general, the inequalities we obtain are not sharp. In this chapter we shall present an approach which can be used to derive the optimal constants and review the results in this direction. Recall that if f is a martingale taking values in a Banach space $\mathcal{B}$ with norm $|\cdot|$, then its square function is given by

$$
S(f) = \left[\sum_{k=0}^{\infty} |df_k|^2 \right]^{1/2}.
$$

We will also use the notation

$$
S_n(f) = \left[\sum_{k=0}^{n} |df_k|^2 \right]^{1/2},
$$

for the truncated square function, $n = 0, 1, 2, \ldots$.

8.1 Modification of Burkholder's method

First we introduce a version of Burkholder's technique which allows to study the estimates involving a martingale, its maximal and square function simultaneously. Let $\mathcal{B}$ be a separable Banach space, set $D = \mathcal{B} \times [0, \infty) \times [0, \infty)$ and fix $V : D \to \mathbb{R}$, which satisfies the condition

$$
V(x, y, z) = V(x, y, |x| \vee z) \qquad \text{for all } (x, y, z) \in D.
$$

We are interested in showing that

$$
\mathbb{E} V(f_n, S_n(f), |f_n|^*) \leq 0, \qquad n = 0, 1, 2, \ldots \tag{8.1}
$$

for any simple discrete-time martingale f taking values in $\mathcal{B}$. Suppose that $U : D \to \mathbb{R}$ enjoys the following four properties:

$1°$ $U(x, y, z) \geq V(x, y, z)$ for any $(x, y, z) \in D$,

$2°$ $\mathbb{E}U(x + d, \sqrt{y^2 + |d|^2}, z) \leq U(x, y, z)$ for any $(x, y, z) \in D$ such that $|x| \leq z$ and any simple centered random variable d taking values in $\mathcal{B}$,

$3°$ $U(x, |x|, |x|) \leq 0$ for all $x \in \mathcal{B}$,

$4°$ $U(x, y, z) = U(x, y, |x| \vee z)$ for all $(x, y, z) \in D$.

Theorem 8.1. *Let V be as above. Then there is a function U satisfying $1°$–$4°$ if and only if (8.1) is valid.*

Proof. The underlying concept is the same. If U satisfies the conditions $2°$ and $4°$, then the process $(U(f_n, S_n(f), |f_n|^*))_{n \geq 0}$ is a supermartingale: indeed, for any $n \geq 1$,

$$\mathbb{E}\big[U(f_n, S_n(f), |f_n|^*)|\mathcal{F}_{n-1}\big]$$
$$= \mathbb{E}\big[U(f_{n-1} + df_n, \sqrt{S_{n-1}^2(f) + |df_n|^2}, |f_{n-1}|^*)|\mathcal{F}_{n-1}\big]$$
$$\leq U(f_{n-1}, S_{n-1}(f), |f_{n-1}|^*).$$

Thus, by $1°$ and $3°$,

$$\mathbb{E}V(f_n, S_n(f), |f_n|^*) \leq \mathbb{E}U(f_n, S_n(f), |f_n|^*) \leq \mathbb{E}U(f_0, S_0(f), |f_0|^*) \leq 0.$$

We turn to the reverse implication. The desired special function is given by

$$U^0(x, y, z) = \sup\{\mathbb{E}V(f_n, \sqrt{y^2 - |x|^2 + S_n^2(f)}, |f_n| \vee z)\},$$

where the supremum is taken over all n and all simple martingales f starting from x which are given on the probability space $([0, 1], \mathcal{B}([0, 1]), |\cdot|)$. Here the filtration may vary. To see that U^0 majorizes V, it suffices to observe that for $n = 0$ the expression in the parentheses equals $V(x, y, |x| \vee z) = V(x, y, z)$. To show $2°$, we make use of the "splicing argument". Let d be as in the statement and let $K = \{x_1, x_2, \ldots, x_k\}$ be the set of its values: $\mathbb{P}(d = x_j) = p_j > 0$ and $\sum_{j=1}^k p_j = 1$. For any $\varepsilon > 0$ and $1 \leq j \leq k$, let f^j be a simple sequence starting from $x + x_j$, such that

$$\mathbb{E}V(f_\infty^j, \sqrt{y^2 + |x_j|^2 - |x + x_j|^2 + S_\infty^2(f^j)}, |f^j|^* \vee z) \geq U(x + x_j, \sqrt{y^2 + |x_j|^2}, z) - \varepsilon.$$

Let $a_0 = 0$ and $a_j = \sum_{\ell=1}^j p_\ell$, $j = 1, 2, \ldots, k$. We define a simple sequence f by $f_0 \equiv x$ and

$$f_n(\omega) = f_{n-1}^j((\omega - a_{j-1})/(a_j - a_{j-1})), \qquad n \geq 1,$$

when $\omega \in (a_{j-1}, a_j]$. Then f is a simple martingale with respect to its natural filtration. Consequently,

$$U(x, y, z) \geq \mathbb{E}V(f_\infty, \sqrt{y^2 - |x|^2 + S_\infty^2(f)}, |f|^* \vee z).$$

Since $|x| \leq z$, we have $|f|^* \vee z = \sup_{n \geq 1} |f_n| \vee z$ and the right-hand side above equals

$$\sum_{j=1}^{k} \int_{a_{j-1}}^{a_j} V(f_\infty(\omega), \sqrt{y^2 - |x|^2 + S_\infty^2(f)(\omega)}, |f|^*(\omega) \vee z) d\omega$$

$$= \sum_{j=1}^{k} p_j \mathbb{E} V(f_\infty^j, \sqrt{y^2 + |x_j|^2 - |x + x_j|^2 + S_\infty^2(f^j)}, |f^j|^* \vee z)$$

$$\geq \sum_{j=1}^{k} p_j U(x + x_j, \sqrt{y^2 + |x_j|^2}, z) - k\varepsilon = \mathbb{E} U(x + d, \sqrt{y^2 + |d|^2}, z) - k\varepsilon.$$

Since ε was arbitrary, this gives $2°$. Condition $3°$ follows immediately from (8.1) and the definition of U^0. Finally, $4°$ is a direct consequence of the formula for U^0. This completes the proof. □

Remark 8.1. Obviously, the above theorem can be used when the estimate involves only f and $S(f)$, simply by omitting the variable z. Then the condition $4°$ is "empty" and need not be taken into consideration.

While studying square function inequalities, it is often of interest to determine optimal constants for conditionally symmetric martingales. Then the above approach works and the only modification is that in $2°$ one has to consider symmetric random variables d. By easy induction on the number of values of d, this condition can be simplified to the case when these random variables take only two values. That is, it can be rewritten in the equivalent form

$2°$ for any $(x, y, z) \in D$ such that $|x| \leq z$ and any $d \in \mathcal{B}$,

$$\frac{1}{2} \left[U(x + d, \sqrt{y^2 + |d|^2}, z) + U(x - d, \sqrt{y^2 + |d|^2}, z) \right] \leq U(x, y, z).$$

This observation shows the following general fact: the optimal constants for conditionally symmetric martingales are equal to those coming from the dyadic case. However it should be stressed that typically they do differ from those of the general case. We shall see this below.

Remark 8.2. The method described above deals only with discrete-time martingales. However, in general the passage to the continuous-time analogues requires no additional effort: in most cases it is just a matter of standard approximation; see, e.g., Lemma 2.2 in Burkholder's paper [38]. Thus, in what follows, we shall concentrate on martingales indexed by $\{0, 1, 2, \ldots\}$.

8.2 Weak type estimates

8.2.1 Formulation of the result

As usual, we begin our study by reviewing the facts concerning weak type estimates. We have the following statement for general martingales.

Theorem 8.2. *If f is an $\mathcal{H}$-valued martingale, then*

$$||S(f)||_{1,\infty} \leq \sqrt{e}||f||_1. \tag{8.2}$$

The constant is the best possible, even if $\mathcal{H} = \mathbb{R}$.

It turns out that the estimate above characterizes Hilbert spaces.

Theorem 8.3. *If $\mathcal{B}$ is not a Hilbert space, then there is a $\mathcal{B}$-valued martingale f such that*

$$\mathbb{P}(S(f) \geq 1) > \sqrt{e}||f||_1.$$

The best constants in the corresponding weak type (p,p) inequalities, with $p > 1$ and $p \neq 2$, are not known.

In the case when f is assumed to be conditionally symmetric, we shall prove the following two results.

Theorem 8.4. *If f is a conditionally symmetric $\mathcal{H}$-valued martingale, then*

$$||S(f)||_{1,\infty} \leq K||f||_1, \tag{8.3}$$

where

$$K = \exp\left(-\frac{1}{2}\right) + \int_0^1 \exp\left(-\frac{t^2}{2}\right) dt \approx 1,4622.$$

The constant K is the best possible.

The second result is the following weak type estimate in the reverse direction.

Theorem 8.5. *For any $\mathcal{H}$-valued conditionally symmetric martingale f we have*

$$||f||_{p,\infty} \leq C_p||S(f)||_p, \qquad 1 \leq p \leq 2, \tag{8.4}$$

where

$$C_p = \frac{2^{1-p/2}\pi^{p-3/2}\Gamma((p+1)/2)}{\Gamma(p+1)} \cdot \frac{1 + \frac{1}{3^2} + \frac{1}{5^2} + \frac{1}{7^2} + \cdots}{1 - \frac{1}{3^{p+1}} + \frac{1}{5^{p+1}} - \frac{1}{7^{p+1}} + \cdots}.$$

The constant C_p is the best possible.

For the remaining values of the parameter p, the question about the best constants in the corresponding estimates is open.

8.2.2 Proof of Theorem 8.2

Proof of (8.2). By standard approximation, it suffices to show that for any simple martingale f we have

$$\mathbb{P}(S_n(f) \geq 1) \leq \sqrt{e}\,\mathbb{E}|f_n|, \qquad n = 0, 1, 2, \dots.$$

This estimate has the form for which Burkholder's method is directly applicable. Let $V(x, y) = 1_{\{y \geq 1\}} - \sqrt{e}|x|$ for $x \in \mathcal{H}$ and $y \in [0, \infty)$. The corresponding special function $U : \mathcal{H} \times [0, \infty) \to \mathbb{R}$ is given by

$$U(x, y) = \begin{cases} 1 - (1 - y^2)^{1/2} \exp\left[\frac{|x|^2}{2(1-y^2)}\right] & \text{if } |x|^2 + y^2 < 1, \\ 1 - \sqrt{e}|x| & \text{if } |x|^2 + y^2 \geq 1. \end{cases} \tag{8.5}$$

We shall verify the properties 1°, 2° and 3°. To show the majorization, we may assume that $|x|^2 + y^2 < 1$; then the inequality takes the form

$$\exp\left[\frac{|x|^2}{2(1-y^2)}\right] \leq \sqrt{e}\frac{|x|}{\sqrt{1-y^2}} + \frac{1}{\sqrt{1-y^2}}$$

and follows immediately from the elementary bound $\exp(s^2/2) \leq \sqrt{e}\,s + 1$, $s \in [0, 1]$, applied to $s = |x|/\sqrt{1-y^2}$. To check 2°, let

$$A(x, y) = \begin{cases} -x(1 - y^2)^{-1/2} \exp\left[\frac{|x|^2}{2(1-y^2)}\right] & \text{if } |x|^2 + y^2 < 1, \\ -\sqrt{e}\,x' & \text{if } |x|^2 + y^2 \geq 1. \end{cases}$$

We shall establish the pointwise inequality

$$U(x + d, \sqrt{y^2 + d^2}) \leq U(x, y) + A(x, y) \cdot d \tag{8.6}$$

for all $x, d \in \mathcal{H}$ and $y \geq 0$. This will clearly yield 2°, by taking expectation. We start by showing that $U(x, y) \leq 1 - \sqrt{e}|x|$ for all $x \in \mathcal{H}$ and $y \geq 0$. This is trivial for $|x|^2 + y^2 \geq 1$, while for remaining x, y can be transformed into an equivalent inequality

$$\frac{|x|^2}{1 - y^2} \leq \exp\left(\frac{|x|^2}{1 - y^2} - 1\right),$$

which is obvious. Consequently, when $|x|^2 + y^2 \geq 1$, we have

$$U(x+d, \sqrt{y^2 + d^2}) \leq 1 - \sqrt{e}|x+d| \leq 1 - \sqrt{e}|x| + A(x, y) \cdot d = U(x, y) + A(x, y) \cdot d.$$

Now suppose that $|x|^2 + y^2 < 1$. Observe that for $X, D \in \mathcal{H}$ with $|D| < 1$ we have

$$\exp\left[\frac{|D|^2|X|^2 + 2X \cdot D + |D|^2}{1 - |D|^2}\right] \geq \exp\left[\frac{(X \cdot D)^2 + 2X \cdot D + |D|^2}{1 - |D|^2}\right]$$
$$\geq \frac{(X \cdot D)^2 + 2X \cdot D + |D|^2}{1 - |D|^2} + 1$$
$$= \frac{(1 + X \cdot D)^2}{1 - |D|^2}.$$

Plugging $X = x/\sqrt{1-y^2}$ and $D = d/\sqrt{1-y^2}$ we get (8.6) in the case when $|x + d|^2 + y^2 + d^2 \leq 1$. Finally, if $|x|^2 + y^2 < 1 < |x + d|^2 + y^2 + d^2$, then substituting X and D as previously, we have $|X| < 1$, $|X + D|^2 + |D|^2 > 1$ and (8.6) can be written in the form

$$\exp\left(\frac{|X|^2 - 1}{2}\right)(1 + X \cdot D) \leq |X + D|$$

or

$$\exp\left(\frac{|X|^2 - 1}{2}\right)\left(1 + \frac{|X + D|^2 - |X|^2 - |D|^2}{2}\right) \leq |X + D|.$$

Now we fix $|X|$, $|X + D|$ and maximize the left-hand side over D. Consider two cases. If $|X + D|^2 + (|X + D| - |X|)^2 < 1$, then there is $D' \in \mathcal{H}$ satisfying $|X + D| = |X + D'|$ and $|X + D'|^2 + |D'|^2 = 1$. Consequently,

$$\exp\left(\frac{|X|^2 - 1}{2}\right)\left(1 + \frac{|X + D|^2 - |X|^2 - |D|^2}{2}\right)$$

$$\leq \exp\left(\frac{|X|^2 - 1}{2}\right)\left(1 + \frac{|X + D'|^2 - |X|^2 - |D'|^2}{2}\right) \leq |X + D'| = |X + D|.$$

Here the first inequality follows from the fact that $|D'| < |D|$, while to obtain the second we have applied (8.6) to $x = X$, $y = 0$ and $d = D'$ (for these x, y and d we have already established the bound). Suppose then, that $|X + D|^2 + (|X + D| - |X|)^2 \geq 1$. This inequality is equivalent to

$$|X + D| \geq \frac{1 - |X|^2}{\sqrt{2 - |X|^2} - |X|}$$

and hence

$$\exp\left(\frac{|X|^2 - 1}{2}\right)\left(1 + \frac{|X + D|^2 - |X|^2 - |D|^2}{2}\right) - |X + D|$$

$$\leq \exp\left(\frac{|X|^2 - 1}{2}\right)\left(1 + \frac{|X + D|^2 - |X|^2 - (|X + D| - |X|)^2}{2}\right) - |X + D|$$

$$= \exp\left(\frac{|X|^2 - 1}{2}\right)(1 - |X|^2) + \left[\exp\left(\frac{|X|^2 - 1}{2}\right)|X| - 1\right]|X + D|$$

$$\leq \frac{1 - |X|^2}{\sqrt{2 - |X|^2} - |X|}\left[\exp\left(\frac{|X|^2 - 1}{2}\right)\sqrt{2 - |X|^2} - 1\right].$$

It suffices to observe that the last expression in the square brackets is nonpositive, which follows from the estimate $\exp(1 - |X|^2) \geq 2 - |X|^2$. This completes the proof of 2°. Finally, 3° is a consequence of (8.6): $U(x, |x|) \leq U(0, 0) + A(0, 0) \cdot x = 0$. This completes the proof of (8.2). $\qquad\square$

Sharpness. We shall show that the constant $\sqrt{e}$ is the best possible even if we restrict ourselves to the real-valued martingales f that satisfy $S(f) \geq 1$ almost surely. To do this, let us introduce the special function

$$U^0(x, y) = \inf\{\mathbb{E}|f_n|\},$$

where the infimum is taken over all n and all martingales from $M(x, y)$. This class consists of all simple martingales starting from x and satisfying

$$|y|^2 - x^2 + S^2(f) \geq 1 \qquad \text{almost surely.} \tag{8.7}$$

Obviously, U^0 is symmetric: $U^0(-x, y) = U^0(x, y)$. Let us show that $U^0(x, y) = |x|$ when $x^2 + |y|^2 \geq 1$. The inequality $U^0(x, y) \geq |x|$ is trivial. The reverse estimate is also immediate when $y \geq 1$ (we take a constant martingale), and for $y < 1$, we consider $f \in M(x, y)$ such that $df_1 = \pm\sqrt{1 - y^2}$ and $df_2 = df_3 = \cdots \equiv 0$.

The next move is the following. Reasoning as in Theorem 8.1, we show that U^0 satisfies the convexity condition

2° For any $x \in \mathbb{R}$, $y \geq 0$ and any centered simple random variable d we have

$$\mathbb{E}U^0(x + d, \sqrt{y^2 + d^2}) \geq U^0(x, y).$$

Furthermore, the restriction to f satisfying (8.7) has the advantage that the function U^0 has an additional homogeneity-type property. Namely, for all $x \in \mathbb{R}$ and $y \in [0, 1)$ we have

$$U^0(x, y) = \sqrt{1 - y^2}U^0\left(\frac{x}{\sqrt{1 - y^2}}, 0\right), \tag{8.8}$$

which follows immediately from the fact that $f \in M(x, y)$ if and only if

$$f/\sqrt{1 - y^2} \in M(x/\sqrt{1 - y^2}, 0).$$

Now apply 2° to some $x \in (0, 1)$, $y = 0$ and a centered random variable d which takes two values: $\delta > 0$ and $-2x/(1 + x^2)$. We get

$$U^0(x, 0) \leq \frac{\delta(1 + x^2)}{2x + \delta(1 + x^2)}U^0\left(-\frac{x(1 - x^2)}{1 + x^2}, \frac{2x}{1 + x^2}\right) + \frac{2x}{2x + \delta(1 + x^2)}U^0\left(x + \delta, \delta\right)$$

$$\leq \frac{\delta(1 - x^2)}{2x + \delta(1 + x^2)}U^0(x, 0) + \frac{2x\sqrt{1 - \delta^2}}{2x + \delta(1 + x^2)}U^0\left(x + \delta, 0\right),$$

by (8.8) and the inequality $U^0(x + \delta, \delta) \leq U^0(x + \delta, 0)$, which follows directly from the definition of U^0. This implies

$$\frac{U^0(x + \delta, 0)}{U^0(x, 0)} \geq \frac{1 + \delta x}{\sqrt{1 - \delta^2}} > 1 + \delta x. \tag{8.9}$$

Now fix a (large) positive integer N and apply this inequality to $x = n/N$ and $\delta = 1/N$, where $n = 1, 2, \ldots, N-1$. Multiplying the obtained estimates, we get

$$\frac{U^0(1,0)}{U^0(1/N,0)} > \prod_{n=1}^{N-1} \left(1 + \frac{n}{N^2}\right).$$

The right-hand side converges to $e^{1/2}$ as $N \to \infty$. Furthermore, we have $U^0(1,0) = 1$ (see the beginning) and, by $2°$,

$$U^0(0,0) \le (U^0(1/N,1/N) + U^0(-1/N,1/N))/2 = U^0(1/N,1/N) \le U^0(1/N,0).$$

This yields $U^0(0,0) \le e^{-1/2}$, which is precisely the claim.   □

On the search of the suitable majorant. In fact, much of the work in this direction has been already carried out in the proof of sharpness. For U^0 defined there, we know that $U^0(x,y) = |x|$ when $x^2 + y^2 \ge 1$; for the remaining x and y, (8.8) and (8.9) hold. Reasoning as above, the latter estimate yields $U^0(x,0) \le \exp(x^2 - 1)$, so, by the homogeneity property,

$$U^0(x,y) \ge \sqrt{1-y^2} \exp\left(\frac{x^2}{1-y^2} - 1\right).$$

It is natural to conjecture that we have equality above; then the formula for the special function follows immediately. Namely, we take

$$U(x,y) = 1 - CU^0(x,y),$$

which is precisely (8.5).

For an alternative approach using the method of moments, see Cox [53].   □

8.2.3   Proof of Theorem 8.3

Let $(\mathcal{B}, ||\cdot||)$ be a separable Banach space and let $\beta(\mathcal{B})$ be the least extended positive number β such that for any $\mathcal{B}$-valued martingale f we have

$$\mathbb{P}(S(f) \ge 1) \le \beta ||f||_1. \tag{8.10}$$

For $x \in \mathcal{B}$ and $y \ge 0$, let $M(x,y)$ denote the class of all simple $\mathcal{B}$-valued martingales f given on the probability space $([0,1], \mathcal{B}(0,1), |\cdot|)$, such that $f_0 \equiv x$ and

$$y^2 - ||x||^2 + S^2(f) \ge 1 \qquad \text{almost surely.}$$

Here the filtration may vary. The key object is the function $U^0 : \mathcal{B} \times [0,\infty) \to \mathbb{R}$, given by

$$U^0(x,y) = \inf\{\mathbb{E}||f_n||\},$$

where the infimum is taken over all n and all $f \in M(x,y)$. Repeating the arguments from Section 8.1, we show the following fact.

Lemma 8.1. *The function U^0 enjoys the following properties.*

$1°'$ *For any $x \in B$ and $y \geq 0$ we have $U^0(x,y) \geq ||x||$.*

$2°'$ *For any $x \in B$, $y \geq 0$ and any simple centered B-valued random variable T,*

$$\mathbb{E}U^0(x + T, \sqrt{y^2 + ||T||^2}) \geq U^0(x,y).$$

$3°'$ *For any $x \in B$ we have $U^0(x, ||x||) \geq \beta(B)^{-1}$.*

Further properties are described in the next lemma.

Lemma 8.2.

(i) *The function U^0 satisfies the symmetry condition*

$$U^0(x,y) = U^0(-x,y)$$

for all $x \in B$ and $y \geq 0$.

(ii) *The function U^0 has the homogeneity-type property*

$$U^0(x,y) = \sqrt{1 - y^2}U^0\left(\frac{x}{\sqrt{1-y^2}}, 0\right)$$

for all $x \in B$ and $y \in [0,1)$.

(iii) *If $z \in B$ satisfies $||z|| = 1$ and $0 \leq s < t \leq 1$, then*

$$U^0(sz, 0) \leq U^0(tz, 0) \exp((s^2 - t^2)||z||^2/2). \tag{8.11}$$

Proof. (i) It suffices to use the equivalence $f \in M(x,y)$ if and only if $-f \in M(-x,y)$.

(ii) This follows immediately from the fact that $f \in M(x,y)$ if and only if $f/\sqrt{1-y^2} \in M(x/\sqrt{1-y^2}, 0)$.

(iii) Fix $x \in B$ with $0 < ||x|| < 1$ and $\delta > 0$ such that $||x + \delta x|| \leq 1$. Apply $2°'$ to $y = 0$ and a centered random variable T which takes two values: δx and $-2x/(1 + ||x||^2)$. We get

$$U^0(x,0) \leq \frac{\delta||x||(1 + ||x||^2)}{2||x|| + \delta||x||(1 + ||x||^2)} U^0\left(-\frac{x(1 - ||x||^2)}{1 + ||x||^2}, \frac{2||x||}{1 + ||x||^2}\right)$$
$$+ \frac{2||x||}{2||x|| + \delta||x||(1 + ||x||^2)} U^0\left(x + \delta x, \delta||x||\right).$$

By (i) and (ii), the first term on the right equals

$$\frac{\delta||x|| |1 - ||x||^2|}{2||x|| + \delta||x||(1 + ||x||^2)} U^0(x, 0).$$

The second summand can be bounded from above by

$$\frac{2||x||}{2||x|| + \delta||x||(1 + ||x||^2)} U^0(x + \delta x, 0),$$

because $M(x + \delta x, 0) \subset M(x + \delta x, \delta||x||)$. Using these two facts into the inequality above and recalling that $||x|| \leq 1$ (so $|1 - ||x||^2| = 1 - ||x||^2$) we get

$$\frac{U^0(x + \delta x, 0)}{U^0(x, 0)} \geq 1 + \delta||x||^2.$$

This gives

$$\frac{U^0(x(1 + k\delta), 0)}{U^0(x(1 + (k-1)\delta), 0)} \geq 1 + \delta(1 + (k-1)\delta)||x||^2,$$

provided $||x(1 + k\delta)|| \leq 1$. Consequently, if N is an integer such that $||x(1 + N\delta)|| \leq 1$, then

$$\frac{U^0(x(1 + N\delta), 0)}{U^0(x, 0)} \geq \prod_{k=1}^{N}(1 + \delta(1 + (k-1)\delta)||x||^2). \tag{8.12}$$

Now we turn to (8.11). Assume first that $s > 0$. Put $x = sz$, $\delta = (t/s - 1)/N$ and let $N \to \infty$ in the inequality above to obtain

$$\frac{U^0(tz, 0)}{U^0(sz, 0)} \geq \exp\left(\frac{1}{2}||z||^2(t^2 - s^2)\right),$$

which is the claim. Next, suppose that $s = 0$. For any $0 < s' < t$ we have, by $2^{\circ\prime}$,

$$U^0(0, 0) \leq \frac{1}{2}U^0(s'z, ||s'z||) + \frac{1}{2}U^0(-s'z, ||s'z||) = U^0(s'z, ||s'z||) \leq U^0(s'z, 0),$$

where in the latter passage we have used the inclusion $M(s'z, 0) \subset M(s'z, ||s'z||)$. Thus,

$$\frac{U^0(tz, 0)}{U^0(0, 0)} \geq \frac{U^0(tz, 0)}{U^0(s'z, 0)} \geq \exp\left(\frac{1}{2}||z||^2(t^2 - (s')^2)\right)$$

and it remains to let $s' \to 0$. $\square$

Next, let us assume that $\beta(\mathcal{B}) = \sqrt{e}$. This allows us to derive the explicit formula for U^0.

Lemma 8.3. *If* $\beta(\mathcal{B}) = \sqrt{e}$, *then*

$$U^0(x, y) = \begin{cases} \sqrt{1 - y^2} \exp\left(\frac{||x||^2}{2(1-y^2)} - \frac{1}{2}\right) & \text{if } ||x||^2 + y^2 < 1, \\ ||x|| & \text{if } ||x||^2 + y^2 \geq 1. \end{cases}$$

Proof. First let us focus on the set $\{(x, y) : ||x||^2 + y^2 \geq 1\}$. By $1°$, we have $U^0(x, y) \leq ||x||$. To get the reverse estimate, consider a martingale f such that $f_0 \equiv x$, df_1 takes the values $-x$ and x, and $df_2 = df_3 = \cdots \equiv 0$. Then $y^2 - ||x||^2 + S^2(f) = y^2 + ||x||^2 \geq 1$ (so $f \in M(x, y)$) and $||f||_1 = ||x||$. Now suppose that $||x||^2 + y^2 < 1$. Using items (ii) and (iii) of Lemma 8.2, we have

$$U^0(x, y) = \sqrt{1 - y^2} U^0 \left(\frac{x}{\sqrt{1 - y^2}}, 0 \right) \geq U^0(0, 0) \sqrt{1 - y^2} \exp \left(\frac{||x||^2}{2(1 - y^2)} \right),$$

so, by $3°'$ in Lemma 8.1,

$$U^0(x, y) \geq \sqrt{1 - y^2} \exp \left(\frac{||x||^2}{2(1 - y^2)} - \frac{1}{2} \right).$$

To get the reverse bound, we use the homogeneity of U^0 and (8.11) again:

$$U^0(x, y) = \sqrt{1 - y^2} U^0 \left(\frac{x}{\sqrt{1 - y^2}}, 0 \right)$$
$$\leq \sqrt{1 - y^2} U^0 \left(\frac{x}{|x|}, 0 \right) \exp \left(\frac{1}{2} \left(\frac{||x||^2}{1 - y^2} - 1 \right) \right)$$
$$= \sqrt{1 - y^2} \exp \left(\frac{||x||^2}{2(1 - y^2)} - \frac{1}{2} \right),$$

where in the last line we have used the equality $U^0(\bar{x}, 0) = ||\bar{x}||$ valid for $\bar{x}$ of norm 1 (we have just established this in the first part of the proof). For completeness, let us mention here that if $x = 0$, then $x/|x|$ should be replaced above by any vector of norm one. $\square$

Lemma 8.4. *Suppose that $\beta(\mathcal{B}) = \sqrt{e}$ and assume that x, $y \in \mathcal{B}$ and $\alpha > 0$ satisfy*

$$||x|| < 1, \quad ||x + \alpha x + y||^2 + ||\alpha x + y||^2 < 1 \text{ and } ||x + \alpha x - y||^2 + ||\alpha x - y||^2 < 1.$$

Then

$$2 + 2\alpha ||x||^2 \leq \sqrt{1 - ||\alpha x + y||^2} \exp \left[\frac{||x + \alpha x + y||^2}{2(1 - ||\alpha x + y||^2)} - \frac{||x||^2}{2} \right]$$
$$+ \sqrt{1 - ||\alpha x - y||^2} \exp \left[\frac{||x + \alpha x - y||^2}{2(1 - ||\alpha x - y||^2)} - \frac{||x||^2}{2} \right]. \tag{8.13}$$

Proof. Consider a random variable T such that

$$\mathbb{P} \left(T = -\frac{2x}{1 + ||x||^2} \right) = p, \quad \mathbb{P}(T = \alpha x + y) = \mathbb{P}(T = \alpha x - y) = \frac{1 - p}{2},$$

where $p \in (0, 1)$ is chosen so that $\mathbb{E}T = 0$. That is,

$$p = \frac{\alpha(1 + ||x||^2)}{2 + \alpha(1 + ||x||^2)}.$$

By $2^{\circ\prime}$, we have $U^0(x,0) \leq \mathbb{E}U^0(x+T, ||T||)$. Since $||x+T||^2 + ||T||^2 < 1$ almost surely, the previous lemma implies that this can be rewritten in the equivalent form

$$
\exp\left(\frac{||x||^2}{2}\right) \leq p\sqrt{1 - \left(\frac{2||x||}{1+||x||^2}\right)^2} \exp\left(\frac{\left|\left|x\left(\frac{-1+||x||^2}{1+||x||^2}\right)\right|\right|^2}{2\left(1 - \left(\frac{2||x||}{1+||x||^2}\right)^2\right)}\right)
$$
$$
+ \frac{1-p}{2}\sqrt{1 - ||\alpha x + y||^2}\exp\left(\frac{||x + \alpha x + y||^2}{2(1 - ||\alpha x + y||^2)}\right)
$$
$$
+ \frac{1-p}{2}\sqrt{1 - ||\alpha x - y||^2}\exp\left(\frac{||x + \alpha x - y||^2}{2(1 - ||\alpha x - y||^2)}\right).
$$

However, the first term on the right equals

$$
\frac{\alpha(1 - ||x||^2)}{2 + \alpha(1 + ||x||^2)}\exp\left(\frac{||x||^2}{2}\right)
$$

and, in addition, $(1-p)/2 = (2 + \alpha(1 + ||x||^2))^{-1}$. Consequently, it suffices to multiply both sides of the inequality above by $(2 + \alpha(1 + ||x||^2))\exp\left(-||x||^2/2\right)$; the claim follows. $\square$

Now we are ready to complete the proof of the characterization. Suppose that a, b belong to the unit ball K of $\mathcal{B}$ and take $\varepsilon \in (0, 1/2)$. Applying (8.13) to $x = \varepsilon a$, $y = \varepsilon^2 b$ and $\alpha = \varepsilon$ gives

$$
\begin{aligned}
2 + 2\varepsilon^3||a||^2 \leq &\sqrt{1 - \varepsilon^4||a+b||^2}\exp(m(a,b)) \\
&+ \sqrt{1 - \varepsilon^4||a-b||^2}\exp(m(a,-b)),
\end{aligned}
\tag{8.14}
$$

where

$$
\begin{aligned}
m(a,b) &= \frac{\varepsilon^2||a + \varepsilon(a+b)||^2}{2(1 - \varepsilon^4||a+b||^2)} - \frac{\varepsilon^2||a||^2}{2} \\
&= \frac{\varepsilon^2}{2}(||a + \varepsilon(a+b)||^2 - ||a||^2) + \frac{\varepsilon^6||a + \varepsilon(a+b)||^2||a+b||^2}{2(1 - \varepsilon^4||a-b||^2)}.
\end{aligned}
$$

It is easy to see that there exists an absolute constant M_1 such that

$$
\sup_{a,b \in K} |m(a,b)| \leq M_1\varepsilon^3.
$$

Consequently, there is a universal $M_2 > 0$ such that if ε is sufficiently small, then

$$
\exp(m(a,b)) \leq 1 + m(a,b) + m(a,b)^2
$$
$$
\leq 1 + \frac{\varepsilon^2}{2}(||a + \varepsilon(a+b)||^2 - ||a||^2) + M_2\varepsilon^6
$$

for any $a, b \in K$. Since $\sqrt{1-x} \leq 1 - x/2$ for $x \in (0,1)$, the inequality (8.14) implies

$$2 + 2\varepsilon^3 ||a||^2 \leq (1 - \varepsilon^4 ||a+b||^2/2) \left(1 + \frac{\varepsilon^2}{2}(||a+\varepsilon(a+b)||^2 - ||a||^2) + M_2\varepsilon^6\right)$$
$$+ (1 - \varepsilon^4 ||a-b||^2/2) \left(1 + \frac{\varepsilon^2}{2}(||a+\varepsilon(a-b)||^2 - ||a||^2) + M_2\varepsilon^6\right).$$

This, after some manipulations, leads to

$$||a + \varepsilon(a+b)||^2 + ||a + \varepsilon(a-b)||^2 - 2||a(1+\varepsilon)||^2$$
$$\geq \varepsilon^2(||a+b||^2 + ||a-b||^2 - 2||a||^2) - 2\varepsilon^4 M_3,$$

where M_3 is a positive constant not depending on ε, a and b. Equivalently,

$$\left|\left|a + \frac{\varepsilon}{1+\varepsilon}b\right|\right|^2 + \left|\left|a - \frac{\varepsilon}{1+\varepsilon}b\right|\right|^2 - 2||a||^2 - 2\left|\left|\frac{\varepsilon}{1+\varepsilon}b\right|\right|^2$$
$$\geq \frac{\varepsilon^2}{(1+\varepsilon)^2}(||a+b||^2 + ||a-b||^2 - 2||a||^2 - 2||b||^2) - 2\frac{\varepsilon^4}{(1+\varepsilon)^2}M_3.$$

Next, let $c \in \mathcal{B}$, $\gamma > 0$ and substitute $a = \gamma c$; we assume that γ is small enough to ensure that $a \in K$. If we divide both sides by γ^2 and substitute $\delta = \varepsilon(1+\varepsilon)^{-1}\gamma^{-1}$, we obtain

$$||c + \delta b||^2 + ||c - \delta b||^2 - 2||c||^2 - 2||\delta b||^2$$
$$\geq \delta^2(||\gamma c + b||^2 + ||\gamma c - b||^2 - 2||\gamma c||^2 - 2||b||^2) - 2\varepsilon^2\delta^2 M_3$$
$$\geq \delta^2(||\gamma c + b||^2 + ||\gamma c - b||^2 - 2||\gamma c||^2 - 2||b||^2) - 2\delta^4 M_3$$

Let γ and ε go to 0 so that δ remains fixed. We obtain that for any $\delta > 0$, $b \in K$ and $c \in \mathcal{B}$,

$$||c + \delta b||^2 + ||c - \delta b||^2 - 2||c||^2 - 2||\delta b||^2 \geq -2\delta^4 M_3. \tag{8.15}$$

Now let N be a large positive integer and consider a symmetric random walk $(g_n)_{n>0}$ over the integers, starting from 0. Let $\tau = \inf\{n : |g_n| = N\}$. The inequality (8.15), applied to $\delta = N^{-1}$, implies that for any $a \in \mathcal{B}$ and $b \in K$, the process

$$(\xi_n)_{n\geq 0} = \left(\left(\left|\left|a + \frac{bg_{\tau \wedge n}}{N}\right|\right|^2 - \left\{\frac{||b||^2}{N^2} - \frac{M_3}{N^4}\right\}(\tau \wedge n)\right)\right)_{n\geq 0}$$

is a submartingale. Since $\mathbb{E}(\tau \wedge n) = \mathbb{E}g_{\tau \wedge n}^2$, we obtain

$$\mathbb{E}\left[\left|\left|a + \frac{bg_{\tau \wedge n}}{N}\right|\right|^2 - \left\{\frac{||b||^2}{N^2} - \frac{M_3}{N^4}\right\}g_{\tau \wedge n}^2\right] = \mathbb{E}\xi_n \geq \mathbb{E}\xi_0 = ||a||^2.$$

Letting $n \to \infty$ and using Lebesgue's dominated convergence theorem gives

$$\frac{1}{2} \left[||a+b||^2 + ||a-b||^2 \right] - ||b||^2 + \frac{M_3}{N^2} \geq ||a||^2.$$

It suffices to let N go to ∞ to obtain

$$||a+b||^2 + ||a-b||^2 \geq 2||a||^2 + 2||b||^2.$$

We have assumed that b belongs to the unit ball K, but, by homogeneity, the above estimate extends to any $b \in \mathcal{B}$. Putting $a+b$ and $a-b$ in the place of a and b, respectively, we obtain the reverse estimate

$$||a+b||^2 + ||a-b||^2 \leq 2||a||^2 + 2||b||^2.$$

This implies that the parallelogram identity is satisfied and hence $\mathcal{B}$ is a Hilbert space.

8.2.4   Proof of Theorem 8.4

We shall need the following auxiliary fact.

Lemma 8.5. *Assume $\psi : \mathbb{R} \to \mathbb{R}$ is even, of class C^2 and with the further property that ψ, ψ'' are convex. Then for any u, $v \in \mathcal{H}$,*

$$|\psi'(|u|)u' - \psi'(|v|)v'| \leq \frac{|u-v|}{2} \left[\psi''(|u|) + \psi''(|v|) \right]. \qquad (8.16)$$

Proof. Squaring both sides, we see that the inequality above is equivalent to $A \leq B(u', v')$, where A, B depend only on $|u|$ and $|v|$. Therefore we only need to check the inequality for u, v satisfying $(u', v') = \pm 1$. If $(u', v') = 1$, then (8.16) takes the form

$$|\psi'(|u|) - \psi'(|v|)| \leq |u-v| \left[\frac{\psi''(|u|) + \psi''(|v|)}{2} \right],$$

and follows from

$$|\psi'(|u|) - \psi'(|v|)| = \left| \int_{|v|}^{|u|} \psi''(s) ds \right| \leq ||u| - |v|| \cdot \frac{\psi''(|u|) + \psi''(|v|)}{2}.$$

The last inequality is a consequence of the convexity of ψ''. If $(u', v') = -1$, then, since ψ' is odd,

$$|\psi'(|u|)u' - \psi'(|v|)v'| = |\psi'(|u|) + \psi'(|v|)| = |\psi'(|u|) - \psi'(-|v|)|$$

$$= \left| \int_{-|v|}^{|u|} \psi''(s) ds \right| \leq ||u| + |v|| \cdot \frac{\psi''(|u|) + \psi''(|v|)}{2} = |u-v| \cdot \frac{\psi''(|u|) + \psi''(|v|)}{2}.$$

The proof is complete. $\square$

Proof of (8.3). We must show that for any dyadic martingale f.

$$\mathbb{P}(S_n(f) \geq 1) \leq K\mathbb{E}|f_n|, \qquad n = 0, 1, 2, \ldots.$$

Let $V(x, y) = 1_{\{y \geq 1\}} - K|x|$ for $x \in \mathcal{H}$ and $y \geq 0$. To introduce the corresponding special function, define $\psi : [-1, 1] \to \mathbb{R}$ by

$$\psi(s) = K^{-1}\left(e^{-s^2/2} + s \int_0^s e^{-t^2/2}\mathrm{dt}\right). \tag{8.17}$$

It is easy to check that ψ satisfies the differential equation

$$\psi''(s) + s\psi'(s) - \psi(s) = 0 \qquad \text{for } s \in (-1, 1). \tag{8.18}$$

Let $U : \mathcal{H} \times [0, \infty) \to \mathbb{R}$ be given by

$$U(x, y) = \begin{cases} 1 - K\sqrt{1 - y^2}\psi\left(\dfrac{|x|}{\sqrt{1-y^2}}\right) & \text{if } |x|^2 + y^2 < 1, \\ 1 - K|x| & \text{if } |x|^2 + y^2 \geq 1. \end{cases}$$

Let us prove that U and V satisfy the conditions $1°$, $2°$ and $3°$. The majorization is obvious when $|x|^2 + y^2 \geq 1$, and for the remaining x, y it can be written in the form

$$\psi(s) \leq s + \frac{1}{K\sqrt{1 - y^2}}, \qquad \text{where } s = \frac{|x|}{\sqrt{1 - y^2}} < 1.$$

However, ψ is convex, so it suffices to check the above estimate for $s = 0$ and $s = 1$; in both cases the bound holds and $1°$ follows. We turn to $2°$. Observe first that

$$U(x, y) \leq 1 - K|x| \qquad \text{for all } x \in \mathcal{H}, y \geq 0. \tag{8.19}$$

This is trivial for $|x|^2 + y^2 \geq 1$, and for the remaining x, y follows immediately from the estimate $\psi(s) \leq s$, which can be verified by standard analysis of the derivative. Consequently, when $|x|^2 + y^2 \geq 1$,

$$U(x, y) = 1 - K|x| \geq \frac{1}{2}\left[(1 - K|x + d|) + (1 - K|x - d|)\right]$$

$$\geq \frac{1}{2}\left[U(x + d, \sqrt{y^2 + d^2}) + U(x - d, \sqrt{y^2 + d^2})\right].$$

Suppose then that $|x|^2 + y^2 < 1$. We shall prove $2°$ only in the case when $|x \pm d|^2 + y^2 \leq 1$, for the remaining values of d the reasoning is similar. Let

$$G(s) = \frac{1}{2}\left[U(x + ds, \sqrt{y^2 + d^2s^2}) + U(x - d, \sqrt{y^2 + d^2s^2})\right], \qquad s \in [0, 1]$$

and note that we must prove that $G(0) \geq G(1)$. Of course we will be done if we show that G is nonincreasing. We may and do assume that $y = 0$: otherwise we divide throughout by $\sqrt{1-y^2}$ and substitute $X = x/\sqrt{1-y^2}$, $D = d/\sqrt{1-y^2}$. Setting $x_\pm = (x \pm ds)/\sqrt{1-|d|^2 s^2}$ and using (8.18), we compute that

$$G'(s) = -\left(\psi'(|x_+|)x'_+ - \psi'(|x_-|)x'_-\right) \cdot \frac{d}{2} + \frac{|d|^2 s}{2\sqrt{1-|d|^2 s^2}}(\psi''(x_-) + \psi''(x_+)).$$

By (8.16), applied to $-\psi$, and the equality

$$\frac{|x_+ - x_-|}{2} = \frac{|d|s}{\sqrt{1-|d|^2 s^2}},$$

we obtain

$$G'(s) \le -\left(\psi'(|x_+|)x'_+ - \psi'(|x_-|)x'_-\right) \cdot \frac{d}{2} - |\psi'(|x_+|)x'_+ - \psi'(|x_-|)x'_-| \frac{|d|}{2} \le 0,$$

and 2° follows. Finally, 3° is an immediate consequence of 2°. The inequality is established. $\square$

Sharpness. We shall show that the estimate is sharp for continuous-time martingales (we interpret the square function as a quadratic covariance process). This approach, frequently used below for other inequalities in the dyadic setting, has the advantage that we may use Brownian motion, for which the calculations are easier to handle. So, let $B = (B_t)_{t \ge 0}$ be a standard Brownian motion starting from 0 and ε be a Rademacher random variable independent of B. Introduce the stopping time $\tau = \inf\{t : B_t^2 + t \ge 1\}$, satisfying $\tau \le 1$ almost surely, and let the process $X = (X_t)_{t \ge 0}$ be given by

$$X_t = B_{\tau \wedge t} + \varepsilon B_\tau I_{\{t \ge 1\}}.$$

The process X is a Brownian motion, which stops at the moment τ, and then at time 1 jumps to one of the points 0, $2B_\tau$ with probability $1/2$ and stays there forever. Clearly, it is a martingale with respect to its natural filtration. Its square bracket process satisfies

$$[X, X]_1 = [B]_\tau + |B_\tau|^2 = \tau + B_\tau^2 = 1 \quad \text{almost surely,}$$

and, as we shall prove now, $||X||_1 = 1/K$. Observe that $||X||_1 = ||X_1||_1 = ||X_\tau||_1 = ||B_\tau||_1$. Let $U : \mathbb{R} \times \mathbb{R}_+ \to \mathbb{R}$ be given by

$$U(x,t) = \sqrt{1-t} \exp\left(-\frac{x^2}{2(1-t)}\right) + |x| \int_0^{|x|/\sqrt{1-t}} \exp(-s^2/2)ds,$$

if $t + x^2 < 1$, and $U(x,t) = K|x|$ otherwise. It can be readily verified that U is continuous and satisfies the heat equation $U_t + \frac{1}{2}U_{xx} = 0$ on the set $\{(x,t) : t + x^2 < 1\}$. This implies that $(U(B_{\tau \wedge t}, \tau \wedge t))_{t \ge 0}$ is a martingale adapted to $\mathcal{F}^B$ and therefore

$$K||B_\tau||_1 = \mathbb{E}U(B_\tau, \tau) = U(0,0) = 1.$$

This shows the sharpness of (8.2) in the continuous-time setting. Now the passage to the discrete-time case can be carried out using standard approximation techniques. However, our proof will be a bit different. Suppose that the best constant in the inequality (8.2) for dyadic martingales equals K_0. Let $U^0 : \mathbb{R} \times [0, \infty) \to \mathbb{R}$ be given by

$$U^0(x, y) = \sup\{\mathbb{P}(y^2 - x^2 + S^2(f) \geq 1) - K_0||f||_1\},$$

where the supremum is taken over all the simple martingales starting from x and dyadic differences df_n, $n = 1, 2, \ldots$, and set $W(x, t) = U^0(x, \sqrt{t})$. It is not difficult to see that W is continuous. Indeed, let f, $f_0 \equiv x$, be as in the definition of W. Fix x' and let $f' = f + x' - x$. Then $x^2 - S^2(f) = (x')^2 - S^2(f')$ and, for any $t \geq 0$,

$$\mathbb{P}(t - x^2 + S^2(f) \geq 1) - K_0||f||_1 \leq \mathbb{P}(t - (x')^2 + S^2(f') \geq 1) - K_0||f'||_1 + K_0|x - x'|,$$

which implies $W(x, t) \leq W(x', t) + K_0|x - x'|$ so for fixed t, $W(\cdot, t)$ is K_0-Lipschitz. Hence, applying (iii), for any $s < t$ and any x,

$$W(x, s) \geq \frac{1}{2}[W(x - \sqrt{t - s}, t) + W(x + \sqrt{t - s}, t] \geq W(x, t) - K_0\sqrt{t - s}.$$

On the other hand, $W(x, s) \leq W(x, t)$, by the definition of W. Therefore, for any x, $W(x, \cdot)$ is continuous. This yields the continuity of W.

Now extend W to the whole $\mathbb{R}^2$ by setting $W(x, t) = W(x, 0)$ for $t < 0$. Let $\delta > 0$ and convolve W with a nonnegative smooth function g^δ satisfying $||g^\delta||_1 = 1$ and supported on the ball centered at $(0, 0)$ and of radius δ. We obtain a smooth function W^δ, for which $2°$ is still valid. Dividing this inequality by d^2 and letting $d \to 0$ gives $W_t^\delta + \frac{1}{2}W_{xx}^\delta \leq 0$ and hence, by Itô's formula, $\mathbb{E}W^\delta(B_\tau, \tau) \leq W^\delta(0, 0)$. Now let $\delta \to 0$ and use the continuity of W and Lebesgue's dominated convergence theorem to conclude that $\mathbb{E}W(B_\tau, \tau) \leq W(0, 0)$. The final step is that, by $2°$,

$$W(B_\tau - B_\tau, \tau + B_\tau^2) + W(B_\tau + B_\tau, \tau + B_\tau^2) \leq 2W(B_\tau, \tau) \quad \text{almost surely,}$$

which yields $\mathbb{E}W(X_1, [X, X]_1) \leq W(0, 0)$ and, by the first part, $K_0 \geq K$. $\square$

On the search of the suitable majorant. As in the general case, we define

$$U^0(x, y) = \inf\{\mathbb{E}|f_n|\},$$

where the infimum is taken over all n and all simple martingales f with dyadic kth differences, $k = 1, 2, \ldots$. Furthermore, let $W(x, y^2) = U^0(x, y)$. Arguing as above, we obtain that

$$W(x, t) = |x| \quad \text{when } x^2 + t \geq 1.$$

Now, suppose that

(A1) W is of class C^2 on $\{(x, t) : x^2 + t < 1\}$.

Repeating the proof of Theorem 8.1, we see that the function U^0 satisfies the convexity property: in particular, for $x^2 + t < 1$ and $s > 0$,

$$\frac{W(x + s, t + s^2) + W(x - s, t + s^2) - W(x, t)}{2s^2}$$
$$= \frac{U(x + s, \sqrt{t + s^2}) + U(x - s, \sqrt{t + s^2}) - 2U(x, \sqrt{t})}{2s^2} \geq 0.$$

Letting $s \to 0$ gives $W_t + \frac{1}{2}W_{xx} \geq 0$ and leads to the assumption

(A2) W satisfies $W_t + \frac{1}{2}W_{xx} = 0$ on $\{(x, t) : x^2 + t < 1\}$.

This yields the special function used above. To see this, note that (8.8) holds and gives

$$W(x, t) = U^0(x, \sqrt{t}) = \sqrt{1 - t}\, U^0\left(\frac{x}{\sqrt{1 - t}}, 0\right).$$

Let $\psi(s) = U^0(s, 0)$ for $s \in [-1, 1]$. Then the function ψ is of class C^2, is even and we have $\psi(1) = 1$. Furthermore, (A2) implies that ψ satisfies (8.18) and we easily compute that this function must be given by (8.17). $\square$

8.2.5   Proof of Theorem 8.5

Here the reasoning is similar to that presented in the proof of the weak type inequality for orthogonal martingales. Let $H = \mathbb{R} \times (0, \infty)$, $S = \mathbb{R} \times (-1, 1)$ and $S^+ = (0, \infty) \times (-1, 1)$. For $1 \leq p \leq 2$, introduce the harmonic function $A = A_p : H \to \mathbb{R}$ given by the Poisson integral

$$A(\alpha, \beta) = \frac{1}{\pi} \int_{-\infty}^{\infty} \frac{\beta \left|\frac{2}{\pi} \log |t|\right|^p}{(\alpha - t)^2 + \beta^2}\, dt.$$

It is easy to see that the function A satisfies

$$\lim_{(\alpha, \beta) \to (z, 0)} A(\alpha, \beta) = \left(\frac{2}{\pi}\right)^p |\log |z||^p, \qquad z \neq 0. \qquad (8.20)$$

Consider the conformal mapping φ given by $\varphi(z) = ie^{\pi z/2}$, or, in the real coordinates,

$$\varphi(x, y) = \left(-e^{\pi x/2} \sin\left(\frac{\pi}{2}y\right), e^{\pi x/2} \cos\left(\frac{\pi}{2}y\right)\right), \qquad (x, y) \in \mathbb{R}^2,$$

which maps S onto H. Let $A = A_p$ be defined on the strip S by $A(x, y) = A(\varphi(x, y))$. Then the function A is harmonic on S, since it is the real part of an analytic function. By (8.20), we can extend A to the continuous function on the closure $\overline{S}$ of S by $A(x, \pm 1) = |x|^p$. One easily checks that for $(x, y) \in S$,

$$A(x, y) = \frac{1}{\pi} \int_{\mathbb{R}} \frac{\cos\left(\frac{\pi}{2} y\right) \left|\frac{2}{\pi} \log|s| + x\right|^p}{(s - \sin(\frac{\pi}{2} y))^2 + \cos^2(\frac{\pi}{2} y)} ds \qquad (8.21)$$

for $|y| < 1$. Substituting $s := 1/s$ and $s := -s$ above, we see that A satisfies

$$A(x, y) = A(-x, y) = A(x, -y) \qquad \text{for } (x, y) \in S. \qquad (8.22)$$

Finally, let $U = U_p : [0, \infty) \times \mathbb{R} \to \mathbb{R}$ be given by $U(x, y) = x^p$ for $|y| > 1$ and

$$U(x, y) = c_p \int_{\mathbb{R}} A(ux, y) \exp(-u^2/2) du$$

otherwise; here $c_p = \left(\int_{\mathbb{R}} |u|^p \exp(-u^2/2) du\right)^{-1} = \left(2^{(p+1)/2} \Gamma\left(\frac{p+1}{2}\right)\right)^{-1}$. Clearly, U is continuous and, by (8.22), we have

$$U(x, y) = 2c_p \int_0^\infty A(ux, y) \exp(-u^2/2) du \qquad \text{on } \overline{S^+}. \qquad (8.23)$$

Introduce $V : [0, \infty) \times \mathcal{H} \to \mathbb{R}$ by $V(x, y) = U(0, 0)1_{\{|x| \geq 1\}} - x^p$. The key properties of U are the following. We omit the proof, analogous argumentation can be found in Chapter 6.

Lemma 8.6.

(i) *The function U satisfies the differential equation*

$$U_x(x, y) + x U_{yy}(x, y) = 0 \qquad \text{on } S^+. \qquad (8.24)$$

(ii) *U is superharmonic on S^+.*

(iii) *$U(0, 0) = C_p^{-p}$.*

(iv) *For any $(x, y) \in [0, \infty) \times \mathbb{R}$,*

$$x^p \leq U(x, y) \leq x^p + U(0, 0)1_{\{|y| < 1\}}. \qquad (8.25)$$

(v) *$U_x(x, y) \geq 0$ on $(0, \infty) \times \mathbb{R}$ and $U_y(x, y) \leq 0$ on $(0, \infty) \times ((0, \infty) \setminus \{1\})$.*

In the lemmas below, we will use the following notation. For $a \in \mathcal{H}$, let $a' = a/|a|$ if $a \neq 0$ and $a' = 0$ otherwise; furthermore, $a^* = a$ for $|a| \leq 1$ and $a^* = a/|a|$ otherwise.

Lemma 8.7. *For any $a, b \in \mathcal{H}$ we have $|a^* + b^*| \leq |a + b|$.*

Proof. If both $|a|, |b| \leq 1$, the claim is obvious. If $|a| > 1 \geq |b|$,

$$|a + b|^2 - |a^* + b^*|^2 = |a|^2 - 1 + 2\langle a \cdot b \rangle (1 - |a|^{-1}) \geq |a|^2 - 1 - 2|a|(1 - |a|^{-1}) \geq 0$$

and similarly for $|b| > 1 \geq |a|$. Finally, if $|a| > 1$ and $|b| > 1$, then

$$|a + b|^2 - |a^* + b^*|^2 = |a|^2 + |b|^2 + 2\langle a \cdot b \rangle (1 - (|a||b|)^{-1}) - 2 \geq |a|^2 + |b|^2 - 2|a||b| \geq 0,$$

as desired. $\qquad \square$

Lemma 8.8. *For any* $(x, y) \in [0, \infty) \times \mathcal{H}$ *and any* $d \in \mathcal{H}$,

$$2U(x, |y|) \leq U\big((x^2 + |d|^2)^{1/2}, |y + d|\big) + U\big((x^2 + |d|^2)^{1/2}, |y - d|\big). \qquad (8.26)$$

Proof. It is convenient to split the proof into three parts.

   Case 1: $|y| \geq 1$. Then the estimate is trivial: indeed, by part (iv) of Lemma 8.6,

$$2U(x, |y|) \leq 2(x^2 + |d|^2)^{p/2} \leq U\big((x^2 + |d|^2)^{1/2}, |y + d|\big) + U\big((x^2 + |d|^2)^{1/2}, |y - d|\big).$$

   Case 2: $|y| < 1$, $|y \pm d| \leq 1$. For $t \in [0, 1]$, let $\psi(t) = U\big(x_+^t, |y_+^t|\big) + U\big(x_+^t, |y_-^t|\big)$, where $x_+^t = (x^2 + t^2|d|^2)^{1/2}$ and $y_\pm^t = y \pm td$. We have that $\psi'(t)/|d|$ equals

$$\frac{t|d|}{x_+^t}\left[U_x\big(x_+^t, |y_+^t|\big) + U_x\big(x_+^t, |y_-^t|\big)\right] + \left(U_y\big(x_+^t, |y_+^t|\big)(y_+^t)' - U_y\big(x_+^t, |y_-^t|\big)(y_-^t)'\right) \cdot d'$$

(when $|y + td| = 0$, the differentiation is allowed since $U_y(x, 0) = 0$). We will prove that this is nonnegative, which will clearly yield the claim. It suffices to show that

$$\frac{|y_+^t - y_-^t|}{2x_+^t}\left[U_x\big(x_+^t, |y_+^t|\big) + U_x\big(x_+^t, |y_-^t|\big)\right] \geq \left|U_y\big(x_+^t, |y_+^t|\big)(y_+^t)' - U_y\big(x_+^t, |y_-^t|\big)(y_-^t)'\right|.$$

Note that if we square both sides, the estimate becomes $A \leq B \cdot (y_+^t)' \cdot (y_-^t)'$, where A and B depend only on $|y_+^t|$ and $|y_-^t|$. Thus it suffices to prove it for $(y_+^t)' = \pm(y_-^t)'$. When $(y_+^t)'$ and $(y_-^t)'$ are equal, we use (8.24) and conclude that the inequality reads

$$-\frac{\big||y_+^t| - |y_-^t|\big|}{2}\left[U_{yy}\big(x_+^t, |y_+^t|\big) + U_{yy}\big(x_+^t, |y_-^t|\big)\right] \geq \left|\int_{|y_-^t|}^{|y_+^t|} U_{yy}(x_+^t, s)\mathrm{d}s\right|.$$

This follows from the fact that U_{yy} is nonpositive and concave: by Lemma 7.69, we have $x^2 U_{yyyy}(x, y) = U_{xx}(x, y) + U_{yy}(x, y) \leq 0$ for $(x, y) \in S^+$. The case $(y_+^t)' = -(y_-^t)'$ is dealt with in the same manner.

   Case 3: $|y| < 1$, $d > 1 - |y|$. This can be reduced to the previous case. Set $y_+ = (y + d)^*$, $y_- = (y - d)^*$ and $\tilde{y} = (y_+ + y_-)/2$, $\tilde{d} = (y_+ - y_-)/2$, $\tilde{x} = (x^2 + |d|^2 - \tilde{d}^2)^{1/2}$. By Lemma 8.7, $|\tilde{y}| \leq |y|$ and $|\tilde{d}| \leq |d|$, so $\tilde{x} \geq x$. Now, using part (v) of Lemma 8.6 and the fact that $\tilde{y}, d$ satisfy the assumptions of Case 2, we obtain

$$2U(x, y) \leq 2U(\tilde{x}, \tilde{y}) \leq U\big((\tilde{x}^2 + \tilde{d}^2)^{1/2}, y_+\big) + U\big((\tilde{x}^2 + \tilde{d}^2)^{1/2}, y_-\big)$$

and the latter sum is precisely the right-hand side of (8.26). $\square$

Remark 8.3. The choice $x = 0$, $y = 0$ in (8.26) gives $U(0, 0) \leq U(d, d)$ for all $d \geq 0$.

   Thus, the functions $(x, y) \to U(0, 0) - U(x, y)$ and V satisfy the properties $1°$, $2°$ and $3°$. The result follows.

Sharpness. Suppose that γ_p is the optimal constant in (8.4) for real-valued dyadic martingales. Arguing as in the inequality above, this yields a corresponding weak type inequality

$$\mathbb{P}(|B_\tau| \geq 1) \leq \gamma_p^p \mathbb{E}\tau^{p/2}, \tag{8.27}$$

where B is a standard Brownian motion and τ is any stopping time of B. On the other hand, let $\eta = \inf\{t : |B_t| = 1\}$ and consider the process $(U(\sqrt{\eta \wedge t}, B_{\eta \wedge t}))_{t \geq 0}$. By (8.24) and Itô's formula, it is a martingale with expectation equal to $U(0,0)$. By (8.25) and the exponential integrability of η, this martingale converges almost surely and in L^1 to $\eta^{p/2}$, which, by Lemma 8.6, yields $C_p^{-p} = \mathbb{E}\eta^{p/2}$ and, consequently, $1 = \mathbb{P}(|B_\eta| \geq 1) = C_p^p \mathbb{E}\eta^{p/2}$. By (8.27), this implies $\gamma_p \geq C_p$ and completes the proof. $\qquad\square$

On the search of the suitable majorant. Consider the function

$$U^0(x,y) = \inf\{\mathbb{E}(y^2 - x^2 + S_n^2(f))^{p/2}\},$$

where the infimum is taken over all n and all simple dyadic martingales f starting from x and satisfying $\mathbb{P}(|f_\infty| \geq 1) = 1$. We easily see that $U^0(x,y) = y^p$ when $|x| \geq 1$. Passing to continuous-time martingales raises the problem of finding U on the strip S satisfying the heat equation. This leads to the function above. $\qquad\square$

8.3 Moment estimates

8.3.1 Formulation of the result

Burkholder's L^p estimate for differentially subordinate martingales yields the following result.

Theorem 8.6. *Suppose that f is a Hilbert-space-valued martingale. Then for $1 < p < \infty$,*

$$(p^* - 1)^{-1}||f||_p \leq ||S(f)||_p \leq (p^* - 1)||f||_p. \tag{8.28}$$

The left-hand inequality is sharp for $p \geq 2$ and the right-hand inequality is sharp for $1 < p \leq 2$. In the remaining case, the optimal constants are not known.

When $p = 1$, then the right-hand inequality above does not hold with any finite constant. However, the left-hand inequality does. We shall prove the following.

Theorem 8.7. *Suppose that f is a Hilbert-space-valued martingale. Then*

$$||f||_1 \leq 2||S(f)||_1 \tag{8.29}$$

and the constant is the best possible.

When f is assumed to be conditionally symmetric, the constants change and involve the special functions studied in Appendix.

Theorem 8.8. *Let f be a conditionally symmetric martingale taking values in a Hilbert space $\mathcal{H}$. Let ν_p be the smallest positive zero of the confluent hypergeometric function and z_p be the largest positive zero of the parabolic cylinder function of parameter p. Then we have*

$$\|f\|_p \leq \nu_p \|S(f)\|_p \qquad \text{for } 0 < p \leq 2, \tag{8.30}$$
$$\|f\|_p \leq z_p \|S(f)\|_p \qquad \text{for } p \geq 3 \tag{8.31}$$

and

$$\nu_p \|S(f)\|_p \leq \|f\|_p \qquad \text{for } p \geq 2. \tag{8.32}$$

The inequalities are sharp. They are already sharp when f is assumed to be a dyadic martingale taking values in $\mathbb{R}$.

8.3.2   Proof of Theorem 8.6

As already mentioned, the validity of (8.28) follows immediately from moment estimate for differentially subordinate martingales. Thus all we need is to deal with the optimality of the constants.

The left-hand inequality in (8.28) *is sharp for $p \geq 2$.* Suppose that the inequality holds for all real-valued martingales with some constant C and let

$$U^0(x, y) = \sup\{\mathbb{E}|f_n|^p - C^p \mathbb{E}(y^2 - |x|^2 + S_n^p(f))^{p/2}\},$$

where the supremum is taken over the corresponding parameters. It is easy to see that U^0 is homogeneous of order p and, by Theorem 8.1, satisfies the properties $1°$, $2°$ and $3°$. Fix a small $\delta > 0$ and apply $2°$ with $x = \delta$, $y = 1$ and a mean zero random variable d taking the values

$$s = \frac{2\delta}{1 - \delta^2} \quad \text{and} \quad t = -\frac{-1 + (p-1)\sqrt{1 - p(p-2)\delta^2}}{p(p-2)}.$$

We obtain

$$U^0(1, \delta) \geq \frac{s}{s - t} U^0(1 + t, \sqrt{\delta^2 + t^2}) + \frac{-t}{s - t} U^0(1 + s, \sqrt{\delta^2 + s^2}). \tag{8.33}$$

Now, by $1°$,

$$U^0(1 + t, \sqrt{\delta^2 + t^2}) \geq (1 + t)^p - C^p(\delta^2 + t^2)^{p/2}$$
$$= (\delta^2 + t^2)^{p/2}\left[(p-1)^p - C^p\right]$$
$$= (\delta^2 + t^2)^{p/2}\left[(p^* - 1)^p - C^p\right]$$

and, by homogeneity,

$$U^0(1 + s, \sqrt{\delta^2 + s^2}) = U^0\left(\frac{1 + \delta^2}{1 - \delta^2}, \frac{\delta(1 + \delta^2)}{1 - \delta^2}\right) = \left(\frac{1 + \delta^2}{1 - \delta^2}\right)^p U^0(1, \delta).$$

Using these two facts in (8.33) and subtracting $U^0(1, \delta)$ from both sides gives

$$0 \geq \frac{s(\delta^2 + t^2)^{p/2}}{s - t} \left[(p^* - 1)^p - C^p \right] + U^0(1, \delta) \left[\frac{-t}{s - t} \left(\frac{1 + \delta^2}{1 - \delta^2} \right)^p - 1 \right].$$

Now divide throughout by δ and let $\delta \to 0$. Then the second term above converges to 0 and we shall obtain

$$0 \geq 2p^{1-p} \left[(p^* - 1)^p - C^p \right].$$

That is, we have $C \geq p^* - 1$ and we are done. $\qquad\square$

The right-hand inequality in (8.28) is sharp for $1 < p \leq 2$. This can be established in a similar manner. The details are left to the reader. $\qquad\square$

8.3.3 Proof of Theorem 8.7

First we prove some auxiliary inequalities that will be needed later.

Lemma 8.9. *Let $x, d \in \mathcal{H}$, $y \in \mathbb{R}_+$, $y < |x|$. Then*

$$\sqrt{y^2 + |d|^2} - y \geq \sqrt{|x|^2 + |d|^2} - |x|. \tag{8.34}$$

If, in addition, $\sqrt{y^2 + |d|^2} \geq |x + d|$, then

$$\sqrt{2y^2 + 2|d|^2 - |x + d|^2} - 2y \geq \sqrt{2|x|^2 + 2|d|^2 - |x + d|^2} - 2|x|. \tag{8.35}$$

Proof. The inequality (8.34) is equivalent to

$$|x| - y \geq \sqrt{|x|^2 + |d|^2} - \sqrt{y^2 + |d|^2} = \frac{|x|^2 - y^2}{\sqrt{|x|^2 + |d|^2} + \sqrt{y^2 + |d|^2}},$$

or

$$\sqrt{|x|^2 + |d|^2} + \sqrt{y^2 + |d|^2} \geq |x| + y,$$

which is obvious.

Now we turn to (8.35). We may write it as follows:

$$2|x| - 2y \geq \sqrt{2|x|^2 + 2|d|^2 - |x + d|^2} - \sqrt{2y^2 + 2|d|^2 - |x + d|^2}$$
$$= \frac{2|x|^2 - 2y^2}{\sqrt{2|x|^2 + 2|d|^2 - |x + d|^2} + \sqrt{2y^2 + 2|d|^2 - |x + d|^2}},$$

which can be transformed into

$$\sqrt{2|x|^2 + 2|d|^2 - |x + d|^2} + \sqrt{2y^2 + 2|d|^2 - |x + d|^2} \geq |x| + y.$$

The left-hand side of the inequality above is equal to

$$|x - d| + \sqrt{|x + d|^2 + 2(y^2 + |d|^2 - |x + d|^2)}$$

and, due to the assumption $\sqrt{y^2 + |d|^2} \geq |x + d|$, is bounded from below by

$$|x - d| + |x + d| \geq 2|x| > |x| + y.$$

This is precisely what we need.   □

Proof of (8.29). Let $V : \mathcal{H} \times [0, \infty) \to \mathbb{R}$ be given by $V(x, y) = |x| - 2y$. The corresponding special function $U : \mathcal{H} \times [0, \infty) \to \mathbb{R}$ is given by

$$U(x, y) = \begin{cases} -\sqrt{2y^2 - |x|^2} & \text{if } y \geq |x|, \\ |x| - 2y & \text{if } y < |x|. \end{cases} \qquad (8.36)$$

We shall verify the conditions $1°$–$4°$; in fact, the last condition is empty, since there is no maximal function in the studied estimate. To prove the majorization, we may assume that $y \geq |x|$. The inequality takes the form

$$2y - |x| \geq \sqrt{2y^2 - |x|^2}$$

and can be rewritten as $2(y - |x|)^2 \geq 0$, after squaring both sides. To establish $2°$, assume first that $y \geq |x|$. If $y = 0$, then $x = 0$ and the inequality is trivial: it reduces to the inequality $\mathbb{E}|d| \geq 0$. Suppose then, that $y > 0$. It suffices to show that for any $d \in \mathcal{H}$,

$$U(x + d, \sqrt{y^2 + d^2}) \leq U(x, y) + \frac{x \cdot d}{\sqrt{2y^2 - |x|^2}}.$$

Then $2°$ follows by taking expectations of both sides. We have

$$2(\sqrt{y^2 + |d|^2})^2 - |x + d|^2 \geq 2|x|^2 + 2|d|^2 - |x|^2 - 2(x \cdot d) - |d|^2 = |x - d|^2 \geq 0$$

and

$$U(x + d, \sqrt{y^2 + |d|^2}) \leq -\sqrt{2(y^2 + |d|^2)} - |x + d|^2,$$

see the proof of $1°$ above. Hence it suffices to check the inequality

$$\sqrt{2(y^2 + |d|^2) - |x + d|^2} \geq \sqrt{2y^2 - |x|^2} - \frac{x \cdot d}{\sqrt{2y^2 - |x|^2}}, \qquad (8.37)$$

or

$$\sqrt{2y^2 - |x|^2}\sqrt{2y^2 - |x|^2 - 2(x \cdot d) + |d|^2} \geq 2y^2 - |x|^2 - x \cdot d.$$

But we have

$$(2y^2 - |x|^2)(2y^2 - |x|^2 - 2x \cdot d + |d|^2)$$
$$\geq (2y^2 - |x|^2)^2 - 2(2y^2 - |x|^2)(x \cdot d) + |x|^2|d|^2$$
$$\geq (2y^2 - |x|^2 - x \cdot d)^2$$

and the inequality follows.

Now suppose that $y < |x|$ and let $d \in \mathcal{H}$. Again, we will be done if we show that

$$U(x + d, \sqrt{y^2 + |d|^2}) \leq U(x, y) + \frac{x \cdot d}{|x|}.$$

If $\sqrt{y^2 + d^2} < |x + d|$, then this is equivalent to

$$2\sqrt{y^2 + d^2} - |x + d| \geq 2y - |x| - \frac{x \cdot d}{|x|},$$

or

$$2\sqrt{y^2 + |d|^2} - 2y \geq |x + d| - |x| - \frac{x \cdot d}{|x|}. \tag{8.38}$$

By (8.34), we may bound the left-hand side of the above inequality from below by $2\sqrt{|x|^2 + |d|^2} - 2|x|$ and, therefore, it suffices to prove that

$$2\sqrt{|x|^2 + |d|^2} - |x + d| \geq |x| - \frac{x \cdot d}{|x|}. \tag{8.39}$$

Setting $y = |x|$ in (8.37), we get

$$\sqrt{2(|x|^2 + |d|^2) - |x + d|^2} \geq |x| - \frac{x \cdot d}{|x|} \tag{8.40}$$

and it suffices to note that

$$2\sqrt{|x|^2 + |d|^2} - |x + d| \geq \sqrt{2(|x|^2 + |d|^2) - |x + d|^2}.$$

This establishes (8.39). Let us now consider the case $\sqrt{y^2 + |d|^2} \geq |x + d|$. We must prove that

$$\sqrt{2(y^2 + |d|^2) - |x + d|^2} \geq 2y - |x| - \frac{x \cdot d}{|x|},$$

or

$$\sqrt{2(y^2 + |d|^2) - |x + d|^2} - 2y \geq -|x| - \frac{x \cdot d}{|x|}.$$

By inequality (8.35), the left-hand side is not smaller than

$$\sqrt{2(|x|^2 + |d|^2) - |x + d|^2} - 2|x|,$$

which, with the aid of (8.40), yields the desired inequality. The last step is to verify that U satisfies 3°. But this is obvious: $U(x, |x|) = -|x| \leq 0$. This completes the proof of (8.29). $\qquad \square$

Sharpness of (8.29). Suppose that the best constant in (8.29) for real-valued martingales f equals C. Let $V(x, y) = |x| - C|y|$ and

$$U^0(x, y) = \sup\{\mathbb{E}V(f_n, \sqrt{y^2 - |x|^2 + S_n^2(f)})\},$$

where the supremum is taken over all n and all simple martingales f starting from x. For a fixed nonnegative integer n, apply $2°$ to $x = n$, $y = \sqrt{n}$ and a mean zero random variable d taking the values $s < 0$ and 1. We obtain

$$\frac{s}{s-1} U(n+1, \sqrt{n+1}) + \frac{1}{1-s} U(n+s, \sqrt{n+s^2}) \leq U(n, \sqrt{n}),$$

which, by $1°$, implies

$$\frac{s}{s-1} U(n+1, \sqrt{n+1}) + \frac{1}{1-s} V(n+s, \sqrt{n+s^2}) \leq U(n, \sqrt{n}).$$

Now we let $s \to -\infty$ and get

$$U(n+1, \sqrt{n+1}) + 1 - C \leq U(n, \sqrt{n}).$$

This, by induction, implies that for any nonnegative integer n,

$$U(n, \sqrt{n}) \leq U(0,0) + n(C-1),$$

so, by $3°$ and the further use of $1°$, $V(n, \sqrt{n}) \leq n(C-1)$. Equivalently,

$$C \geq \frac{2n}{n + \sqrt{n}}$$

and letting $n \to \infty$ yields the result. $\square$

8.3.4   Proof of Theorem 8.8

We shall only focus on the second estimate; the remaining ones can be established in a similar manner. See [197].

Proof of (8.31). Suppose that $p \geq 3$. Clearly, if suffices to prove that for any $\varepsilon > 0$,

$$\|f_n\|_p \leq z_p \|S(f) + \varepsilon\|_p$$

and, consequently, we may consider functions U_p, V_p defined only on $\mathcal{H} \times (0, \infty)$ (that is, we may assume that $y \neq 0$). Let V_p be given by $V_p(x, y) = |x|^p - z_p^p y^p$, $x \in \mathcal{H}$, $y > 0$. Let ϕ_p be defined by (A.4) in the Appendix. The special function $U_p : \mathcal{H} \times (0, \infty) \to \mathbb{R}$ is given as follows:

$$U_p(x, y) = \begin{cases} \alpha_p y^p \phi_p(|x|/y) & \text{if } |x| \geq z_p y, \\ |x|^p - z_p y^p & \text{if } |x| < z_p y. \end{cases}$$

Here

$$\alpha_p = \frac{p z_p^{p-1}}{\phi_p'(z_p)} = \frac{z_p^{p-1}}{\phi_{p-1}(z_p)}.$$

To show that $U_p \geq V_p$, it suffices to consider x, y satisfying $|x| \geq z_p y$, for which the estimate becomes

$$F(s) := \alpha_p \phi_p(s) - s^p + z_p^p \geq 0, \qquad s \geq z_p.$$

We see that $F(z_p) = 0$, so it suffices to show that

$$F'(s) = \alpha_p \phi_p'(s) - ps^{p-1} \geq 0 \qquad s > z_p.$$

Since we have equality when $s \downarrow z_p$, we will be done if we prove that the function $s \mapsto \phi_p'(s)/s^{p-1}$ is nondecreasing on $[z_p, \infty)$. Differentiating, this is equivalent to proving that $\phi_p''(s)s - (p-1)\phi_p'(s) \geq 0$ or $\phi_p'''(s) \geq 0$ on $[z_p, \infty)$. This is proved in part (v) of Lemma A.11 in the Appendix.

The proof of the concavity property $2°$ is very elaborate, so we will only sketch it and refer the interested reader to [197] for details. First one shows the condition for $\mathcal{H} = \mathbb{R}$. For a fixed $x \in \mathbb{R}$, $y > 0$, introduce the function

$$G(d) = \frac{1}{2}\left[U_p(x+d, \sqrt{y^2+d^2}) + U_p(x-d, \sqrt{y^2+d^2})\right] - U_p(x,y)$$

$$= \frac{1}{2}(y^2+d^2)^{p/2}\left[U_p\left(\frac{x+d}{\sqrt{y^2+d^2}}, 1\right) + U_p\left(\frac{x-d}{\sqrt{y^2+d^2}}, 1\right)\right] - y^p U_p\left(\frac{x}{y}, 1\right).$$

One can check that $G'(d) \leq 0$ for all $d > 0$, considering separately six cases, depending on the positions of the points (x,y), $(x \pm d, \sqrt{y^2+d^2})$. This gives $G(d) \leq G(0) = 0$ for all $d \geq 0$ and $2°$ follows. In the case when $\mathcal{H}$ is a general Hilbert space, let x, $d \in \mathcal{H}$, $y \geq 0$, substitute $X = |x|$, $D = |d|$ and let $x \cdot d = XD\cos\theta$. One writes the condition in the form

$$\frac{1}{2}(y^2+D^2)^{p/2}\left[U_p\left(\sqrt{\frac{X^2+D^2+2XD\cos\theta}{y^2+D^2}}, 1\right)\right. \tag{8.41}$$

$$\left. + U_p\left(\sqrt{\frac{X^2+D^2-2XD\cos\theta}{y^2+D^2}}, 1\right)\right] - y^p U_p\left(\frac{X}{y}, 1\right) \leq 0.$$

The real-valued case considered earlier implies that the estimate holds when $\cos\theta = 1$. Consequently, if we define $G : [0,1] \to \mathbb{R}$ by

$$G(t) = U_p\left(\sqrt{\frac{X^2+D^2+2XDt}{y^2+D^2}}, 1\right) + U_p\left(\sqrt{\frac{X^2+D^2-2XDt}{y^2+D^2}}, 1\right),$$

then it suffices to show that $G'(t) > 0$ on $(0,1)$. Again, this is done by tedious verification of six cases depending on the values of X, y, D and t. We omit the details. Finally, the condition $3°$ follows immediately from $2°$. This completes the proof. $\qquad\square$

Remark 8.4. One can check that when $2 < p < 3$ and we define U_p, V_p by the above formulas, then the condition $2°$ *does not* hold, even in the real-valued case. It is not difficult to see that this implies that for $2 < p < 3$ the optimal constant in (8.31) is strictly larger than z_p.

Sharpness. Fix $p \geq 3$. Let $C > 0$ be a constant such that for any dyadic martingale f we have $||f||_p \leq C||S(f)||_p$. Thus, using approximation, if $B = (B_t)_{t\geq 0}$ is a standard Brownian motion and τ is a stopping time of B, then

$$||B_\tau||_p \leq C||\tau||_p.$$

However, as shown by Davis [59], the optimal constant in this inequality equals z_p. Consequently, $C \geq z_p$ and we are done.   □

On the search of the suitable majorant. As already mentioned above, while studying estimates for dyadic martingales and their square functions, it is often fruitful to look at corresponding estimates for a stopped Brownian motion. In other words, instead of investigating the properties of

$$U^0(x, y) = \sup\{\mathbb{E}|f_n|^p - C^p\mathbb{E}(y^2 - x^2 + S_n^2(f))^{p/2}\},$$

we study the following optimal stopping problem:

$$U(x, y) = \sup\{\mathbb{E}|B_\tau|^p - C^p\mathbb{E}(y^2 - x^2 + [B, B]_\tau)^{p/2}\}$$
$$= \sup\{\mathbb{E}|B_\tau|^p - C^p\mathbb{E}(y^2 + \tau)^{p/2}\},$$

where the Brownian motion starts from x and the supremum is taken over all stopping times of B satisfying $\mathbb{E}\tau^{p/2} < \infty$. This leads to $C = z_p$ and the function U_p studied above: see Pedersen and Peskir [167]. See also the search corresponding to the inequality (6.12), presented in Subsection 6.3.3 of Chapter 6.   □

8.4 Moment estimates for nonnegative martingales

8.4.1 Formulation of the result

If a martingale is nonnegative, then the inequalities (8.28) hold also for $p < 1$. Let

$$C_p = \left(\int_1^\infty (1 + t^2)^{p/2} \frac{dt}{t^2}\right)^{1/p}, \quad \text{if } p \neq 0,$$

$$C_0 = \lim_{p \to 0} C_p = \exp\left(\int_1^\infty \frac{1}{2}\log(1 + t^2)\frac{dt}{t^2}\right).$$

We shall prove the following fact.

Theorem 8.9. *If f is a nonnegative martingale, then*

$$||f||_p \leq ||S(f)||_p \leq C_p||f||_p, \quad \text{if } p < 1, \tag{8.42}$$

and the inequality is sharp.

8.4.2 Proof of Theorem 8.9

Proof of (8.42). Since $p < 1$ and the martingale is nonnegative, we have that $||f||_p = ||f_0||_p$. Thus all we need is the double inequality

$$||f_0||_p \leq ||S_n(f)||_p \leq C_p ||f_0||_p, \qquad n = 0, 1, 2, \ldots. \qquad (8.43)$$

First note that the left inequality is obvious, since $||f_0||_p = ||S_0(f)||_p \leq ||S_n(f)||_p$. Furthermore, it clearly is sharp. Hence, it remains to prove the right inequality in (8.43). We focus on the case $p \neq 0$. Let us introduce the functions $V_p, U_p :$ $(0, \infty)^2 \to \mathbb{R}$ by

$$V_p(x, y) = p y^p \quad \text{and} \quad U_p(x, y) = px \int_x^\infty (y^2 + t^2)^{p/2} \frac{dt}{t^2}.$$

The desired bound can be stated in the form

$$\mathbb{E} V_p(f_n, S_n(f)) \leq \mathbb{E} U_p(f_0, f_0)$$

and hence all we need is to verify 1° and 2°. The majorization follows from

$$U_p(x, y) - V_p(x, y) = px \int_x^\infty \left[(y^2 + t^2)^{p/2} - y^p \right] \frac{dt}{t^2}.$$

To prove 2°, we shall show a bit stronger statement: for any positive x and any number $d > -x$,

$$U(x + d, \sqrt{y^2 + d^2}) \leq U(x, y) + U_x(x, y) d.$$

Integration by parts yields

$$U_p(x, y) = p(y^2 + x^2)^{p/2} + p^2 x \int_x^\infty (y^2 + t^2)^{p/2-1} dt \qquad (8.44)$$

and

$$U_{px}(x, y) = p \int_x^\infty (y^2 + t^2)^{p/2} \frac{dt}{t} - p \frac{(y^2 + x^2)^{p/2}}{x} = p^2 \int_x^\infty (y^2 + t^2)^{p/2-1} dt.$$

Hence, we must prove that

$$p(y^2 + d^2 + (x + d)^2)^{p/2} + p^2 (x + d) \int_{x+d}^\infty (y^2 + d^2 + t^2)^{p/2-1} dt$$

$$- p(y^2 + x^2)^{p/2} - p^2 x \int_x^\infty (y^2 + t^2)^{p/2-1} dt - p^2 d \int_x^\infty (y^2 + t^2)^{p/2-1} d \leq 0,$$

or, equivalently,

$$F(x) := p \frac{(y^2 + d^2 + (x + d)^2)^{p/2} - (y^2 + x^2)^{p/2}}{x + d}$$

$$- p^2 \left[\int_x^\infty (y^2 + t^2)^{p/2-1} dt - \int_{x+d}^\infty (y^2 + d^2 + t^2)^{p/2-1} dt \right] \leq 0.$$

We have

$$F'(x)(x+d)^2 = p^2(y^2+x^2)^{p/2-1}(x+d)d$$
$$- p[(y^2+d^2+(x+d)^2)^{p/2} - (y^2+x^2)^{p/2}], \qquad (8.45)$$

which is nonnegative due to the mean value property of the function $t \mapsto t^{p/2}$. Hence

$$F(x) \le \lim_{s \to \infty} F(s) = 0$$

and the proof is complete.   □

Sharpness. Fix $\varepsilon > 0$ and define f by $f_n = (1+n\varepsilon)\,(0,(1+n\varepsilon)^{-1}]$, $n = 0, 1, 2, \ldots$. Then it is easy to check that f is a nonnegative martingale, $df_0 = (0,1]$,

$$df_n = \varepsilon\,(0,(1+n\varepsilon)^{-1}] - (1+(n-1)\varepsilon)\,((1+n\varepsilon)^{-1},(1+(n-1)\varepsilon)^{-1}],$$

for $n = 1, 2, \ldots$, and

$$S(f) = \sum_{n=0}^{\infty}(1+n\varepsilon^2+(1+n\varepsilon)^2)^{1/2}\,((1+(n+1)\varepsilon)^{-1},(1+n\varepsilon)^{-1}].$$

Furthermore, for $p < 1$ we have $\|f\|_p = 1$ and, if $p \neq 0$,

$$\|S(f)\|_p^p = \varepsilon\sum_{n=0}^{\infty}\frac{(1+n\varepsilon^2+(1+n\varepsilon)^2)^{p/2}}{(1+(n+1)\varepsilon)(1+n\varepsilon)},$$

which is a Riemann sum for C_p^p. Finally, the case $p = 0$ is dealt with by passing to the limit; this is straightforward, as the martingale f does not depend on p.   □

On the search of the suitable majorant. It is not difficult to obtain the above special function U. For a fixed $p < 1$, $p \neq 0$, let

$$U^0(x,y) = \sup\{\mathbb{E}(y^2 - x^2 + S_n^2(f))^{p/2}\},$$

where the supremum is taken over all simple nonnegative martingales f starting from x. For a given x, $y > 0$, what is the choice for f which yields the largest expectation in the parentheses? It is instructive to look at the corresponding moment inequality (4.37) for nonnegative martingales and their differential subordinates. There the optimal processes (f, g) had the following property: if $(f_0, g_0) = (x, y)$, then at the next step the process moved either to the y-axis or "a bit to the right". It is natural to conjecture that here the situation is similar. That is, the optimal martingale satisfies the condition $df_n \in \{\varepsilon, -f_{n-1}\}$ on the set $\{f_{n-1} > 0\}$ and $df_n = 0$ on $\{f_{n-1} = 0\}$. This leads to the function U: see the example above and apply the homogeneity of U.   □

8.5 Ratio inequalities

8.5.1 Formulation of the result

We start with the following negative result.

Theorem 8.10. *Let p, q, M be positive numbers. Then there is a simple martingale f such that*

$$\mathbb{E}\frac{|f_\infty|^p}{(1+S^2(f))^q} > M.$$

In other words, the expectation of the ratio $|f_\infty|^p/(1+S^2(f))^q$ cannot be bounded by an absolute constant for any p, q. However, if we narrow the class of martingales to the conditionally symmetric ones, such a bound exists. We have the following statement.

Theorem 8.11. *Let p, q be fixed positive numbers. If $p < 2q$, then for any conditionally symmetric martingale f,*

$$\sup_n \mathbb{E}\frac{|f_n|^p}{(1+S_n^2(f))^q} \leq z_*^p \left(M\left(q - \frac{p}{2}, \frac{1}{2}, \frac{z_*^2}{2}\right)\right)^{-1}, \qquad (8.46)$$

where M is the Kummer function (see Appendix) and $z_* = z_*(p,q)$ is the unique solution to the equation

$$pM\left(q - \frac{p}{2}, \frac{1}{2}, \frac{z^2}{2}\right) = z^2(2q-p)M\left(q - \frac{p}{2}+1, \frac{3}{2}, \frac{z^2}{2}\right). \qquad (8.47)$$

The constant on the right in (8.46) *is the best possible even in the case of real-valued dyadic martingales.*

If $p \geq 2q$, then there is no finite universal $C_{p,q} < \infty$ for which

$$\sup_n \mathbb{E}\frac{|f_n|^p}{(1+S_n^2(f))^q} \leq C_{p,q}$$

holds for any conditionally symmetric martingale f.

We also establish the following nonsymmetric ratio inequality for real-valued conditionally symmetric martingales.

Theorem 8.12. *Let $N = 1, 2, \ldots$ and q be a positive number. If $q > N - 1/2$, then for any real-valued conditionally symmetric martingale f we have*

$$\sup_n \mathbb{E}\frac{f_n^{2N-1}}{(1+S_n^2(f))^q} \leq z_*^{2N-1}\exp(-z_*^2/4)(D_{2(N-q)-1}(-z_*))^{-1}. \qquad (8.48)$$

Here D is a parabolic cylinder function (see Subsection 8.5.2 below) and $z_* = z_*(N,q)$ is the unique solution to the equation

$$(2N-1)D_{2(N-q)-1}(-z) = z(2(q-N)+1)D_{2(N-q-1)}(-z). \qquad (8.49)$$

*The constant appearing on the right in (8.48) is the best possible even in the case
of dyadic martingales.*

If $q \leq N - 1/2$, then there is no universal $C'_{N,q}$ for which

$$\sup_n \mathbb{E} \frac{f_n^{2N-1}}{(1 + S_n^2(f))^q} \leq C'_{N,q}$$

is valid for any conditionally symmetric martingale f.

We shall only present the proofs of Theorems 8.10 and 8.11. The proof of
Theorem 8.12 is completely analogous and we refer the interested reader to the
original paper [143] for details.

8.5.2   Lack of the ratio inequalities in the general case

Let p, $q > 0$ be fixed and let $U^0 : \mathbb{R} \times [0, \infty) \to (-\infty, \infty]$ be defined by

$$U^0(x, y) = \sup \left\{ \mathbb{E} \frac{|f_n|^p}{(1 + y^2 - |x|^2 + S_n^2(f))^q} \right\},$$

where the supremum is taken over all n and all simple martingales starting from x.
If there is $(x, y) \in \mathbb{R} \times (0, \infty)$ such that $U^0(x, y) = \infty$, then we are done. Indeed,
it is easy to see that there is a simple martingale f starting from 0 such that
$\mathbb{P}(f_\infty = x, S(f) = y) > 0$ and hence, by the splicing argument, f can be extended
so that the ratio $|f|^p/(1 + S^2(f))^q$ has arbitrarily large expectation. So, assume
that U^0 is finite. Let us apply 2° to some $x \in \mathbb{R}$, $y > 0$ and a centered random
variable d taking values $d_1 > 0 > d_2$. We get

$$U^0(x, y) \geq \frac{d_1}{d_1 - d_2} U^0 \left(x + d_2, \sqrt{y^2 + d_2^2} \right) + \frac{-d_2}{d_1 - d_2} U^0 \left(x + d_1, \sqrt{y^2 + d_1^2} \right)$$

$$\geq \frac{-d_2}{d_1 - d_2} U^0 \left(x + d_1, \sqrt{y^2 + d_1^2} \right),$$

since U^0 is nonnegative. Letting $d_2 \to -\infty$ we obtain the inequality $U^0(x, y) \geq
U \left(x + d_1, \sqrt{y^2 + d_1^2} \right)$, which, by induction, leads to

$$U^0(x, y) \geq U \left(x + N d_1, \sqrt{y^2 + N d_1^2} \right),$$

for any nonnegative integer N. Take $x = 0$, $y = 1$ and apply 1° to get

$$\frac{(N d_1)^p}{(2 + N d_1^2)^q} \leq U^0(0, 1).$$

It suffices to take $d_1 = 1/\sqrt{N}$ and let $N \to \infty$ to obtain $U^0(0, 1) = \infty$, a contradiction. This proves the claim.

8.5.3 Proof of Theorem 8.11

Proof of (8.46). Introduce the function $V : \mathcal{H} \times (0, \infty) \to \mathbb{R}$ by $V(x, y) = |x|^p / y^{2q}$. The special function $U : \mathcal{H} \times (0, \infty) \to \mathbb{R}$ corresponding to our problem is given by

$$U(x, y) = z_*^p y^{p-2q} M \left(q - \frac{p}{2}, \frac{1}{2}, \frac{|x|^2}{2y^2} \right) \Big/ M \left(q - \frac{p}{2}, \frac{1}{2}, \frac{z_*^2}{2} \right).$$

To show 1°, denote $a = |x|/y$ and observe that the majorization is equivalent to

$$F(a) := \frac{M(q - \frac{p}{2}, \frac{1}{2}, \frac{a^2}{2})}{a^p} \geq \frac{M(q - \frac{p}{2}, \frac{1}{2}, \frac{z_*^2}{2})}{z_*^p}.$$

We have, by (A.1),

$$F'(a) = a^{-p-1} \left[a^2 (2q - p) M \left(q - \frac{p}{2} + 1, \frac{3}{2}, \frac{a^2}{2} \right) - pM \left(q - \frac{p}{2}, \frac{1}{2}, \frac{a^2}{2} \right) \right],$$

so $F'(a) < 0$ for $a < z_*$ and $F'(a) > 0$ for $a > z_*$. Thus $F(a) \geq F(z_*)$, which is 1°. To prove 2°, let $\psi : \mathbb{R} \to \mathbb{R}$ be defined by $\psi(s) = M(q - p/2, 1/2, s^2/2)$. Fix $x, d \in \mathcal{H}$, $y > 0$ and introduce $G = G_{x,y,d} : [0, \infty) \to \mathbb{R}$ by the formula

$$G(s) = (y^2 + |d|^2 s^2)^{p/2-q} \left[\psi \left(\frac{|x - ds|}{\sqrt{y^2 + |d|^2 s^2}} \right) + \psi \left(\frac{|x + ds|}{\sqrt{y^2 + |d|^2 s^2}} \right) \right].$$

Since 2° is equivalent to the inequality $G(1) \leq G(0)$, we will be done if we show that G is nonincreasing. Clearly, we may assume that $d \neq 0$. Denote

$$s_1 = \frac{x - ds}{\sqrt{y^2 + |d|^2 s^2}}, \quad s_2 = \frac{x + ds}{\sqrt{y^2 + |d|^2 s^2}}$$

and observe that

$$\frac{G'(s)}{(y^2 + |d|^2 s^2)^{p/2-q-1}} = (p - 2q)|d|^2 s \left[\psi(|s_1|) + \psi(|s_2|) \right]$$

$$+ \psi'(|s_1|)[-(x - ds)' \cdot d\sqrt{y^2 + |d|^2 s^2} - |s_1||d|^2 s]$$

$$+ \psi'(|s_2|)[(x + ds)' \cdot d\sqrt{y^2 + |d|^2 s^2} - |s_2||d|^2 s].$$

Now divide throughout by $\sqrt{y^2 + |d|^2 s^2}|d|$ and note that

$$\frac{|d|s}{\sqrt{y^2 + |d|^2 s^2}} = \frac{|s_1 - s_2|}{2}, \quad (x - ds)' = s_1' \text{ and } (x + ds)' = s_2'.$$

Furthermore, as $(p - 2q)\psi(z) = -\psi''(z) + z\psi'(z)$ for all z, the above equality transforms into

$$\frac{G'(s)}{(y^2 + |d|^2 s^2)^{p/2-q-1/2}|d|} = -\frac{|s_1 - s_2|}{2} [\psi''(|s_1|) + \psi''(|s_2|)]$$

$$- (\psi'(|s_1|)s_1' - \psi'(|s_2|)s_2') \cdot d'.$$

The right-hand side does not exceed

$$-\frac{|s_1 - s_2|}{2}[\psi''(|s_1|) + \psi''(|s_2|)] + |\psi'(|s_1|)s_1' - \psi'(|s_2|)s_2'|,$$

which is nonpositive due to (8.16); it is straightforward to check that ψ has all the required properties. The initial condition is of the form

$3°$ $U(x, \sqrt{1 + |x|^2}) \leq C_{p,q}$ for all $x \in \mathcal{H}$.

Since $C_{p,q} = U(0, 1)$, this follows immediately from $2°$:

$$U(0, 1) \geq \frac{1}{2}[U(x, \sqrt{1 + |x|^2}) + U(-x, \sqrt{1 + |x|^2})] = U(x, \sqrt{1 + |x|^2}).$$

The proof is complete. □

Sharpness. Arguing as in the proof of the weak type inequality, we see that it suffices to prove that $C_{p,q}$ is optimal in the inequality

$$\mathbb{E}\left[\frac{|B_\tau|^p}{(1 + \tau)^q}\right] \leq C_{p,q},$$

where B is a standard one-dimensional Brownian motion and τ is a finite stopping time of B. For the proof of this, see [167]. □

On the search of the suitable majorant. We proceed as in the above estimates for conditionally symmetric martingales and begin by studying the corresponding inequality for the stopped Brownian motion. The optimal stopping problem we obtain is the following: for given $p, q > 0$, determine

$$W(t, x) = \sup \mathbb{E}\frac{|x + B_\tau|^p}{(t + \tau)^q},$$

where the supremum is taken over all finite stopping times of B. Pedersen and Peskir [167] showed that the value function W is infinite if $q \leq p/2$, while for $q > p/2$ it is given by

$$W(t, x) = \begin{cases} z_*^p t^{p/2-q} M(q - \frac{p}{2}, \frac{1}{2}, \frac{x^2}{2t})/M(q - \frac{p}{2}, \frac{1}{2}, \frac{z_*^2}{2}) & \text{if } |x|/\sqrt{t} < z_*, \\ |x|^p/t^q & \text{if } |x|/\sqrt{t} \geq z_*, \end{cases}$$

where z_* is defined in (8.47). This function is the key to (8.46) and leads to the corresponding special function $U : \mathcal{H} \times (0, \infty) \to \mathbb{R}$; namely, a natural guess is to put $U(x, y) = W(y^2, |x|)$. However, this choice is not convenient: due to the existence of two different formulas in the definition, the verification of the properties $1°$, $2°$ and $3°$ turns out to be very elaborate. To avoid these technicalities, we use one of these formulas on the whole domain: fortunately the function we obtain still enjoys he necessary properties. □

8.6 Maximal inequalities

We turn to the estimates involving the square and the maximal function of a martingale.

8.6.1 Formulation of the results

The first result we shall prove is the following.

Theorem 8.13. *For any Hilbert-space-valued martingale f we have*

$$||S(f)||_1 \leq \sqrt{3}\,||\,|f|^*||_1 \tag{8.50}$$

and the constant is the best possible.

The inequality in the reverse direction also holds with some finite constant, but its optimal value is unknown. The only result in this direction is the following bound for the one-sided maximal function of a conditionally symmetric martingale. Let $s_0 = -0.8745\ldots$ be the unique negative solution to the equation

$$\int_0^{s_0} \exp\left(\frac{u^2}{2}\right) du + 1 = 0 \tag{8.51}$$

and set

$$\beta = \exp\left(\frac{s_0^2}{2}\right) = 1.4658\ldots.$$

Theorem 8.14. *If f is a conditionally symmetric martingale taking values in $\mathbb{R}$, then*

$$||f^*||_1 \leq \beta||S(f)||_1. \tag{8.52}$$

The constant is the best possible. It is already the best possible for dyadic martingales.

Theorem 8.15. *For any Hilbert-space-valued martingale f we have*

$$||\,|f|^*||_p \leq p||S(f)||_p, \qquad p \geq 2. \tag{8.53}$$

The constant p is the best possible.

Finally, we shall establish the following fact concerning nonnegative martingales.

Theorem 8.16. *If f is a nonnegative martingale, then*

$$||S(f)||_p \leq \sqrt{2}||f^*||_p, \qquad \text{if } p \leq 1, \tag{8.54}$$

and the constant $\sqrt{2}$ is the best possible.

8.6.2   Proof of Theorem 8.13

Proof of (8.50). The claim is equivalent to

$$\mathbb{E}V(f_n, S_n(f), |f_n|^* \vee \varepsilon) \leq 0, \qquad \varepsilon > 0, \, n = 0, 1, 2, \ldots,$$

where $V : \mathcal{H} \times [0, \infty) \times (0, \infty) \to \mathbb{R}$ is given by $V(x, y, z) = y - \sqrt{3}(|x| \vee z)$. The special function U is given by

$$U(x, y, z) = \frac{y^2 - |x|^2 - 2(|x| \vee z)^2}{2\sqrt{3}(|x| \vee z)},$$

for $x \in \mathcal{H}$, $y \geq 0$ and $z > 0$. Let us verify that U and V have the necessary properties. The majorization $U \geq V$ is straightforward:

$$\frac{y^2 - |x|^2 - 2(|x| \vee z)^2}{2\sqrt{3}(|x| \vee z)} \geq \frac{y^2 - 3(|x| \vee z)^2}{2\sqrt{3}(|x| \vee z)}$$
$$= \frac{(y - \sqrt{3}(|x| \vee z))^2}{2\sqrt{3}(|x| \vee z)} + |y| - 2\sqrt{3}(|x| \vee z)$$
$$\geq |y| - 2\sqrt{3}(|x| \vee z).$$

To show 2° we may assume that $z = 1$, due to homogeneity. As usual, we show a slightly stronger, pointwise statement: for all x, $d \in \mathcal{H}$ and $y \geq 0$ with $|x| \leq 1$,

$$U(x + d, \sqrt{y^2 + |d|^2}, 1) \leq U(x, y, 1) - \frac{x \cdot d}{\sqrt{3}}.$$

When $|x+d| \leq 1$, the two sides are equal. If $|x+d|$ exceeds 1, the estimate becomes

$$(y^2 + |d|^2 - |x + d|^2 + 2|x + d|)(1 - |x + d|) \leq 0$$

and follows from

$$y^2 + |d|^2 - |x + d|^2 + 2|x + d| \geq |d|^2 - |x + d|^2 + 2|x + d|$$
$$= |d|^2 + 1 - (|x + d| - 1)^2$$
$$\geq |d|^2 + 1 - (|x| + |d| - 1)^2$$
$$= (|x| + 2|d|)(1 - |x|) + |x| \geq 0.$$

Finally, the conditions 3° and 4° are obvious. This proves (8.50).   □

Sharpness. Suppose that the best constant in (8.50), restricted to real-valued martingales, equals C. Let

$$U^0(x, y, z) = \sup\{\mathbb{E}\sqrt{y^2 - x^2 + S_n^2(f)} - C\mathbb{E}(|f_n|^* \vee z)\},$$

where the supremum is taken over the usual parameters. Then U satisfies $1°$–$4°$ and is homogeneous of order 1. Applying $2°$ to $x = z = 1$, $y < 1$ and a centered random variable d taking the values -1 and $N = 2y^2/(1 - y^2)$, gives

$$U^0(1, y, 1) \geq \frac{N}{N+1} U^0(0, \sqrt{y^2 + 1}, 1) + \frac{1}{N+1} U^0(1 + N, \sqrt{y^2 + N^2}, 1)$$

$$= \frac{N}{N+1} U^0(0, \sqrt{y^2 + 1}, 1) + U^0\left(1, \frac{\sqrt{y^2 + N^2}}{1 + N}, 1\right)$$

$$= \frac{N}{N+1} U^0(0, \sqrt{y^2 + 1}, 1) + U^0(1, y, 1).$$

where in the second passage we used homogeneity. This implies $U^0(0, y, 1) \leq 0$ for any $y \in [1, \sqrt{2})$. Applying $2°$ again and then using $1°$ gives

$$U^0(0, y, 1) \geq \frac{1}{2}\left[U^0(-1, \sqrt{y^2 + 1}, 1) + U^0(1, \sqrt{y^2 + 1}, 1)\right]$$

$$= U^0(1, \sqrt{y^2 + 1}, 1) \geq \sqrt{y^2 + 1} - C.$$

Thus, by the preceding arguments, $C \geq \sqrt{y^2 + 1}$ if $y \in [1, \sqrt{2})$; this yields $C \geq \sqrt{3}$ and shows that no constant smaller than $\sqrt{3}$ suffices in (8.13). $\qquad\square$

8.6.3 Proof of Theorem 8.14

Proof of (8.52). The argumentation is similar to that presented in the proofs of (8.46) and (8.31), so we shall only provide the formula for the special function. Let

$$\phi(s) = -e^{s^2/2} + s \int_0^s e^{t^2/2} dt.$$

Then ϕ is a confluent hypergeometric function of parameter 1 (see [1]). Let

$$V(x, y, z) = (x \vee z) - \beta y \quad \text{and} \quad U(x, y, z) = x + y\phi\left(\frac{x - (x \vee z)}{y}\right).$$

Conditions $1°$, $2°$ and $3°$ are verified in the usual way, and in the proof of the concavity property one exploits the differential equation $\phi''(s) - s\phi(s) + \phi(s) = 0$. We omit the details: see [148]. $\qquad\sqcup$

Sharpness. By approximation, it suffices to prove that the constant β is optimal for continuous-path martingales. Let us first recall a result of Shepp [185]. Suppose that $B^0 = (B_t^0)$ is a Brownian motion starting from 0 and consider the stopping time $T = T_a$ defined by

$$T_a = \inf\{t > 0 : |B_t^0| = a\sqrt{1 + t}\}.$$

Shepp has shown that if $a < \nu_1$, then the stopping time T_a belongs to $L^{1/2}$.

Now let $B^1 = (B^1_t)$ be a Brownian motion starting from 1 and let

$$\tau = \inf\{t > 0 : (B^1_t)^* - B^1_t = -s_0\sqrt{1+t}\}. \tag{8.55}$$

In view of Levy's theorem, the distributions of τ and T_{-s_0} coincide, so in particular we have $\mathbb{E}\tau^{1/2} < \infty$ (since $-s_0 < \nu_1$). Let $X = (B^1_{\tau \wedge t})$ and consider the process $Y = (U(X_t, [X, X]_t, X^*_t))$. Using Itô's formula, it is easy to check that Y is a martingale which converges in L^1. Now it suffices to note that $\mathbb{E}Y_0 = 0$ and $Y_t \to X^* - \beta[X, X]^{1/2}$ almost surely as $t \to \infty$. This shows that

$$\||X^*\||_1 = \beta\||[X, X]^{1/2}\||_1 \tag{8.56}$$

and we are done. $\square$

8.6.4   Proof of Theorem 8.15

Proof of (8.53). This follows immediately from Doob's maximal inequality and (8.28). $\square$

Sharpness. Suppose that for any real-valued martingale f we have $\|\, |f|^*\,\|_p \leq C\|S(f)\|_p$, where $C > 0$ is an absolute constant. Let

$$U^0(x, y, z) = \sup\{\mathbb{E}(|f_n|^* \vee z)^p - C^p\mathbb{E}(y^2 - x^2 + S_n^2(f))^{p/2}\},$$

where the supremum is taken over all n and all simple real-valued martingales starting from x. Then U^0 satisfies 1°–4° and is homogeneous of order p. Fix $\varepsilon > 0$ and apply 2° to $x = z = 1$, $y = \varepsilon$ and to a mean-zero variable d taking values $-1/p$ and $\delta = 2\varepsilon^2/(1 - \varepsilon^2)$. We obtain

$$U^0(1, \varepsilon, 1) \geq \frac{p\delta}{1 + p\delta}U^0\left(1 - \frac{1}{p}, \sqrt{\varepsilon^2 + \frac{1}{p^2}}, 1\right) + \frac{1}{1 + p\delta}U^0(1 + \delta, \sqrt{\varepsilon^2 + \delta^2}, 1).$$

Using 1° and 4°, this implies

$$U^0(1, \varepsilon, 1) \geq \frac{p\delta}{1 + p\delta}\left[1 - \left(C\sqrt{\varepsilon^2 + \frac{1}{p^2}}\right)^p\right] + \frac{1}{1 + p\delta}U^0(1 + \delta, \sqrt{\varepsilon^2 + \delta^2}, 1 + \delta).$$

Finally, by homogeneity,

$$U^0(1 + \delta, \sqrt{\varepsilon^2 + \delta^2}, 1 + \delta) = \left(\frac{1 + \varepsilon^2}{1 - \varepsilon^2}\right)^p U^0(1, \varepsilon, 1),$$

so plugging this above and working a little bit yields

$$U^0(1, \varepsilon, 1)\left[1 - \frac{1}{1 + p\delta}\left(\frac{1 + \varepsilon^2}{1 - \varepsilon^2}\right)^p\right] \geq \frac{p\delta}{1 + p\delta}\left[1 - \left(C\sqrt{\varepsilon^2 + \frac{1}{p^2}}\right)^p\right].$$

Now divide both sides by ε^2 and let $\varepsilon \to 0$. Then the left-hand side converges to 0 and we obtain $0 \geq 2p(1 - (C/p)^p)$, that is, $C \geq p$. This completes the proof. $\square$

8.6.5 Proof of Theorem 8.16

The inequality (8.54) may be proved using a special function involving three variables. However, this function seems to be difficult to construct and we have managed to find it only in the case $p = 1$ (see Remark 8.5 below). To overcome this problem, we need an extension of Burkholder's method allowing to work with other operators: we will establish the stronger result

$$||T(f)||_p \le \sqrt{2}||f^*||_p, \quad \text{if } p \le 1. \tag{8.57}$$

Here, given a martingale f, we define the sequence $(T_n(f))$ by

$$T_0(f) = |f_0|, \quad T_{n+1}(f) = (T_n^2(f) + df_{n+1}^2)^{1/2} \vee f_{n+1}^*, \quad n = 0, 1, 2, \ldots,$$

and $T(f) = \lim_{n \to \infty} T_n(f)$. Observe that $T_n(f) \ge S_n(f)$ for all n, which can be easily proved by induction. Thus (8.57) implies (8.54).

Theorem 8.17. *Suppose that U and V are functions from $\{(x, y, z) \in (0, \infty)^3 : y \ge x \vee z\}$ into $\mathbb{R}$ satisfying*

1° $V(x, y, z) \le U(x, y, z)$,

2° *if $0 < x \le z \le y$ and d is a simple mean zero random variable with $\mathbb{P}(x+d > 0) = 1$, then*

$$\mathbb{E}U(x + d, \sqrt{y^2 + d^2} \vee (x + d), z) \le U(x, y, z).$$

3° $U(x, x, x) \le 0$ *for any $x \ge 0$.*

4° $U(x, y, z) = U(x, y, x \vee z)$ *for all (x, y, z).*

Then $\mathbb{E}V(f_n, T_n(f), |f_n|^*) \le 0$ for all nonnegative integers n and simple positive martingales f.

The proof of this theorem is analogous to that of Theorem 8.1 and is omitted. We start with some auxiliary technical results.

Lemma 8.10.

(i) *If $z \ge d > 0$ and $y > 0$, then*

$$p\left[(y^2 + d^2 + z^2)^{p/2} - (y^2 + (z - d)^2)^{p/2}\right] - p^2 z \int_{z-d}^{z} (y^2 + t^2)^{p/2-1} dt \le 0. \tag{8.58}$$

(ii) *If $-z < d \le 0$ and $Y > 0$, then*

$$p\frac{(Y + (z + d)^2)^{p/2} - (Y^2 - d^2 + z^2)^{p/2}}{z + d} + p^2 \int_{z+d}^{z} (Y + t^2)^{p/2-1} dt \le 0. \tag{8.59}$$

(iii) *If $y \geq z \geq x > 0$, then*

$$p\left[(y^2 + x^2)^{p/2} - 2^{p/2} z^p\right] + p^2 \frac{x^2 + y^2}{2x} \int_x^z (y^2 + t^2)^{p/2-1} dt \geq 0. \qquad (8.60)$$

(iv) *If $D \geq z \geq x > 0$, $y \geq z$, then*

$$p\left[(y^2 + (D-x)^2 + D^2)^{p/2} - (y^2 + x^2)^{p/2} + 2^{p/2}(z^p - D^p)\right] \qquad (8.61)$$

$$- p^2 D \int_x^z (y^2 + t^2)^{p/2-1} dt \leq 0.$$

Proof. Denote the left-hand sides of (8.58)–(8.61) by $F_1(d)$, $F_2(d)$, $F_3(x)$ and $F_4(x)$, respectively. The inequalities will follow by simple analysis of the derivatives.

(i) We have

$$F_1'(d) = p^2 d[(y^2 + d^2 + z^2)^{p/2-1} - (y^2 + (z-d)^2)^{p/2-1}] \leq 0,$$

as $(z-d)^2 \leq d^2 + z^2$. Hence $F_1(d) \leq F_1(0+) = 0$.

(ii) The expression $F_2'(d)(z+d)^2$ equals

$$p\left[(Y - d^2 + z^2)^{p/2} - (Y + (z+d)^2)^{p/2} + \frac{p}{2}(Y - d^2 + z^2)^{p/2-1} \cdot 2d(z+d)\right] \geq 0,$$

due to the mean value property. This yields $F_2(d) \leq F_2(0) = 0$.

(iii) We have

$$F_3'(x) = \frac{p^2}{2}\left(1 - \frac{y^2}{x^2}\right)\left[(y^2 + x^2)^{p/2-1} x + \int_x^z (y^2 + t^2)^{p/2-1} dt\right] \leq 0$$

and $F_3(x) \geq F_3(z) = p[(y^2 + z^2)^{p/2} - 2^{p/2} z^p] \geq 0$.

(iv) Finally,

$$F_4'(x) = p^2(D-x)\left[-(y^2 + (D-x)^2 + D^2)^{p/2-1} + (y^2 + x^2)^{p/2-1}\right] \geq 0$$

and hence

$$F_4(x) \leq F_4(z) = p\left[(y^2 + (D-z)^2 + D^2)^{p/2} - (y^2 + z^2)^{p/2}\right] - p2^{p/2}(D^p - z^p).$$

The right-hand side decreases as y increases. Therefore

$$F_4(z) \leq p\left[(z^2 + (D-z)^2 + D^2)^{p/2} - 2^{p/2} D^p\right] \leq 0,$$

as $z^2 + (D-z)^2 + D^2 \leq 2D^2$. $\qquad \square$

Now we are ready to establish (8.54). Let

$$V_p(x, y, z) = p\left(y^p - 2^{p/2}(x \vee z)^p\right)$$

and

$$U_p(x, y, z) = p^2 x \int_x^{x \vee z} (y^2 + t^2)^{p/2 - 1} dt + p(y^2 + x^2)^{p/2} - p2^{p/2}(x \vee z)^p. \quad (8.62)$$

We need to check 1°–4°. The majorization is a consequence of the identity

$$U_p(x, y, z) - V_p(x, y, z) = p[(y^2 + x^2)^{p/2} - y^p] + p^2 x \int_x^{x \vee z} (y^2 + t^2)^{p/2 - 1} dt.$$

To show 2°, we shall prove a slightly stronger statement: for any $0 < x \leq z \leq y$ and any $d > -x$,

$$U(x + d, \sqrt{y^2 + d^2} \vee (x + d), z) \leq U(x, y, z) + Ad, \quad (8.63)$$

where

$$A = A(x, y, z) = \begin{cases} U_x(x, y, z), & \text{if } x < z, \\ \lim_{t \uparrow z} U_x(t, y, z) & \text{if } x = z. \end{cases}$$

We consider two cases.

The case $x + d \leq z$. Then (8.63) reads

$$p(y^2 + d^2 + (x + d)^2)^{p/2} + p^2(x + d) \int_{x+d}^z (y^2 + d^2 + t^2)^{p/2 - 1} dt$$

$$\leq p(y^2 + x^2)^{p/2} + p^2(x + d) \int_x^z (y^2 + t^2)^{p/2 - 1} dt,$$

or, in equivalent form,

$$p \frac{(y^2 + d^2 + (x + d)^2)^{p/2} - (y^2 + x^2)^{p/2}}{x + d}$$
$$- p^2 \left[\int_x^z (y^2 + t^2)^{p/2 - 1} dt - \int_{x+d}^z (y^2 + d^2 + t^2)^{p/2 - 1} dt \right] \leq 0.$$

Denote the left-hand side by $F(x)$ and observe that (8.45) is valid; this implies $F(x) \leq F((z - d) \wedge z)$. If $z - d < z$, then $F(z - d) \leq 0$, which follows from (8.58). If conversely, $z \leq z - d$, then $F(z) \leq 0$, which is a consequence of (8.59) (with $Y = y^2 + d^2$).

The case $x + d > z$. If $x + d \geq \sqrt{y^2 + d^2}$, then (8.63) takes the form

$$p[(y^2 + x^2)^{p/2} - 2^{p/2} z^p] + p^2(x + d) \int_x^z (y^2 + t^2)^{p/2 - 1} dt \geq 0.$$

The left-hand side is an increasing function of d, hence, if we fix all the other parameters, it suffices to show the inequality for the least d, which is determined by the condition $x + d = \sqrt{y^2 + d^2}$, that is, $d = (y^2 - x^2)/(2x)$; however, then the estimate is exactly (8.60). Finally, assume $x + d < \sqrt{y^2 + d^2}$. Then (8.63) becomes

$$p(y^2 + d^2 + (x+d)^2)^{p/2} - p 2^{p/2}(x+d)^p$$

$$\leq p(y^2 + x^2)^{p/2} + p^2(x+d)\int_x^z (y^2 + t^2)^{p/2-1} dt - p 2^{p/2} z^p,$$

which is (8.61) with $D = x + d$.

Thus, $2°$ follows. The remaining conditions $3°$ and $4°$ are straightforward. The proof of (8.54) is complete.

Remark 8.5. If $p = 1$, then (8.54) can be shown directly without the use of the operators T_n. Namely, we take $V(x, y, z) = y - \sqrt{2}(x \vee z)$ and then the corresponding special function is

$$U(x, y, z) = \frac{1}{2\sqrt{2}} \frac{y^2 - x^2 - (x \vee z)^2}{x \vee z}.$$

See [144] for details.

Sharpness of (8.54). We shall construct an appropriate example. Fix $M > 1$, an integer $N \geq 1$ and let $f = f^{(N,M)}$ be given by

$$f_n = M^n \, (0, M^{-n}], \quad n = 0, 1, 2, \ldots, N, \quad \text{and} \quad f_N = f_{N+1} = f_{N+2} = \cdots.$$

Then f is a nonnegative martingale,

$$f^* = M^N (0, M^{-N}] + \sum_{n=1}^N M^{n-1}(M^{-n}, M^{-n+1}],$$

$$df_0 = (0, 1], \quad df_n = (M^n - M^{n-1}) \, (0, M^{-n}] - M^{n-1} \, (M^n, M^{-n+1}],$$

for $n = 1, 2, \ldots, N$, and $df_n = 0$ for $n > N$. Hence the square function is equal to

$$\left(1 + \sum_{k=1}^N (M^k - M^{k-1})^2\right)^{1/2} = \left(1 + \frac{M-1}{M+1}(M^{2N} - 1)\right)^{1/2}$$

on the interval $(0, M^{-N}]$, and is given by

$$\left(1 + \sum_{k=1}^{n-1} (M^k - M^{k-1})^2 + M^{2n-2}\right)^{1/2} = \left(1 + \frac{M-1}{M+1}(M^{2n-2} - 1) + M^{2n-2}\right)^{1/2}$$

on the set $(M^{-n}, M^{-n+1}]$, for $n = 1, 2, \ldots, N$.

Now, if $M \to \infty$, then $\|S(f)\|_1 \to 1 + \sqrt{2}N$ and $\|f^*\|_1 \to 1 + N$, therefore, for M and N sufficiently large, the ratio $\|S(f)\|_1 / \|f^*\|_1$ can be made arbitrarily close to $\sqrt{2}$. Similarly, for $p < 1$, $\|S(f)\|_p / \|f^*\|_p \to \sqrt{2}$ as $M \to \infty$ (here we may keep N fixed). Thus the constant $\sqrt{2}$ is the best possible. $\qquad\square$

8.7 Inequalities for the conditional square function

Recall that for any martingale f taking values in a Banach space $\mathcal{B}$ with norm $|\cdot|$, the conditional square function of f is given by

$$s(f) = \left[\sum_{k=0}^{\infty} \mathbb{E}(|df_k|^2|\mathcal{F}_{(k-1)\vee 0})\right]^{1/2}.$$

We will also use the notation

$$s_n(f) = \left[\sum_{k=0}^{n} \mathbb{E}(|df_k|^2|\mathcal{F}_{(k-1)\vee 0})\right]^{1/2},$$

for $n = 0, 1, 2, \ldots$.

8.7.1 Formulation of the results

We start with weak type estimates.

Theorem 8.18. *Let f be an $\mathcal{H}$-valued martingale.*

(i) *If $0 < p \le 2$, then*

$$\|S(f)\|_{p,\infty} \le (\Gamma(p/2+1))^{-1/p}\|s(f)\|_p \tag{8.64}$$

and

$$\|f\|_{p,\infty} \le (\Gamma(p/2+1))^{-1/p}\|s(f)\|_p. \tag{8.65}$$

The inequalities are sharp, even if $\mathcal{H} = \mathbb{R}$.

(ii) *If $p \ge 2$, then*

$$\|s(f)\|_{p,\infty} \le \left(\frac{p}{2}\right)^{1/2-1/p}\|S(f)\|_p \tag{8.66}$$

and

$$\|s(f)\|_{p,\infty} \le \left(\frac{p}{2}\right)^{1/2-1/p}\|f\|_p. \tag{8.67}$$

All the inequalities are sharp, even if $\mathcal{H} = \mathbb{R}$.

We shall also establish the following moment estimates.

Theorem 8.19. *Let f be an $\mathcal{H}$-valued martingale.*

(i) *If $0 < p \le 2$, then*

$$\|f\|_p \le \sqrt{\frac{2}{p}}\|s(f)\|_p,$$
$$\|S(f)\|_p \le \sqrt{\frac{2}{p}}\|s(f)\|_p, \tag{8.68}$$

(ii) *If $p \geq 2$, then*

$$\|s(f)\|_p \leq \sqrt{\frac{p}{2}} \|f\|_p,$$

$$\|s(f)\|_p \leq \sqrt{\frac{p}{2}} \|S(f)\|_p. \tag{8.69}$$

All the estimates are sharp, even if $\mathcal{H} = \mathbb{R}$.

Now we turn to the bounded case. We have the following tail estimates for $s(f)$.

Theorem 8.20. *Let f be an $\mathcal{H}$-valued martingale.*

(i) *If $\|S(f)\|_\infty \leq 1$, then for any $\lambda > 0$ we have*

$$\mathbb{P}(s(f) \geq \lambda) \leq \min(e^{1-\lambda^2}, 1). \tag{8.70}$$

(ii) *If $\|f\|_\infty \leq 1$, then for any $\lambda > 0$ we have*

$$\mathbb{P}(s(f) \geq \lambda) \leq \min(e^{1-\lambda^2}, 1). \tag{8.71}$$

Both bounds are sharp, even if $\mathcal{H} = \mathbb{R}$.

Finally, we shall establish the following Φ-estimates.

Theorem 8.21. *Let f be an $\mathcal{H}$-valued martingale and let $\Phi : [0, \infty) \to \mathbb{R}$ be an increasing convex function.*

(i) *If $\|S(f)\|_\infty \leq 1$, then*

$$\mathbb{E}\Phi(s^2(f)) \leq \int_0^\infty \Phi(t)e^{-t}dt. \tag{8.72}$$

(ii) *If $\|f\|_\infty \leq 1$, then*

$$\mathbb{E}\Phi(s^2(f)) \leq \int_0^\infty \Phi(t)e^{-t}dt. \tag{8.73}$$

Both estimates are sharp, even if $\mathcal{H} = \mathbb{R}$.

8.7.2 On the method

First we shall modify Burkholder's approach so that it works for conditional square functions.

Theorem 8.22. *Let I be a subinterval of $[0, \infty)$ such that $0 \in I$ and suppose that U, V are functions from $I \times [0, \infty)$ to $\mathbb{R}$ satisfying*

$1°$ $V(x, y) \leq U(x, y)$ *for $x \in I$ and $y \geq 0$,*

$2°$ *For any $y \geq 0$, the function $U(\cdot, y)$ is concave on I and*

$$U(x + d, y + d) \leq U(x, y), \quad \text{for } d \geq 0, \ y \geq 0 \text{ and } x, \ x + d \in I. \tag{8.74}$$

$3°$ $U(0, 0) \leq 0$ *for all $x \in I$.*

Let f is a simple $\mathcal{H}$-valued martingale.

(i) *If $S(f) \in I$ almost surely, then for any nonnegative integer n,*

$$\mathbb{E}V(S_n^2(f), s_n^2(f)) \le 0. \tag{8.75}$$

(ii) *If $|f| \in I$ almost surely, then for any nonnegative integer n,*

$$\mathbb{E}V(|f_n|^2, s_n^2(f)) \le 0. \tag{8.76}$$

Proof. By 1°, it suffices to show the assertions with V replaced by U.

(i) Let $F_n = S_n^2(f)$, $G_n = s_n^2(f)$ and set $F_{-1} = G_{-1} \equiv 0$, $\mathcal{F}_{-1} = \{\emptyset, \Omega\}$. Since for any $n \ge 0$ the variable G_n is $\mathcal{F}_{n-1}$-measurable, we have, by the first part of 2° and the conditional Jensen inequality,

$$\mathbb{E}U(F_n, G_n) = \mathbb{E}\big[\mathbb{E}(U(F_n, G_n)|\mathcal{F}_{n-1})\big] \le \mathbb{E}\big[\mathbb{E}U\big(\mathbb{E}(F_n|\mathcal{F}_{n-1}), G_n\big)\big].$$

The process $(U(\mathbb{E}(F_n|\mathcal{F}_{n-1}), G_n))_{n=0}^\infty$ is a supermartingale with respect to $(\mathcal{F}_{n-1})$. Indeed, for $n \ge 0$,

$$\begin{aligned}
&\mathbb{E}[U(\mathbb{E}(F_{n+1}|\mathcal{F}_n), G_{n+1})|\mathcal{F}_{n-1}] \\
&= \mathbb{E}\big[U\big(\mathbb{E}(F_n|\mathcal{F}_n) + \mathbb{E}(|df_{n+1}|^2|\mathcal{F}_n), G_n + \mathbb{E}(|df_{n+1}|^2|\mathcal{F}_n)\big)\big|\mathcal{F}_{n-1}\big] \\
&\le \mathbb{E}[U(\mathbb{E}(F_n|\mathcal{F}_n), G_n)|\mathcal{F}_{n-1}] \\
&\le U(\mathbb{E}(F_n|\mathcal{F}_{n-1}), G_n),
\end{aligned} \tag{8.77}$$

where the first estimate uses (8.74) and the second one follows from 2° and the conditional Jensen inequality. Thus

$$\mathbb{E}U(F_n, G_n) \le \mathbb{E}U(\mathbb{E}(F_0|\mathcal{F}_{-1}), G_0) = \mathbb{E}U(\mathbb{E}(|df_0|^2|\mathcal{F}_{-1}), \mathbb{E}(|df_0|^2|\mathcal{F}_{-1})) \le U(0,0),$$

in view of (8.74), and the needed estimate follows.

(ii) We repeat the proof of (i), word by word, this time with the processes $F_n = \|f_n\|^2$ and $G_n = s_n^2(f)$, $n = 0, 1, 2, \ldots$. The only fact we need is that $\mathbb{E}(F_{n+1}|\mathcal{F}_n) = F_n + \mathbb{E}(|df_{n+1}|^2|\mathcal{F}_n)$; therefore, the chain (8.77) is still valid and the claim follows. $\qquad\square$

8.7.3   Proof of Theorem 8.18

Proof of (8.64)–(8.67). (i) By straightforward approximation, it suffices to show that if f is a simple martingale, then for $p \in (0, 1]$,

$$\mathbb{P}(S_n^2(f) \ge 1) \le \Gamma(p+1)^{-1}\mathbb{E}(s_n^2(f))^p$$

and

$$\mathbb{P}(|f_n|^2 \ge 1) \le \Gamma(p+1)^{-1}\mathbb{E}(s_n^2(f))^p,$$

for $n = 0, 1, 2, \ldots$. Thus we are in the position where Theorem 8.22 can be applied. For $x, y \geq 0$, let $V_p(x, y) = 1_{\{x \geq 1\}} - \Gamma(p+1)^{-1}y^p$ and

$$U_p(x, y) = \begin{cases} 1 - \Gamma(p+1)^{-1}\left[(1-x)e^y \int_y^\infty t^p e^{-t}dt + xy^p\right] & \text{if } x < 1, \\ 1 - \Gamma(p+1)^{-1}y^p & \text{if } x \geq 1. \end{cases}$$

It suffices to show that U_p and V_p satisfy 1°, 2° and 3°. To establish the majorization $V_p \leq U_p$, observe first that we have equality if $x \geq 1$. If $x < 1$, then the inequality can be written in the equivalent form

$$(1-x)\left(e^y \int_y^\infty t^p e^{-t}dt - y^p\right) \leq \Gamma(p+1).$$

This holds true, since $1 - x \leq 1$, the function $H : [0, \infty) \to [0, \infty)$, given by

$$H(y) := e^y \int_y^\infty t^p e^{-t}dt - y^p = pe^y \int_y^\infty t^{p-1}e^{-t}dt$$

is nonincreasing: $H'(y) = p(p-1)e^y \int_y^\infty t^{p-2}e^{-t}dt < 0$, and $H(0) = \Gamma(p+1)$. The first part of 2° is guaranteed by the fact that $U_p(\cdot, y)$ continuous, linear and increasing on $[0, 1]$, and constant on $[1, \infty)$. To check (8.74), it suffices to show that $U_{px}(x, y) + U_{py}(x, y) \leq 0$ for $x \neq 1$, $y > 0$. This is clear for $x > 1$, while for $x < 1$ we have that

$$U_{px}(x, y) + U_{py}(x, y) = Cp(p-1)xe^y \int_y^\infty t^{p-2}e^{-t}dt \leq 0.$$

Finally, 3° holds: $U(0, 0) = 0$. Thus (8.64) and (8.65) are established.

(ii) Here the special function is entirely different. Let $\gamma : [1 - 1/p, 1) \to \mathbb{R}$ be given by

$$\gamma(y) = \frac{1}{p}(p(1-y))^{-1/(p-1)}.$$

Consider the following subsets of $[0, \infty) \times [0, \infty)$:

$$D_1 = [0, \infty) \times [0, 1 - 1/p],$$
$$D_2 = \{(x, y) : x > \gamma(y) + y - 1, 1 - 1/p < y < 1\},$$
$$D_3 = ([0, \infty) \times [0, 1)) \setminus (D_1 \cup D_2),$$
$$D_4 = [0, \infty) \times [1, \infty).$$

Define $U_p, V_p : [0, \infty) \times [0, \infty) \to \mathbb{R}$ by

$$U_p(x, y) = \begin{cases} p^{p-1}(\frac{y}{p-1})^{p-1}[y - px] & \text{on } D_1, \\ \frac{1}{p(1-y)}[(p-1)p^{-p/(p-1)}(1-y)^{-1/(p-1)} - px] & \text{on } D_2, \\ 1 - p^{p-1}(1 + x - y)^p & \text{on } D_3, \\ 1 - p^{p-1}x^p & \text{on } D_4 \end{cases}$$

and $V_p(x, y) = 1_{[1,\infty)}(y) - p^{p-1}x^p$. In view of Theorem 8.22, it suffices to check that U_p, V_p satisfy the conditions 1°, 2° and 3°. To prove the majorization, observe that if $(x, y) \in D_1$, then

$$U_{px}(x, y) - V_{px}(x, y) = p^p \left[x^{p-1} - \left(\frac{y}{p-1} \right)^{p-1} \right],$$

which implies that

$$U_p(x, y) - V_p(x, y) \geq U_p(y/(p-1), y) - V_p(y/(p-1), y) = 0.$$

If $(x, y) \in D_2$, then

$$U_{px}(x, y) - V_{px}(x, y) = -\frac{1}{1-y} + p^p x^{p-1}.$$

Setting

$$x_0 = \frac{1}{p}[p(1-y)]^{-1/(p-1)},$$

we see that $(x_0, y) \in D_2$ and $U_p(x, y) - V_p(x, y) \geq U_p(x_0, y) - V_p(x_0, y) = 0$. If (x, y) belongs to D_3, then $U_{px}(x, y) - V_{px}(x, y) = p^p[x^{p-1} - (1+x-y)^{p-1}] < 0$, so

$$U_p(x, y) - V_p(x, y) \geq U_p(\gamma(y) + y - 1, y) - V_p(\gamma(y) + y - 1, y),$$

and the right-hand side is nonnegative, as $(\gamma(y) + y - 1, y)$ belongs to the closure of D_2, where we have already established the majorization. We complete the proof of 1° by noting that $U_p = V_p$ on D_4. The first part of 2° is clear. To establish (8.74) it suffices to prove that $U_{px} + U_{py} \leq 0$ in the interiors of the sets D_i, $i = 1, 2, 3, 4$. The direct calculation shows that $U_{px}(x, y) + U_{py}(x, y)$ equals

$$\begin{cases} -p^p(p-1)^{2-p}xy^{p-2} & \text{if } (x, y) \in D_1^{\circ}, \\ (1-y)^{-2}[-x+y-1+\gamma(y)] & \text{if } (x, y) \in D_2^{\circ}, \\ 0 & \text{if } (x, y) \in D_3^{\circ}, \\ -p^p x^{p-1} & \text{if } (x, y) \in D_4^{\circ} \end{cases}$$

and it is evident that all the expressions are nonpositive on the corresponding sets. Finally, 3° is obvious. The proof is complete. □

Sharpness of (8.64) and (8.65). Let $\delta \in (0, 1)$ be fixed and let $(X_n)_{n=0}^{\infty}$ be a sequence of independent random variables sharing the same distribution given by

$$\mathbb{P}(X_n = 1) = \delta = 1 - \mathbb{P}(X_n = 0).$$

Furthermore, let (ε_n) be a sequence of independent Rademacher variables, independent also of (X_n). Introduce the stopping time $\tau = \inf\{n : X_n = 1\}$, set $df_n = \varepsilon_n X_n 1_{\{\tau \geq n\}}$, $n = 0, 1, 2, \ldots$ and let $(\mathcal{F}_n)$ be the natural filtration of f. Then f is a martingale (which is even conditionally symmetric), for which $|f_n| \uparrow |f_\infty| \equiv 1$ and $S(f) \equiv 1$ almost surely; hence $||f||_{p,\infty} = ||s(f)||_{p,\infty} = 1$. Furthermore, as $\mathbb{E}(df_n^2|\mathcal{F}_{n-1}) = 1_{\{\tau \geq n\}} \mathbb{E}X_n = \delta 1_{\{\tau \geq n\}}$, we have $s^2(f) = \delta(\tau + 1)$. Since τ has geometric distribution, we have, for $0 < p \leq 2$,

$$||s(f)||_p^p = ||s^2(f)||_{p/2}^{p/2} = \delta \sum_{n=1}^{\infty} (\delta n)^{p/2} (1 - \delta)^{n-1}$$

and we see that the right-hand side, by choosing δ sufficiently small, can be made arbitrarily close to $\int_0^\infty t^{p/2} e^{-t} dt = \Gamma(p/2 + 1)$. This implies that the constant in (8.64) and (8.65) is indeed the best possible. $\square$

Sharpness of (8.66) *and* (8.67). If $p = 2$, then the constant martingale $f_0 = f_1 = f_2 = \cdots \equiv 1$ gives equality in (8.66) and (8.67). Suppose then, that $p > 2$. Let $\delta \in (0, 1 - 2/p)$, N be a positive integer and set

$$r = \left(\frac{p-2}{p\delta}\right)^{1/N} > 1. \tag{8.78}$$

Furthermore, assume that N is large enough so that $q := (r - 1)(p - 2)/(2r) < 1$. Consider a sequence $(X_n)_{n=0}^{N}$ of independent random variables such that

$$\mathbb{P}\left(X_n = \frac{2r^n \delta}{p-2}\right) = q = 1 - \mathbb{P}(X_n = 0), \quad n = 0, 1, 2, \ldots,$$

and $X_N \equiv 2/p$. Let $(\varepsilon_n)_{n=0}^{N}$ be a sequence of independent Rademacher variables, independent also of (X_n). Set $\tau = \inf\{n : X_n \neq 0\}$ and let $df_n = \varepsilon_n \sqrt{X_n} 1_{\{\tau \geq n\}}$, $n = 0, 1, 2, \ldots, N$. We easily see that f is a conditionally symmetric martingale satisfying $|f_N| = S_N(f) = \sqrt{X_\tau}$. Therefore

$$||f||_p^p = ||S(f)||_p^p = \sum_{n=0}^{N-1} \left(\frac{2r^n \delta}{p-2}\right)^{p/2} (1-q)^n q + \left(\frac{2}{p}\right)^{p/2} (1-q)^N$$

$$= \left(\frac{2\delta}{p-2}\right)^{p/2} q \frac{1 - (r^{p/2}(1-q))^N}{1 - r^{p/2}(1-q)} + \left(\frac{2}{p}\right)^{p/2} (1-q)^N. \tag{8.79}$$

On the other hand, it can be easily verified that $\mathbb{E}(df_n^2|\mathcal{F}_{n-1}) = (r-1)r^{n-1}\delta 1_{\{\tau \geq n\}}$ for $n < N$ and $\mathbb{E}(df_N^2|\mathcal{F}_{N-1}) = (2/p)1_{\{\tau=N\}}$, so

$$s^2(f) = \sum_{n=0}^{\tau \wedge (N-1)} (r-1)r^{n-1}\delta + \frac{2}{p}1_{\{\tau=N\}} = \frac{\delta}{r}(r^{(\tau+1)\wedge N} - 1) + \frac{2}{p}1_{\{\tau=N\}}.$$

On the set $\{\tau = N\}$ we have, by (8.78),

$$s^2(f) = \frac{p-2}{pr} - \frac{\delta}{r} + \frac{2}{p} \geq \frac{1-\delta}{r},$$

so

$$\mathbb{P}\left(s(f) \geq \left(\frac{1-\delta}{r}\right)^{1/2}\right) \geq \mathbb{P}(\tau = N) = (1-q)^N.$$

Hence

$$\frac{\|f\|_p^p}{\|s(f)\|_{p,\infty}^p} = \frac{\|S(f)\|_p^p}{\|s(f)\|_{p,\infty}^p} \leq \left(\frac{r}{1-\delta}\right)^{p/2} \frac{\|f\|_p^p}{(1-q)^N}.$$

Now let $N \to \infty$ (so r tends to 1). We have

$$\lim_{N \to \infty} \frac{q}{1 - r^{p/2}(1-q)} = \lim_{r \to 1} \left(\frac{1 - r^{p/2}}{(r-1)(p-2)} \cdot 2r + r^{p/2}\right)^{-1} = -\frac{p-2}{2}$$

and, by (8.78),

$$\lim_{N \to \infty} (1-q)^N = \lim_{r \to 1} \left(1 - \frac{(r-1)(p-2)}{2r}\right)^{\log \frac{p-2}{p\delta} / \log r} = \left(\frac{p\delta}{p-2}\right)^{(p-2)/2}.$$

Therefore, again using (8.78),

$$\lim_{N \to \infty} \frac{\|f\|_p^p}{(1-q)^N} = \left(\frac{2\delta}{p-2}\right)^{p/2} \cdot \left(-\frac{p-2}{2}\right) \cdot \left(\frac{p-2}{p\delta}\right)^{(p-2)/2}$$

$$+ \left(\frac{2}{p-2}\right)^{p/2} \cdot \frac{p-2}{2} \cdot \left(\frac{p-2}{p}\right)^{p/2} + \left(\frac{2}{p}\right)^{p/2}$$

$$= -\delta \left(\frac{2}{p}\right)^{(p-2)/2} + \left(\frac{2}{p}\right)^{p/2-1}.$$

If δ is taken sufficiently small, we see that for any $\kappa > 0$ the ratios $\|f\|_p^p/\|s(f)\|_p$ and $\|S(f)\|_p^p/\|s(f)\|_p$ can be made smaller than $(2/p)^{p/2-1} + \kappa$. This shows that the constant $(p/2)^{1/2-1/p}$ is indeed the best possible in (8.66) and (8.67). $\square$

8.7.4 Proof of Theorem 8.19

Proof of (8.68) and (8.69). We must show that

$$\mathbb{E}V(S_n^2(f), s_n^2(f)) \leq 0, \qquad \mathbb{E}V(|f_n|^2, s_n^2(f)) \leq 0, \qquad n = 0, 1, 2, \ldots,$$

where $V : [0, \infty) \times [0, \infty) \to \mathbb{R}$ is given by $V(x,y) = x^p - (y/p)^p$, $0 < p \leq 1$. The corresponding special function $U : [0, \infty) \times [0, \infty) \to \mathbb{R}$ is given by

$$U(x,y) = p(y/p)^{p-1}(x - y/p).$$

By Theorem 8.22, it suffices to check 1°, 2° and 3°. The majorization $U \geq V$ follows immediately from the mean-value property of the concave function $t \mapsto t^p$. The first part of 2° is obvious. To prove (8.74), note that

$$(y+d)^p \left(\frac{x+d}{y+d} - \frac{1}{p} \right) - y^p \left(\frac{x}{y} - \frac{1}{p} \right)$$
$$= x[(y+d)^{p-1} - y^{p-1}] + \left\{ d(y+d)^{p-1} - \frac{1}{p}[(y+d)^p - y^p] \right\} \leq 0,$$

since both terms are nonpositive: this is due to the assumption $p \leq 1$ and the mean value property. Finally, 3° is evident. This proves (8.68) and (8.69).

(ii) This is shown exactly in the same manner, using the functions $V(x,y) = -x^p + (y/p)^p$ and

$$U(x,y) = -p(y/p)^{p-1}(x - y/p),$$

for $p \geq 1$. We omit the details. $\square$

Sharpness. We shall only focus on (8.68) and (8.69). Fix $0 < p < 2$, take $\sqrt{p/2} < b < 1$ and let $a_0 = 1$, $a_n = 1 - b^2/n$ for $n \geq 1$. Define a conditionally symmetric martingale f on $[0,1)$ by setting $df_0 \equiv 0$ and

$$df_n = \sqrt{n} \left[\prod_{i=0}^{n} a_i, \frac{1+a_n}{2} \prod_{i=0}^{n-1} a_i \right) - \sqrt{n} \left[\frac{1+a_n}{2} \prod_{i=0}^{n-1} a_i, \prod_{i=0}^{n-1} a_i \right)$$

(as usual, we identify a set with its indicator function). We easily see that

$$|f_n|^2 = S_n^2(f) = \sum_{k=1}^{n} k \left[\prod_{i=0}^{k} a_i, \prod_{i=0}^{k-1} a_i \right)$$

and, after some computations,

$$\mathbb{E}(df_n^2|\mathcal{F}_{n-1}) = b^2 \left[0, \prod_{i=0}^{n-1} a_i \right),$$

so

$$s_n^2(f) = b^2 n \left[0, \prod_{i=0}^{n} a_i \right) + \sum_{k=1}^{n} b^2 k \left[\prod_{i=0}^{k} a_i, \prod_{i=0}^{k-1} a_i \right).$$

Now, we have

$$\mathbb{E}s_n^p(f) \left[0, \prod_{i=0}^{n} a_i \right) = b^p n^{p/2} \prod_{i=0}^{n-1} \left(1 - \frac{b^2}{n} \right) \sim b^p n^{p/2} n^{-b^2} \to 0$$

as $n \to \infty$, so the ratio $||s_n(f)||_p/||f||_p = ||s_n(f)||_p/||S(f)||_p$ can be made arbitrarily close to b. This shows that the constant $\sqrt{2/p}$ is indeed the best in (8.68) and (8.69). $\square$

8.7.5 Proof of Theorem 8.20

Proof of (8.70) *and* (8.71). For $\lambda \leq 1$ the inequality is trivial, so we may assume that $\lambda > 1$. Let $U, V : [0,1] \times [0,\infty) \to \mathbb{R}$ be given by $V(x,y) = 1_{\{y \geq \lambda^2\}} - e^{1-\lambda^2}$ and

$$U(x,y) = \begin{cases} (1-x)\exp(y+1-\lambda^2) - e^{1-\lambda^2} & \text{if } y \leq \lambda^2 - 1, \\ (1-x)(\lambda^2 - y)^{-1} - e^{1-\lambda^2} & \text{if } \lambda^2 - 1 < y < x + \lambda^2 - 1, \\ 1 - e^{1-\lambda^2} & \text{if } y \geq x + \lambda^2 - 1. \end{cases}$$

It is straightforward to verify $1°$ and the first part of $2°$. Furthermore, (8.74) follows from the continuity of U on $[0,1) \times \mathbb{R}$ and the fact that

$$U_x(x,y) + U_y(x,y) = \begin{cases} -x\exp(y+1-\lambda^2), & \text{if } y < \lambda^2 - 1, \\ (1-x-\lambda^2+y)(\lambda^2-y)^{-2} & \text{if } \lambda^2 - 1 < y < x + \lambda^2 - 1, \\ 0, & \text{if } y > x + \lambda^2 - 1 \end{cases}$$

is nonpositive. Finally, $U(0,0) = 0$. Hence, by Theorem 8.22,

$$\mathbb{P}(s^2(f) \geq \lambda^2) \leq e^{1-\lambda^2},$$

provided $||f||_\infty \leq 1$ or $||S(f)||_\infty \leq 1$. This yields (8.70) and (8.71). □

Sharpness. If $\lambda \leq 1$ then we have equalities if we take $e_0 = e_1 = e_2 = \cdots \equiv 1$ and $f_0 = f_1 = f_2 = \cdots \equiv 1$. Suppose that $\lambda > 1$, take a positive integer N and set $\delta = (\lambda^2 - 1)/N$. The example is similar to the one used in Section 5.1. Let $(X_n)_{n=0}^N$ be a sequence of independent random variables such that

$$\mathbb{P}(X_n = 1) = \delta = 1 - \mathbb{P}(X_n = 0), \qquad n = 0, 1, 2, \ldots, N-1$$

and $X_N \equiv 1$. Finally, let $\tau = \inf\{n : X_n = 1\}$, (ε_n) be a sequence of independent Rademacher variables and $df_n = \varepsilon_n X_n 1_{\{\tau \geq n\}}$, $n = 0, 1, 2, \ldots, N$, $df_n \equiv 0$ for $n > N$.

We easily check that $||f||_\infty = ||S(f)||_\infty = 1$ (in fact, for any $0 \leq n \leq N-1$, $S_n(f), |f_n| \in \{0,1\}$ and $S_N(f) = |f_N| = 1$ with probability 1). Moreover, we see that $\mathbb{E}(df_n^2 | \mathcal{F}_{n-1}) = \delta 1_{\{\tau \geq n\}}$ almost surely for $n < N$ and hence $s^2(f) = (\tau + 1)\delta \leq \lambda - 1 < \lambda$ on $\tau < N$; on the other hand, as $\mathbb{E}(df_N^2 | \mathcal{F}_{N-1}) = 1$ with probability 1, we have $s(f) = \lambda$ on $\{\tau = N\}$ and hence

$$\mathbb{P}(s(f) \geq \lambda) = (1 - \delta)^N.$$

It suffices to note that the right-hand side converges to $e^{1-\lambda^2}$ as $N \to \infty$. Therefore (8.70) and (8.71) are sharp. □

8.7.6 Proof of Theorem 8.21

Proof of (8.72) *and* (8.73). With no loss of generality we may assume that Φ is of class C^1. The functions $U, V : [0, 1] \times [0, \infty) \to \mathbb{R}$ corresponding to our problem are given by $V(x, y) = \Phi(y) - \int_0^\infty \Phi(t)e^{-t}dt$ and

$$U(x, y) = x\Phi(y) + (1-x)e^y \int_y^\infty e^{-t}\Phi(t)dt - \int_0^\infty \Phi(t)e^{-t}dt.$$

Since Φ is nondecreasing, we have

$$e^y \int_y^\infty e^{-t}\Phi(t)dt \geq e^y \int_y^\infty e^{-t}\Phi(y)dt = \Phi(y),$$

which gives the majorization $U \geq V$. For a fixed y, the function $U(\cdot, y)$ is linear, so the first part of $2°$ holds. To check (8.74), observe that if x, y, d are as assumed, then $U_x(x, y) + U_y(x, y)$ equals

$$x\left[\Phi'(y) + \Phi(y) - e^y \int_y^\infty e^{-z}\Phi(z)dz\right] = x\left[\Phi'(y) - e^y \int_y^\infty e^{-z}\Phi'(z)dz\right] \leq 0,$$

where we have used integration by parts and the fact that Φ' is nondecreasing. Finally, $U(0, 0) = 0$. Hence, by Theorem 8.22, for any n,

$$\mathbb{E}\Phi\left(s_n^2(f)\right) \leq \int_0^\infty \Phi(t)e^{-t}dt,$$

which is what we need.									□

Sharpness. For $\delta \in (0, 1)$, let f be a martingale as in Subsection 5.1. We have $\|f\|_\infty = \|S(f)\|_\infty = 1$,

$$\mathbb{E}\Phi(s^2(f)) = \delta \sum_{n=1}^\infty \Phi(\delta n)(1 - \delta)^n,$$

which, if n is chosen sufficiently large, can be made arbitrarily close to $\int\limits_0^\infty \Phi(t)e^{-t}dt$. This shows that the bounds in (8.72) and (8.73) are optimal.									□

8.8 Notes and comments

The inequalities comparing various sizes of a martingale and its square function are classical and go back to the works of many mathematicians: we mention here Burkholder and Gundy [40], Cox [53], Cox and Kertz [54], Doob [70], Davis [57], Dellacherie and Meyer [67], Fefferman [78], Garsia [81], [82], Klincsek [108], Pittenger [175], Wang [197], [199] and others. Related results appear in many areas of mathematics: in classical harmonic analysis, see Stein [187], [188], noncommutative probability theory, see Carlen and Kree [43], Pisier and Xu [174], Randrianantoanina [178].

Section 8.1. The modification of Burkholder's method for square-function inequalities was introduced in [38]. A similar approach (though more technical and more exact), called the method of moments, was used by Cox [53] in his proof of the weak type estimate (8.2).

Section 8.2. The weak type estimate for square function (in the general setting) was obtained by Burkholder in [17] without an explicit estimate of the constant C. In [18] he proved that the inequality holds with a constant $C = 3$ and decreased it to $C = 2$ in [19]. The paper [13] by Bollobás contains a different proof of the inequality with the constant 2 as well as an example which shows that the optimal value must be at least $3/2$. In addition, Bollobás conjectured that the best constant equals $\sqrt{e}$ and constructed "a piece" of Burkholder's special function used above. Finally, the sharp version of the inequality (for real-valued martingales) was obtained by Cox [53]. In fact, he established the following much more exact estimate

$$\mathbb{P}(S_n(f) \geq 1) \leq \left(\frac{n}{n-1}\right)^{(n-1)/2} ||f_n||_1, \qquad n = 2, 3, \ldots,$$

and proved it is sharp for each n. The vector-valued version presented above is new, but exploits the well-known objects discovered by Bollobás and Cox.

The weak type estimate (8.3) for real and dyadic martingales is due to Bollobás. The sharpness of this estimate and the extension to conditionally symmetric martingales were established by the author in [136]. The above extension to Hilbert-space-valued martingales as well as the characterization of Hilbert spaces can be found in [157]. Finally, the contents of Theorem 8.5 are taken from [139].

Section 8.3. Moment inequalities for martingale square functions appeared for the first time in Burkholder's paper [17]. Pittenger [175] showed the estimate

$$||f||_p \leq (p-1)||S(f)||_p, \qquad p \geq 3,$$

in the real-valued setting, but the proof can be easily extended to cover martingales taking values in Hilbert spaces. The double inequality (8.28) follows from Burkholder's results in [24]. The limit case $p = 1$, presented in Theorem 8.7 above, was studied by the author in [123]. Theorem 8.8 was established by Wang [197], as a part of his Ph.D. Thesis. For the construction of the corresponding special functions and related results concerning optimal stopping problems, see Davis [59], Novikov [122], Pedersen and Peskir [167], Peskir [169] and Shepp [185].

Section 8.4. The material in this section comes from the author's paper [144].

Section 8.5. Ratio inequalities appear naturally in many situations, for example while studying self-normalized processes and their applications in statistics. See de la Peña [61], de la Peña, Klass and Lai [62]–[64], de la Peña, Lai and Shao

[65], de la Peña and Pang [66]. The above ratio inequalities for martingale square functions (taken from [143]) are based on a corresponding results in continuous time, coming from optimal stopping: see Pedersen and Peskir [167].

Section 8.6. The double inequality comparing $||f^*||_1$ and $||S(f)||_1$ is due to Davis [57], who established it using his famous decomposition. The refinement described in Theorem 8.13 comes from Burkholder's paper [38]. For (suboptimal) results in the opposite direction, see Garsia [82] and the author [123]. The L^1 estimate for one-sided maximal function stated in Theorem 8.14 is taken from [148]. The inequality (8.53) was established by Klincsek [108] for $p = 3, 4, 5, \ldots$; furthermore, he conjectured that it holds for all $p \geq 2$. Finally, Theorem 8.16 comes from the author's paper [144].

Section 8.7. The inequalities for conditional square functions were studied by many mathematicians. A famous result in the martingale case is the existence of absolute constants c_p, C_p, such that, for any martingale f, we have

$$c_p||f||_p \leq \max\left\{ ||s(f)||_p, \left\|\left(\sum_{k=0}^{\infty} |df_k|^p\right)^{1/p}\right\|_p \right\} \leq C_p||f||_p, \tag{8.80}$$

for $p \geq 2$. The version for the sums of independent mean zero random variables was first studied by Rosenthal [183], and later improved by Johnson, Schechtman and Zinn [104] (see also Kwapień and Woyczyński [112]). The general version for martingales was established by Burkholder [18], with C_p of the optimal order $O(\sqrt{p})$ and c_p of the suboptimal order $O(1/\sqrt{p})$. Then Hitczenko [90], [91] determined the optimal order of C_p and showed it is the same as in the case of sums of independent random variables: $C_p = O(p/\ln p)$. The problem of finding the optimal values of the constants c_p and C_p seems to be open. There is an interesting dual version of (8.80), established by Junge and Xu in [106]. It asserts that if $1 < p < 2$ and f is real-valued, then, for some absolute C_p,

$$C_p^{-1}||f||_p \leq \inf\left\{ ||s(g)||_p + \left\|\left(\sum_{k=0}^{\infty} |dh_k|^p\right)^{1/p}\right\|_p \right\} \leq C_p||f||_p,$$

where the infimum runs over all possible decompositions of f as a sum $f = g + h$ of two martingales.

The modification of Burkholder's method presented above appears in [141], but in fact it can be extracted from Wang's paper [199]. The weak type estimates presented in Theorem 8.18 above are taken from [141]. In fact, the author determined in [165] the optimal values of the weak (p, q)-constants. The moment inequalities are due to Wang [199]; see also Garsia [82]. Finally, the estimates for bounded f or $S(f)$ come from [141].

Appendix

In this part we introduce a family of special functions and present some of their properties. Much more information on this subject can be found in [1].

A.1 Confluent hypergeometric functions and their properties

We start with the definition of Kummer's function $M(a, b, z)$. It is a solution of the differential equation

$$zy''(z) + (b - z)y'(z) - ay'(z) = 0$$

and its explicit form is given by

$$M(a, b, z) = 1 + \frac{a}{b}z + \frac{a(a+1)}{b(b+1)}\frac{z^2}{2!} + \cdots .$$

The *confluent hypergeometric function* M_p is defined by the formula

$$M_p(x) = M(-p/2, 1/2, x^2/2), \qquad x \in \mathbb{R}.$$

If p is an even positive integer: $p = 2n$, then M_p is a constant multiple of the Hermite polynomial of order $2n$ (where the constant depends on n). Note also that

$$M'(a, b, z) = \frac{a}{b}M(a + 1, b + 1, z). \tag{A.1}$$

A.2 Parabolic cylinder functions and their properties

The *parabolic cylinder* functions (also known as Whittaker's functions) are closely related to the confluent hypergeometric functions. They are solutions of the differential equation

$$y''(x) + (ax^2 + bx + c)y(x) = 0.$$

We will be particularly interested in the special case

$$y''(x) - \left(\frac{1}{4}x^2 - p - \frac{1}{2}\right) y(x) = 0. \qquad (A.2)$$

There are two linearly independent solutions of this equation, given by

$$y_1(x) = e^{-x^2/4} M\left(-\frac{p}{2}, \frac{1}{2}, \frac{x^2}{2}\right)$$

and

$$y_2(x) = x e^{-x^2/4} M\left(-\frac{p}{2} + \frac{1}{2}, \frac{3}{2}, \frac{x^2}{2}\right).$$

The parabolic cylinder function D_p is defined by

$$D_p(x) = A_1 y_1(x) + A_2 x y_2(x),$$

where

$$A_1 = \frac{2^{p/2}}{\sqrt{\pi}} \cos(p\pi/2) \Gamma((1+p)/2) \quad \text{and} \quad A_2 = \frac{2^{(1+p)/2}}{\sqrt{\pi}} \sin(p\pi/2) \Gamma(1+p/2).$$

$$(A.3)$$

Denote

$$\phi_p(s) = e^{s^2/4} D_p(s), \qquad s \in \mathbb{R}, \qquad (A.4)$$

and let z_p stand for the largest positive root of D_p. If D_p has no positive roots, we set $z_p = 0$.

Later on, we will need the following properties of ϕ_p.

Lemma A.11. *Let p be a fixed number.*

(i) *For all $s \in \mathbb{R}$,*

$$p\phi_p(s) - s\phi_p'(s) + \phi_p''(s) = 0 \qquad (A.5)$$

and

$$\phi_p'(s) = p\phi_{p-1}(s). \qquad (A.6)$$

(ii) *We have the asymptotics*

$$\phi_p(s) = s^p \left(1 - \frac{p(p-1)}{2s^2} + \frac{p(p-1)(p-2)(p-3)}{8s^4} + o(s^{-5})\right) \qquad \text{as } s \to \infty.$$

$$(A.7)$$

(iii) *If $p \le 1$, then ϕ_p is strictly positive on $(0, \infty)$.*
(iv) *$z_p = 0$ for $p \le 1$ and $z_p > z_{p-1}$ for $p > 1$.*
(v) *For $p > 1$,*

$$\phi_p(s) \ge 0, \quad \phi_p'(s) > 0 \quad \text{and} \quad \phi_p''(s) > 0 \qquad \text{on } [z_p, \infty). \qquad (A.8)$$

Furthermore, if $1 < p \le 2$, then $\phi_p'''(s) \le 0$ on $[z_p, \infty)$, while for $p \ge 2$ we have $\phi_p'''(s) \ge 0$ on $[z_p, \infty)$; the inequalities are strict unless $p = 2$.

Proof. (i) follows immediately from (A.2) and the definition of ϕ_p.

(ii) See 19.6.1 and 19.8.1 in [1].

(iii) If $p = 1$, then the assertion is clear, since $h_1(s) = s$. If $p < 1$ then, by (i) and 19.5.3 in [1], we have

$$\phi_p'(s) = p\phi_{p-1}(s) = \frac{p}{\Gamma(1-p)} \int_0^\infty u^{-p} \exp(-su - u^2/2) du > 0.$$

It suffices to use $\phi_p(0) = A_1 > 0$ (see (A.3)) to obtain the claim.

(iv) The first part is an immediate consequence of (iii). To prove the second, we use induction on $\lceil p \rceil$. When $1 < p \le 2$, we have $\phi_p(0) = A_1 < 0$ (see (A.3)) and, by (ii), $\phi_p(s) \to \infty$ as $s \to \infty$, so the claim follows from the Darboux property. To carry out the induction step, take $p > 2$ and write (A.5) in the form

$$\phi_p(z_{p-1}) = pz_{p-1}\phi_{p-1}(z_{p-1}) - p\phi_{p-1}'(z_{p-1}).$$

But, by the hypothesis, $z_{p-1} > 0$: this implies that z_{p-1} is the largest root of ϕ_{p-1}. Therefore, by the asymptotics (A.7), we obtain $\phi_{p-1}'(z_{p-1}) \ge 0$. Plugging this above yields $\phi_p(z_{p-1}) \le 0$, so, again by (A.7), we have $z_{p-1} \le z_p$. However, the inequality is strict, since otherwise, by (i), we would have $\phi_{p-n}(z_p) = 0$ for all integers n. This would contradict (iii).

(v) This follows immediately from (iii), (iv) and the equalities $\phi_p' = p\phi_{p-1}$, $\phi_p'' = p(p-1)\phi_{p-2}$ and $\phi_p''' = p(p-1)(p-2)\phi_{p-3}$. $\qquad\square$

Further properties of ϕ_p are described in the following lemma.

Lemma A.12. *For $p > 1$, let $F_p : [0, \infty) \to \mathbb{R}$ be given by*

$$F_p(s) = p(p-2)\phi_p(s) - (2p-3)s\phi_p'(s) + s^2\phi_p''(s). \tag{A.9}$$

(i) *If $0 < p \le 1$, then F_p is nonpositive.*
(ii) *If $1 < p \le 2$, then F_p is nonnegative.*
(iii) *If $2 \le p \le 3$, then F_p is nonpositive.*

Proof. (i) By (A.5), we have

$$F_p(s) = p(p-2-s^2)\phi_p(s) + (s^2 - 2p + 3)s\phi_p'(s). \tag{A.10}$$

A little calculation gives that for $s > 0$,

$$F_p'(s) = -2ps\phi_p(s) + (p^2 - 4p + 3 - (p-3)s^2)\phi_p'(s) + (s^3 - (2p-3)s)\phi_p''(s),$$

which, by (A.5), can be recast as

$$F_p'(s) = ps(-s^2 + 2p - 5)\phi_p(s) + (s^4 - 3(p-2)s^2 + p^2 - 4p + 3)\phi_p'(s). \tag{A.11}$$

After lengthy but simple manipulations, this can be written as

$$F_p'(s) = F_p(s)\frac{s^4 - 3(p-2)s^2 + p^2 - 4p + 3}{s(s^2 - 2p + 3)} + \frac{p(p-1)(p-2)(3-p)\phi_p(s)}{s(s^2 - 2p + 3)}.$$

(A.12)

The second term above is nonnegative for $0 < p \le 1$ (see Lemma A.11 (iii)). Furthermore,

$$\frac{s^4 - 3(p-2)s^2 + p^2 - 4p + 3}{s(s^2 - 2p + 3)} - s = \frac{(3-p)(s^2 - p + 1)}{s(s^2 - 2p + 3)} \ge 0.$$

Now suppose that $F_p(s_0) > 0$ for some $s_0 \ge 0$. Then, by (A.12) and the above estimates, $F_p'(s) \ge F_p(s)s$ for $s > s_0$, which yields $F_p(s) \ge F_p(s_0)\exp((s^2 - s_0^2)/2)$ for $s \ge s_0$. However, by (A.7), the function F_p has polynomial growth. A contradiction, which finishes the proof of (i).

(ii) It can be easily verified that $F_p'(s) = pF_{p-1}(s)$ for all p and s. Consequently, by the previous part, we have that F_p is nonincreasing and it suffices to prove that $\lim_{s\to\infty} F_p(s) \ge 0$. In fact, the limit is equal to 0, which can be justified using (A.10) and (A.7): F_p is of order at most s^{p+2} as $s \to \infty$, and one easily checks that the coefficients at s^p and s^{p+2} vanish.

(iii) We proceed in the same manner as in the proof of (ii). The function F_p is nondecreasing and $\lim_{s\to\infty} F_p(s) = 0$, thanks to (A.10) and (A.7) (this time one also has to check that the coefficient at s^{p-2} is equal to 0).   □

Let us mention here that the arguments presented in the proof of the above lemma (equations (A.10) and (A.11)) lead to some interesting bounds for the roots z_p, $1 \le p \le 3$. For example, if $1 \le p \le 2$, then, as we have shown, the function F_p is nonnegative: thus, putting $s = z_p$ in (A.10) and exploiting (A.8) yields $z_p^2 \ge 2p - 3$ (which is nontrivial for $p > 3/2$). Furthermore, F_p is nonincreasing, so taking $s = z_p$ in (A.11) gives

$$z_p^4 - 3(p-2)z_p^2 + p^2 - 4p + 3 \le 0,$$

which can be rewritten in the more explicit form

$$z_p^2 \le \frac{3(p-2) + \sqrt{9(p-2)^2 - 4(p-1)(p-3)}}{2}.$$

Note that the bound is quite tight: we have equality for $p \in \{1, 2\}$. Similarly, in the case when $2 \le p \le 3$ we obtain the following estimates:

$$\frac{3(p-2) + \sqrt{9(p-2)^2 - 4(p-1)(p-3)}}{2} \le z_p^2 \le 2p - 3$$

and we have the (double) equality for $p \in \{2, 3\}$.

In particular, the above inequalities yield

Corollary 3. *We have $z_p^2 \le p - 1$ for $1 < p \le 2$ and $z_p^2 \ge p - 1$ for $2 \le p \le 3$.*

Bibliography

[1] M. Abramowitz and I.A. Stegun, editors, *Handbook of Mathematical Functions with Formulas, Graphs and Mathematical Tables*, Reprint of the 1972 edition, Dover Publications, Inc., New York, 1992.

[2] D.J. Aldous, *Unconditional bases and martingales in $L_p(F)$*, Math. Proc. Cambridge Phil. Soc. **85** (1979), 117–123.

[3] A. Baernstein II, *Some sharp inequalities for conjugate functions*, Indiana Univ. Math. J. **27** (1978), 833–852.

[4] R. Bañuelos and K. Bogdan, *Lévy processes and Fourier multipliers*, J. Funct. Anal. **250** (2007) no. 1, 197–213.

[5] R. Bañuelos and P. Janakiraman, *L_p-bounds for the Beurling-Ahlfors transform*, Trans. Amer. Math. Soc. **360** (2008), 3603–3612.

[6] R. Bañuelos and P.J. Méndez-Hernandez, *Space-time Brownian motion and the Beurling-Ahlfors transform*, Indiana Univ. Math. J. **52** (2003), no. 4, 981–990.

[7] R. Bañuelos and A. Osękowski, *Burkholder inequalities for submartingales, Bessel processes and conformal martingales*, to appear in Amer. J. Math.

[8] R. Bañuelos and G. Wang, *Sharp inequalities for martingales with applications to the Beurling-Ahlfors and Riesz transformations*, Duke Math. J. **80** (1995), 575–600.

[9] R. Bañuelos and G. Wang, *Orthogonal martingales under differential subordination and application to Riesz transforms*, Illinois J. Math. **40** (1996), 678–691.

[10] R. Bañuelos and G. Wang, *Davis's inequality for orthogonal martingales under differential subordination*, Michigan Math. J. **47** (2000), 109–124.

[11] K. Bichteler, *Stochastic integration and L^p-theory of semimartingales*, Ann. Probab. **9** (1981), 49–89.

[12] S. Bochner and A.E. Taylor, *Linear functionals on certain spaces of abstractly-valued functions*, Ann. of Math. (2) **39** (1938), no. 4, pp. 913–944.

[13] B. Bollobás, *Martingale inequalities*, Math. Proc. Cambridge Phil. Soc. **87** (1980), 377–382.

[14] A. Borichev, P. Janakiraman and A. Volberg, *Subordination by orthogonal martingales in L^p and zeros of Laguerre polynomials*, arXiv:1012.0943v3.

[15] A. Borichev, P. Janakiraman and A. Volberg, *On Burkholder function for orthogonal martingales and zeros of Legendre polynomials*, to appear in Amer. J. Math., arXiv:1002.2314v3.

[16] J. Bourgain, *Some remarks on Banach spaces in which martingale difference sequences are unconditional*, Ark. Mat. **21** (1983), no. 2, 163–168.

[17] D.L. Burkholder, *Martingale transforms*, Ann. Math. Statist. **37** (1966), 1494–1504.

[18] D.L. Burkholder, *Distribution function inequalities for martingales*, Ann. Probab. **1** (1973), 19–42.

[19] D.L. Burkholder, *A sharp inequality for martingale transforms*, Ann. Probab. **7** (1979), 858–863.

[20] D.L. Burkholder, *A geometrical characterization of Banach spaces in which martingale difference sequences are unconditional*, Ann. Probab. **9** (1981), 997–1011.

[21] D.L. Burkholder, *Martingale transforms and geometry of Banach spaces*, Proceedings of the Third International Conference on Probability in Banach Spaces, Tufts University, 1980, Lecture Notes in Math. 860 (1981), 35–50.

[22] D.L. Burkholder, *A nonlinear partial differential equation and the unconditional constant of the Haar system in L^p*, Bull. Amer. Math. Soc. **7** (1982), 591–595.

[23] D.L. Burkholder, *A geometric condition that implies the existence of certain singular integrals of Banach-space-valued functions*, Conference on Harmonic Analysis in Honor of Antoni Zygmund, Chicago, 1981, Wadsworth, Belmont, CA (1983), pp. 270–286.

[24] D.L. Burkholder, *Boundary value problems and sharp inequalities for martingale transforms*, Ann. Probab. **12** (1984), 647–702.

[25] D.L. Burkholder, *An elementary proof of an inequality of R.E.A.C. Paley*, Bull. London Math. Soc. **17** (1985), 474–478.

[26] D.L. Burkholder, *Martingales and Fourier analysis in Banach spaces*, Probability and Analysis (Varenna, 1985) Lecture Notes in Math. **1206**, Springer, Berlin (1986), pp. 61–108.

[27] D.L. Burkholder, *An extension of a classical martingale inequality*, Prob. Theory and Harmonic Analysis (J.-A. Chao and A.W. Woyczyński, eds.), Marcel Dekker, New York, 1986, pp. 21–30.

[28] D.L. Burkholder, *Sharp inequalities for martingales and stochastic integrals*, Colloque Paul Lévy (Palaiseau, 1987), Astérisque **157–158** (1988), 75–94.

[29] D.L. Burkholder, *Differential subordination of harmonic functions and martingales*, Harmonic Analysis and Partial Differential Equations (El Escorial, 1987), Lecture Notes in Math. **1384** (1989), pp. 1–23.

[30] D.L. Burkholder, *A proof of Pełczyński's conjecture for the Haar system*, Studia Math. **91** (1988), 268–273.

[31] D.L. Burkholder, *On the number of escapes of a martingale and its geometrical significance*, in "Almost Everywhere Convergence", edited by Gerald A. Edgar and Louis Sucheston. Academic Press, New York, 1989, pp. 159–178.

[32] D.L. Burkholder, *Explorations in martingale theory and its applications*, École d'Été de Probabilités de Saint-Flour XIX–1989, pp. 1–66, Lecture Notes in Math. **1464**, Springer, Berlin, 1991.

[33] D.L. Burkholder, *Sharp probability bounds for Ito processes*, Current Issues in Statistics and Probability: Essays in Honor of Raghu Raj Bahadur (edited by J.K. Ghosh, S.K. Mitra, K.R. Parthasarathy and B.L.S. Prakasa) (1993), Wiley Eastern, New Delhi, pp. 135–145.

[34] D.L. Burkholder, *Strong differential subordination and stochastic integration*, Ann. Probab. **22** (1994), pp. 995–1025.

[35] D. L. Burkholder, *Sharp norm comparison of martingale maximal functions and stochastic integrals*, Proceedings of the Norbert Wiener Centenary Congress, 1994 (East Lansing, MI, 1994), pp. 343–358, Proc. Sympos. Appl. Math., 52, Amer. Math. Soc., Providence, RI, 1997.

[36] D.L. Burkholder, *Some extremal problems in martingale theory and harmonic analysis*, Harmonic analysis and partial differential equations (Chicago, IL, 1996), 99–115, Chicago Lectures in Math., Univ. Chicago Press, Chicago, IL, 1999.

[37] D.L. Burkholder, *Martingales and singular integrals in Banach spaces*, Handbook of the Geometry of Banach Spaces, Vol. 1, 2001, pp. 233–269.

[38] D.L. Burkholder, *The best constant in the Davis inequality for the expectation of the martingale square function*, Trans. Amer. Math. Soc. **354** (2002), 91–105.

[39] D.L. Burkholder, *Martingales and singular integrals in Banach spaces*, Handbook of the Geometry of Banach Spaces, Vol. 1, 2001, pp. 233–269.

[40] D.L. Burkholder and R.F. Gundy, *Extrapolation and interpolation of quasilinear operators on martingales*, Acta Math. **124** (1970), 249–304.

[41] A.P. Calderón and A. Zygmund, *On the existence of certain singular integrals*, Acta Math. **88** (1952), 85–139.

[42] A.P. Calderón and A. Zygmund, *On singular integrals*, Amer. J. Math. **78** (1956), 289–309.

[43] E.A. Carlen and P. Krée, *On martingale inequalities in non-commutative stochastic analysis*, J. Funct. Anal. **158** (1998), 475–508.

[44] S.D. Chatterji, *Les martingales et leurs applications analytiques*, Lecture Notes in Mathematics **307** (1973), 27–164.

[45] C. Choi, *A submartingale inequality*, Proc. Amer. Math. Soc. **124** (1996), 2549–2553.

[46] C. Choi, *A norm inequality for Itô processes*, J. Math. Kyoto Univ. **37** Vol. 2 (1997), 229–240.

[47] C. Choi, *A weak-type submartingale inequality*, Kobe J. Math. **14** (1997), 109–121.

[48] C. Choi, *An inequality of subharmonic functions*, J. Korean Math. Soc. **34** (1997), 543–551.

[49] C. Choi, *A weak-type inequality of subharmonic functions*, Proc. Amer. Math. Soc. **126** (1998), 1149–1153.

[50] C. Choi, *A weak-type inequality for differentially subordinate harmonic functions*, Trans. Amer. Math. Soc. **350** (1998), 2687–2696.

[51] K.P. Choi, *Some sharp inequalities for martingale transforms*, Trans. Amer. Math. Soc. **307** (1988), 279–300.

[52] K.P. Choi, *A sharp inequality for martingale transforms and the unconditional basis constant of a monotone basis in $L^p(0,1)$*, Trans. Amer. Math. Soc. **330** (1992) no. 2, 509–529.

[53] D.C. Cox, *The best constant in Burkholder's weak-L^1 inequality for the martingale square function*, Proc. Amer. Math. Soc. **85** (1982), 427–433.

[54] D.C. Cox and R.P. Kertz, *Common strict character of some sharp infinite-sequence martingale inequalities*, Stochastic Processes Appl. **20** (1985), 169–179.

[55] S. Cox and M. Veraar, *Some remarks on tangent martingale difference sequences in L^1-spaces*, Electron. Comm. Probab. **12** (2007), 421–433.

[56] B. Davis, *A comparison test for martingale inequalities*, Ann. Math. Statist. **40** (1969), 505–508.

[57] B. Davis, *On the integrability of the martingale square function*, Israel J. Math. **8** (1970), 187–190.

[58] B. Davis, *On the weak type $(1,1)$ inequality for conjugate functions*, Proc. Amer. Math. Soc. **44** (1974), 307–311.

[59] B. Davis, *On the L^p norms of stochastic integrals and other martingales*, Duke Math. J. **43** (1976), 697–704.

[60] R.D. DeBlassie, *Stopping times of Bessel processes*, Ann. Probab. **15** (1987), 1044–1051.

[61] V.H. de la Pena, *A general class of exponential inequalities for martingales and ratios*, Ann. Probab. **27** (1999), no. 1, 537–564.

[62] V.H. de la Pena, M.J. Klass, T.L. Lai, *Moment bounds for self-normalized martingales*, High-dimensional probability, II (Seattle, WA, 1999), 3–11, Progr. Probab., 47, Birkhäuser Boston, Boston, MA, 2000.

[63] V.H. de la Pena, M.J. Klass, T.L. Lai, *Self-normalized processes: exponential inequalities, moment bounds and iterated logarithm laws*, Ann. Probab. **32** (2004), no. 3A, 1902–1933.

[64] V.H. de la Pena, M.J. Klass, T.L. Lai, *Pseudo-maximization and self-normalized processes*, Probab. Surv. **4** (2007), 172–192.

[65] V.H. de la Pena, T.L. Lai, Q.-M. Shao, *Self-normalized Processes. Limit Theory and Statistical Application*, Probability and its Applications (New York). Springer-Verlag, Berlin, 2009.

[66] V.H. de la Pena, G. Pang, *Exponential inequalities for self-normalized processes with applications*, Electron. Commun. Probab. **14** (2009), 372–381.

[67] C. Dellacherie and P.A. Meyer, *Probabilities and Potential* B: *Theory of Martingales*, North-Holland, Amsterdam, 1982.

[68] M. Dindoš and T. Wall, *The sharp A_p constant for weights in a reverse-Hölder class*, Rev. Mat. Iberoam. **25** (2009), no. 2, 559–594.

[69] C. Doléans, *Variation quadratique des martingales continues à droite*, Ann. Math. Statist. **40** (1969), 284–289.

[70] J.L. Doob, *Stochastic Processes*, Wiley, New York, 1953.

[71] P. Enflo, *Banach spaces which can be given an equivalent uniformly convex norm*, Israel J. Math. **13** (1972), 281–288.

[72] M. Essén, *A superharmonic proof of the M. Riesz conjugate function theorem*, Ark. Mat. **22** (1984), 241–249.

[73] M. Essén, *Some best constant inequalities for conjugate harmonic functions*, International Series of Numerical Math. **103**, Birkhäuser Verlag, Basel (1992), pp. 129–140.

[74] M. Essén, D.F. Shea and C.S. Stanton, *Some best constant inequalities of $L(\log L)^\alpha$ type*, Inequalities and Applications 3 (ed. R.P. Agarwal), World Scientific Publishing (1994), pp. 233–239.

[75] M. Essén, D.F. Shea and C.S. Stanton, *Best constant inequalities for conjugate functions*, J. Comput. Appl. Math. **105** (1999), 257–264.

[76] M. Essén, D.F. Shea and C.S. Stanton, *Best constants in Zygmund's inequality for conjugate functions*, A volume dedicated to Olli Martio on his 60th birthday (ed. Heinonen, Kilpeläinen and Koskela), report **83** (2001), Department of Mathematics, University of Jyväskila, pp. 73–80.

[77] M. Essén, D.F. Shea and C.S. Stanton, *Sharp $L \log^\alpha L$ inequalities for conjugate functions*, Ann. Inst. Fourier, Grenoble **52**, (2002), no. 623–659.

[78] C. Fefferman, *Characterization of bounded mean oscillation*, Bull. Amer. Math. Soc. **77** (1971), 587–588.

[79] Gamelin, T.W., *Uniform Algebras and Jensen Measures*, Cambridge University Press, London, 1978.

[80] L.E. Dubins, D. Gilat and I. Meilijson, *On the expected diameter of an L_2-bounded martingale*, Ann. Probab. **37** (2009), 393–402.

[81] A.M. Garsia, *The Burgess Davis inequalities via Fefferman's inequality*, Ark. Mat. **11** (1973), 229–237.

[82] A.M. Garsia, *Martingale Inequalities: Seminar Notes on Recent Progress*, Benjamin, Reading, Massachusetts, 1973.

[83] R.K. Getoor and M.J. Sharpe, *Conformal martingales,* Invent. Math. **16** (1972), 271–308.

[84] D. Gilat, *On the ratio of the expected maximum of a martingale and the L_p-norm of its last term*, Israel J. Math. **63** (1988), 270–280.

[85] R.F. Gundy, *A decomposition for L^1-bounded martingales*, Ann. Math. Statist. **39** (1968), 134–138.

[86] R.F. Gundy and N.Th. Varopoulos, *Les transformations de Riesz et les intégrales stochastiques*, C. R. Acad. Sci. Paris Sér. A-B **289** (1979), A13–A16.

[87] W. Hammack, *Sharp inequalities for the distribution of a stochastic integral in which the integrator is a bounded submartingale*, Ann. Probab. **23** (1995), 223–235.

[88] W. Hammack, *Sharp maximal inequalities for stochastic integrals in which the integrator is a submartingale*, Proc. Amer. Math. Soc. **124**, Vol. 3 (1996), 931–938.

[89] P. Hitczenko, *Comparison of moments for tangent sequences of random variables*, Probab. Theory Related Fields **78** (1988), no. 2, 223–230.

[90] P. Hitczenko, *Best constants in martingale version of Rosenthal's inequality*, Ann. Probab. **18** (1990), no. 4, 1656–1668.

[91] P. Hitczenko, *On a domination of sums of random variables by sums of conditionally independent ones*, Ann. Probab. **22** 1 (1994), no. 2, 453-468.

[92] P. Hitczenko, *Sharp inequality for randomly stopped sums of independent non-negative random variables*, Stoch. Proc. Appl. **51** (1994), no. 1, 63–73.

[93] B. Hollenbeck, N.J. Kalton and I.E. Verbitsky, *Best constants for some operators associated with the Fourier and Hilbert transforms*, Studia Math. **157** (3) (2003), 237–278.

[94] B. Hollenbeck and I.E. Verbitsky, *Best constants for the Riesz projection*, J. Funct. Anal. **175** (2000), 370–392.

[95] B. Hollenbeck and I.E. Verbitsky, *Best constant inequalities involving the analytic and co-analytic projections*, Operator Theory: Adv. Appl. vol. 202. Birkhäuser (2010), pp. 285–296.

[96] V.I. Istrăţescu, *Inner Product Structures*, Reidel, Boston, MA, 1987.

[97] T. Iwaniec and G. Martin, *Riesz transforms and related singular integrals*, J. Reine Angew. Math. **473** (1996), 25–57.

[98] S.D. Jacka, *Optimal stopping and best constants for Doob-like inequalities I: The case $p = 1$*, Ann. Probab. **19** (1991), 1798–1821.

[99] R.C. James, *Some self-dual properties of normed linear spaces*, Ann. of Math. Studies **69** (1972), 159–175.

[100] R.C. James, *Super-reflexive Banach spaces*, Canad. J. Math. **24** (1972), 896–904.

[101] R.C. James, *Super-reflexive spaces with bases*, Pacific J. Math. **41** (1972), 409–419.

[102] R.C. James, *Bases in Banach spaces*, Amer. Math. Monthly **89** (1982), 625–640.

[103] P. Janakiraman, *Best weak-type (p,p) constants, $1 < p < 2$, for orthogonal harmonic functions and martingales*, Illinois J. Math. **48** (2004), 909–921.

[104] W.B. Johnson, G. Schechtman, and J. Zinn, *Best constants in moment inequalities for linear combinations of independent and exchangeable random variables*, Ann. Probab. **13** (1985), no. 1, 234–253.

[105] P. Jordan and J. von Neumann, *On inner products in linear metric spaces*, Ann. of Math. **36** (1935), 719–723.

[106] M. Junge and Q. Xu, *Noncommutative Burkholder/Rosenthal inequalities*, Ann. Probab. **31** (2003), 948–995.

[107] Y-H. Kim and B-I. Kim, *A submartingale inequality*, Comm. Korean Math. Soc. **13** (1998), No. 1, pp. 159-170.

[108] G. Klincsek, *A square function inequality*, Ann. Probab. **5** (1977), 823–825.

[109] A.N. Kolmogorov, *Sur les fonctions harmoniques conjugées et les séries de Fourier*, Fund. Math. **7** (1925), 24–29.

[110] S. Kwapień, *Isomorphic characterizations of inner product spaces by orthogonal series with vector-valued coefficients*, Studia Math. **44** (1972), 583–595.

[111] S. Kwapień, W.A. Woyczyński, *Tangent Sequences of Random Variables: Basic Inequalities and Their Applications*, in "Almost Everywhere Convergence", edited by Gerald A. Edgar and Louis Sucheston. Academic Press, New York, 1989, pp. 237–266.

[112] S. Kwapień, W.A. Woyczyński, *Random Series and Stochastic Integral: Single and Multiple*, Birkhäuser, 1992.

[113] J.M. Lee, *On Burkholder's biconvex-function characterization of Hilbert spaces*, Proc. Amer. Math. Soc. **118** (1993), no. 2, 555–559.

[114] J.M. Lee, *Biconcave-function characterisations of UMD and Hilbert spaces*, Bull. Austral. Math. Soc. **47** (1993), no. 2, 297–306.

[115] J. Marcinkiewicz, *Quelques théorèmes sur les séries orthogonales*, Ann. Soc. Polon. Math. **16** (1937), 84–96.

[116] B. Maurey, *Système de Haar*, Séminaire Maurey-Schwartz, 1974–1975, Ecole Polytechnique, Paris, 1975.

[117] T.R. McConnell, *A Skorohod-like representation in infinite dimensions*, Probability in Banach Spaces V, Lecture Notes in Math. **1153** (1985), 359–368.

[118] T.R. McConnell, *Decoupling and stochastic integration in UMD Banach spaces*, Probab. Math. Statist. **10** (1989), no. 2, 283–295.

[119] F.L. Nazarov and S.R. Treil, *The hunt for a Bellman function: applications to estimates for singular integral operators and to other classical problems of harmonic analysis*, St. Petersburg Math. J. **8** (1997), 721–824.

[120] F.L. Nazarov, S.R. Treil and A. Volberg, *Bellman function in stochastic control and harmonic analysis (how our Bellman function got its name)*, Oper. Theory: Adv. Appl. **129** (2001), 393–424.

[121] J. Neveu, *Martingales à temps discret*, Masson et Cie, Paris, 1972.

[122] A.A. Novikov, *On stopping times for the Wiener process* (Russian, English summary), Teor. Veroyatnost. i Primenen. **16** (1971), 458–465.

[123] A. Osękowski, *Two inequalities for the first moment of a martingale, its square and maximal function*, Bull. Polish Acad. Sci. Math. **53** (2005), 441–449.

[124] A. Osękowski, *Inequalities for dominated martingales*, Bernoulli **13** no. 1 (2007), 54–79.

[125] A. Osękowski, *Sharp Norm Inequalities for Martingales and their Differential Subordinates*, Bull. Polish Acad. Sci. Math. **55** (2007), 373–385.

[126] A. Osękowski, *Sharp LlogL inequality for differentially subordinated martingales*, Illinois J. Math. **52**, Vol. 3 (2008), 745–756.

[127] A. Osękowski, *Sharp inequality for bounded submartingales and their differential subordinates*, Electr. Commun. Probab. **13** (2008), 660–675.

[128] A. Osękowski, *Weak type inequality for noncommutative differentially subordinated martingales*, Probab. Theory Related Fields **140** (2008), 553–568.

[129] A. Osękowski, *Sharp norm inequality for bounded submartingales*, Journal of Inequalities in Pure and Applied Mathematics (JIPAM) **9** Vol. 4 (2008), art. 93.

[130] A. Osękowski, *On Φ-inequalities for bounded submartingales and subharmonic functions*, Commun. Korean Math. Soc. **23**, Vol. 2 (2008), pp. 269–277.

[131] A. Osękowski, *Sharp maximal inequality for stochastic integrals*, Proc. Amer. Math. Soc. **136** (2008), 2951–2958.

[132] A. Osękowski, *Strong differential subordination and sharp inequalities for orthogonal processes*, J. Theoret. Probab. **22**, Vol. 4 (2009), 837–855.

[133] A. Osękowski, *Sharp weak type inequalities for differentially subordinated martingales*, Bernoulli **15**, Vol. 3 (2009), 871–897.

[134] A. Osękowski, *Sharp norm comparison of the maxima of a sequence and its predictable projection*, Statist. Probab. Lett. **79** Vol. 16 (2009), 1784–1788.

[135] A. Osękowski, *Weak type inequality for the square function of a nonnegative submartingale*, Bull. Polish Acad. Sci. Math. **57**, Vol. 1 (2009), 81–89.

[136] A. Osękowski, *On the best constant in the weak type inequality for the square function of a conditionally symmetric martingale*, Statist. Probab. Lett. **79** (2009), 1536–1538.

[137] A. Osękowski, *Sharp norm inequalities for stochastic integrals in which the integrator is a nonnegative supermartingale*, Probab. Math. Statist. **29**, Vol. 1 (2009), 29–42.

[138] A. Osękowski, *Sharp maximal inequality for martingales and stochastic integrals*, Electr. Commun. in Probab. **14** (2009), 17–30.

[139] A. Osękowski, *Weak type inequalities for conditionally symmetric martingales*, Statist. Probab. Lett. **80**, Vol. 8 (2010), 2009–2013.

[140] A. Osękowski, *Sharp tail inequalities for nonnegative submartingales and their strong differential subordinates*, Electr. Commun. in Probab. **15** (2010), 508–521.

[141] A. Osękowski, *Sharp inequalities for sums of nonnegative random variables and for a martingale conditional square function*, ALEA: Latin American Journal of Probability and Mathematical Statistics **7** (2010), 243–256.

[142] A. Osękowski, *Sharp maximal bound for continuous martingales*, Statist. Probab. Lett. **80**, Vol. 4 (2010), 1405–1408.

[143] A. Osękowski, *Sharp ratio inequalities for a conditionally symmetric martingale*, Bull. Polish Acad. Sci. Math. **58**, Vol. 1 (2010), 65–77.

[144] A. Osękowski, *Sharp inequalities for the square function of a nonnegative martingale*, Probab. Math. Statist. **30**, Vol. 1 (2010), 61–72.

[145] A. Osękowski, *Logarithmic estimates for nonsymmetric martingale transforms*, Statist. Probab. Lett. **80**, Vol. 4 (2010), 678–682.

[146] A. Osękowski, *Sharp moment inequalities for differentially subordinated martingales*, Studia Math. **201** (2010), 103–131.

[147] A. Osękowski, *Maximal inequalities for stochastic integrals*, Bull. Polish Acad. Sci. Math. **58** (2010), 273–287.

[148] A. Osękowski, *Sharp maximal inequalities for the martingale square bracket*, Stochastics **82** (2010), 589–605.

[149] A. Osękowski, *Logarithmic estimates for submartingales and their differential subordinates*, J. Theoret. Probab. **24** (2011), 849–874.

[150] A. Osękowski, *A sharp weak-type bound for Itô processes and subharmonic functions*, Kyoto J. Math. **51** (2011), 875–890.

[151] A. Osękowski, *On relaxing the assumption of differential subordination in some martingale inequalities*, Electron. Commun. Probab. **16** (2011), 9–21.

[152] A. Osękowski, *Sharp maximal inequality for nonnegative martingales*, Statist. and Probab. Lett. **81** (2011), 1945–1952.

[153] A. Osękowski, *Maximal inequalities for continuous martingales and their differential subordinates*, Proc. Amer. Math. Soc. **139** (2011), 721–734.

[154] A. Osękowski, *A maximal inequality for nonnegative sub- and supermartingales*, Math. Ineq. Appl. **14** (2011), 595–604.

[155] A. Osękowski, *Sharp and strict L^p-inequalities for Hilbert-space-valued orthogonal martingales*, Electron. J. Probab. **16** (2011), 531–551.

[156] A. Osękowski, *Sharp maximal inequalities for the moments of martingales and nonnegative submartingales*, Bernoulli **17** (2011), 1327–1343.

[157] A. Osękowski, *Weak type inequality for the martingale square function and a related characterization of Hilbert spaces*, Probab. Math. Statist. **31** (2011), 227–238.

[158] A. Osękowski, *A weak-type inequality for orthogonal submartingales and subharmonic functions*, Bull. Polish Acad. Sci. Math. **59** (2011), 261–274.

[159] A. Osękowski, *Sharp weak type inequalities for the Haar system and related estimates for non-symmetric martingale transforms*, Proc. Amer. Math. Soc. **140** (2012), 2513–2526.

[160] A. Osękowski, *Weak type (p, q)-inequalities for the Haar system and differentially subordinated martingales*, Math. Nachr. **285** (2012), 794–807.

[161] A. Osękowski, *Sharp inequalities for differentially subordinate harmonic functions and martingales*, to appear in Canadian Math. Bull.

[162] A. Osękowski, *Sharp inequality for martingale maximal functions and stochastic integrals*, Illinois J. Math. **54** (2010), 1133–1156.

[163] A. Osękowski, *Best constant in Zygmund's inequality and related estimates for orthogonal harmonic functions and martingales*, J. Korean Math. Soc. **49** (2012), 659–670.

[164] A. Osękowski, *Sharp maximal inequalities for continuous-path semimartingales*, to appear in Math. Ineq. Appl.

[165] A. Osękowski, *Best constants in the weak type inequalities for a martingale conditional square function*, Statist. Probab. Lett. **82** (2012), 885–893.

[166] R.E.A.C. Paley, *A remarkable series of orthogonal functions I*, Proc. London Math. Soc. **34** (1932) 241–264.

[167] J.L. Pedersen and G. Peskir, *Solving non-linear optimal stopping problems by the method of time-change*, Stochastic Anal. Appl., **18** (2000), 811–835.

[168] A. Pelczynski, *Norms of classical operators in function spaces*, in "Colloque en l'honneur de Laurent Schwartz", Vol. 1, Astérisque **131** (1985), 137–162.

[169] G. Peskir, *Optimal stopping of the maximum process: The maximality principle*, Ann. Probab. **26** (1998), 1614–1640.

[170] G. Peskir and A. Shiryaev, *Optimal Stopping and Free Boundary Problems*, Lect. Notes in Math., ETH Zürich, Birkhäuser Basel, 2006.

[171] S.K. Pichorides, *On the best values of the constants in the theorems of M. Riesz, Zygmund and Kolmogorov*, Studia Math. **44** (1972), 165–179.

[172] G. Pisier, *Un exemple concernant la super-réflexivité* (French), Séminaire Maurey-Schwartz 1974–1975: Espaces L^p applications radonifiantes et géométrie des espaces de Banach, Annexe No. 2, 12 pp. Centre Math. École Polytech., Paris, 1975.

[173] G. Pisier, *Martingales with values in uniformly convex spaces*, Israel J. Math. **20** (1975), 326–350.

[174] G. Pisier and Q. Xu, *Non-commutative martingale inequalities*, Commun. Math. Phys. **189** (1997), 667–698.

[175] A.O. Pittenger, *Note on a square function inequality*, Ann. Probab. **7** (1979), 907–908.

[176] S. Petermichl and J. Wittwer, *A sharp estimate for the weighted Hilbert transform via Bellman functions*, Michigan Math. J. **50** (2002), 71–87.

[177] N. Randrianantoanina, *Noncommutative martingale transforms*, J. Funct. Anal. **194** (2002), 181–212.

[178] N. Randrianantoanina, *Square function inequalities for non-commutative martingales*, Israel J. Math. **140** (2004), 333–365.

[179] M. Riesz, *Les fonctions conjugées et les séries de Fourier*, C. R. Acad. Paris, **178** (1924), 1464–1467.

[180] M. Riesz, *Sur les fonctions conjugées*, Math. Z. **27** (1927), 218–244.

[181] M. Rao, *Doob decomposition and Burkholder inequalities*, Séminaire de Probabilités VI (Univ. Strasbourg, 1970–71), Lecture Notes in Math. **258**, Springer-Verlag, Berlin, 1972, pp. 198–201.

[182] D. Revuz and M. Yor, *Continuous Martingales and Brownian Motion*, 3rd edition, Springer Verlag, 1999.

[183] H.P. Rosenthal, *On the subspaces of L^p ($p > 2$) spanned by sequences of independent random variables*, Israel J. Math. **8** (1970) 273–03.

[184] Q.-M. Shao, *A comparison theorem on maximal inequalities between negatively associated and independent random variables*, J. Theoret. Probab. **13** (2000), 343–356.

[185] L.A. Shepp, *A first passage time for the Wiener process*, Ann. Math. Statist. **38** (1967), 1912–1914.

[186] L. Slavin and V. Vasyunin, *Sharp results in the integral-form John-Nirenberg inequality*, Trans. Amer. Math. Soc. **363** (2011), 4135–4169.

[187] E.M. Stein, *The development of square functions in the work of A. Zygmund*, Bull. Amer. Math. Soc. **7** (1982), 359–376.

[188] E.M. Stein, *Harmonic Analysis: Real-Variable Methods, Orthogonality, and Oscillatory Integrals*, Princeton University Press, Princeton, NJ, 1993.

[189] Y. Suh, *A sharp weak type (p,p) inequality $(p > 2)$ for martingale transforms and other subordinate martingales*, Trans. Amer. Math. Soc. **357** (2005), no. 4, 1545–1564

[190] B. Tomaszewski, *The best constant in a weak-type H^1-inequality, complex analysis and applications*, Complex Variables Theory and Appl. **4** (1984), 35–38.

[191] B. Tomaszewski, *Some sharp weak-type inequalities for holomorphic functions on the unit ball of $\mathbb{C}^n$*, Proc. Amer. Math. Soc. **95** (1985), 271–274.

[192] N.Th. Varopoulos, *Aspects of probabilistic Littlewood-Paley theory*, J. Funct. Anal **38** (1980), 25–60.

[193] V. Vasyunin, *The exact constant in the inverse Hölder inequality for Muckenhoupt weights* (Russian), Algebra i Analiz **15** (2003), 73–117; translation in St. Petersburg Math. J. **15** (2004), no. 1, 49–79.

[194] V. Vasyunin and A. Volberg, *The Bellman function for the simplest two-weight inequality: an investigation of a particular case* (Russian), Algebra i Analiz **18** (2006), 24–56; translation in St. Petersburg Math. J. **18** (2007), no. 2, 201–222.

[195] V. Vasyunin and A. Volberg, *Bellman functions technique in harmonic analysis* (sashavolberg.wordpress.com).

[196] I.E. Verbitsky, *Estimate of the norm of a function in a Hardy space in terms of the norms of its real and imaginary part*, Mat. Issled. **54** (1980), 16–20 (Russian); English transl. in Amer. Math. Soc. Transl. (2) **124** (1984), 11–15.

[197] G. Wang, *Sharp square-function inequalities for conditionally symmetric martingales*, Trans. Amer. Math. Soc., **328** (1991), no. 1, 393–419.

[198] G. Wang, *Sharp maximal inequalities for conditionally symmetric martingales and Brownian motion*, Proc. Amer. Math. Soc. **112** (1991), 579–586.

[199] G. Wang, *Sharp inequalities for the conditional square function of a martingale*, Ann. Probab. **19** (1991), 1679–1688.

[200] G. Wang, *Differential subordination and strong differential subordination for continuous-time martingales and related sharp inequalities*, Ann. Probab. 23 (1995), no. 2, 522–551.

[201] J. Wittwer, *A sharp estimate on the norm of the martingale transform*, The University of Chicago. ProQuest LLC, Ann Arbor, MI, 2000. 43 pp.

[202] J. Wittwer, *A sharp estimate on the norm of the continuous square function*, Proc. Amer. Math. Soc. **130** (2002), no. 8, 2335–2342.

[203] J. Wittwer, *Bellman functions and MRA wavelets*, Rocky Mountain J. Math. **37** (2007), 343–358.

[204] A. Zygmund, *Sur les fonctions conjugées*, Fund. Math. **13** (1929), 284–303.

[205] A. Zygmund, *Trigonometric Series*, Vol. 1 and 2, Cambridge University Press, London, 1968.

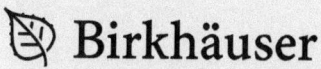 Birkhäuser | **www.birkhauser-science.com**

Monografie Matematyczne, New Series (MMNS)

Starting in the 1930s with volumes written by such distinguished mathematicians as Banach, Saks, Kuratowski, and Sierpinski, the original series grew to comprise 62 excellent monographs up to the 1980s. In cooperation with the Institute of Mathematics of the Polish Academy of Sciences (IMPAN), Birkhäuser now resumes this tradition to publish high quality research monographs in all areas of pure and applied mathematics.

Edited by
Przemysław Wojtaszczyk, IMPAN and Warsaw University, Poland

■ **Vol. 71: Bukovský, L.**, The Structure of the Real Line
2011. 550 pages. Hardcover.
ISBN 978-3-0348-0005-1

The rapid development of set theory in the last fifty years, mainly in obtaining plenty of independence results, strongly influenced an understanding of the structure of the real line. This book is devoted to the study of the real line and its subsets taking into account the recent results of set theory. Whenever possible the presentation is done without the full axiom of choice. Since the book is intended to be self-contained, all necessary results of set theory, topology, measure theory, descriptive set theory are revisited with the purpose to eliminate superfluous use of an axiom of choice. The duality of measure and category is studied in a detailed manner. Several statements pertaining to properties of the real line are shown to be undecidable in set theory. The metamathematics behind it is shortly explained in the appendix. Each section contains a series of exercises with additional results.

■ **Vol. 70: Positselski, L.**, Homological Algebra of Semimodules and Semicontramodules. Semi-infinite Homological Algebra of Associative Algebraic Structures
2010. 274 pages. Hardcover.
ISBN 978-3-0346-0435-7

This monograph deals with semi-infinite homological algebra. Intended as the definitive treatment of the subject of semi-infinite homology and cohomology of associative algebraic structures, it also contains material on the semi-infinite (co)homology of Lie algebras and topological groups, the derived comodulecontramodule correspondence, its application to the duality between representations of infinitedimensional Lie algebras with complementary central charges, and relative non-homogeneous Koszul duality.

The book explains with great clarity what the associative version of semi-infinite cohomology is, why it exists, and for what kind of objects it is defined. Semialgebras, contramodules, exotic derived categories, Tate Lie algebras, algebraic Harish-Chandra pairs, and locally compact totally disconnected topological groups all interplay in the theories developed in this monograph. Contramodules, introduced originally by Eilenberg and Moore in the 1960s but almost forgotten for four decades, are featured prominently in this book, with many versions of them introduced and discussed.

Rich in new ideas on homological algebra and the theory of corings and their analogues, this book also makes a contribution to the foundational aspects of representation theory. In particular, it will be a valuable addition to the algebraic literature available to mathematical physicists.

■ **Vol. 69: Panchapagesan, T.V.**, The Bartle-Dunford-Schwartz Integral
2008. 318 pages. Hardcover.
ISBN 978-3-7643-8601-6

■ **Vol. 68: Grigoryan, S.A.**, Shift-invariant Uniform Algebras on Groups
2006. 294 pages. Hardcover.
ISBN 978-3-7643-7606-2

■ **Vol. 67: Zoladek, H.**, The Monodromy Group
2006. 592 pages. Hardcover.
ISBN 978-3-7643-7535-5

■ **Vol. 66: Müller, P.F.X.**, Isomorphisms between H^1 Spaces
2005. 472 pages. Hardcover.
ISBN 978-3-7643-2431-5

■ **Vol. 65: Badescu, L.**, Projective Geometry and Formal Geometry
2004. 228 pages. Hardcover.
ISBN 978-3-7643-7123-4

■ **Vol. 64: Walczak, P.**, Dynamics of Foliations, Groups and Pseudogroups
2004. 240 pages. Hardcover.
ISBN 978-3-7643-7091-6